Springer Series in
Surface Sciences

19

Editor: Robert Gomer

Springer Series in **Surface Sciences**

Editors: G. Ertl and R. Gomer   Managing Editor: H. K. V. Lotsch

Volume 1: **Physisorption Kinetics** By H. J. Kreuzer, Z. W. Gortel

Volume 2: **The Structure of Surfaces** Editors: M. A. Van Hove, S. Y. Tong

Volume 3: **Dynamical Phenomena at Surfaces, Interfaces and Superlattices**
Editors: F. Nizzoli, K.-H. Rieder, R. F. Willis

Volume 4: **Desorption Induced by Electronic Transitions, DIET II**
Editors: W. Brenig, D. Menzel

Volume 5: **Chemistry and Physics of Solid Surfaces VI**
Editors: R. Vanselow, R. Howe

Volume 6: **Low-Energy Electron Diffraction**
Experiment, Theory and Surface Structure Determination
By M. A. Van Hove, W. H. Weinberg, C.-M. Chan

Volume 7: **Electronic Phenomena in Adsorption and Catalysis on Semiconductors and Dielectrics**
By V. F. Kiselev, O. V. Krylov

Volume 8: **Kinetics of Interface Reactions** Editors: M. Grunze, H. J. Kreuzer

Volume 9: **Adsorption and Catalysis on Transition Metals and Their Oxides**
By V. F. Kiselev, O. V. Krylov

Volume 10: **Chemistry and Physics of Solid Surfaces VII**
Editors: R. Vanselow, R. F. Howe

Volume 11: **The Structure of Surfaces II**
Editors: J. F. van der Veen, M. A. Van Hove

Volume 12: **Diffusion at Interfaces: Microscopic Concepts**
Editors: M. Grunze, H. J. Kreuzer, J. J. Weimer

Volume 13: **Desorption Induced by Electronic Transitions, DIET III**
Editors: R. H. Stulen, M. L. Knotek

Volume 14: **Solvay Conference on Surface Science** Editor: F. W. de Wette

Volume 15: **Surfaces and Interfaces of Solids** By H. Lüth

Volume 16: **Theory of the Atomic and Electronic Structure of Surfaces**
By M. Lannoo and P. Friedel

Volume 17: **Adhesion and Friction** Editors: M. Grunze and H. J. Kreuzer

Volume 18: **Auger Spectroscopy and Electronic Structure**
Editors: G. Cubiotti, G. Mondio, K. Wandelt

Volume 19: **Desorption Induced by Electronic Transitions, DIET IV**
Editors: G. Betz and P. Varga

G. Betz P. Varga (Eds.)

Desorption Induced by Electronic Transitions DIET IV

Proceedings of the Fourth International Workshop, Gloggnitz, Austria, October 2–4, 1989

With 209 Figures

Springer-Verlag
Berlin Heidelberg New York London
Paris Tokyo Hong Kong Barcelona

Dr. Gerhard Betz
Dr. Peter Varga
Institut für Allgemeine Physik, Technische Universität, Wiedner Hauptstrasse 8–10,
A-1040 Wien, Austria

Series Editors

Professor Dr. Gerhard Ertl
Fritz-Haber-Institut der Max-Planck-Gesellschaft, Faradayweg 4–6,
D-1000 Berlin 33, Germany

Professor Robert Gomer, Ph. D.
The James Franck Institute, The University of Chicago, 5640 Ellis Avenue,
Chicago, IL 60637, USA

Managing Editor: Dr. Helmut K. V. Lotsch
Springer-Verlag, Tiergartenstrasse 17,
D-6900 Heidelberg, Fed. Rep. of Germany

ISBN-13: 978-3-642-84147-7 e-ISBN-13: 978-3-642-84145-3
DOI: 10.1007/978-3-642-84145-3

Softcover reprint of the hardcover 1st edition 1990

2154/3150(3011)-543210 – Printed on acid-free paper

Preface

This volume contains the full proceedings of the Fourth International Workshop on "Desorption Induced by Electronic Transitions" (DIET IV), held October 2–4, 1989, at the Burghotel Kranichberg, Gloggnitz, Austria. It followed earlier workshops held in Williamsburg, USA (1982), Schloß Elmau, FRG (1984), and Shelter Island, USA (1987).

The workshop was attended by 77 participants from 9 countries, and 66 papers (8 invited, 24 oral contributions and 34 poster contributions) were presented. The proceedings contain 53 separate papers and focus on 8 main topics: (1) Basic Mechanisms and Theory, (2) Photophysics and Photochemistry, (3) Laser Induced Desorption, (4) Adsorbed Layers, (5) Condensed Gases, (6) Insulators, (7) Gas Phase Processes and (8) Applications and Related Processes.

The workshop was opened by N. Tolk, who initiated this series of meetings in 1982, and by D. Menzel, who mentioned the 25th birthday of one of the basic processes in DIET, the so-called MGR model, which was first published in 1964. One of the goals of the meeting was to achieve a better understanding and further improved theoretical descriptions of desorption by electronic transitions.

Besides the "classic" studies on electron- and photon-induced desorptions from adsorbates, a relatively large number of papers, compared to the previous meeting, dealt with laser-induced processes (laser ablation), and increasing interest in this subject can be expected for the next meeting because of many still unanswered questions. A further focal point was "sputtering" of alkali halides, where the lively and enthusiastic discussion that characterized the whole workshop reached a peak. Last but not least, it should be mentioned that photochemistry, which has been covered for the first time within the frame of this workshop, is a new and very interesting part of DIET with a number of promising applications.

Finally we acknowledge the assistance given by T.E. Madey, D. Menzel and N.H. Tolk in developing the workshop program and we would also like to extend our appreciation to all the participants, who made this meeting such a successful and fruitful one.

Vienna, February 1990

G. Betz
P. Varga

Photo by W. Husinsky

Conference Photo

1 P. Wurz
2 P. Varga
3 R.W. Dreyfus
4 N. Itoh
5 M. Sakurai
6 T. Yasue
7 G. Rocker
8 T. Gotoh
9 C.T. Reimann
10 J. Delwiche
11 G. Dujardin
12 U. Diebold
13 R.A. Baragiola
14 M.R. Schneider
15 E. Hunger
16 G. Betz
17 R.E. Walkup
18 R. Kelly
19 M. Szymonski
20 Zhen Wu
21 L. Hellner
22 Y. Nakashima
23 F. Träger
24 H. Sambe
25 M.J. Besnard-Ramage
26 A.R. Burns
27 K. Ueda
28 J.L. de Segovia
29 M. Bernheim
30 L.T. Hudson
31 P. Feulner
32 P.H. Bunton
33 B. Bourguignon
34 C. Jedrzejek
35 H. Antonsson
36 P.M. Ossi
37 C.E. Young
38 J.A. Yarmoff
39 H. Zacharias
40 H.B. Nielsen
41 A. Klekamp
42 T.E. Madey
43 L. Calliari
44 J.M. White
45 Y. Baba
46 R.A. Rosenberg
47 P. Nordlander
48 M. Suhren
49 N.H. Tolk
50 E. Hasselbrink
51 W. Ho
52 S.L. Bennett
53 D. Weide
54 T.A. Green
55 R. Franchy
56 J. Sarnthein
57 M. Reichling
58 D.E. Ramaker
59 D. Feldmann
60 D. Menzel
61 M. Wolf
62 M. Christiansen
63 J. Onsgaard
64 I. Arakawa
65 J.W. Gadzuk
66 F. Marchetti
67 A. Rosen
68 N. Martensson
69 Th. Mull
70 D. Fick
71 V. Kempter

Contents

Part IV **Adsorbed Layers**

Part V **Condensed Gases**

Part VI Insulators

Part I

Basic Mechanisms and Theory

New Theoretical Aspects in DIET

J.W. Gadzuk

Center for Atomic, Molecular, and Optical Physics,
National Institute of Standards and Technology, Gaithersburg, MD20899, USA

Abstract. A number of stimulated desorption issues that have been addressed in terms of wave packet propagation over image-potential augmented, intermediate state potential curves will be discussed. Specifically, giant enhancements in ESD yields of O(a)/Pd(111) and laser-induced desorption for NO/Pt(111) will be considered.

1. INTRODUCTION

Desorption induced by electronic transitions (DIET)[1,2] is initiated by creation of an electronically excited state of the coupled "desorbate" surface system, usually with either an energetic electronic or photon beam in which case one refers to electron or photon stimulated desorption (ESD or PSD), respectively. The electronic excitation can involve both valence and/or core electrons. The important point is that within a Born-Oppenheimer separation of electronic and nuclear (ion core) motion, the electronic state of the system specifies the potential energy functions entering the nuclear equations of motion. Thus the electronically excited states that are relevant for stimulated desorption are those (such as a steeply repulsive dissociative/desorptive curve) that not only impose forces on the adsorbed atom or molecule initially in an equilibrium geometry, but also survive long enough for the desorbate to respond to these forces and displace in a consequential manner.

Theoretical modeling and conceptualizing on stimulated desorption has shown a double faceted emphasis which frequently occurs in studies of time-dependent quantum molecular processes[3,4]. On the one hand have been those studies focusing on the microscopic mechanisms responsible for the repulsive potential surfaces[5] leading to desorption, while on the other have been the studies keyed into the desorption dynamics executed over generic potential energy surfaces characterized by a given topology, independent of the system-specific chemical properties responsible for that topology[6]. Based on the success of wave packet modeling in providing intuitively appealing accounts of gas phase photodissociation dynamics due to electronic excited states[3,4], a similar level of success can be anticipated in the closely related process of stimulated desorption[6].

From the point of view of simplified one-dimensional nuclear dynamics, the situation in stimulated desorption is modeled in terms of "desorbate"-surface potential energy curves (PEC), as shown in Figure 1. The initial state of the particle that will be desorbed is thermally distributed among the bound states of the atom/molecule-surface PEC. According to both the long-standing Menzel-Gomer-Redhead (MGR) model[5a] and the Knotek-Feibelman (KF) model[5b], desorption is initiated by first creating an electronically excited state of the surface complex, with either an electron or photon beam. If the PEC associated with the excited state is both repulsive and sufficiently long-lived to survive surface quenching[7], then "standard" desorption can occur and this is independent of the specific nature of the electronic state responsible for the PEC.

Springer Series in Surface Sciences, Vol. 19
Desorption Induced by Electronic Transitions, DIET IV
Editors: G. Betz · P. Varga

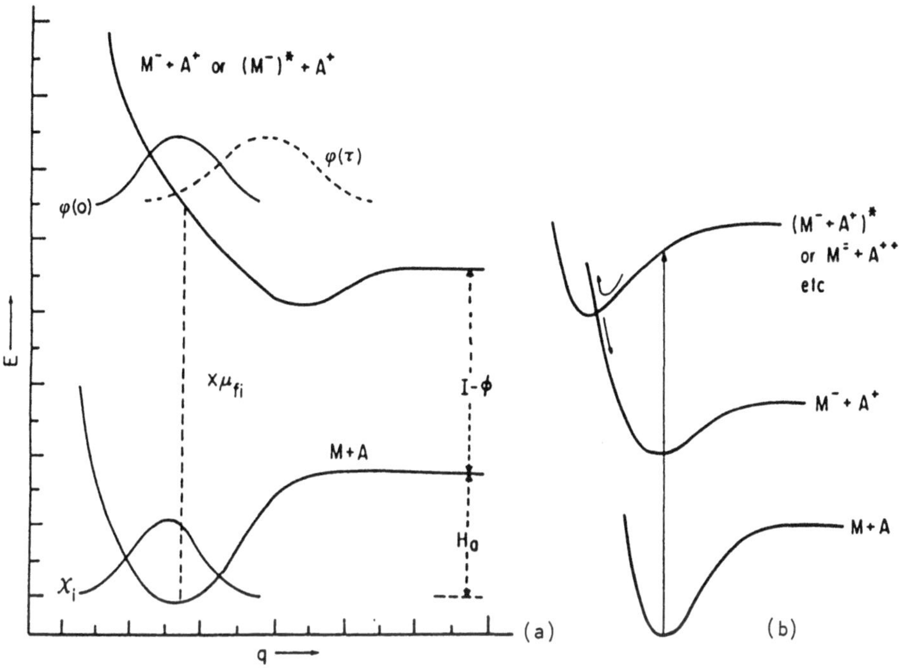

Figure 1. Potential curves illustrating various mechanisms for stimulated desorption. (a) Direct excitation onto a repulsive curve such as MGR or KF. (b) Excitation onto an intermediate bound ion curve prior to desorption à la Antoniewicz.

Another interesting stimulated desorption scenario, shown in Figure 1b, was suggested by Antoniewicz[8]. He realized that if the initially excited state was ionic, then, due to image potential screening[9], the forces determining initial wave packet propagation could be directed toward the surface. Closer to the surface additional electronic transitions and thus curve crossings could be experienced onto repulsive states, in analogy with predissociative processes in gas phase. For various reasons, the Antoniewicz mechanism has been enjoying a recent popularity[6c,10-12] and as such, will form the basis for the new issues in DIET which will be the focus of this paper. In Section 2, a quick summary of theoretical work published since the last DIET workshop, which pertains to stimulated desorption via an Antoniewicz-related process, will be offered. Two specific examples will then be considered in Sections 3 and 4. In 3, the giant enhancement recently observed[13] in the ESD yields for O(a)/Pd(111) and in 4, (non-thermal) laser induced desorption observed for NO/Pt(111)[14], will both be interpreted in terms of a wave packet model involving a temporary image-potential-augmented intermediate state.

2. RECENT DEVELOPMENTS

At present, systems of atoms or molecules physisorbed on metal surfaces, have received the most attention with respect to the possible role of transient image potentials in desorption dynamics. A particularly seminal experiment, due to Feulner and Menzel[15], reported on the desorption characteristics for N_2O/Ru(100). In brief, their observations which "stimulated" subsequent theoretical modeling were:

(i) only neutral N_2O desorbed;

(ii) the total ESD yield was ≈0.1;

(iii) The "skewed-bell shaped" kinetic energy distribution of the desorbed molecules was peaked at 112 meV, with half maximum values at 34 meV and 312 meV.

Qualitatively, the Antoniewicz mechanism provides a credible zeroth-order explanation for these results. However, attempts to reconcile both the high yield and the kinetic energy distribution on a quantitative basis have not yet been totally satisfactory[10,11]. Whether this is a result of inadequacies in the dynamics models or the PEC is still unresolved.

The study of ESD of chemisorbed oxygen[6c,13] has benefited from the insights provided by Avouris and coworkers in their work on fluorine desorption[5c]. In both cases, the large electronegativity of the adsorbate with respect to any "good" metal substrate assures that both O and F would absorb as (fractionally) charged negative ions. For the theoretically studied system of F/Al, the lowest lying electronic excited states responsible for desorption involve "electron detachment" from the F^-, thus turning off the attractive image potential contribution to the adsorbate-surface interaction. This being done, the more-or-less neutral F finds itself on a repulsive potential curve and thus attempts to desorb before electron capture returns the system to its stable bound ground state. Presumably, a similar path to ESD is followed by adsorbed O, but with an important and intriguing exception, to be discussed in Section 3. For present purposes we just note that the picture provided by the density functional calculations is consistent with the intuitive image potential mechanism.

3. RESONANCE ENHANCED ESD

Experiments have been carried out by the Pittsburgh group under Yates on the system of atomic oxygen adsorbed on Pd(111) in which the total ESD cross section as a function of incident electron energy in the ≈5-500 eV range was inferred[13]. A striking enhancement in the observed cross section by a factor ≈five within a very narrow incident energy range (≈10±2eV), was observed which was very different from conventional desorption, suggesting the occurrence of some sort of resonance-assisted desorption process near 10eV. It has been proposed that this behavior could be a consequence of an adsorbed oxygen shape resonance in which the incident electron is temporarily trapped. The plausibility of such resonance formation in the 10eV range has been supported by Hartree-Fock calculations of the total energies for various charge states of atomic oxygen[6c]. The type of enhancement mechanism we have in mind is pictured in Fig. 2. Let the potential energy function for the O/Pd(111) system be the attractive potential labeled $V_a(z)$. Initially the chemisorbed O is taken to be in the Gaussian harmonic oscillator vibrational ground state associated with $V_a(z)$. Now suppose that an incident electron with kinetic energy ϵ_{in} is captured by $O^{-\delta}$ forming a temporary negative $O^{-(1+\delta)}$ ion resonance where δ is the noninteger charge number of the chemisorbed O. When the incident electron becomes trapped in the resonance state, the additional image-potential type of attraction is turned on, resulting in the augmented negative ion/surface potential curve

$$V_-(z) \simeq \epsilon^* + V_a(z) - e^2/4z, \tag{1}$$

where ϵ^* is a constant whose value is of order 10eV, and z is the distance between the effective image plane and the position of the $O^{-\delta}$.

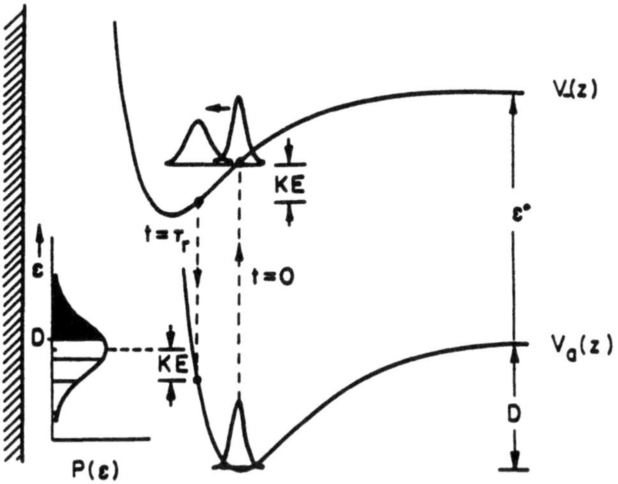

Fig. 2. Potential energy curves for chemisorbed and negative ion oxygen, showing wavepacket propagation throughout the time sequence involving the negative ion shape resonance. The distribution of final oxygen states, between vibrationally excited bound and desorptive continuum states, is shown as $P(\epsilon)$ vs ϵ.

It is hypothesized that the actual resonance-assisted desorption event could proceed as follows. Upon capture of the incident electron, the wave packet characterizing the nuclear motion of the adsorbed O suddenly finds itself as a nonstationary state on $V_-(z)$ as in a Franck-Condon transition, and thus accelerates inwards towards the surface. After a time interval τ_R, the negative ion resonance lifetime, the trapped electron on $O^{-(1+\delta)}$ departs, and the moving, displaced, and distorted wave packet is returned to $V_a(z)$. As is apparent from Fig. 2, the dynamics on $V_-(z)$, the image-potential-augmented intermediate state potential, can easily lead to an enormous enhancement in the electron stimulated desorption yield represented by the area under the cross labeled region of the final state distribution labeled $P(\epsilon)$.

The process just outlined has been quantified[6c] using semiclassical Gaussian wave packet dynamics[2b,3]. In this picture, the "two-transition process" views the combined sequence of: (i) initial preparation of $|\phi(t=0)\rangle=|\phi_0\rangle$, the harmonic oscillator ground state of V_a as a nonstationary state on V_- at $t=0$; (ii) time evolution of $|\phi(t)\rangle$ under the Hamiltonian $H_-=T+V_-$ (T=kinetic energy operator) such that $|\phi(t)\rangle=\exp(-iH_-t/\hbar)|\phi(t=0)\rangle$; (iii) return of $|\phi(t=\tau_R)\rangle$, the moving and distorted wave packet, to V_a at $t=\tau_R$, as the preparation of a nonstationary state $|\phi(t'=0;\tau_R)\rangle$, on the ground electronic state PES. Subsequent time evolution of this nonstationary state, called the "Raman wave function" in other obvious contexts,[3,16] is given by $|\phi(t';\tau_R)\rangle=\exp(-iH_a t'/\hbar)\exp(-iH_-\tau_R/\hbar)|\phi_0\rangle$, where $t'=0$ coincides with $t=\tau_R$. From this perspective, the final transition at $t=\tau_R$, "looks" like a "standard" Franck-Condon type of transition, but one involving a moving rather than fixed initial state wave packet[2b,6c]. The resulting energy distribution of excited states, equivalent to a Franck-Condon type of distribution, is given as

$$P(\epsilon;\tau_R)=\frac{1}{2\pi}\int_{-\infty}^{\infty}dt'\, e^{i\epsilon t'/\hbar}\langle\phi(t'=0;\tau_R)|\phi(t';\tau_R)\rangle \qquad (2)$$

which is the Fourier transform of the time-dependent overlap of the initially prepared state of V_a at $t=\tau_R$ (as a result of the temporary negative ion resonance) with the subsequent time evolved state[2b,3,16].

If the wave packets required in Eq.2 are obtained using the Heller method of Gaussian wave packet propagation in the short-time-limit[3], then it has been demonstrated that Eq.2 yields the Gaussian distribution (for the continuum desorptive states)

$$P(\epsilon;\tau_R)=\frac{1}{\Delta(\tau_R)\sqrt{2\pi}}\exp\left[-\frac{1}{2}\left(\frac{\epsilon-\epsilon_f(\tau_R)}{\Delta(\tau_R)}\right)^2\right] \qquad (3)$$

with

$$\Delta(\tau_R)\equiv\bar{u}V_a'(z_{\tau_R})[1+(\omega\tau_R)^2]^{1/2}, \qquad (4)$$

$$\epsilon_f(\tau_R)=p_{\tau_R}^2/2m + V_a(z_{\tau_R}),$$

$\bar{u}=(\hbar/2m\omega)^{1/2}$, the spread of the original (t=0) oscillator probability distribution, and p_{τ_R} and z_{τ_R} the momentum and position of the wave packet center at time τ_R when it is returned to $V_a(z)$ (with slope $V_a'(z)$). From Eq. 1, it is apparent that both p_{τ_R} and z_{τ_R} depend upon the image charge forces acting throughout the life of the temporary negative ion. The resonance enhanced ESD probability is given by those (continuum) states in the distribution [Eq. 3] with energy in excess of D, the binding energy of the adsorbate; that is

$$P_{des}(\tau_R)=\int_D^\infty P(\epsilon;\tau_R)d\epsilon$$
$$=0.5\left[1-\mathrm{erf}\left(\frac{D-\epsilon_f(\tau_R)}{\sqrt{2}\Delta(\tau_R)}\right)\right] \qquad (5)$$

which is the final result to be presented here. In effect, P_{des} given by Eq. 5 is a function of incident electron energy through the implicit energy dependence of τ_R which is non-zero only for energies compatible with the shape resonance. The theoretical enhanced ESD probability given by Eqs. 3-5 has been evaluated with parameters appropriate to O(a)/Pd(111) (that is D≃3.0eV, $\hbar\omega$≃0.1eV, and $\bar{u}$≃0.04Å). The results shown in Fig. 3

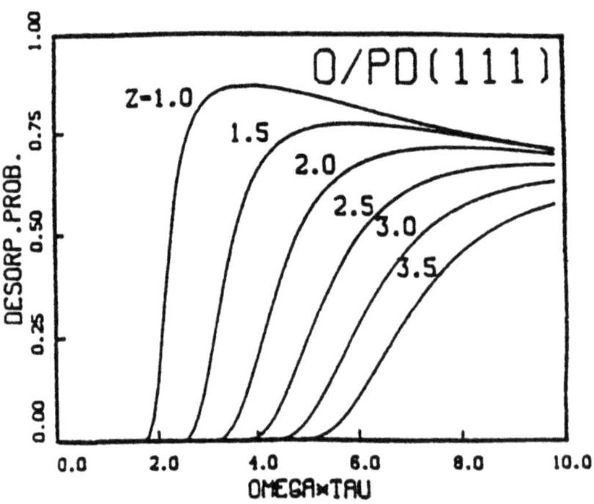

Fig. 3. Resonance enhanced desorptive probability vs resonance lifetime (in units of ω^{-1}) for atomic oxygen on Pd(111) treating the adsorption equilibrium location with respect to the image plane parametrically.

are given in the form of P_{des} as a function of resonance lifetime (in units of ω^{-1}) with z_{eq}, the location of the minimum in $V_a(z)$ with respect to the image plane varied parametrically. Note that $\omega\tau_R=2\pi$ corresponds to the vibrational period of the ground state of $V_a(z)$, which for the present system is $\simeq 4\times10^{-14}$s. The main point of course is that the proposed mechanism does lead to significant enhancement in the desorption yield, for all reasonable values of z_{eq}, when a negative ion resonance is formed for a time duration $\simeq 10^{-14}$ sec. Further implications have been discussed elsewhere[6c,13].

4. LASER DRIVEN ESD

Recent studies have sought to utilize lasers to drive surface reactions in some way other than local heating[14,17]. A prototypical example of a non-thermal, optically driven surface reaction is the NIST final-state-resolved experiment in which laser-induced molecular desorption of NO from Pt(111) was shown to be due to a non-thermal mechanism[14]. Furthermore, there were compelling reasons to believe that laser excitation of substrate electron-hole pairs was the initial step in the desorption process. The observations which pertain to issues addressed here are the following. Laser light with $h\nu=0.65$, 1.17, 2.33, or 3.49 eV was focused onto the surface, the polarization and fluence varied parametrically and various properties of the desorbed NO were measured[14]. With regard to translational dynamics, it was found that: (i) NO did not desorb when $h\nu<D\simeq 1$ eV, the NO desorption energy; (ii) For $h\nu=2.33$ and 3.49 eV, the average kinetic energy measured in the time-of-flight (ToF) translational energy distribution was found to exceed the peak surface temperature by a factor of 4 and for $h\nu=1.17$ eV, by a somewhat smaller factor; (iii) Although the ToF distribution, hence <KE>, were independent of laser fluence, the yield varied linearly with it; (iv) Desorption was strongly forward peaked ($\approx\cos^{11}\theta$); (v) Desorption yields with s- and p- polarized light scaled with the amount of radiation absorbed by the Pt and not with the magnitude of the electric field normal or parallel to the surface. These observations are suggestive of desorption induced by photogenerated excited carriers (henceforth taken to be electrons) produced in the near surface region which interact with the adsorbed NO by inelastic resonance scattering through the $2\pi^*$ level. The orbital bonding for NO chemisorbed on a metal surface is characterized by broadened and mostly occupied 4σ, 5σ, 1π, and $2\pi^*$ orbitals with energies below the metal Fermi level, and unoccupied $2\pi^*$ orbitals above the Fermi level, as sketched in the "energy level" diagram of Fig. 4. Inverse photoemission studies have confirmed this picture, both for CO and NO adsorbed on transition metals[18].

The proposed optically driven, non-thermal desorption proceeds in the following way. The incident radiation with photon energy $h\nu$ creates a distribution of excited conduction band electrons with energy ϵ_{in} in the range $\epsilon_{fermi}<\epsilon_{in}\leq\epsilon_{fermi}+h\nu$, which produces a flux of photoelectrons $\equiv j_{el}(\mathbf{k}(\epsilon_{in});h\nu)$ from within the substrate, incident upon the adsorbed NO. This distribution should be reasonably smoothly varying with energy on the scale of Δ^*, the $2\pi^*$ level half width. For those photo-excited electrons resonant with the $2\pi^*$ "level", the incident electron with energy ϵ_{in} may hop onto the adsorbed NO, reside in this resonance for a time τ_R, and then scatter back into an unoccupied conduction band state with energy $\epsilon_{fin}\leq\epsilon_{in}$, leaving behind some part of the adsorption system excited with energy $\epsilon=\epsilon_{in}-\epsilon_{fin}$. Just as for resonance enhanced ESD of O discussed in Section 3, the formation of a temporary negative molecular ion (NO^-) can result in extremely τ_R-dependent desorption of NO if $h\nu$ is large enough for ϵ to exceed D, the desorption energy. The basic image-potential-assisted model shown in Fig. 2 is therefore equally relevant here. Additional excitation of internal vibrational and/or rotational modes of

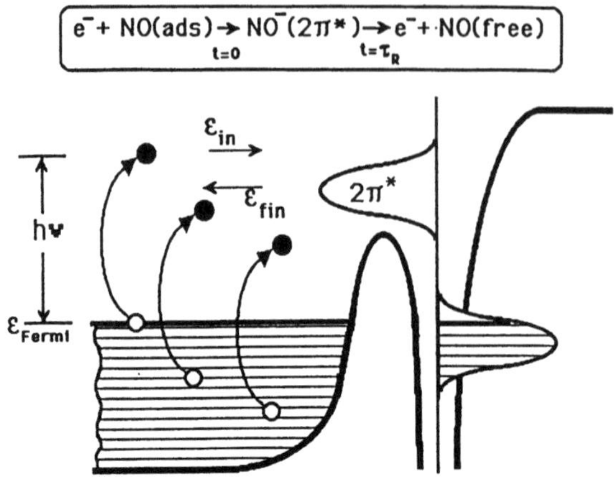

Fig. 4. **Energy level diagram for inelastic hot electron scattering through 2π* shape resonance of adsorbed NO.**

the desorbed molecule is expected[19,20] which should be a sensitive function of τ_R[2b]. As detailed elsewhere[12], this dependence can be used as an internal self-consistency test in which observed desorption characteristics and internal vibrational state populations are necessarily related when the electron-mediated desorption mechanism proposed here is applicable.

Proceeding towards a theory, the energy-resolved rate of production of NO molecules with total energy ϵ, due to resonance scattering (involving temporary occupation of the $2\pi^*$ shape resonance for a duration τ_R) of optically excited electrons from $\mathbf{k}_i$, energy $\epsilon_{in}(\mathbf{k}_i)$, to state $\mathbf{k}_f$, energy $\epsilon_{fin}=\epsilon_{in}-\epsilon\geq\epsilon_{fermi}$ is given by an expression of the form

$$\frac{d\dot{N}(\epsilon;\tau_R,h\nu)}{d\epsilon}=\sum_{\mathbf{k}_i} j_{el}(\mathbf{k}_i;h\nu)\sigma_{Tot}(\epsilon_{in}(\mathbf{k}_i);\epsilon;\tau_R) \ . \qquad (6)$$

In Eq. 6, the total cross section for inelastic resonance scattering in which energy ϵ is transferred from the excited electron to the molecular system is

$$\sigma_{Tot}\approx P(\epsilon;\tau_R)\sigma_{res}(\epsilon_{in};\epsilon), \qquad (7)$$

where σ_{res} is the electron scattering cross section, evaluated at some appropriate fixed molecular geometry and $P(\epsilon;\tau_R)$ given by Eq. 3, is the total probability distribution for exciting the molecule via a negative ion resonance of lifetime τ_R. Equations 6 and 7 can be brought to a product form

$$\frac{d\dot{N}(\epsilon;\tau_R,h\nu)}{d\epsilon}\equiv P(\epsilon;\tau_R)Q_{2\pi^*}(\epsilon;h\nu), \qquad (8)$$

where $Q_{2\pi^*}$, the electronic factor, determines the overall scale of the effect in a number of physically distinct ways: (i) the direct proportionality of $d\dot{N}/d\epsilon$ to j_{el}, the incident flux of hot electrons; (ii) phase space factors for incident and scattered electrons; (iii) tuning $h\nu$ for optimal utilization of the $2\pi^*$ resonance.

A number of analytic expressions related to possible observables in the laser-induced desorption experiments follow from Eqs. 3 and 8. To begin with, the desorption probability for the adsorbed particle known to have

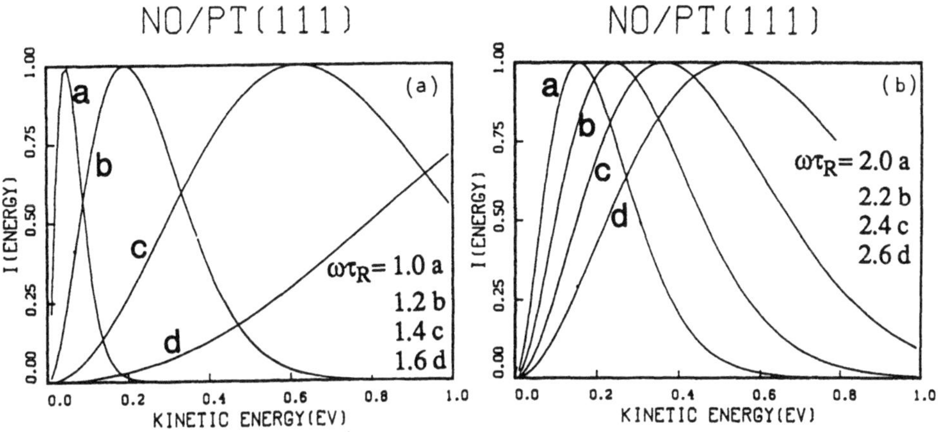

Fig. 5. **Kinetic energy distributions for desorbed NO, with parameters $\omega\tau_R$** as labeled. (a) z_{eq}=1Å, (b) z_{eq}=2Å.

experienced the effects of the temporary negative ion resonance (i.e. $Q_{2\pi^*}=1$ in Eq. 8) is given by Eq. 5, and the observed desorption yield is independent of $h\nu$ other than through the multiplicative scale factor $Q_{2\pi^*}$. Evaluating Eq. 5, now with NO/Pt parameters (D≈1.08 eV, $\hbar\omega$=0.056 eV, $\bar{u}$≈0.035Å), the resulting $P_{des}(\tau_R)$ as a function of $\omega\tau_R$ look similar enough to the curves in Fig. 3 that they need not be shown here. The main point that the proposed resonance mechanism does in fact lead to desorption of course remains intact. To the extent that one can identify a desorption threshold (determined by the classical motion of the wave packet center), said threshold is determined by the condition $D=\epsilon_f(\tau_R)$.

The measured ToF distributions are of the form

$$I(v)\approx v^n P(\epsilon;\tau_R)\; Q_{2\pi^*}(\epsilon;h\nu) \qquad (9)$$

where the velocity of the desorbed molecule is given by $v=[2(\epsilon-D)/m]^{1/2}$ and the value of the exponent n depends upon the details of the experiment.

Theoretical kinetic energy distributions deduced from Eq. 9 with $Q_{2\pi^*}$=constant and n=5 are shown in Fig. 5 with the choices z_{eq}=1 and 2Å, treating $\omega\tau_R$ parametrically. For the system NO/Pt(111), the main body of the kinetic energy distribution of the desorbed molecules moves through the observed 0-to-1 eV range for that combination of parameters z_{eq} and τ_R for which P_{des} rises from threshold to saturation as in Fig. 3. Indeed the basic features of the experimental spectra are qualitatively well simulated by theory, at least for short resonance lifetimes in which $\epsilon_f(\tau_R)\approx D$.

Finally, it is appropriate to comment on the relationship between the present mechanism and one proposed by Burns, Stechel, and Jennison[20] for higher energy ESD, also for the system NO/Pt(111). They considered desorption which followed upon ionization of the NO by creation of a hole in an initially occupied 5σ orbital of the NO/Pt complex, this occurring for incident electron energies in excess of an ≈6 eV threshold. Invoking an ingenious "equivalent-binding" arguement, the PES for the intermediate state NO^+ was related to that of the weakly bound, free-rotor state of physisorbed O_2. It was demonstrated how the release of the hindered rotor potential[19] associated with the chemisorbed NO^+ while in the intermediate long-lived (hopefully!) $NO^+(=O_2)$ state, allowed the NO^+ to access portions of the final state PES which led to desorption. Presumably their mechanism could also play a role in electron-mediated laser-induced desorption for photon energies high enough to produce internally generated electrons in excess of the required threshold energy. Nonetheless, at the wavelengths pertinent to the present experiments, the mechanism is of no consequence.

ACKNOWLEGEMENTS

I am grateful for lively, pleasurable, and stimulating discourse with Rich Cavanagh, Charles (Bruce) Clark, Lee Richter, and John Yates on topics related, and otherwise, to issues aired here.

REFERENCES

1. Excellent introductions to both the historical background and state-of-the-art of the field are given in the proceedings of the DIET workshops: (a) DIET I, edited by N.H. Tolk, M.M. Traum, J.C. Tully, and T.E. Madey (Springer, Berlin, 1983). (b) DIET II, edited by W. Brenig and D. Menzel (Springer, Berlin, 1985). (c) DIET III, edited by R.H. Stulen and M.L. Knotek (Springer, Berlin, 1988).
2. a) M.L. Knotek, Rep, Prog. Phys. 47, 1499 (1984);
b) J.W. Gadzuk, Ann. Rev. Phys. Chem. 39, 395 (1988);
c) P. Avouris and R.E. Walkup, Ann. Rev. Phys. Chem. 40, 173 (1989).
3. E.J. Heller, J. Chem. Phys. 62, 1544 (1975): Acc. Chem. Res. 14, 368 (1981); in Potential Energy Surfaces and Dynamics Calculations, edited by D.G. Truhlar (Plenum, New York, 1981).
4. (a) R. Kosloff, J. Phys. Chem. 92, 2087 (1987). (b) S. Sawada. R. Heather, B. Jackson, and H. Metiu, J. Chem. Phys. 83, 3009 (1985); (c) R.D. Coalson and J.L. Kinsey, J. Chem. Phys. 85, 4322 (1986); (d) V. Mohan and N. Sathyamurthy, Comp. Phys. Rep. 7, 213 (1988). (e) D.J. Tannor and S.A. Rice, Adv. Chem. Phys. 70, 441 (1988).
5. Three examples which emphasize the most common possibilities are: (a) Bonding-to-antibonding-orbital transition of valence electrons, usually synonymous with the Menzel, Gomer, Redhead model: D. Menzel and R. Gomer, J. Chem. Phys. 41, 3311 (1964); P.E. Redhead, Can. J. Phys. 42, 886 (1964). (b) Core-hole-initiated Auger processes: P.J. Feibelman and M.L. Knotek, Phys. Rev. B 18, 6531 (1978). (c) Neutralization of electronegative adsorbates, thus switching off the image potential attraction: Ph. Avouris, R. Kawai, N.D. Lang, and D.M. Newns, J. Chem. Phys. 89, 2388 (1988).
6. Examples based on time-dependent wave packet propagation, the method of choice in this paper, include: (a) W. Brenig, Z. Phys. B 23, 361 (1976); J. Phys. Soc. Jpn. 51, 1914 (1982); Ref. 1(a), p.30. (b) E.B. Stechel, Ref. 1(b), p.32; E.B. Stechel, D.R. Jennison, and A.R. Burns, Ref. 1 (c), p.136. (c) J.W. Gadzuk and C.W. Clark, J. Chem. Phys. 91, 3174 (1989).
7. J.W. Gadzuk, Comm. At. Mol. Phys. 16, 219 (1985); R. Brako and D.M. Newns, Rep. Prog. Phys. 52, 655 (1989).
8. P.R. Antoniewicz, Phys. Rev. B 21, 3811 (1980).
9. P.J. Jennings and R.O. Jones, Adv. in Phys. 37, 341 (1988).
10. a) Z.W. Gortel, H.J. Kreuzer, P. Feulner, and D. Menzel, Phys. Rev. B 35, 8951 (1987-I);
b) Z.W. Gortel, R. Teshima, and H.J. Kreuzer, Phys. Rev. B 37, 3183 (1988).
11. W. Hübner and W. Brenig, Z. Phys. B 74, 361 (1989).
12. J.W. Gadzuk, L.J. Richter, S.A. Buntin, R.R. Cavanagh, and D.S. King, (to be published).

13. A. Hoffman, X. Guo, J.T. Yates,Jr., J.W. Gadzuk, and C.W. Clark, J. Chem. Phys. 90, 5793 (1989).
14. S.A. Buntin, L.J. Richter, R.R. Cavanagh, and D.S. King, Phys. Rev. Letters 61, 1321 (1988); D.S. King and R.R. Cavanagh, Adv. Chem. Phys. 76, 45 (1989).
15. P. Feulner, D. Menzel, H.J. Kreuzer, and Z.W. Gortel, Phys. Rev. Letters 53, 671 (1984).
16. E.J. Heller, R.L. Sundberg, and D. Tannor, J. Phys. Chem. 86, 1822 (1982); S.O. Williams and D.G. Imre, J. Phys. Chem. 92, 3363 (1988).
17. T.J. Chuang, Surface Sci. Repts. 3, 1 (1983); W. Ho, Comments Cond. Mat. Phys. 13, 293 (1988).
18. V. Dose, Surface Science Repts. 5, 337 (1985); P.D. Johnson and S.L. Hulbert, Phys. Rev. B 35, 9427 (1987-II); B. Gumhalter, K. Wandelt, and Ph. Avouris, Phys. Rev. B 37, 8048 (1988-I).
19. J.W. Gadzuk, U. Landman, E.J. Kuster, C.L. Cleveland, and R.N. Barnett, Phys. Rev. Letters 49, 426 (1982).
20. A.R. Burns, E.B. Stechel, and D.R. Jennison, Phys. Rev. Letters 58, 250 (1987).

Lifetimes of Excited Atoms near Surfaces

P. Nordlander

Department of Physics and Rice Quantum Institute,
Rice University, Houston, TX 77030, USA

Abstract. First principles calculations of the shifts and lifetimes of atomic energy levels outside clean and impurity covered surfaces are presented. It is shown that in the presence of coadsorbed impurities both the widths and the shift of atomic levels with position can be strongly modified compared to the clean metal case. Such changes can significantly alter the probability for charge transfer between an atom and a surface. This effect is illustrated by an application to a series of recent ESD experiments of excited H from K covered Pt(100).

1. Introduction

Charge transfer processes play an important role in surface science. Many surface reactions such as dissociation and sticking are believed to depend crucially on the probability for an electron to transfer from the surface into excited atomic or molecular levels[1]. Electron transfer processes also play a very important role in quenching certain dynamical responses that could be induced by electronic transitions. One example of this is stimulated desorption[2]. The stimulated desorption process can occur if an atom is electronically excited into an antibonding state sufficiently longlived to enable the atom to acquire enough kinetic energy to leave the surface[3, 4, 5]. In the same way, photochemistry at surfaces depends on the enhanced reactivity of electronically excited species and the time the photon induced excitation survives[6].

The occurence of charge transfer at surfaces can also be used to probe important microscopic features of the surface. For instance, in the Scanning Tunneling Microscope, (STM), both the electronic and geometrical structure of the surface is probed by measuring the tunneling current between an external tip and the surface as a function of lateral position and voltage[7]. The state selective study of desorbing or sputtered particles can in principle reveal the same type of information but also provide details of the electronic structure of the surface not accessible with STM[8].

The purpose of this paper is to present a theoretical scheme for the description of charge transfer processes between atoms and surfaces. Such a method will be presented in section 2. Results for various excited atoms outside both clean and impurity covered metal surfaces will be presented in section 3. In section 4, the calculations will be combined with a dynamical theory for the calculation of the desorption yield of excited H in ESD from K covered Pt(100).

Springer Series in Surface Sciences, Vol. 19
Desorption Induced by Electronic Transitions, DIET IV
Editors: G. Betz · P. Varga

2. Theory

In this section, the theoretical method used in the present calculation will be presented. The origin of the shift and broadening of atomic levels near surfaces are changes in the electron potential around the atom due to the presence of the surface. It is therefore important to model the surface electronic structure accurately. Section 2.1 contains a description of how the surface electron potential can be calculated using density functional methods. From the surface potential, the broadening and shift of the atomic resonances can be calculated. In section 2.2 the details of the resonance calculation are presented.

2.1 Electron potential outside surfaces

In the following, we will assume a one-electron description of the surface and atomic electrons. Atomic units will be used throughout the text except when otherwise indicated. The coordinate system will be cylindrical with the positive z-axis oriented towards vacuum. The coordinate origin, z=0, corresponds to the actual location of the surface. The radial coordinate ρ refers to a surface normal through the atom. Upper case letters will be used to describe atomic coordinates and lower case letters refer to the electron coordinates.

For the surface we adopt the jellium model[9]. The jellium approximation amounts to treating the conduction electrons in the solid as an electron gas. The positive background in the solid is smeared out as a constant attractive potential. The surface is modeled by abruptly truncating this potential at z=0. The jellium model is entirely specified by the valence electron density and the corresponding r_s describing the average distance between the electrons. In what follows, we will consider jellium with r_s=2.0 corresponding to aluminum.

In the jellium model, the total potential for the electron at coordinates (ρ,z) in the presence of an atom at distance Z from the surface can be written as

$$V^{eff}(\rho, z; Z) = V_0^s(z) + V^A(r) + \Delta V_A^s(\rho, z; Z) \tag{1}$$

The first part of the potential describes the bare electron-surface interaction. This potential can be calculated within the non-local density functional scheme. In the present case we have adopted the weighted density approximation[10]. This particular many body approach describes both the image interaction and the potential in the bulk. For large z, $V_0^s \longrightarrow \frac{-1}{4(z-z_{im})}$, where z_{im} is the image plane defined as the first moment of the charge distribution induced by an external electric field[11]. The atomic potential $V^A(r)$ is modeled using a pseudo potential[12](Alkali atoms) or a bare Coulomb potential(Hydrogen). The $\Delta V_A^s(\rho, z; Z)$ term describes how the bare surface electron potential is modified when an adsorbate is present. For large z this term approaches $\frac{1}{\sqrt{\rho^2+(z+Z-2z_{im})^2}}$. For intermediate distances this term is estimated using a linear response approach. The electron charge induced by an external perturbation is distributed in a thin layer around the surface. The thickness of this layer Δ as well as z_{im} depends on the r_s of the metal and has been calculated within the local density approximation[11].

From the proton-induced surface electron charge distribution, the induced electrostatic potential as well as the exchange correlation potential can be calculated using Poisson's equation and a proper exchange correlation functional.

In fig. 1, we show the electron potential for an adsorbate outside a jellium surface modeled using the present many body approach. We note that the bare surface potential, V_0^s, provides a force towards the surface on the electron while the proton induced potential, ΔV_A^s, is repulsive and therefore reduces the tunneling rates.

Close to the atom, the total surface induced electron potential is repulsive. The atomic states will thus shift upwards. This means that it will be easier to ionize an atom close to the surface. From Fig. 1, it is also clear why the atomic states will broaden, i.e., become resonances. Inside the surface the electron potential is attractive and electrons can tunnel between the atom and the surface.

Coadsorption of alkali atoms on metal surfaces is often used as a means of changing the work function of the metal, thus to modify the role the surface can play in charge transfer processes. In order to obtain a microscopic understanding of the charge transfer dynamics between atoms and alkali promoted surfaces, it is important to investigate the behaviour of resonances near coadsorbed alkali atoms.

In order to model the impurity/metal systems, calculations of the shifts and broadening of atomic levels will be performed for hydrogen levels outside a potassium atom chemisorbed on jellium. Due to the reduced symmetry when two different atoms are present on the surface, the hydrogen will always be placed along the surface normal through the chemisorbed impurity so that cylindrical symmetry of the potential is maintained.

The chemisorbed potassium is modeled using a K pseudopotential[12], and a negative image charge located at $2z_{im} - Z_K$. The chemisorption distance, Z_K, of potassium on Al jellium is 4 a.u. Due to the electro-positivity of potassium, the induced electrostatic potential is negative in the surface region. In fig. 2, a contour plot of the induced electrostatic potential is shown. The present modeling of the alkali induced electrostatic fields quantitatively agrees with the results from ab initio calculations[13].

The impurity induced changes in the surface exchange correlation potential are neglected, since these are confined to the immediate vicinity of the impurity, and therefore not expected to influence the shift or width of the hydrogen resonances. The calculated impurity induced potential is added to the expression (1) for the total potential V^{eff}.

It is clear that both the shifts of atomic levels and the tunneling rates will be influenced by such impurities. In the next section, an accurate calculation of these quantities will be presented.

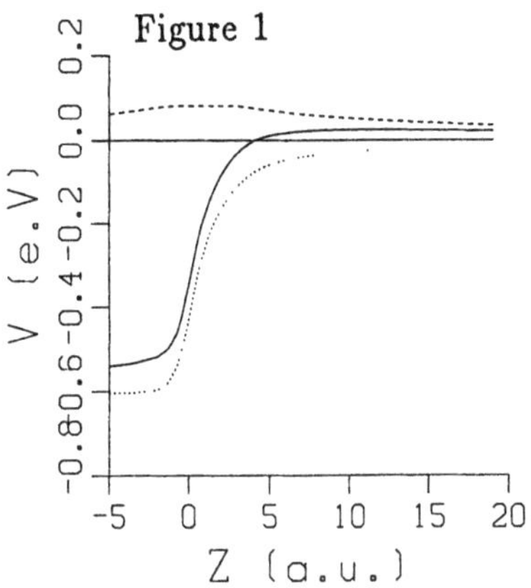

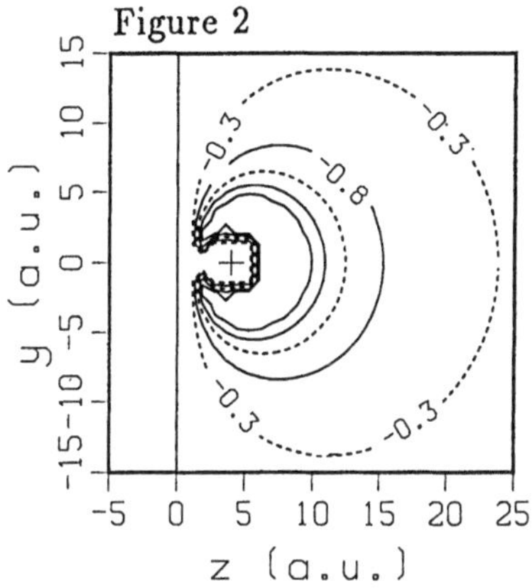

Figure 1: The different contributions to the surface potential outside a metal surface. The atom is placed at Z=10 a.u. The dotted line is the bare surface potential, $V_0^s(z)$, the dashed line is the atom induced potential, ΔV_a^s, and the solid line is the total potential. The substrate is Al and the distance z is measured from the jellium edge.

Figure 2: Contour plot of the induced electrostatic potential for a potassium atom chemisorbed outside a perfectly conducting surface. The K is placed at 4 a.u. outside the metal.The energy unit is eV.

2.2 Calculations of the shifts and broadening of atomic levels near surfaces

In order to calculate the energy of the atomic resonances, the Schrödinger equation for the electrons must be solved,

$$[-\nabla^2 + V^{eff}(\rho, z; Z)]\psi = \epsilon\psi \tag{2}$$

with the proper boundary conditions. Resonances satisfy the so-called Siegert boundary conditions:

$$\Psi(r) \longrightarrow \exp(ik_R r + k_I r) f(\Omega) \tag{3}$$

where k_I is positive. The energy is related to the complex wavenumber, k through $\epsilon = -\frac{1}{2}(k_R + ik_I)^2$. The energy thus becomes complex: $\epsilon = \epsilon_R - i\epsilon_I$. The real part of the energy, ϵ_R, describes the energy of the level and the imaginary part, ϵ_I, describes the width of the resonance.

It can be seen that these boundary conditions diverge at infinity. This is because the number of electrons has to be conserved. The time evolution of the resonance wavefunction is

$$|\Psi(r,t)|^2 = \exp(-2\epsilon_I t)|\Psi(r,0)|^2. \tag{4}$$

At infinite time, the integral over all space of this expression has to be finite.

From this expression it can also be seen that the lifetime, Γ, is related to the imaginary part of the resonance energy through, $\Gamma = 2\epsilon_I$.

A convenient solution to the Schrödinger equation is provided by the so-called complex scaling method[14, 15]. The idea here is to introduce a complex variable substitution in the radial coordinate r,

$$r \longrightarrow \exp(i\theta)r. \quad (5)$$

Upon this variable transformation the resonance boundary condition is changed to

$$\Psi(r) \longrightarrow \exp(i(k_R cos\theta + k_I sin\theta)r + (k_I cos\theta - k_R sin\theta)r). \quad (6)$$

If θ is chosen larger than arctan$\frac{k_I}{k_R}$, this boundary condition goes to zero for large r. This means that the resulting Hamiltonian can be diagonalized using an integrable basis. The advantage of simpler boundary conditions is at the expense of having to invert a complex nonhermitian Hamiltonian. This lengthens the computation time somewhat but is not a serious problem. The wave functions are expanded in a finite basis set consisting of generalized Laguerre polynomials. The Hamiltonian is then diagonalized. The accuracy of the calculations can be checked by investigating the dependence of the calculated eigenvalues on the parameter θ. For a complete set of basis functions there should be no θ dependence provided $\theta > arctan\frac{k_I}{k_R}$. Further details have been published elsewhere[16, 17, 18]

3. Results

In fig. 3 we show how the lowest excited H levels shift and broaden with distance from an Al surface(r_s=2). All levels shift upwards and become broader with decreasing atom-surface separation. We note that the degeneracy of the n=2 and n=3 states is lifted. The n=2 states are initially four-fold degenerate. As the atom comes close to the surface, three distinct states are formed that asymptotically derive from the atomic Stark states. These states are denoted $\psi_2^- = \psi_{2s} + \psi_{2p_z}$, $\psi_2^0 = \psi_{2p_{x,y}}$, and $\psi_2^+ = \psi_{2s} - \psi_{2p_z}$, and oriented away, along and towards the surface respectively. As the atom comes closer to the surface, the atomic wavefunction contains an increasing amount of surface and bulk states. The different levels show a complicated behavior with distance. The state oriented towards the surface increases in energy fastest due to the overlap with the surface electrons and eventually crosses the two other states. We note that the different states have very different lifetimes. The state oriented towards the surface, ψ_2^+, has about two orders of magnitude larger width than the state oriented towards vacuum ψ_2^-. This is because tunneling rates are exponential functions of distance and therefore strongly dependent on the orientation of the electronic state. The n=3 levels are initially nine-fold degenerate. The surface lifts some of the degeneracy and six distinct states are formed. At a given distance, the most longlived of the n=3 states are three to four orders of magnitude narrower than the broadest level.

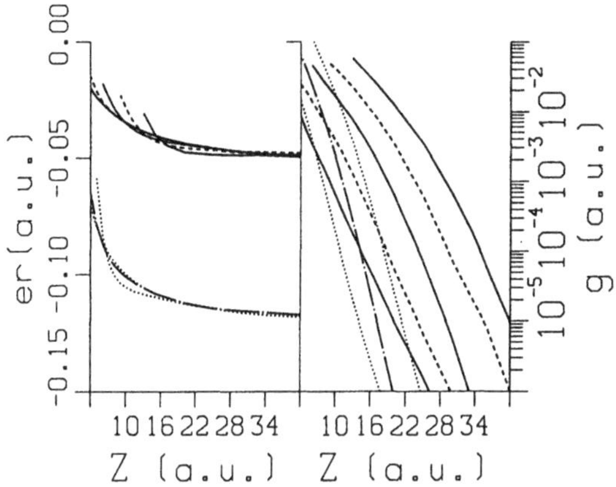

Figure 3: Calculated energy shifts (to the left) and widths (to the right) of the lowest excited hydrogen states as a function of distance outside an Aluminium surface. The dotted lines are the H(n=2,m=0) states. The chaindotted line is the H(n=2,m=1) state. The solid lines are the H(n=3,m=0) states and the dashed lines are the H(n=3,m=1) states. The energy is in a.u.

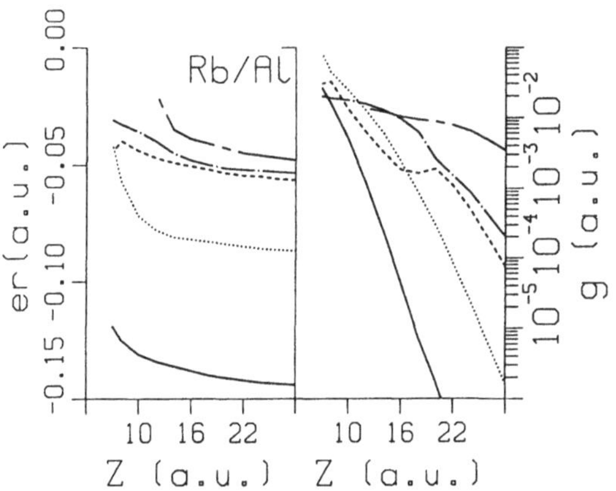

Figure 4: Calculated energy shifts (to the left) and widths (to the right) of the lowest Rb atomic states, m=0, as function of distance outside an Al surface. The states are labelled by their asymptotic limits. Atomic units are used.

In fig. 4 we show how the lowest excited levels of rubidium shift and broaden with distance on an Al surface. The shifts are qualitatively similar to those obtained for hydrogen but the lifetimes behave very differently. The widths of the 6s and 4d levels show considerable structure around Z=16 a.u.. This is due to hybridization. The spatial extent of the ψ_{6s} and ψ_{4d} wave functions is different and consequently these levels will shift differently with distance. For Z=16 a.u., the energies of these states are relatively close and the states can hybridize. It can be seen that the state that derives from the atomic 4d levels actually becomes narrower with decreasing atom-surface separation.

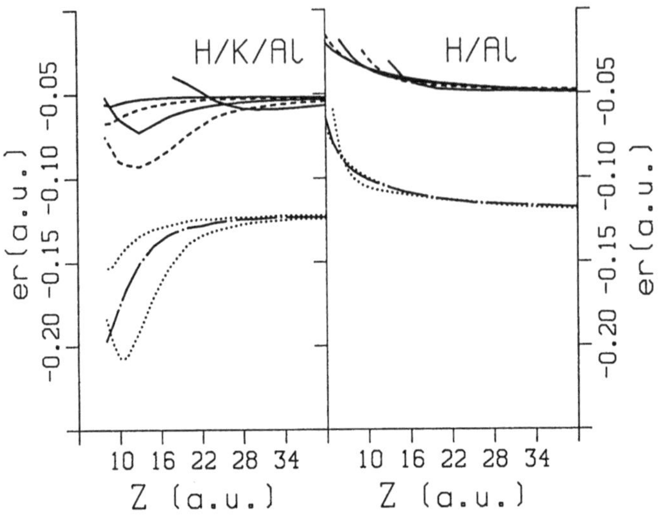

Figure 5: Calculated energy shifts for H outside a K atom chemisorbed on an Al surface (to the left) and outside a clean Al surface (to the right). The notation and units are as in fig. 3.

The variation of the width with distance is obviously non-exponential. Similar effects show up for all alkali atoms[17].

In fig. 5, we compare the calculated shifts as a function of distance for H(n=2) and H(n=3) states outside a clean Al and outside a potassium atom chemisorbed on Al. It can be seen that in the latter case the hydrogen levels shift downwards toward the surface. We also note that the splittings between the different n=2 and n=3 states is strongly increased. These effects are induced by the strong dipole field induced by the K. Such a large downshift of atomic levels near chemisorbed impurities means that charge transfer from the metal to the atom is facilitated. These topics will be further discussed in the next section.

Surface scattering or desorption experiments involve the motion of an atom relative to the surface. The probability for charge transfer will thus depend on the velocity, tunneling rates, and the relative position between occupied and unoccupied levels of the surface and the atom.

The standard theory for charge transfer involves the solution of the time-dependent Anderson Hamiltonian. This was first done in this context by Blandin et al.[19]. Subsequent applications of this method to a variety of dynamical processes have established the usefulness of this formalism[20, 21, 22, 23]. The Hamiltonian has the form

$$H(t) = \epsilon_a(t)n_a + \sum_k \epsilon_k c_k^+ c_k + \sum_k V_{ak}(t) c_k^+ c_a + h.c., \tag{7}$$

where h.c. is the hermitian conjugate. This Hamiltonian can be solved using the equations of motion for the operators c_a and c_k. Assuming finite temperature and a low constant normal velocity, $v_\perp$, and introducing the width, $\Delta(Z)$, of the atomic resonance, $|a>$,

$$\Delta(Z) = 2\pi \sum_k |V_{ak}(Z)|^2 \delta(\epsilon_a(Z) - \epsilon_k), \tag{8}$$

the solution can be obtained using the semi-classical approximation[18]:

$$\begin{aligned} < n_a(\infty) > &= < n_a(Z_0) > \exp(-\chi(Z_0)) \\ &+ \int_{Z_0}^{\infty} dZ f(\epsilon_a(Z), T) \frac{\Delta(Z)}{v_\perp} \exp(-\chi(Z)). \end{aligned} \tag{9}$$

In this expression, $n_a(Z_0)$ is the initial population of $|a>$ and f(ϵ) is the Fermi-Dirac distribution. $\chi(Z)$ is related to $\Delta(Z)$ through,

$$\chi(Z) = \int_Z^{\infty} \frac{dZ_1}{v_\perp} \Delta(Z_1). \tag{10}$$

This expression has a simple interpretation. The first term describes the memory of the initial configuration at Z_0. The second term is the non-adiabatic term describing the charge transfer.

4. ESD of excited H from alkali promoted metals

The detection of excited atoms in stimulated desorption provides a particularly interesting means of studying charge transfer reactions between atoms and surfaces. In these experiments, the initial state of the desorbing species can be studied separately and therefore be relatively well characterized.

In recent ESD experiments[24, 25], the desorption of hydrogen from Ni and Pt was studied as a function of work function (alkali coverage). The formation of H(n=2) and H(n=3) was detected from the intensity of Lyman-α and Lyman-β radiation emitted from the desorbing particles. It was found that the formation of H(n=2) and H(n=3) increases monotonically with alkali coverage even for work functions as high as 5 eV. The onsets of H(n=2) and H(n=3) occur over a much broader region of work functions and for much larger work functions than what would be predicted for desorption from a clean surface. On a clean metal, the H levels shift relatively uniformly upwards. The thresholds for formation of H(n=2) would be expected to be relatively narrow and located at a work function of around 1 eV. From the results in section 3, it is obvious that if the desorbing H atoms originate from the vicinity of the coadsorbed alkali atoms, their charge transfer dynamics will be modified. Since in this region the atomic levels are downshifted and the splittings between the different hybrids have increased one would expect larger and smoother work function thresholds.

The adsorption and desorption properties of hydrogen on alkali covered metal surfaces has recently been the subject of an extensive study[26], in which is was concluded that the desorbing hydrogen originates from alkali-hydride-like surface complexes on the metal, with the H terminally bound to the the alkali impurity.

The excited state desorption yield for a H/K/Metal system can be calculated using eq. (9). The only unknown parameters relate to the trajectory of the desorbing H. The bond length between H and K is taken as 4 a.u., corresponding to the free KH molecular bond length. The motion of H is assumed to be normal to the surface in agreement with the ESDIAD studies[26]. The kinetic energy of the H is taken to be 1 eV. These assumptions are entirely realistic and the results of the calculation do not depend sensitively on these parameters.

For hydrogen, the formation of the different excited states is in principle complicated by intra atomic correlation effects. An exact treatment of the time dependent Anderson Hamiltonian including such effects is complicated. For the work functions in the present experiment, such problems are avoided however. The downward shifts of the H(n=3) excited state are insufficient to allow tunneling from the metal into these states. The occurence of any H(n=3) excited state would therefore arise purely from the memory term. Only the most longlived of the H(n=3) states, ψ_3^-, can survive the passage out from the surface. With the above trajectory parameters, this state has a survival probability of 25% if formed initially. The remaining H(n=3) states decay immediately if formed near the surface. The ψ_2^- state is long lived and only contributes to the H(n=2) yield through the memory term. The survival probability for this state is 60%. For the description of the formation of the remaining H(n=2) states, we note that the ψ_2^+ level and the ψ_2^0 level have such short lifetimes that they will be populated immediately if their energy is below the Fermi level near the surface. These levels can therefore effectively be treated as a threefold degenerate ψ_2^0 level. The width of this level is so large that the memory term for this state can be neglected. The ψ_2^0 and ψ_2^+ states are the only states that can be formed by resonant tunneling from the metal. Since all other excited states except ψ_2^- and ψ_3^- will decay in the vicinity of the surface, the yield of positive ions, P^+, desorbing from the surface will be determined by the probability of formation of ψ_2^0 and ψ_2^+. Since desorption can only occur from a KH surface complex, the total yield of desorbing particles will be proportional to the alkali coverage, Θ_K. The resulting expressions for the population of the H(n=2) and H(n=3) states are

$$\begin{aligned} n_3^-(\infty)_\phi &= k\Theta_K P_3^- \exp(-\chi_3^-(Z_0)) \\ n_2^-(\infty)_\phi &= k\Theta_K P_2^- \exp(-\chi_2^-(Z_0)) \\ n_2^0(\infty)_\phi &= [k\Theta_K - n_3^-(\infty) - n_2^-(\infty)] \\ &\times \int_{Z_0}^{\infty} dZ f(\epsilon_a(Z),T)\frac{\Delta(Z)}{v_\perp}\exp(-\chi_2^0(Z)) \\ P_\phi^+ &= k\Theta_K - n_2^{tot}(\infty)_\phi - n_3^{tot}(\infty)_\phi \end{aligned} \tag{11}$$

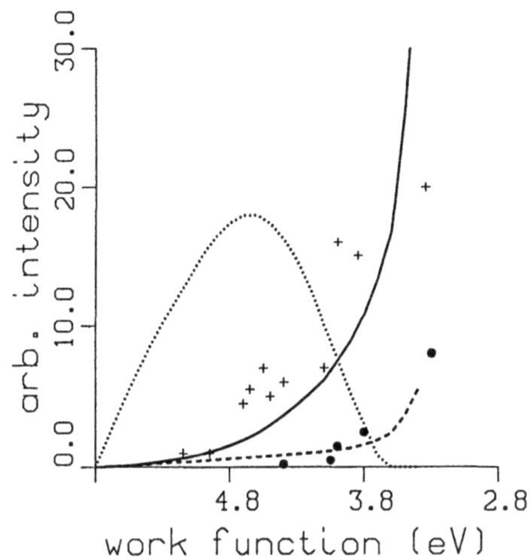

Figure 6: Calculated Lyman-α intensity (solid line) and Lyman-β intensity (dashed line) as function of work function for ESD of H from potassium covered Pt(001). The dash dotted line is the positive ion intensity, P^+, multiplied by a factor of 5. The experimental points are from ref[25]. The intensity unit is arbitrary.

The relation between work function and alkali coverage has been measured[25], T is the surface temperature (300K). $\Delta(Z), \epsilon_a(Z)$ and $\chi(Z)$ are calculated for the different states respectively. We assume that all states are populated initially with equal probability, P.

These expressions can readily be evaluated as function of alkali coverage. The results of the calculation are shown in fig. 6 along with the experimentally determined points.

It can be seen that the agreement between theory and experiment is excellent. Both the H(n=2) and H(n=3) intensity increases linearly at small coverages (large work functions). This is due to the memory terms for the respective states. We also note that the yield of H(n=3) is much lower than the yield of H(n=2). This is due to the differences in lifetimes between the levels. At a work function of around 5 eV, the intensity of H(n=2) starts to increase more rapidly. This is due to the possibilities of formation of H(n=2) at smaller work functions. At a work function of around 3.5 eV, the slope of the work function versus coverage relation decreases, leading to a correspondingly larger increase in alkali coverage. This leads to a more rapid increase of the yield of excited H. We have here neglected the proposed increase in bond length[27] between adsorbed alkali atoms and the metal surface below this work function. A proper inclusion of this effect would result in a slightly larger slope of the H(n=3) intensity curve for work functions below 3.5 eV, since the survival probability for an excited state is larger the further out from the surface the atom starts.

The decrease in the total ion yield for work functions smaller than 4.5 eV is in good agreement with the findings in ref.[26]. The reason for this behavior is the efficient formation of H(n=2). This process could not occur on the clean metal surface since the H levels would be image shifted towards the vacum level.

5. Discussion and conclusions

It has been shown how accurate calculations of atomic level shifts and broadening can be performed outside a realistic model of a surface. In particular it has been demonstrated that chemisorbed impurities can strongly influence both the shifts and broadening of the atomic levels. Finally it has beeen demonstrated that these calculations can be combined with a dynamical theory to describe the formation and survival of excited H desorbing from an alkali promoted metal surface. The good agreement between the theory and the experiment support the proposed theoretical model. The author acknowledges many useful discussions with J.C. Tully. This work was supported by the National Science Foundation under grant no. DMR-88-01027.

References

[1] B. Lundqvist, in *Many-body phenomena at surfaces, Eds. D.C. Langreth and H. Suhl*, Academic Press, New York, 1984.

[2] T. E. Madey, Science **234**, 316 (1986).

[3] D. Menzel and R. Gomer, J. Chem. Phys. **41**, 3311 (1964).

[4] P. E. Redhead, Can. J. Phys. **42**, 886 (1964).

[5] P. J. Feibelman and M. L. Knotek, Phys. Rev. B **18**, 531 (1978).

[6] W. Ho, Comments Cond. Mat. Phys. **13**, 293 (1988).

[7] R. Wolkow and P. Avouris, Phys. Rev. Lett. **60**, 1049 (1988).

[8] J. N. M. van Wunnik, R. Brako, K. Makoshi, and D. Newns, Surf. Sci. **126**, 618 (1983).

[9] N. D. Lang, Solid State Physics **28**, 225 (1985).

[10] S. Ossicini, C. M. Bertoni, and P. Gies, Europhys. Lett. **1**, 661 (1986).

[11] N. D. Lang, Phys. Rev. B **7**, 3541 (1973).

[12] J. N. Bardsley, Case Studies in Atomic Physics **4**, 299 (1974).

[13] N. D. Lang, S. Holloway, and J. K. Norskov, Surf. Sci. **150**, 24 (1985).

[14] W. P. Reinhardt, Ann. Rev. Phys. Chem. **33**, 223 (1982).

[15] B. R. Junker, Adv. At. Mol. Phys. **18**, 207 (1982).

[16] P. Nordlander and J. C. Tully, Phys. Rev. Lett. **61**, 990 (1988).

[17] P. Nordlander and J. C. Tully, Submitted to Phys. Rev. B (1989).

[18] P. Nordlander and J. C. Tully, Surf. Sci **211**, 207 (1989).

[19] A. Blandin, A. Nourtier, and D. Hone, J. Physique **37**, 369 (1976).

[20] J. C. Tully, Phys. Rev. B **16**, 4324 (1977).

[21] J. K. Norskov and B. I. Lundqvist, Phys. Rev. B **19**, 5661 (1979).

[22] R. Brako and D. M. Newns, Surf. Sci. **108**, 253 (1981).

[23] N. D. Lang, Phys. Rev. B **27**, 2019 (1983).

[24] P. D. Johnson, X. Pan, J. Tranquada, S. L. Hulbert, and E. Johson, in *Proc. DIET-III*, Springer-Verlag, Berlin Heidelberg, 1988.

[25] P. D. Johnson, A. J. Viescas, P. Nordlander, and J. C. Tully, To be published (1989).

[26] A. M. Lanzillotto, M. J. Dresser, M. D. Alvey, and J. Y. Jr., J. Chem. Phys. **89**, 570 (1988).

[27] G. Lamble, R. Brooks, D. King, and D. Norman, Phys. Rev. Lett. **61**, 1112 (1988).

Image-Screening of Positive Ions at Metal Surfaces, and Electronically Stimulated Desorption of Noble Gas Atoms

R.E. Walkup, Ph. Avouris, and N.D. Lang

IBM Research Division, Thomas J. Watson Research Center,
P.O. Box 218, Yorktown Heights, NY 10598, USA

Abstract. Diabatic potential energy curves are reported for ions interacting with metal surfaces. The calculations are based on a local density functional approach, which fully accounts for screening by the electrons within the metal valence band. Results for Ar^+ are used to predict yields and energy distributions for Ar atoms which desorb upon ionization of physisorbed Ar. The calculated potentials provide a substantially improved description of the desorption process compared to previous models based on a classical image potential for the ion-surface interaction. Classical and quantal treatments of desorption are compared, and quantum effects in the dynamics of desorption are discussed.

1. Introduction

The potential energy surface plays a fundamental role in desorption induced by electronic transitions. In this article we discuss potential energy surfaces for simple model systems consisting of an atom adsorbed on a metal surface. Starting from the ground electronic state of the system, we consider electronically excited states that are formed by electron or photon impact, resulting in the creation of a hole in a level located below the bottom of the metal valence band. For such excited states, the hole can only be filled by Auger neutralization, thus the lifetime of the excited state should be long enough for substantial nuclear motion to occur [1]. Depending on the nature of the potential energy surfaces, such an electronic transition may result in the desorption of ions, excited-state neutral atoms, or ground-state atoms. First we discuss the potential energy surfaces for atom-metal systems, and then we discuss nuclear motion leading to desorption, with particular emphasis on the desorption of noble gas atoms.

2. Discussion

We have calculated potential energy (PE) curves for a variety of positive ion - metal surface systems using the local-density-functional (LDF) approach of Lang and Williams [2]. In this treatment, the metal is regarded as a semi-infinite medium with a uniform positive background charge density. An atomic nucleus is placed a distance, d, from the positive background edge of the metal, and the resulting electron density is calculated self-consistently, using the Kohn-Sham density-functional theory. This method is useful for studies of the ground electronic state of chemisorbed [2] and physisorbed [3] atoms. The method also applies for excited-states formed by creation of a hole in a level of the adsorbate located below the bottom of the metal

Springer Series in Surface Sciences, Vol. 19
Desorption Induced by Electronic Transitions, DIET IV
Editors: G. Betz · P. Varga

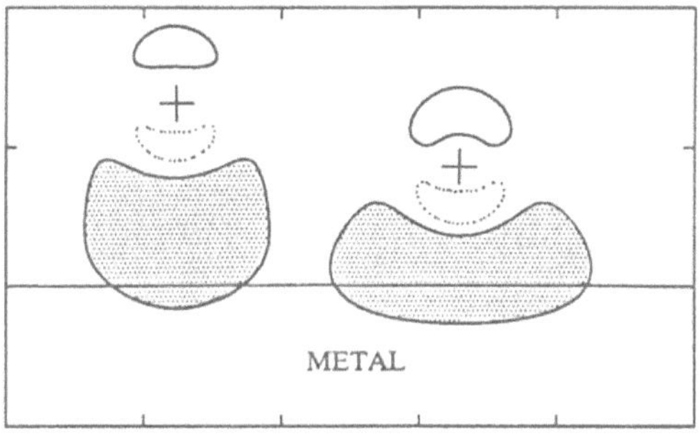

Figure 1. Charge density difference contours, $\rho_{ion+metal}$ - $(\rho_{free\ ion} + \rho_{bare\ metal})$, are shown for Ar^+ at distances of 6.5 bohr (left) and 4.25 bohr (right) from the positive background edge of the metal (indicated by the horizontal line). The position of the Ar nucleus is shown (+), and the densities shown are $\pm$ 0.001 electrons $bohr^{-3}$. The charge bound to the Ar^+ ion is polarized away from the metal, and there is an induced image-like charge density on the metal (shaded area). As the ion approaches the surface, the center-of-gravity of the image-like density moves toward the metal.

valence band. In this case the LDF calculations converge to the lowest energy state with the specified core-hole occupancy. The important feature of this method is that screening by electrons within the metal valence band is fully accounted for. Thus image effects and charge-transfer screening are automatically included. Previous calculations for such diabatic excited states have been reported for Cl adsorbed on metallic sodium [4] and F adsorbed on aluminum [5].

Consider a singly ionized atom, with a hole in a level located below the bottom of the metal valence band. At large distances, the ion-surface interaction will be attractive due to the induced image charge density, which forms at the metal surface. For distances that are very large on the atomic scale, the interaction is given by the classical image potential, $V_i(x) = -e^2/[4(x - x_i)]$, where x_i is the location of the image plane [6] (i.e. the center of gravity of the induced surface charge distribution). However, when the positive ion is in the near-surface region, there is overlap between the metal and ion charge densities, and the ion-surface potential will differ from the classical image form. We have calculated PE curves for a number of atomic ions including H^+, O^+, F^+, He^+, Ne^+, Ar^+, and Kr^+; and we have found that the classical image potential systematically over-estimates the attractive force in the near-surface region. The reason for this can be seen from examination of the charge densities, as shown in Fig. 1 for Ar^+ interacting with a metal characterized by $r_s = 3$ bohr. When a positive ion approaches the surface region, the induced image-like charge density is first attracted toward the ion. Thus the center-of-gravity of the image-like charge is shifted outward compared to the asymptotic location of the image plane. There is also a repulsive interaction between electrons bound to the positive ion and electrons in the image-like distribution. As a result, the image-like charge distribution is excluded from the immediate vicinity of the ion. As the ion moves toward the surface, the image-like distribution is pushed toward the metal - see Fig. 1. Thus the center-of-gravity of the image-like charge moves inward as the ion moves toward the surface. This is illustrated in Fig. 2, which shows the center-of-gravity of the induced image-like charge for Ar^+ as a function of ion position. The shift of the effective image plane results in a reduction of the attractive force compared to the classical image

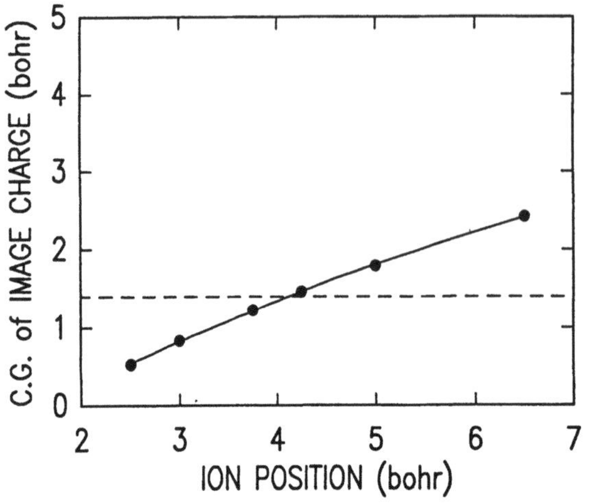

Figure 2. The center-of-gravity (C.G.) of the induced image-like charge is shown vs ion position for Ar^+ interacting with a metal characterized by r_s = 3 bohr. The asymptotic location of the image plane is indicated by the horizontal dashed line at 1.4 bohr.

formula with a fixed image plane. There is also some transfer of charge from the metal into large-radius orbitals of the adsorbate. This reduces the effective ion charge, and this also tends to reduce the attractive image-like force.

The role of charge-transfer into large radius orbitals is controlled by the relationship between the binding energy of the excited neutral atom and the work function of the metal. This relationship determines the location of the A*-M asymptote relative to the $A^+ - M^-$ asymptote, where A* and A^+ refer to excited and ionized atomic states and M represents the metal - see Fig. 3. For many ion-metal systems, the LDF potential curves approach an excited neutral asymptote because the binding energy of the lowest excited atomic state is often greater than the metal work function. However, in the near-surface region an ionic configuration is generally favored due to the attractive image-like interaction. For Ar^+ and a metal characterized by $r_s = 3$ bohr, the $Ar^*(3p^5 4s)$- M asymptote is ≃0.8 eV below the Ar^+ - M^- asymptote. According to the classical image formula, the ionic curve should cross below the excited neutral curve at ≃ 10 bohr from the positive background edge. Thus, in the near-surface region, the ionic curve is substantially below the excited neutral curve. In this case, there is little charge transfer into large radius orbitals, and the lowest energy state with a core-hole is strongly ionic in character. However, for systems with very strongly bound excited states, such as Ne and F, the lowest energy state with a single core-hole has substantial excited neutral character in the near-surface region. Such excited-state mixing effects were pointed out in an earlier study of fluorine on aluminum [5]. A particularly striking example of charge transfer into large radius orbitals is provided by the states of noble gas atoms resulting from the formation of two holes in the outermost shell. In this case, doubly-excited neutral states are considerably lower in energy than the double positive ion. In particular, for Ar with two holes in the 3p level, the n = 4 levels are pulled down in energy, and they overlap the metal valence band. These levels are occupied by ≃2 electrons due to charge transfer from the metal; thus the lowest state with two holes in the 3p level is essentially an excited

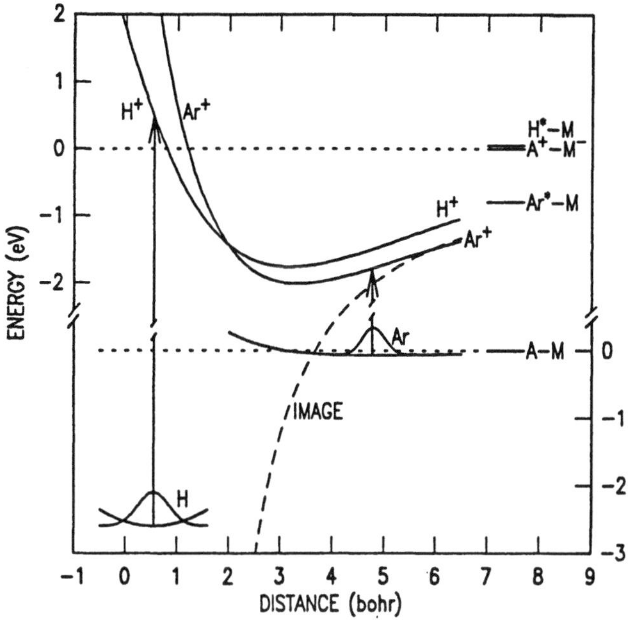

Figure 3. Potential energy curves are shown for chemisorbed H, H^+ , physisorbed Ar, and Ar^+ interacting with a metal characterized by $r_s = 3$ bohr. Distance is measured from the positive background edge of the metal. The ground-state curves (right scale) show the energy measured from the neutral atom - metal asymptote, indicated by A-M. The ionic curves (left scale) show the energy measured from the asymptote corresponding to a free positive ion and an electron bound to the metal, indicated by $A^+ - M^-$. The location of the lowest excited neutral states for H* and Ar* are also shown. The diabatic LDF curve labeled Ar^+ correlates with the excited neutral state (Ar* − M) at very large distances. The classical image potential (dashed curve) is also shown.

neutral state. Such charge transfer processes play a crucial role in the determination of excited-state lifetimes as well as the potential energy surfaces.

In Fig. 3, the ground-state and excited-state potential energy curves are shown for atomic hydrogen and atomic argon [7] adsorbed on a metal substrate characterized by r_s = 3 bohr. Atomic hydrogen exemplifies the behavior of chemisorbed atoms, and atomic Ar exemplifies that of physisorbed atoms. For hydrogen, the minimum in the ground-state PE curve occurs for rather small atom-metal separations. At such small distances, the effect of the discrete metal-ion cores is quite substantial; and the actual PE surfaces will differ from the results of the uniform positive background model [2,8]. Nevertheless, the minimum in the ground-state potential for H will occur for much smaller atom-surface distances than the corresponding minimum in the H^+ PE curve. Thus an electronic excitation that creates a hole in the 1s level of chemisorbed H may correspond to a transition to the repulsive part of the H^+ - metal PE curve, and direct desorption of H^+ is possible. This desorption mechanism corresponds to the model originally proposed by Menzel, Gomer [9], and Redhead [10]. The H^+ - metal PE curve is repulsive at short distances primarily because of Coulomb repulsion between the proton and the positively charged metal ion cores (which are smeared out to form the uniform positive background in the model). At larger distances from the surface, this repulsion is screened, and the induced image-like charge density results in an attractive interaction. For Ar, the form of the ion-surface PE curve is quite similar, but the prospects for desorption are completely different. In this case, the minimum in the ground electronic state is very shallow and occurs at a rather large atom-surface separation. At such large atom-surface distances, the uniform positive background model of the metal is reasonably accurate. Upon ionization of physisorbed Ar (i.e. creation of a hole in the 3p level), the initial force is directed toward the surface. The Franck-Condon transition is to bound regions of the ionic PE curve, thus no Ar^+ desorption is expected. However, if the Ar^+ ion gains sufficient kinetic energy on the ionic curve, then upon Auger neutralization, a neutral Ar atom can recoil from the surface and desorb. This mechanism corresponds to the model proposed by Antoniewicz [11]. It is interesting to note that the PE curves for singly ionized atoms and metal surfaces are very similar for all of the systems that we have investigated. This similarity is quite evident in the H^+ and Ar^+ potentials shown in Fig. 3. These two species have very different prospects for desorption because of the properties of the ground electronic states: the ground-state bonding determines the part of the ionic curve accessed by a Franck-Condon transition, and this, along with neutralization processes, determines the range of motion of the positive ion.

In the desorption of noble gas atoms via the Antoniewicz mechanism, the attractive force acting on the ion plays a very important role. This force determines the kinetic energy gained by the ion during the average lifetime for Auger neutralization. Thus the yield and kinetic energy distribution of desorbed neutral atoms will depend sensitively on the ionic PE curve. In order to make a decisive comparison with experimental data, we have carried out both classical and quantal treatments of noble gas desorption [7], using the LDF potential curves for the physisorbed ground-state as well as the ionic state. Intuitively, one may expect that a quantal treatment of the problem is needed because the wavepacket describing the position of the nucleus is quite broad, owing to the fact that the physisorption well is very soft. In the quantal approach, a term $-iW(x)$ is added to the potential energy in order to describe the net effect of Auger neutralization, as discussed by Brenig and co-workers [12]. $W(x)$ is directly related to the neutralization rate, $\Gamma(x) = 2W(x)/\hbar$, which is allowed to vary with distance from the metal surface. The probability for neutralization to result in a given eigenstate, Ψ_f, of the ground state is [13]: $\mathrm{Prob}(\Psi_f) = \int_0^\infty dt \, | \langle \Psi_f | \Gamma(x)^{1/2} | \phi_i(x,t) \rangle |^2$, where $\phi_i(x,t)$ is the time-dependent wavepacket describing nuclear motion on the ionic PE curve. The wavepacket, $\phi_i(x,t)$, was obtained by direct numerical solution of the time-dependent Schrödinger equation subject to the condition that at $t=0$, $\phi_i(x,t)$ is the lowest bound-state wavefunction of the physisorption well, as appropriate for a Franck-Condon transition to the ionic PE curve at $T=0$ K. The experimental quantities of interest are the yield, Y, and the kinetic-energy (flux) distribution, F(E), of the desorbed neutral atoms. These were evaluated by the relationships $Y = \int_0^\infty dk \, \mathrm{Prob}(\Psi_k)$ and $F(E) \propto k \, \mathrm{Prob}(\Psi_k)$, where Ψ_k is a suitably normalized continuum eigenstate of the ground-state system characterized by a kinetic energy $E = (\hbar k)^2/(2m)$. The only "free" parameters in our treatment are those that characterize the Auger neutralization rate: $\Gamma(x) = \Gamma_0 (x/x_0)^2 \exp[-b(x - x_0)]$, where x is the distance measured from the outermost internuclear plane and x_0 is the location of the bottom of the physisorption well. The Auger matrix element scales as the square of metal wavefunctions averaged

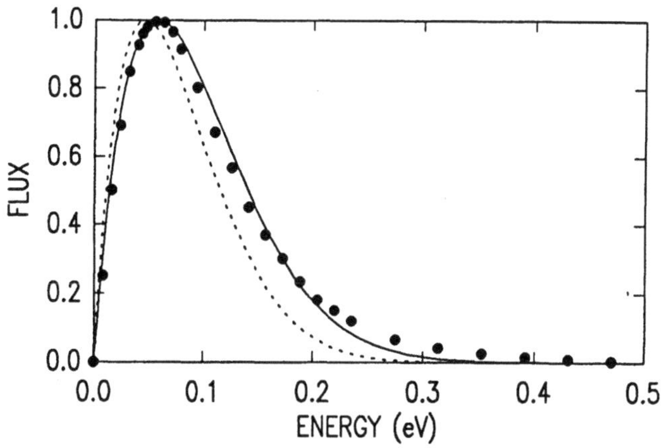

Figure 4. Kinetic energy distributions for neutral argon desorbed via the Antoniewicz mechanism are shown. The experimental data of Steinacker and Feulner [15] are shown (dots) together with quantal (solid line) and classical (dotted line) predictions. The theoretical curves use the same initial ion distribution and Auger decay parameters. All curves are scaled to a peak value of 1.0.

over the adsorbate position; thus $\Gamma(x)$ should be roughly proportional to the square of the metal electron density vs. distance from the surface [14], and the exponential factor b is expected to be $\sim 5\text{Å}^{-1}$.

A comparison of the calculated kinetic energy distribution for Ar with the recent experimental measurements of Steinacker and Feulner [15] is shown in Fig. 4. Good agreement is achieved by choosing $\hbar\Gamma_0 = 0.03$ eV and $b = 6$ Å^{-1}. For these values the net desorption yield is predicted to be $\simeq$ 12%. This is the correct order of magnitude, but it is below the experimental estimate of $\simeq$64% [15]. The discrepancy in the yield is due in part to the neglect of inelastic effects in the model. Using classical trajectory calculations, we estimate that ~20% of the Ar kinetic energy is transferred to metal phonons and/or neighboring adsorbates. If an allowance for inelastic effects of this magnitude is made, similar agreement with the experimental kinetic energy distribution can be attained with yield estimates of ~ 20%. This is still a factor of 3 below the experimental value. The experimental estimate [15] is based on the assertion that the neutral Ar yield is unity when Ar^{+2} is formed. However, our results suggest that doubly ionized Ar should be charge-transfer screened to form an excited neutral state. Thus a fast intra-atomic Auger decay process could occur, resulting in Ar^+ and an ejected electron. Based on data for doubly excited states of gas-phase Ar [16], Auger widths of ~0.02 eV are expected. For rates of this magnitude, there should be a substantial probability for Auger decay to occur before the Ar atom gains much kinetic energy on the excited-state PE curve. If this is the case, then the experimental yield estimate is too large. We expect that better agreement between experiment and theory will come with more accurate absolute yield data and further improvements in the desorption model. In any case it is clear that the LDF potentials provide a much better description than previous models based on the classical image potential. The image potential over-estimates the initial attractive force by a factor of ~2; and the image potential rapidly becomes too attractive. Thus, in order to get the correct peak kinetic energy one must over-estimate the neutralization rate. This results in an

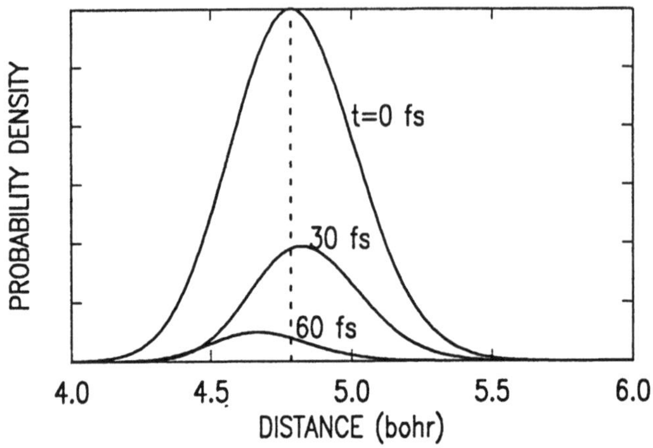

Figure 5. The ionic wavepacket, $|\phi(x,t)|^2$, is shown for ionization of physisorbed Ar at t=0 (upper curve), 30 fs (middle curve), and 60 fs (lower curve). Due to the spatial variation of the neutralization rate, the initial motion of the wavepacket is away from the surface, i.e. opposite to the direction of the classical force.

underestimate of the net desorption yield. For example, using the classical image potential, essentially the same energy distribution as shown in Fig. 4 is obtained with $\hbar\Gamma_0 \simeq 2.0$ eV, and $b \simeq 20$ Å^{-1}. For these parameters, the corresponding yield estimate is ~2.8%. The neutralization parameters obtained by using the classical image potential are not realistic: the width is considerably too large (i.e. much greater than experimental widths [17]), and the variation of the decay rate with distance is too steep to be physically meaningful. Similar difficulties in fitting the neutralization parameters were noted by Moog et al. [18]. In contrast, reasonable neutralization parameters are obtained with the LDF potentials, along with a significantly larger calculated yield. Thus the LDF results provide a much improved description of the desorption process; but a more detailed model accounting for inelastic effects is needed for accurate quantitative predictions.

A more direct probe of the force resulting from ionization of a physisorbed atom is provided by measurements of the vibronic contribution to photoemission linewidths [19]. This width is essentially equal to the slope of the ionic PE curve times the width of the initial coordinate distribution. Experimental data for Ar on Pt(111) [17] are in good accord with the predictions from the LDF PE curves [7], thus providing strong support for the LDF potentials.

The motion of the ion wavepacket has some interesting features due to the spatial variation of the neutralization rate. This is illustrated in Fig. 5, which shows the probability density in the ionic state at t=0, 30 fs, and 60 fs. The part of the wavepacket closest to the surface experiences a larger decay rate, and this results in an outward displacement of the wavepacket at early times, as noted by Hübner and Brenig [20]. In contrast, a classical particle would move toward the surface, in the direction of the classical force. For a general understanding of the desorption process, it is important to consider particle motion in the presence of a spatially varying decay rate. Some physical insight into this problem can be gained by re-writing the time-dependent Schrödinger eqn. in fluid-flow form:

$$\frac{\partial \rho}{\partial t} + \frac{\partial}{\partial x}(\rho \upsilon) = -\Gamma(x)\,\rho$$

$$\frac{\partial J}{\partial t} + \frac{\partial}{\partial x}(J\upsilon) = \frac{\rho}{m}[F_C + F_Q] - \Gamma(x)\,J$$

where ρ and J are the probability and current densities, and $\upsilon = J/\rho$, $F_C = -\frac{\partial V}{\partial x}$, and $F_Q = \frac{\hbar^2}{4m}[\frac{\partial^3 \ln(\rho)}{\partial x^3} + \frac{\partial \ln(\rho)}{\partial x}\frac{\partial^2 \ln(\rho)}{\partial x^2}]$. F_C is the classical force and F_Q is the quantal "force". The time-dependent Schrödinger equation including damping via a local complex potential, is identical to the classical equations of motion for a fluid of non-interacting particles with a spatially varying loss rate, except for the addition of a quantal "force" term, F_Q, which depends on the probability density, ρ. At T=0 K the initial distribution is determined by the ground-state wavefunction, ψ_0, in accordance with the Franck-Condon principle. The same initial conditions can be employed in a classical description by choosing $\rho = |\psi_0|^2$ and $\upsilon = 0$. One can then estimate the difference between classical and quantal dynamics by comparing F_C and F_Q. For a Gaussian distribution, $F_Q = \hbar^2(x - x_c)/[mw^4]$ where x_c is the centroid position and w is the width (half width at 1/e points of the probability density). The quantal "force" tends to spread the distribution. Since F_Q has opposite signs on opposite sides of the distribution there is a partial cancellation of the net quantum correction for spatially averaged quantities. Upon ionization of physisorbed Ar, $F_C \simeq -0.5$ eVÅ^{-1} whereas $F_Q \simeq 0.03(x - x_c)/w$ eVÅ^{-1}. Thus over the bulk of the distribution, the classical force dominates, and the initial motion will be essentially classical. In particular, the initial outward motion of the wavepacket shown in Fig. 5 would also occur for the motion of a classical distribution function. Neutralization preferentially depletes the part of the distribution closest to the surface, thus there is an effective outward motion of the distribution even though the motion of each classical particle within the distribution is inward. Roughly speaking, classical dynamics is good when $|F_C| > |F_Q|$ over the relevant time period. For Ar^+, this criterion is approximately satisfied; thus in this case it is a fair approximation to treat ion motion classically starting with the quantum-mechanically correct initial distribution function, as, for example, in the model of Moog et al. [18]. A comparison of classical and quantal predictions of the Ar kinetic energy distribution is shown in Fig. 4. The quantal prediction is somewhat broader, in better agreement with experiment. The classical prediction for the yield is ≃12%, in agreement with the quantal result.

In general, the importance of quantum effects in electronically-stimulated desorption depends critically on the value of the classical force resulting from the electronic transition. If the transition is to a steeply sloped excited-state PE curve, then the classical force will be large and the motion will be essentially classical. It is seen that a classical treatment of the dynamics can be rather good, provided that the width of the initial distribution of coordinates is accounted for. In fact a large wavepacket width makes the motion more classical since this reduces the quantum "force" in comparison to the classical force. The particle mass enters into the problem only weakly because the wavepacket width decreases as the mass increases, thus partially compensating for the m^{-1} term in the quantal "force". Thus, for example, quantum effects in the dynamics are of comparable importance for Ar and Kr. Quantum effects could be quite important if the neutral and ionic PE curves had closely similar well positions, but different curvatures, as suggested by Jennison et al. [21].

3. Conclusion

In summary, we have carried out self-consistent calculations of diabatic PE curves for ions interacting with metal surfaces. For intermediate distances from the surface, the attractive force is considerably smaller than predicted by the classical image potential. The reduced attractive force is attributed to a displacement of the effective image plane with ion position and (for certain systems) to mixing with excited-neutral states. The resulting PE curves provide an improved description of a variety of dynamical processes such as desorption stimulated by the ionization of physisorbed atoms, and Franck-Condon broadening of lines observed by photoemission spectroscopy. The role of quantum effects on ion motion was also investigated. For noble-gas desorption, a classical treatment of ion motion is qualitatively correct; however, a quantal approach is needed for an accurate description of the kinetic energy distributions.

References

1. Ph. Avouris and R. E. Walkup, Annu. Rev. Phys. Chem. 40 , 173 (1989)
2. N. D. Lang and A. R. Williams, Phys. Rev. B 18 , 616 (1978)
3. N. D. Lang, Phys. Rev. Lett. 46 , 842 (1981)
4. N. D. Lang, J. K. Norskov, and B. I. Lundqvist, Physica Scripta 34 , 77 (1986)
5. Ph. Avouris, R. Kawai, N. D. Lang, and D. M. Newns, Phys. Rev. Lett. 59 , 2215 (1987); and J. Chem. Phys. 89 , 2388 (1988)
6. N. D. Lang and W. Kohn, Phys. Rev. B 7 , 3541 (1973)
7. R. E. Walkup, Ph. Avouris, N. D. Lang, and R. Kawai, to be published
8. H. Hjelmberg, Surface Sci. 81 , 539 (1979)
9. D. Menzel and R. Gomer, J. Chem. Phys. 41 , 3311 (1964)
10. P. A. Redhead, Can. J. Phys. 42 , 886 (1964)
11. P. R. Antoniewicz, Phys. Rev. B 21 , 3811 (1980)
12. W. Brenig, Z. Physik B 23 , 361 (1976); and P. Schuck and W. Brenig, Z. Physik B 46 , 137 (1982)
13. Z. W. Gortel, R. Teshima, and H. J. Kreuzer, Phys. Rev. B 37 , 3183 (1988)
14. N. D. Lang and W. Kohn, Phys. Rev. B 1 , 4555 (1970)
15. E. Steinacker and P. Feulner, Phys. Rev. B (in press)
16. R. P. Madden, D. L. Ederer, and K. Codling, Phys. Rev. 177 , 136 (1969)
17. G. Schönhense, B. Kessler, N. Müller, B. Schmiedeskamp, and U. Heinzmann, Physica Scripta 35 , 541 (1987)

18. E. R. Moog, J. Unguris, and M. B. Webb, Surface Science 134 , 849 (1983)

19. J. W. Gadzuk, S. Holloway, C. Mariani, and K. Horn, Phys. Rev. Lett. 48 , 1288 (1982)

20. W. Hübner and W. Brenig, Z. Phys. B 74 , 361 (1989)

21. D. R. Jennison, E. B. Stechel, and A. R. Burns, pg. 167 Desorption Induced by Electronic Transitions III, ed. R. H. Stulen and M. L. Knotek, Springer-Verlag, Berlin (1988)

Trajectory Calculations of Electron Stimulated Desorption of Positive and Negative Ions

C.Z. Dong, P. Nordlander, and T.E. Madey*

Rutgers, The State University, Department of Physics and
Laboratory for Surface Modification, P.O. Box 849, Piscataway, NJ 08855, USA
*Present address: Rice University, Department of Physics,
Houston, TX 77251, USA

1. Introduction

An important issue in ESDIAD (electron stimulated desorption ion angular distribution) of positive and negative ions from surfaces concerns the influence of final-state effects on the trajectories of desorbed ions. It is believed that the initial repulsive forces propelling ESD ions from surfaces often originate from 2 hole repulsive states which act along the original chemical bond direction[1]. However, final state effects, including image field effects, reneutralization, and coadsorbed impurity ions can influence the measured ion desorption angles by distorting the ion trajectories and thus the angular distributions. Several recent papers have described classical trajectory calculations in which the image force [2,3] and reneutralization [4] were included.

The purpose of the present paper is to present classical trajectory calculations which illustrate the influence of coadsorbed impurity ions on the trajectories of ESD ions from surface molecules. We model the desorption of energetic positive and negative ions from a conducting surface containing either positively or negatively charged impurity ions in close proximity to the site of the desorbing ion. Classical trajectory calculations are performed assuming both screened classical image potential interactions, as well as a more realistic potential (density functional surface potential) for the desorbing ions. Variables include the charge, energy and initial desorption angle of the desorbing ion, and the fractional charge and location of the impurity ion. Neutralization effects are <u>not</u> considered; thus, the results are expected to be most applicable to desorption angles near the normal, where the angular dependence of neutralization is expected to be minimal [4].

The results indicate that ion trajectories can be strongly influenced by adsorbed impurity atoms, particularly for fully ionized impurities, for low ion energy, and for large desorption angle (measured with respect to the surface normal).

2. Calculations Based on Screened Classical Image Potential

The initial conditions for trajectory calculations based on the screened classical image potential are shown in Figure 1. We assume adsorption

Springer Series in Surface Sciences, Vol. 19
Desorption Induced by Electronic Transitions, DIET IV
Editors: G. Betz · P. Varga

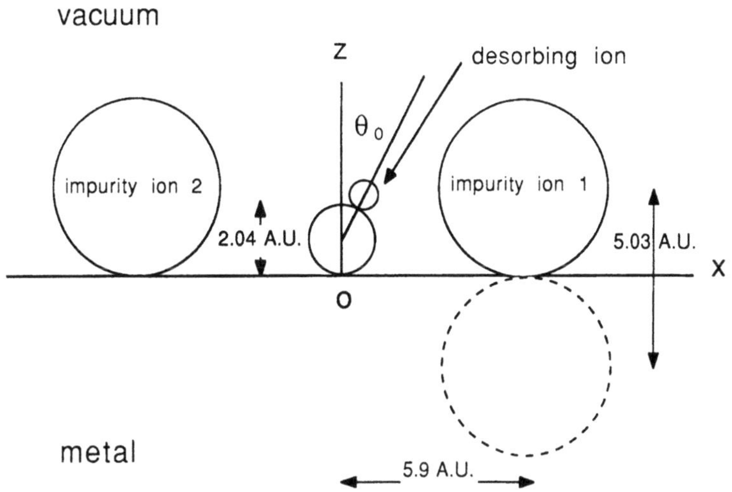

Figure 1. Initial conditions for model calculations

on an ideal planar, conducting surface. The atom sizes and bond distances are appropriate to K atoms (ions) coadsorbed with N-H species (e.g., one of the bonds in NH_3) on a Ru(0001) surface. In most calculations, a single impurity ion and its image are considered (impurity ion 1). In some calculations, both impurity ions shown in Fig.1 are included, along with their images. The quantity δ is the product of the charge of the desorbing ion times the charge of the adsorbed impurity ion. For example, a positive value for δ implies a positive desorbing ion and a positive impurity, or a negative desorbing ion and a negative impurity.

As in previous calculations [3,4], we treat the repulsive potential as an infinitely steep repulsion acting on the ion at a position z_0, where z_0 is the distance of the ion from the image plane at the instant the ion is formed. For ion-surface separations $z > z_0$, the ion moves in an attractive screened image potential $V_I(z)$, approximated by

$$V_I(z) = -e^2/4(z+k^{-1})$$

where k^{-1} is the Thomas-Fermi screening length ($k^{-1} = 0.567$A for Ru). The other forces acting on the desorbing ion are due to the potential established by the impurity ion(s) and image(s). At large distances from the surface, the desorbing ion reaches its final kinetic energy ε_k, chosen to lie between 2 eV and 8 eV. Trajectories are computed directly by integrating Newton's equation of motion.

Figures 2 and 3 show sample trajectories for an ion desorbing in the presence of one and two impurity ions, respectively. In each case, the initial desorption conditions are the same: $\theta_0 = 10°$, $\varepsilon_k = 4$ eV, variable δ. The trajectories are limited to the plane perpendicular to

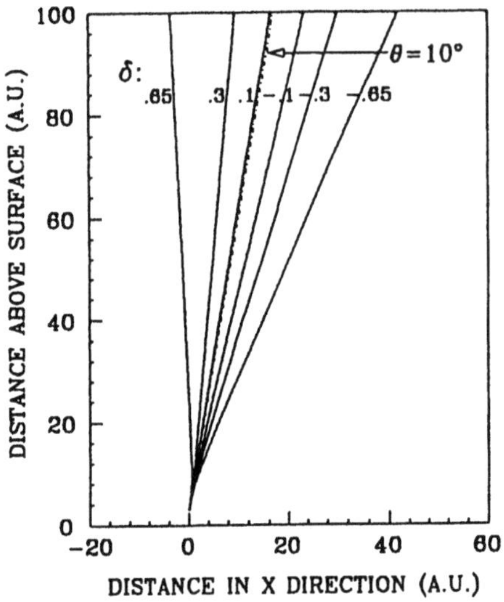

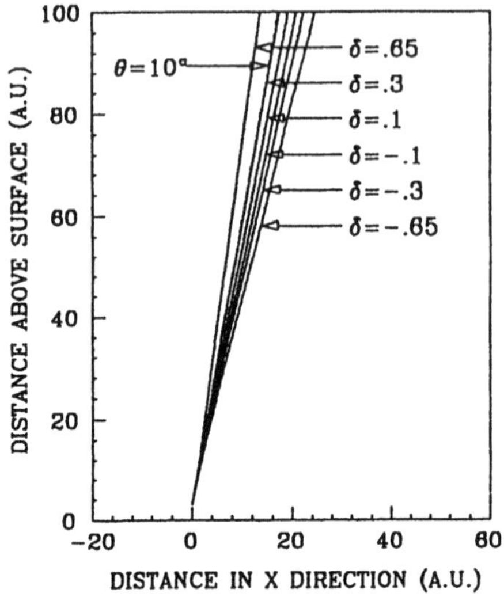

Figure 2. Sample trajectories. Single impurity ion. θ_0 = 10°, ε_k = 4 eV, variable δ.

Figure 3. Sample trajectories. Two impurity ions. θ_0 = 10°, ε_k = 4 eV, variable δ.

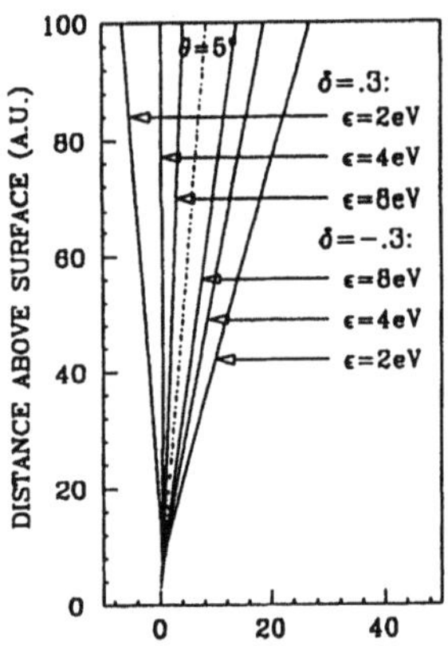

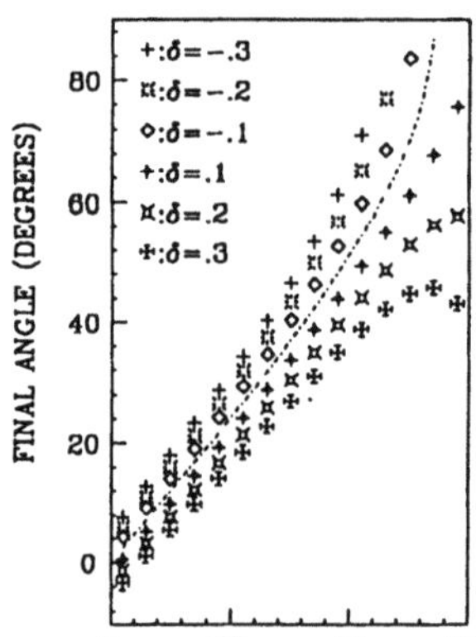

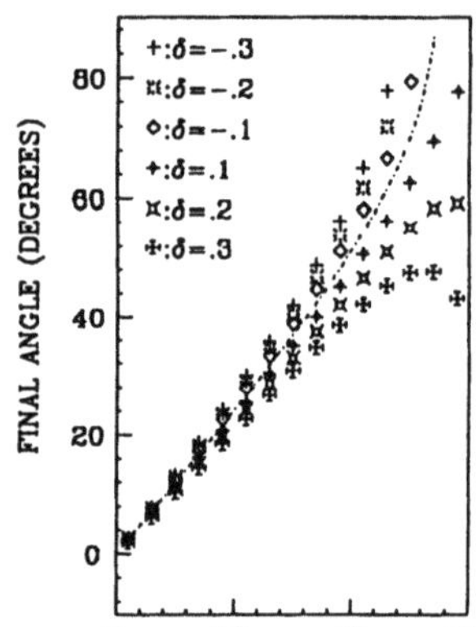

Figure 4. (Left) Influence of kinetic energy on trajectories. θ_0= 5°, δ = 0.3, variable ε_k. Figure 5. (Middle) Final desorption angle vs. initial bond angle for single impurity ion. ε_k = 4 eV, variable δ. Dotdash line indicates no impurity ion. Figure 6. (Right) Final desorption angle vs initial bond angle for two impurity ions. ε_k = 4 eV, variable δ. Dotdash line indicates no impurity ions.

the surface, as defined by the N, H and K species. It is clear that a single ion distorts the in-plane trajectories more than two ions, for which there appears to be a focusing effect. Figure 4 shows that the trajectory distortion is greatest for low values of ion kinetic energy ε_K.

Figures 5 and 6 show the final desorption angle as a function of the initial desorption angle θ_0 for a single impurity ion and two impurity ions, respectively. In each case, $\varepsilon_k = 4$ eV, δ is variable, and the dot-dash line indicates the trajectory in the absence of impurity ions. The focusing effect in Fig. 3 for two impurity ions is seen to be a general phenomenon over a wide range of initial angles, when Fig. 5 and 6 are compared.

3. Calculations Based on Density Functional Surface Potentials

In this section we investigate the sensitivity of desorbing ion trajectories to various approximations for the surface potential. The use of a screened classical image potential for the bare ion surface interaction is a good approximation only at large ion-surface separations. In fig. 7, the screened classical image potential is compared with an accurate density functional surface potential [5]. For large distances, the two different descriptions agree well. It can be seen that for ion-surface separations smaller than 10 a.u., there are marked differences between the two descriptions. The classical description diverges inside the surface while the density functional potential saturates to a finite value.

It is clear that the differences in the description of the bare ion-surface interaction potential will influence the desorbing ion trajectories. In order to quantify these differences we need a more accurate description of the dipole fields induced by the coadsorbed impurities.

This can be obtained from the self-consistent local density calculations of alkali chemisorption by Lang et al. [6]. These calculations employ the so-called jellium model of the surface. In this model, the lateral variations of the bare surface potential are neglected. The jellium model is known to provide an accurate description of the electronic structure of metal surfaces.

For a K atom chemisorbed in its equilibrium position at 4 a.u. from the jellium edge, the induced electrostatic potential can be parametrized using a positive point charge at 4 a.u. outside the surface and a negative point charge at 5 a.u. inside the surface. The charges of these point charges are taken as 0.65. This parameterization should not be mistaken for a real physical situation, but it accurately reproduces the calculated electrostatic potential [6], and is used for the remainder of our calculations.

In Fig. 8, we show a comparison of desorption trajectories calculated using the screened classical surface potential and using the

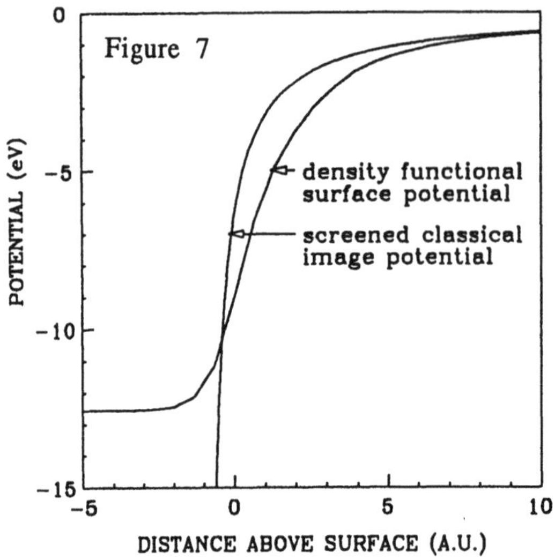

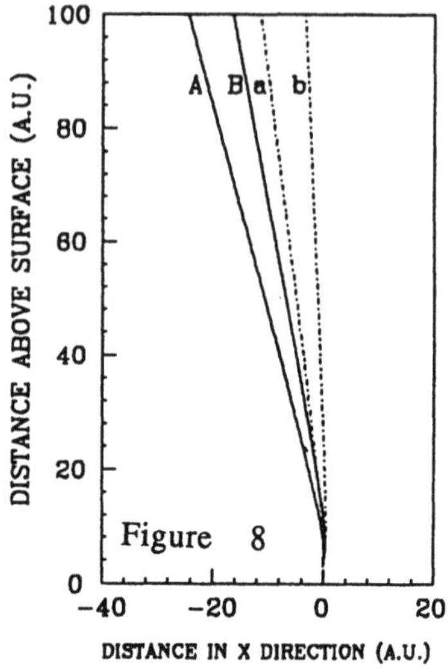

Figure 7. Density functional surface potential used for calculations.
Figure 8. Comparison of trajectories based on screened classical image potential (a,b) and density functional surface potential (A,B).
Single impurity ion. $\delta = 0.65$; $\varepsilon_k = 4$ eV; A, a: $\theta_0 = 5°$; B, b: $\theta_0 = 10°$.

density functional potentials. It can be seen that the classical description underestimates the influence of the surface on the trajectories.

4. Alkali Induced Changes of an ESDIAD Pattern

We now investigate whether the impurity induced ion deflections can be large enough to alter an ESDIAD pattern significantly (Figure 9). We start by considering an H^+ ion desorbing from an NH fragment bonded perpendicular to a clean metal surface (the bond lengths and resulting ion trajectories are appropriate to many low Z diatomic adsorbates, and the trajectories are independent of ion mass). The initial desorption angle is assumed to be distributed around the surface normal according to a gaussian distribution with a full width at half maximum (fwhm) of 40 degrees. Such a distribution accounts qualitatively for the randomness caused by possible frustrated rotation, libration and vibration of a molecule chemisorbed on the surface. The equations of motion are solved for 1000 different trajectories and each plotted point indicates the polar angle and azimuthal angle of a trajectory at large distance from the surface. The results are shown in Fig. 9a. It can be seen that the ESDIAD pattern is cylindrically symmetric as would be expected for normal desorption.

In Fig. 9b, ion desorption trajectories have been calculated for a single K atom chemisorbed at a nearest neighbor site to the NH (cf.

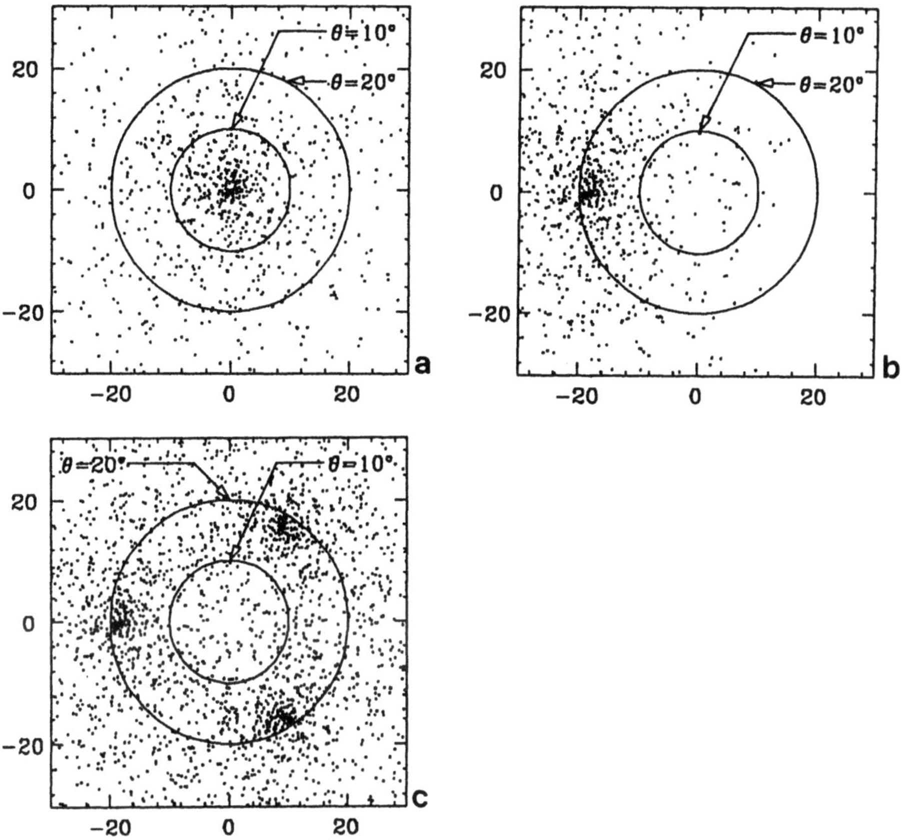

Fig. 9. Model ESDIAD Calculations.

a. Model ESDIAD pattern, $\delta = 0$, $\theta_0 = 0°$(FWHM = 40°).

b. Model ESDIAD pattern, $\delta = 0.65$, $\theta_0 = 0°$(FWHM = 40°). Single impurity at $\phi = 0°$.

c. Pattern b rotated by 120°, 240° to simulate ESD from fcc(111).

Fig.1). The initial conditions are chosen with the same random distribution as for Fig. 9a. It is apparent that the K atom induces a strong deflection of the desorbing ion trajectories. In Fig. 9c, we show the ESDIAD pattern expected for a fcc(111) surface at low alkali coverage. It is assumed that the NH species are chemisorbed at each of three equivalent nearest neighbor sites to the K, and that the distances between the K are so large that the desorbing H^+ trajectories are affected only by the nearest K. The corresponding ESDIAD pattern results from a threefold rotation of the ESDIAD pattern in Fig. 9b. In spite of the fact that the initial ion emission is assumed to be centered on the normal, the impurity-induced deflections are so strong that the apparent ESDIAD pattern suggests a different initial desorption angle. This is a *worst case* scenario involving trajectory perturbation by a highly charged ionic impurity.

5. Conclusion

Changes in the surface potentials induced by impurity atoms can have a large effect on desorbing ion trajectories. The results indicate that caution must be exercised when interpreting ESDIAD data, particularly for coadsorption systems involving ionic or highly polar species. The trajectories measured in an ESDIAD experiment can reveal not only the initial bonding geometry but can also be influenced by coadsorbed impurities. To model such changes it is important to use an accurate description of the surface potential. The so-called screened classical image potential provides only a qualitative potential that can be used for rough estimates of the deflections.

References

[1] T. E. Madey, D. E. Ramaker and R. Stockbauer, Ann. Rev. Phys. Chem. 35, 215 (1984).

[2] W. L. Clinton, Surface Sci. 112, L791 (1981).

[3] Z. Miskovic, J. Vukanic and T. E. Madey, Surface Sci. 141, 285 (1984).

[4] Z. Miskovic, J. Vukanic and T. E. Madey, Surface Sci. 169, 405 (1986).

[5] S. Ossicini, C. M. Bertoni and P. Gies, Europhys. Lett. 1, 661 (1986).

[6] N. D. Lang, S. Holloway and J. K. Norskov, Surface Sci. 150, 24 (1985).

Excitation Lifetimes and Excited-State Forces in Chemisorbed Molecules

D.R. Jennison, A.R. Burns, and E.B. Stechel

Sandia National Laboratories (1151), Albuquerque, NM87185, USA

1. Introduction

Quantum-state populations and yields for the highest probability (neutral desorbate) channels are now available by laser resonance-ionization spectroscopy (RIS) [1,2], thus providing an unprecedented opportunity for theoretical progress in electron-stimulated desorption (ESD). In previous work [1-4], we reported a semi-quantitative explanation of the translational and internal energy distributions of NO^o from Pt(111) based on a detailed microscopic model. More importantly, this work suggested several principles which appear general. After summarizing these, we present threshold data for the stimulated desorption of NO from (NO+O):Pt(111) and for the stimulated dissociation of NO_2:Pt(111), observed by the detection of the NO^o product, $NO_2^*(a) \rightarrow NO(g) + O(a)$. We review the assigned excitations in the NO:Pt(111) case and argue that coadsorbed oxygen greatly increases the excitation lifetime. For NO_2, consideration of the photoelectron spectrum, electronic structure, and excitation lifetimes allows us to assign the two dominant dissociation channels to specific one and two hole excitations. While the holes occupy non-bonding molecular orbitals (MOs), we find charge-transfer screening into a strongly antibonding MO. This screening, while far from complete, strongly influences the excitation lifetimes and forces, and affects the two channels differently.

Valence excitation lifetimes vary by 1-2 orders of magnitude [1,3]; therefore, unless extraordinary differences in the excited state potential energy surfaces occur, the longest-lived excitations dominate neutral yields. Such excitations must decay by valence Auger processes, as they are non-degenerate with substrate valence bands and cannot decay by resonant tunneling (a very rapid process with lifetime widths $\Gamma \sim 1$ eV). A matrix element for valence Auger decay can be written as $\iint d(r_1)d(r_2)\, \phi_i^*(r_1)\phi_j(r_1)\, e^2/r_{12}\, \phi_k^*(r_2)\phi_o(r_2)$. Here an initial-state valence hole in MO i decays through the use of valence electrons in MOs j and k resulting in an outgoing electron in orbital o. The factors which determine the Auger lifetime are the charge distributions of i, j, and k (in a one-electron theory; correlation effects are discussed below) and the momentum transfer between the electrons. Expanding the MOs in atomic orbitals (AOs) and neglecting to first order the interatomic contributions (these are small, including the metal d and s-p shells), angular momentum transfer appears as a factor of $1/(2\ell+1)$ in the AO matrix elements, a result of the expansion of the single-atom terms in spherical coordinates. This transfer is $\ell = 0$, 1 or 2 for decays involving the 2s and 2p shells, corresponding to Auger-rate factors of 1 for intra-orbital decays, 1/9 for s-p decays, and 1/25 for px-py, px-pz, or py-pz decays. The charge-density factor causes the contribution to the Auger rate from a given AO matrix element to scale as the products of the AO Mulliken populations of i, j, and k. The total rate is found by summing over atoms and over the AO Auger matrix elements. Calculations using the actual orbitals determine the relative weights and include the charge

Springer Series in Surface Sciences, Vol. 19
Desorption Induced by Electronic Transitions, DIET IV
Editors: G. Betz · P. Varga

density explicitly. The longest-lived excitations have Γ approaching 0.01 eV [1,3].

The above factors lead to a simple rule for finding the longest-lived valence excitations: holes which are not degenerate with occupied substrate bands and which must be filled using electrons from different AOs. For example, in the case of NO or CO, the $5\sigma^1$, $5\sigma^2$, and $5\sigma^2 4\sigma^1$ hole states should be long lived (electron density in the 2π and 5σ orbitals, respectively, make 1π and 4σ holes short lived). As symmetry is reduced, this rule is less useful because the molecular orbitals are more complex mixtures of different atomic orbitals. In general, two-hole (2h) excitations have lower cross sections for creation but stronger excited-state forces than one hole (1h) excitations. This compensation, and the comparable lifetimes of some 1h and 2h states, allows both channels to be visible in the RIS experiment, albeit with different thresholds, localization probabilities, and energy distributions [2,4]. The 2h excitations are created by shakeoff processes [5] or the decay of deeper 1h states.

The above simple one-electron picture can be in considerable quantitative error due to correlation effects which can substantially alter Auger decay rates [3,6]. For example, if the electrons for the decay both arise in a metal-adsorbate (M-A) hybrid level, the two-electron wavefunction is a linear combination of ionic, M^2 and A^2, and covalent, MA, configurations. Since the excitation is on the adsorbate, the A^2 configuration will dominate the Auger matrix element. One-electron theory is valid only if the weight of this configuration is correctly given by a one-electron charge density. Since those configurations which represent a charged adsorbate are screened by metal image charges (which respond an order of magnitude faster than typical M-A charge fluctuations), one-electron cluster calculations can be in serious quantitative error [3], a point relevant to the lifetime changes produced by oxygen coadsorption, discussed below.

Clearly, an important element in the physics is the rehybridization of the molecular orbitals (MOs) due to chemisorption. Less clearly, but even more important, is the rehybridization in the excited state due to the presence of a valence hole: rehybridization which involves substrate charge density we shall refer to as "screening" the excitation. The short screening length in the metal causes the local $M\text{-}A^*$ complex to remain essentially neutral (i.e., "M" means at most several metal sites), while adsorbate orbitals degenerate with substrate bands create M-A hybrid levels, whose rehybridization in the excited state transfers screening charge onto A if the A level is only partially occupied. The degree to which this occurs depends on the strength of the M-A transfer integral compared with the energy difference between the relevent M and A levels in the presence of the valence hole. This effect alters the charge density available for valence-hole Auger decay and affects the excited state forces and is thus central to understanding both (we find in both the NO and NO_2 cases that the holes are in non-bonding MOs so that the screening charge, in a strongly antibonding MO, has a major effect on the excited state forces). Even though the excited state is relatively short-lived on timescales of atom motion, the high frequency of some vibrations, such as the N-O stretch, causes desorbate vibrational populations to be sensitive to the excited state forces. We suggest this may be a useful probe of charge-transfer screening variations with substrate, site, coverage, or coadsorption.

2. Dissociation Thresholds for NO_2:Pt(111) and (NO+O):Pt(111)

We estimate the dissociation threshold for NO_2:Pt(111) to be ~ 10-14 eV (see Fig. 1), while the desorption threshold for (NO+O):Pt(111) is ~ 8-9 eV (Fig.

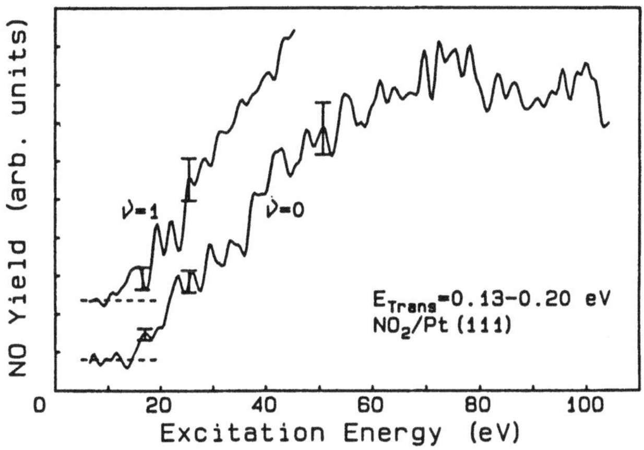

Fig. 1: NO yield from dissociated NO_2 is found for the indicated translational energy windows (from Ref. 8). The excitation energy is referenced to the Fermi level.

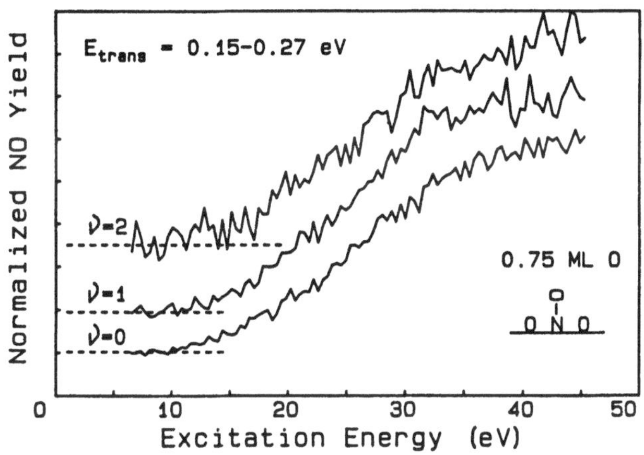

Fig. 2: NO yield from (NO+O):Pt(111). Note different energy scale than Fig. 1.

2; note the different energy scales). Details of the experiment are reported by Burns in this volume [7] and elsewhere [8]. The latter threshold is consistent with the $5\sigma^1$ channel found in the absence of O [1-4]. In addition, there is a hint of a second NO+O channel with a threshold at ~ 18 eV, most visible for the $\nu = 2$ channel. We assign this to the $5\sigma^2$ 2h excitation by lifetime arguments. The $5\sigma^2$ has also been seen in CO neutral and ion desorption by Feulner et al. [9]. (The "resonance" structure previously reported near the NO threshold [1,2] is now known to be an experimental artifact.)

3. Excitation Assignments for NO_2:Pt(111)

The MOs of NO_2 are discussed by Brundle et al. [10] in connection with the gas-phase photoelectron spectrum. A recent photoelectron spectrum of NO_2:Pt(111) by Bartram and Koel [11] is reproduced in Fig. 3. We seek the extent of ground-state MO rehybridization, the magnitude of the transfer integral to the substrate (which sets the amount of excited-state charge-transfer screening), and the longest lived excitations.

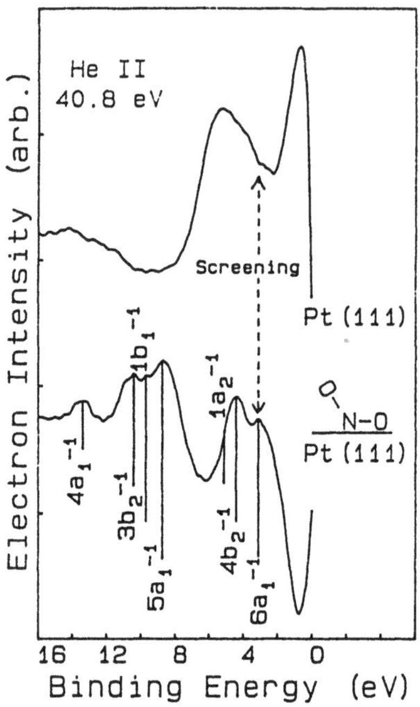

Fig. 3: The photoemission spectrum of NO_2:Pt(111), Ref. 11, is assigned. The "$6a_1$" level is actually a metal-MO hybrid ("$6a_1B$", see text and Ref. 8).

The geometry of NO_2:Pt(111) is a bridge-bonded nitrito complex, bonded side-on with one N-O bond parallel to the surface [12]. The largest M-A transfer integrals (t) are then to MOs made of N and O 2p orbitals in the molecular plane and essentially perpendicular to the N-O bonds; we will refer to these as the $p\pi$ "family" of orbitals which include the $6a_1$, $4b_2$, and the $5a_1$. The $6a_1$ is the shallowest orbital and is half filled in the gas phase. Other families are the $p\sigma$, made up of p-orbitals roughly along the N-O bonds and s-orbitals (forming $3b_2$, $4a_1$ and deep orbitals), and the pz, made of p-orbitals perpendicular to the molecular plane (forming $1a_2$ and $1b_1$). The rehybridization of the $p\pi$ family can be modeled with a tight binding hamiltonian where intra-molecular terms are set by fitting the gas-phase photoemission spectrum and t is then found by fitting the adsorbate spectrum (we neglect the lesser rehybridization of the $p\sigma$ and pz families). This procedure shows us that (1) the $6a_1$ strongly mixes with metal orbitals, splitting into occupied bonding ($6a_1B$) and empty antibonding ($6a_1A$) levels, (2) the deeper $4b_2$ and $5a_1$ MOs are shifted away from E_F by 0.2-0.5 eV but otherwise are not strongly perturbed from their molecular form, and (3) to account for the position of the $6a_1$ orbital relative to E_F, t must be large, ~ 2 eV, causing the $6a_1B$ level to be ~ 50% metal character, so that the chemisorbed molecule is approximately neutral. A large t, however, inhibits excitation charge-transfer screening since t is of the same order as the energy difference between excited state configurations; thus, the $6a_1B$ level only polarizes somewhat in the excited state. This is in contrast with gas-phase bound-to-bound excitations in which an entire electron is placed in the $6a_1$, which explains a difference in chemisorbed vs gas-phase dissociation dynamics [8]. Assignments shown in Fig. 3 use this analysis and arguments of Ref. 10. Gas-phase MO labels indicate the minimal rehybridization of all but $6a_1B$.

The above lifetime discussion causes us to expect that the longest-lived 1h excitation will be the highest member of each orbital family if it is not degenerate with substrate levels. The low energy (< 5 eV) excitations which produce gas phase NO_2 dissociation with unit quantum efficiency are thus too short lived to cause measurable ESD adsorbate dissociation (see Fig. 3). Only the $3b_2^1$ satisfies the 1h guidelines, the other two families having shallow members which can decay by resonant tunneling. The longest lived 2h excitations follow the same rule, having to be non-degenerate with the substrate 2h spectrum. The $3b_2^2$ meets this requirement as will also the $1a_2^2$ if the screened hole-hole interaction is > ~ 2 eV. Because of the limited charge transfer screening, we expect this to be the case. Comparing lifetimes, the $1a_2^2$ is longer lived than the $3b_2^2$ (or $3b_2^1$) by a factor of two to Auger decay (based on the MO populations [8]) because $1a_2$ holes are truly perpendicular to all other shallower orbitals, while $3b_2$ holes are only approximately so. Thus the $1a_2^2$ dominates the 2h contributions.

Comparing 1h and 2h channels, the $1a_2^2$ lifetime compensates for the reduced creation probability of $1a_2^2$ vs $3b_2^1$, and, since both have thresholds of ~10-12 eV as observed, we assign these excitations to the two dominant channels which are observed experimentally in the internal energy distributions of the NO fragment [7,8]. Since these MOs are both non-bonding in the gas phase, screening charge into the highly-antibonding $6a_1$ MO is important to the excited-state forces. The larger screening of the 2h excitation, together with the hole-hole repulsion, causes us to assign the $1a_2^2$ to the higher-energy of the two channels.

4. Discussion of (NO+O):Pt(111)

The lifetime of 5σ holes in NO is dominated by the intensity of the $2\pi^2$ electron configuration in the M-A hybrid $2\pi B$ orbital, which in turn is set in part by the relative energy of the $2\pi^0$, $2\pi^1$, and $2\pi^2$ configurations (for simplicity we omit the $2\pi^3$ and $2\pi^4$). In the absence of oxygen, both the $2\pi^0$ and $2\pi^2$ are screened by a substrate image charge; however, the presence of the negative oxygen favors the $2\pi^0$ positive configuration by electrostatic interaction. We estimate a substantial decrease in $2\pi^2$ excited-state amplitude in the presence of oxygen with the Auger rate scaling as the square of this amplitude. Since this 2π charge-transfer reduction may actually cause a **stronger** excited-state bond [1], we propose a role for the 5σ hole itself in bond weakening. Quantitative details will be reported elsewhere.

Acknowledgements

We thank M. Bartram and B. Koel for making their data available prior to publication. This work was performed at Sandia National Laboratories, supported by the US Department of Energy under contract DE-AC04-76DP00780.

References

1. A. R. Burns, et al., Phys. Rev. Lett. **58**, 250 (1987).
2. A. R. Burns, et al., **Desorption Induced by Electronic Transitions - DIET III**, R. H. Stulen and M. L. Knotek, eds., Springer Verlag (1988), p. 67.
3. D. R. Jennison, et al., Ibid., p. 167.
4. E. B. Stechel, et al., Ibid., p. 136.
5. H. H. Madden, et al., Phys. Rev. B **26**, 896 (1982).
6. D. R. Jennison, Chem. Phys. Lett. **77**, 604 (1981).
7. A. R. Burns, et al., this volume.
8. A. R. Burns, et al., Phys. Rev. B **40**, 9485 (1989).
9. P. Feulner, et al., in Ref. 2, p. 58.
10. C. R. Brundle, et al., J. Chem. Phys. **53**, 705 (1970).
11. M. E. Bartram and B. E. Koel, unpublished.
12. M. E. Bartram, R. G. Windham, and B. E. Koel, Surf. Sci. **184**, 57 (1987).

Part II

Photophysics and Photochemistry

Surface Photochemistry

W. Ho

Laboratory of Atomic and Solid State Physics and Materials Science Center, Cornell University, Ithaca, NY 14853, USA

Abstract. Photochemistry of adsorbed molecules provides a new perspective into our understanding of the interaction of molecules with surfaces and elementary reactions. The photochemical processes which can occur are photodesorption, photodissociation, and photoreaction of the adsorbed molecules. These photo-induced processes have been observed for a range of diatomic and polyatomic molecules on metal, semimetal, semiconductor, and insulator surfaces. Surface photochemistry is selective and can proceed at low temperatures. Although the underlying mechanisms are not completely understood, an increasing number of cross sections have been determined, allowing the beginning of a general understanding of the photo-induced process. The current status of surface photochemistry is summarized in this paper, with the emphasis placed on the effects arising from coadsorbed atoms and molecules.

1. Introduction

Photochemistry at solid surfaces is not a new revelation but a subject which has attracted respectable interest in the past [1-5]. The recent surge in the study of photo-induced processes on solid surfaces has effectively revived the field of surface photochemistry through the use of oriented single crystals, combination of surface probes, ultrahigh vacuum environment, and laser and photon technologies. It has been realized that photons can be used to probe the energetics of the molecules detached from the surface [5,6], and to induce desorption, dissociation, and reaction of adsorbed species [4, 7, 8]. What is perhaps surprising at first glance is the relatively large photoyields observed for molecules adsorbed on metal surfaces. The energy relaxation processes are not prohibitively fast to quench the excited states which are necessary for photochemistry to occur.

Photo-induced processes on solid surfaces include desorption, dissociation, and reaction of adsorbed species. The initial step invariably involves photoabsorption by the adsorbate/substrate, followed by subsequent steps which lead to surface photochemistry. The nature of these elementary steps is currently under active investigation.

The study of the basic mechanisms of photo-induced processes is crucial for understanding the observed photoyields or cross sections. The basic issues which need to be considered are the following: 1. the nature of the initial excitation process, 2. dissipation of the excited state, and 3. the effects on the adsorbate. The main difficulties in deciphering these steps lie in the ultrashort lifetimes and low concentrations of the relevant species. Theoretically, the problem is far from simple due to the complicated electronic structure and dynamics involved [2,7-9]. When all is considered, photoabsorption must lead to an excited state of the *adsorbate-substrate complex* which is unstable towards desorption, dissociation, or reaction of the adsorbate.

2. Established Mechanisms

For molecules adsorbed on surfaces, it is desirable to determine first whether the observed adsorbate transformations are due to thermal or nonthermal mechanisms. The thermal channel arises from irradiative heating and the results are expected to be similar to those obtained by conventional thermal desorption spectroscopy, with proper adjustments for the heating rates. The nonthermal mechanisms give rise to surface photochemistry, which

Springer Series in Surface Sciences, Vol. 19
Desorption Induced by Electronic Transitions, DIET IV
Editors: G. Betz · P. Varga

proceeds independent of the surface temperature. The photochemical products can be either the same or different than those formed by thermal reactions.

In the gas phase, photochemistry is initiated via absorption of visible or UV light by a molecule which makes a transition to an excited electronic state. The excess energy can be dissipated through emission of radiation, nonradiatively by energy transfer to other molecules, or by chemical transformation of the excited molecule.

The presence of the substrate introduces an additional medium for energy dissipation of excited molecules adsorbed on the surface. However, the incident light can also be absorbed predominantly by the substrate and one of the de-excitation channels is via coupling to the adsorbed molecules. The substrate mediated mechanism can be the dominant process by which surface photochemistry occurs.

A number of mechanisms for photo-assisted processes on solid surfaces have been identified within the last six years. In all cases, the most direct and convincing evidence has been obtained from the wavelength dependence of the photoyield. If there are no other competing mechanisms, thermal reactions proceed via substrate heating when the photon power is sufficiently high to induce the required temperature increase. Pulsed laser heating has been used successfully to study molecular desorption [10,11], reaction kinetics [12], and surface diffusion [13]. Thermal desorption has also been observed when a specific adsorbate vibrational mode is initially excited but subsequently decays via substrate phonon excitation (i.e. heating) [14]. However, desorption can indeed occur nonthermally from coupling of a directly excited phonon mode with the adsorbate-substrate complex [15]. Excitation of the plasmon mode of small alkali metal particles deposited on a substrate has also led to nonthermal desorption of individual atoms from the particles [16]. For semiconductors, photogenerated carriers from band gap excitation play an important role in giving rise to nonthermal photodesorption and photodissociation [17-21]. On metal surfaces, hot electrons [22] and electron charge transfer [23] have been observed to play a similar role. Direct photoelectronic excitation of adsorbate is prevalent on metal, semiconductor, and insulator surfaces [24-28].

3. Pre-1980 Surface Photochemistry

The general discussion on *Photo-Effects in Adsorbed Species* held in 1974 illustrates the abundant interest in the field at that time [1]. The substrate was primarily oxide surfaces, such as ZnO and TiO_2, and relatively large molecules were studied. The photochemical processes studied include the oxidation of hydrocarbons (alkanes) [29], reduction and oxidation of alcohols [30], oxidation of CO under ambient mixtures of O_2 and CO [31], polymerization of tetrafluoroethylene [32], and charge transfer interaction with dye molecules [33]. The photogenerated carriers in the oxides are intimately involved in the subsequent surface reactions. In those cases involving oxygen, questions on the charge state and the atomic versus molecular configurations are particularly relevant [34].

Photon irradiation has also been used in the splitting of water and reduction of nitrogen [35] and in isotopic enrichment [36]. The pre-1980 surface photochemistry is characterized predominantly by reactions involving either static or constant flow ambient reactants, and the surfaces are not well characterized in these studies. However, interest in fundamental understanding has motivated experiments to be carried out under better controlled conditions. The very first pulsed laser desorption study used a high power laser to heat a tungsten target [37]. More recently, the interaction of Na and Cs with Ge [38], D_2 with polycrystalline tungsten [10], and H_2 with low index planes of nickel [39] were studied by pulsed laser thermal desorption.

Photodesorption of CO from metal surfaces was observed to have very low cross sections [2, 3]. The importance of the wavelength dependence of the photoyield was realized in the study of CO on Ni foil [40] and polycrystalline W [41]. It is possible that in some of these studies the observed photoyields are due to thermal effects or photodesorption from (locally) oxidized surface [2, 3]. The presence of impurities, in particular carbon, is believed to give rise to photo-production of CO_2 on metal, semiconductor and oxide surfaces [2, 3, 9].

4. Experimental Methodology

The main essential components of an apparatus to study surface photochemistry are the light source, a vacuum chamber with pumping and sample manipulation capabilities, and surface sensitive probes. Both lasers and arc lamps have been used effectively as the excitation source. The arc lamp is frequently operated in the continuous wave (cw) mode [although it is possible to obtain long duration (> μs) pulses] while the laser can be either cw or pulsed with pulse durations lying in the range between microsecond (μs) and femtosecond (fs).

Unless the experimental goal is to measure the internal state distributions of the desorbing molecules, i.e. the translational, rotational, vibrational, and electronic energies, an arc lamp in conjunction with a monochromator is the preferred light source for several reasons. First, it is economical to set up and maintain. Second, it covers a continuous range of wavelengths from ~ 240 nm to 1 μm with sufficiently high flux densities (~ 10^{15} - 10^{17} photons s^{-1} in ~ 10 nm bandwidth). Third, it is easy to operate. Fourth, the cw nature allows direct measurement of the sample temperature during irradiation. The most important implication of these features is that the wavelength dependence of the photoyield can be measured routinely for deciphering the mechanisms of surface photochemistry.

Pulsed lasers are required for measuring the internal state distributions of the desorbing molecules or to excite adsorbate states with narrow energy widths, e.g. the vibrational states [14]. The most prevalent excitation pathway, however, is via electronic excitation of the adsorbate and/or the substrate, and the energy states are generally broad, from fraction of an eV to several eV's. The narrow bandwidth of lasers is thus unnecessary. In addition, direct measurement of the sample temperature with pulsed laser irradiation is not straightforward, although results from calculations are deemed reliable to within 20%.

It is highly desirable to be able to monitor both the adsorbed species and desorbed products in order to fully account for the photo-induced effects. In photon induced desorption spectroscopy (PIDS), the desorbed products are monitored by a mass spectrometer. The adsorbed species before and after irradiation can be characterized by thermal desorption spectroscopy (TDS) and electron energy loss spectroscopy (EELS). These are the three main techniques which have been used in our studies of photo-induced processes. Other surface probes such as infrared absorption-reflection spectroscopy (IRAS) and UV and x-ray photoelectron spectroscopies (UPS and XPS) also reveal incisive information on the nature of the adsorbed species while laser induced fluorescence (LIF) and resonant enhanced multiphoton ionization (REMPI) techniques provide internal state distribution of the desorbed molecules.

The photon sources in our studies consist of an Ar ion laser/dye laser system and xenon (or mercury-xenon) arc lamps in conjunction with monochromators. The arc lamp systems provide continuously turnable radiation from ~ 240 nm to ~ 1 μm. The radiation is focussed onto the sample with approximately the same dimensions as the sample. The thermocouples are attached to the back of the sample and temperature rise is monitored directly under cw conditions. With ~ 10 mW cm^{-2} intensity, the sample temperature rise is < 2 K.

5. Photodesorption, Photodissociation, and Photoreaction Cross Sections

Summaries of cross sections obtained to date for the three photochemical processes are presented in Tables I, II, and III. The photochemical cross section (cm^2 $photon^{-1}$) is the value which multiplied by the photon flux (photons cm^{-2} s^{-1}) gives the inverse of the time constant for first order photochemical processes, or when multiplied by the surface coverage (molecules cm^{-2}) gives the photoyield (molecules $photon^{-1}$). To to be experimentally transparent and for thin adsorbate layers, the cross sections in the tables are computed by considering only the incident flux.

Some trends can be deduced from the compilation given in these tables. The photodesorption of NO is the most extensively studied system. The magnitude of the cross sections for NO photodesorption from different types of substrate is observed to decrease in the order: oxide > semiconductor > coinage metals > platinum metals. This ordering

TABLE I: Summary of Photodesorption Cross Sections

Adsorbate	Substrate	Cross Section (cm^2)	Wavelength (nm)	Light Source	Reference
NO	Ni(100)-O	2×10^{-17}	193	pulsed	42
	Si(111)7×7	1.1×10^{-17}	330	cw	21(e)
		5.2×10^{-18}	430		
	Si(111)7×7-K	1.1×10^{-17}	330	cw	43
		1.6×10^{-18}	430		
	GaAs(110)	2.8×10^{-18}	457	cw	44(f)
	Ag(111)	3.1×10^{-18}	320	cw	45
		7.2×10^{-19}	460		
	Cu(111)	2.2×10^{-18}	320	cw	46
		8.5×10^{-19}	458		
	Pt(111)	2.5×10^{-22}	355	pulsed	22
		1×10^{-22}	532		
		7×10^{-24}	1,064		
		3×10^{-22}	230-1,000	cw	47
		1×10^{-19}(a)	230-1,000		
	Pt(100)	3.5×10^{-18}(b)	193	pulsed	48(g)
		1.0×10^{-19}(c)			
	Pd(111)	$< 1\times10^{-22}$	193	pulsed	49
O_2	Pt(111)	1.8×10^{-20}	230-300	cw	50
	Pd(111)	5×10^{-21}(d)	230-1,000	cw	51
		1.3×10^{-19}	240	cw	52
CO	NiO	5×10^{-18}	300	cw	53

(a) Observed only at saturation coverage.
(b) "Fast" desorption channel.
(c) "Slow" desorption channel.
(d) Include both photodesorption (~ 50%) and photodissociation (~ 50%).
(e) The cross section was obtained by normalizing to the absorbed photo flux, hence the larger value of 1.6×10^{-17} cm^2 reported in Ref. 21 for 330 nm radiation.
(f) The cross section is ill defined since the photodesorption yield depends nonlinearly on the incident power with an exponent of 0.53 ± 0.02. The time decay of the photoyield is algebraic with an exponent of –0.6.
(g) NO^+ is monitored in this work.

can be understood by considering the quenching efficiencies of the different adsorbate-substrate systems. Excited states are quenched more rapidly on metal surfaces compared to semiconductor or oxide surfaces, mainly due to the existence of band gaps in the latter. For the coinage metals Ag and Cu, the interaction with NO is largely with the s band which dominates the density of states near the Fermi level. In contrast, the d band density of states peaks just below the Fermi level for Pt and Pd and the binding of NO is stronger compared to that on Ag and Cu. The five orders of magnitude decrease in cross section for Pt and Pd is attributed to the intimate coupling of NO to the substrate, which provides an efficient pathway for the photo-excited state to decay.

The cross sections for photodesorption of O_2 from Pt and Pd are one to two orders of magnitude larger than those for NO on Pt and Pd, but are still smaller than those for NO on oxide and semiconductor surfaces. The cross section for photodesorption of CO from NiO is increased dramatically compared to the clean surface where the small photoyield (at least four orders of magnitude lower) is most likely due to substrate heating [2, 3]. Again, the oxidized surface removes the efficient quenching channel characteristic of a clean metal surface.

The different photodissociation systems studied to date and their cross sections are listed in Table II. The values of the cross sections vary over five orders of magnitude. Adsorbates on the clean Ni, Pd, and Pt metals give rise to the lowest photodissociation cross sections while those on semiconductor and insulator surfaces yield the largest. The most extensive study was done on $Mo(CO)_6$ where the cross sections on metal and semi-

TABLE II: Summary of Photodissociation Cross Sections

Reaction	Substrate	Cross Section (cm^2)	Wavelength (nm)	Light Source	Reference
$O_2 \rightarrow O + O$	Pd(111)	3.5×10^{-20}	240	cw	52
	Pt(111)	8.5×10^{-21}	230-300	cw	50
$OCS \rightarrow CO + S$	LiF(001)	2×10^{-17}	222	pulsed	54
$CH_3Cl \rightarrow CH_3 + Cl$	Ni(111)	4.3×10^{-20}	193	pulsed	23
		2×10^{-21}	248		
$CH_3Br \rightarrow CH_3 + Br$	Ni(111)-Br	3.5×10^{-20}	193	pulsed	55
	Pt(111)	3×10^{-21}	230-280	cw	26
$CH_3I \rightarrow CH_3 + I$	Pt(111)	2×10^{-22}	320	cw	56
$CH_2I_2 \rightarrow CH_2I + I$	Al_2O_3	4×10^{-19}	308	pulsed	57
$N_2O_4 \rightarrow NO + NO_3$	Pd(111)-NO	3×10^{-18}	193	pulsed	58
$Mo(CO)_6 \rightarrow Mo(CO)_5 + CO$	Si(100)	5×10^{-17}	248	pulsed	59
	Si(111)7×7	6.0×10^{-19}	330	cw	28,60
	Si(111)7×7-K	1.2×10^{-17}	330	cw	60(a)
		8.7×10^{-18}	430		
	Ag(111)	1.8×10^{-17}	285	cw	61
		7.7×10^{-18}	325		
	Cu(111)	1.7×10^{-17}	290	cw	62
		2.2×10^{-18}	325		
	Rh(100)	1.5×10^{-18}	330	cw	63
$Fe(CO)_5 \rightarrow Fe(CO)_4 + CO$	Al_2O_3	1.5×10^{-18}	337	cw	64
	Si(111)7×7	1.2×10^{-16}	249	pulsed	27

(a) The cross section is ill defined since the initial photodissociation yield depends nonlinearly on the incident power with an exponent of ~ 0.5.

TABLE III: Summary of Photoreaction Cross Sections

Reaction	Substrate	Cross Section (cm^2)	Wavelength (nm)	Light Source	Reference
$2HBr \rightarrow H_2 + Br_2$	LiF(001)	6×10^{-19}	193	pulsed	65
$2HCl \rightarrow H_2 + Cl_2$	LiF(001)	1.5×10^{-19}	193	pulsed	65
$2NO \rightarrow N_2O + O$	Ag(111)	1.7×10^{-18}	320	cw	45
		8.4×10^{-19}	430		
	Cu(111)	1.9×10^{-18}	320	cw	46
		5.8×10^{-19}	460		
	GaAs(110)	1.0×10^{-18}	514	cw	44
	Si(111)7×7	1.1×10^{-17}	330	cw	21,43
		5.0×10^{-18}	430		
	Si(111)7×7-K	1.0×10^{-17}	330	cw	43
		1.6×10^{-18}	430		
$2CH_2I_2 \rightarrow C_2H_4 + 4I$	Al Film	-	308	pulsed	66
$O_2 + CO \rightarrow CO_2 + O$	Pt(111)	3.3×10^{-20}	338	cw	68
		2×10^{-21}	443		
$TiCl_4 + Al_2(CH_3)_6 \rightarrow Ti:Cl:Al:CH_3$	Quartz	-	257	cw	67

conductor surfaces are approximately the same order of magnitude. The desorption peak temperatures from TDS are approximately the same on these different surfaces.

A number of photoreaction cross sections have been obtained and are given in Table III. The cross section for CO oxidation on Pt is the lowest in the table. However, the value is similar to that for the O_2 photodissociation on Pt(111) as given in Table II. This is consistent with the mechanism for CO oxidation which involves O_2 photodissociation as one of the key elementary steps.

The value of the photoreaction cross sections may perhaps appear to be larger than expected since the overall chemical reaction involves more than one elementary step. The cross sections, however, may generally be determined by the photodissociation of one of the reactants. When only one type of species is present as the reactants, it is difficult to measure separately the photodissociation step since the subsequent steps occur at rates faster than the probing time scale. However, for two different types of reactants, the photodissociation of one or both reactants can be investigated individually, although the cross section may be affected by the coadsorbate. In this regard, the oxidation of CO on Pt(111) is the only system studied to date which involves two different types of reactants.

For completeness, the photoreactions of CH_2I_2 and $TiCl_4$ + $Al_2(CH_3)_6$ have been included in the table although the cross sections have not been derived. The number of photoreactions studied is less than that of photodesorption or photodissociation since the occurrence of photoreactions is less frequent. The rest of the paper will concentrate mostly on the study of photoreactions, in particular N_2O and CO_2 formation. The photodissociation of $Mo(CO)_6$ on different types of substrates will be discussed first since it illustrates most unambiguously one of the photodissociation mechanisms, namely direct excitation of the adsorbate.

6. Illustrative Systems in Surface Photochemistry

The photodissociation of $Mo(CO)_6$ on Ag(111), Cu(111), basal plane of graphite, and Si(111)7 × 7 is a system which demonstrates most clearly the underlying mechanisms. Coadsorption with potassium has strong effects on the photodissociation process. Subsequent discussions will center on the synthesis of N_2O from adsorbed NO and CO_2 from coadsorbed CO and O_2. Finally, the effects of coadsorbate on surface photochemistry are illustrated via the coadsorption of NO and O_2.

6.1 Photodissociation of $Mo(CO)_6$

Irradiation of metal carbonyls in the gas phase or in solution with UV light (< 360 nm) leads to efficient detachment of CO ligands from the parent molecule [69]. The photodissociation occurs via direct electronic excitation of the metal carbonyl, involving metal-to-ligand charge transfer transition or, less effectively, the ligand field transition. Similar processes are observed to occur on metal, semimetal, and semiconductor surfaces with high efficiencies. The photo-excited electronic state of the metal carbonyl, either free or adsorbed on the surface, is repulsive, leading to photochemical rates which are competitive with substrate quenching rates.

Typical $Mo(CO)_6$ photodissociation spectra are shown in Fig. 1 for monolayer and multilayer coverages on Ag(111). The detached CO is monitored as a function of irradiation time by a mass spectrometer. At $t = 0$ s, the shutter is opened and the CO signal rises rapidly as irradiation is initiated. This initial rapid rise is followed by a monotonic decay which can be fit well with the sum of two or three exponentials, i.e. a fast initial exponential decay followed by slower ones. This time dependence is characteristic of metal carbonyls on different substrate surfaces. The photoyield from the monolayer is about a factor of seven reduced from the multilayer signal while the ratio of the coverages is approximately 1:3.5. The additional reduction beyond that due to the decreased coverage is attributed to the faster quenching rate for adsorbates closer to the surface.

The wavelength dependence of the photoyield gives the most unambiguous evidence for the mechanism involving direct electronic excitation of the adsorbed metal carbonyls. In Fig. 2, the height of the initial jump at $t = 0$ s (see Fig. 1) is plotted versus the irradiation wavelength for monolayer $Mo(CO)_6$ on four different substrates. The observed resonance at ~ 290 nm is due to the metal-to-ligand charge transfer transition and is in good agreement with the same absorption for $Mo(CO)_6$ in cyclohexane solution. The photoyield scales directly with the adsorbate absorption spectrum, a prerequisite for adsorbate mediated mechanism.

The substrate does not have significant influence on the $Mo(CO)_6$ photodissociation cross section (see Table II). However, the width of the resonance appears to be broader for Ag(111) than Cu(111) and graphite, and additionally the smaller peak at ~ 330 nm is

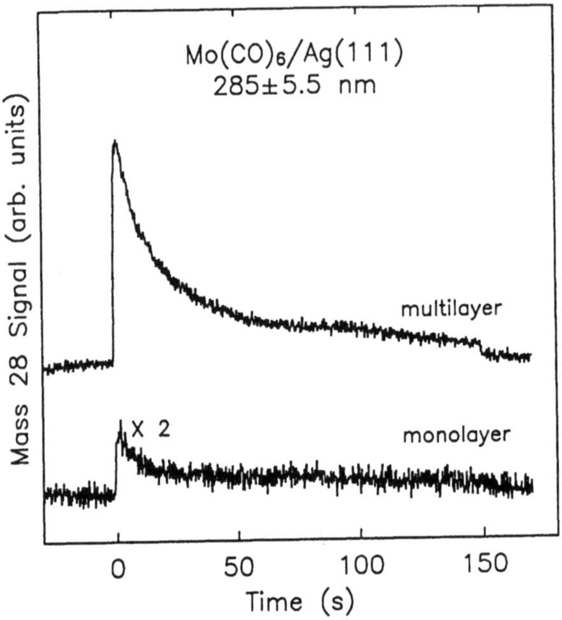

Fig. 1. CO mass signal from photodissociation of $Mo(CO)_6$ on Ag(111) at 80 K. Irradiation is with a 1000 W xenon arc lamp monochromator system.

more prominent. Although, the origin of these effects is not rigorously understood, it is possible that they arise from the interband transition unique to Ag at ~ 330 nm.

When the surface is preadsorbed with K, an additional photodissociation channel of $Mo(CO)_6$ appears. The new channel is substrate mediated, involving photoelectronic excitation of the substrate as the initial step. The wavelength dependence of the photoyield thus tracks the substrate absorptance. Details of the subsequent steps remain to be clarified. However, a likely mechanism involves capture of a photogenerated electron by the adsorbed metal carbonyl, forming a temporary negative ion which is unstable towards ligand detachment.

The most striking effect of K coadsorption lies in the wavelength dependence. In contrast to the limitation of the photoyield to within the UV range for adsorption on the clean surface, photodissociation is observed even in the visible and near IR wavelength regime in the presence of coadsorbed K. This is illustrated by the PID signal with 550 nm irradiation, as shown in Fig. 3. The effects of irradiation are revealed by the differences in the TDS spectra taken before and after irradiation.

The surface photochemistry of a molecule can be affected significantly by coadsorbed species, due to the emergence of additional photochemical mechanisms, new adsorption states, or changes in the electronic structure.

6.2 Photosynthesis of N_2O from NO

The adsorption and reaction of NO are widely studied due to its practical importance and the readiness of obtaining information on internal state distributions by laser based techniques. An interesting and seemingly ubiquitous reaction is the formation of N_2O observed on a number of surfaces either during adsorption, substrate heating, or photon irradiation. It is desirable to investigate whether there are common mechanisms by which N_2O is produced by these different processes.

The results from EELS and TDS recorded before and after irradiation are shown in Fig. 4. From EELS, it can be seen that a number of species are present on the surface upon NO exposures at 80-90 K. Specifically, on Ag and Si the surface is adsorbed with atomic N and O in addition to molecular NO and N_2O. With the exception of N, the other three species are observed on GaAs. For Cu(111), however, N_2O is not present on the surface

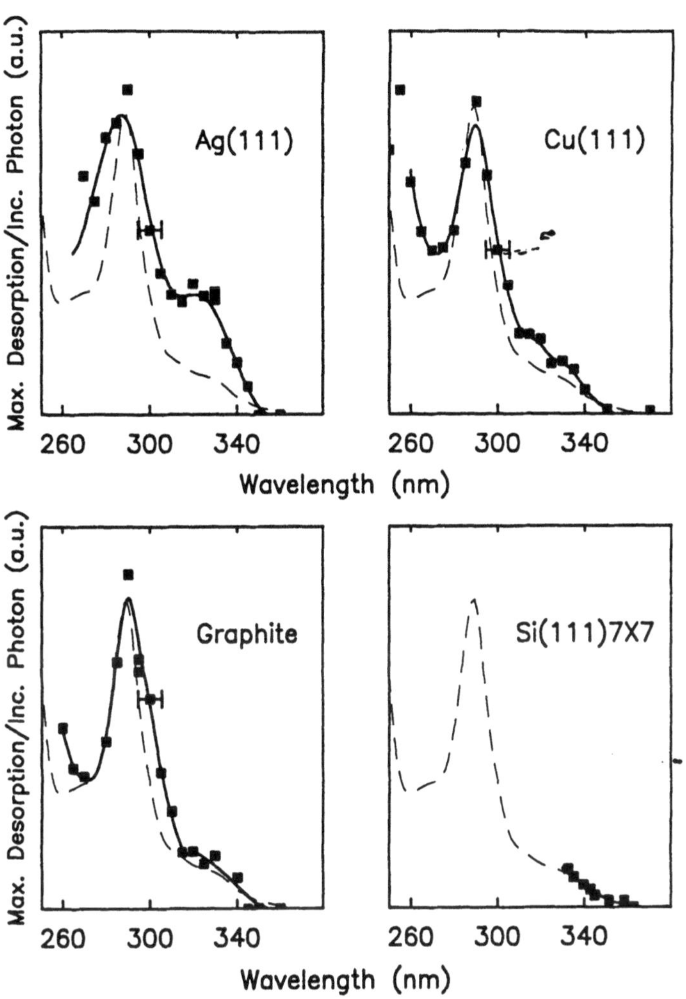

Fig. 2. Wavelength dependence of the initial jump in the CO mass signal immediately after opening the shutter for photodissociation of $Mo(CO)_6$ on four different surfaces at 80-90 K. The dashed curve in each panel is the absorptance of $Mo(CO)_6$ in cyclohexane solution, arbitrarily scaled to the maximum of the adsorbed $Mo(CO)_6$ data. The error bar indicates typical bandwidth used. The wavelength dependence measurements on Si(111)7 × 7 are less complete due to the use of a 150 W xenon arc lamp/filters system (also used for obtaining the rest of the results in this paper for Si) rather than a 1000 W xenon arc lamp/monochromator system.

after NO exposure; the N_2O is formed only by substrate heating or photon irradiation. It is not clear whether N and O are formed during NO adsorption.

As can be seen from the labelled peaks in the EELS spectra, in Fig. 4, photon irradiation leads to the disappearance of the NO (~ 180-239 meV) and N_2O (~ 275 meV) vibrational peaks. The decrease in the NO intensities is due to photodesorption, photodissociation (to form N and O), or photoreaction (to form N_2O). On Ag and Cu, photodissociation of NO does not appear to occur. On Si(111)7 × 7, photodissociation is believed to be the dominant pathway for the decrease in the N_2O coverage.

There are differences in behavior among the four substrates shown in Fig. 4. Photon irradiation does not affect N_2O on Ag(111); the small N_2O desorption peak just below ~

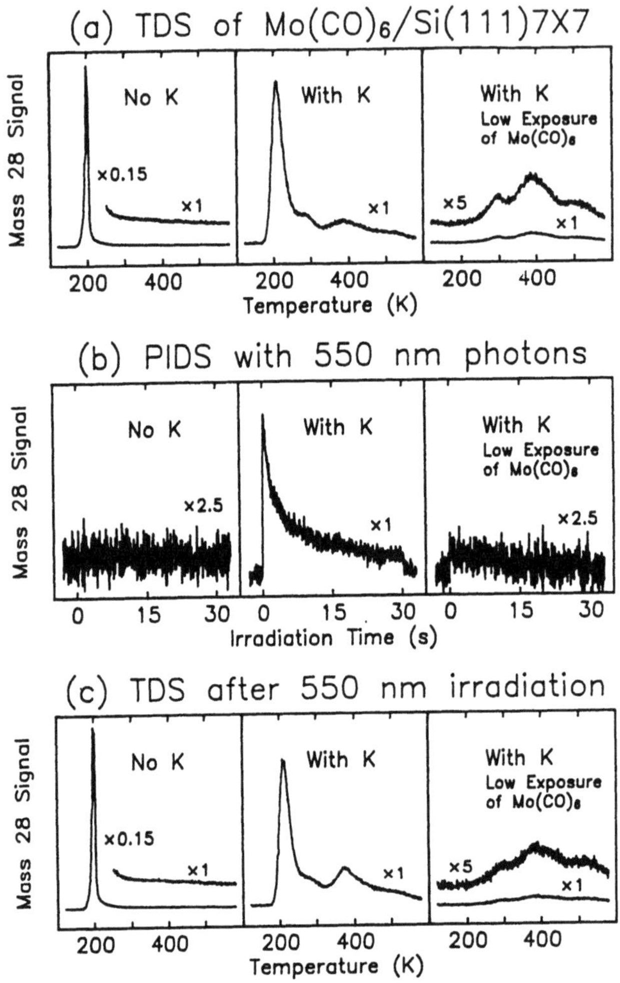

Fig. 3. TDS spectra monitoring mass 28 [CO, a cracking fragment of $Mo(CO)_6$ in the mass spectrometer] of $Mo(CO)_6$ on Si(111)7 × 7 at 90 K, with and without preadsorbed K, recorded (a) before and (c) after photon irradiation. The CO PID signal is shown in (b). Heating rate of 2 K s^{-1} was used. The K precoverage corresponds to ~ 0.5 ML, which gives rise to a maximum decrease in the Si work function. Irradiation is centered at 550 nm with a bandwidth of 65 nm. The shutter is opened at t = 0 s and closed at t = 30 s. The low exposure of $Mo(CO)_6$ is five times less than the higher exposure shown.

100 K is attributed to the formation from NO via substrate heating and disappears after the corresponding NO TDS peak decreases significantly after irradiation. The small N_2O PID signal (approximately a factor of ten less than the NO PID signal) is attributed to photosynthesis from NO.

There are three likely reaction pathways for the formation of N_2O from NO:

$$NO + NO + h\nu \rightarrow O + \{N \rightarrow NO\} \rightarrow O + N_2O, \quad (1)$$

$$NO + NO + h\nu \rightarrow \{NO \rightarrow NO\} \rightarrow O + N_2O, \quad (2)$$

$$NO + N + h\nu \rightarrow \{NO \rightarrow N\} \rightarrow N_2O. \quad (3)$$

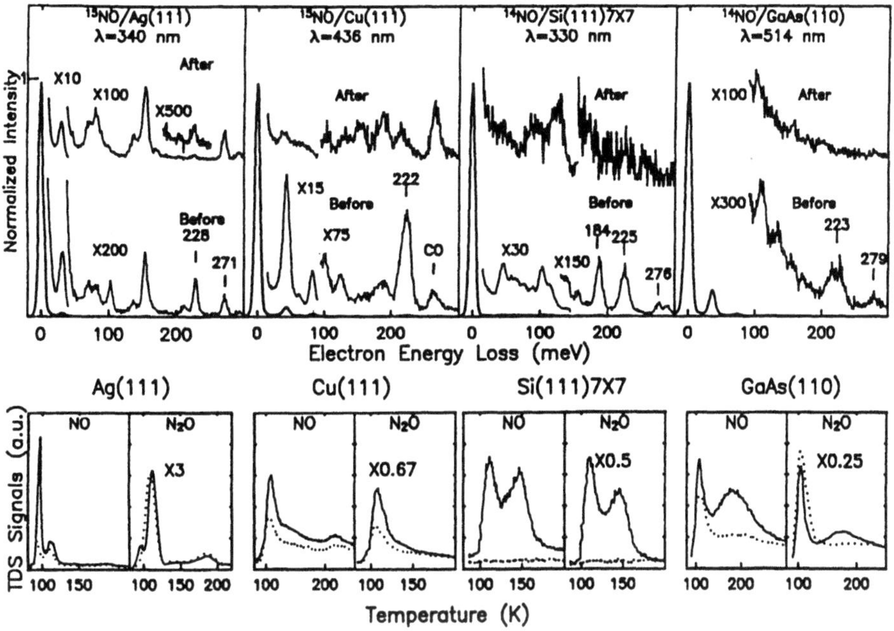

Fig. 4. EELS and TDS spectra recorded before and after (dotted curve for TDS) irradiation at the wavelength indicated. (No correction has been made to the NO TD signal due to cracking of N_2O in the mass spectrometer.) Results for GaAs are obtained with Ar ion laser irradiation with fluence of 160 mW cm^{-2}. The irradiation bandwidths for Ag, Cu, and Si are 11, 11, and 65 nm, with fluence of 7, 23, and 14 mW cm^{-2}, respectively. The samples are at 80-90 K before irradiation and rise by a maximum of ~ 3 K during irradiation. Only the N–O stretch of NO (< 230 meV) and N–N stretch of N_2O (~ 275 meV) are indicated in the spectra.

In the first reaction, photodissociation of a NO molecule leads to a "hot" N atom which collides and reacts with a neighboring NO molecule. Photon irradiation also induces desorption of NO which can collide and react with another NO [Eq. (2)] or a N atom [from the dissociation of NO , Eq. (3)] upon adsorption along its desorption trajectory.

On Cu(111), N_2O is not formed during NO exposures. However, N_2O is produced during both TDS and PID. The N_2O PID signal is approximately half of the NO PID signal. Furthermore, the N_2O TDS lineshape is similar to the NO TDS lineshape, and as the NO TDS intensity decreases, a corresponding decrease is observed for the N_2O TDS intensity. It is also possible that the major origin of the observed NO desorption signal arises from cracking of N_2O in the mass spectrometer.

Irradiation of the Si(111)7 × 7 surface exposed to NO leads to desorption and dissociation of NO and N_2O. However, the N_2O PID signal is mostly due to photosynthesis from NO. The disappearance of N_2O in EELS is mainly caused by photodissociation of N_2O. The lower temperature NO TDS peak (~ 110 K) is composed of NO desorption and the cracking fragment of desorbing N_2O in the mass spectrometer. Direct N_2O thermal desorption accounts for the low temperature N_2O desorption peak (~ 110 K) while the higher temperature peak (~ 150 K) is due entirely to photosynthesis from NO, followed by N_2O desorption.

Further evidence for N_2O photosynthesis rather than photodesorption on Si(111)7 × 7 comes from comparison of the N_2O PID signal with the appropriate NO and N_2O thermal desorption intensities, as shown in Fig. 5. TDS spectra were recorded after different irradiation times. The NO TDS intensity is derived from the higher temperature peak (~ 150 K, see Fig. 4) while the N_2O TDS intensity corresponds to the lower temperature peak (~ 110 K). The N_2O PID signal correlates with the NO TDS intensity, i.e. the amount of

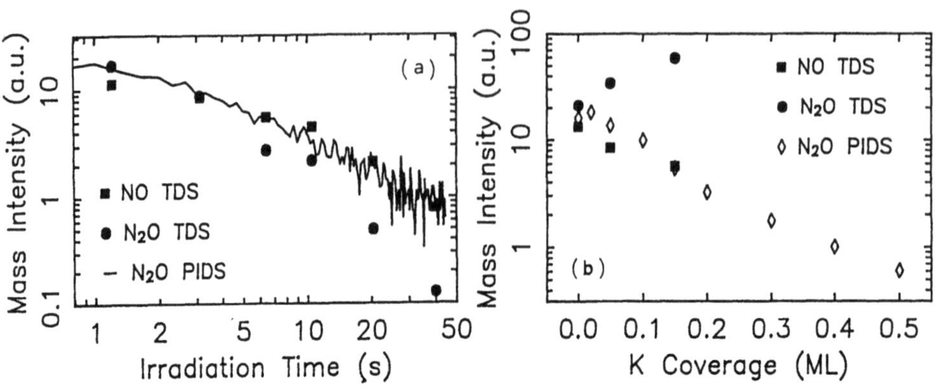

Fig. 5. Comparison of N_2O PID signal to the lower temperature (~ 110 K) N_2O TDS intensity and the higher temperature (~ 150 K) NO TDS intensity as a function of (a) irradiation time and (b) K precoverage on Si(111)7 × 7 at 90 K. In (a), the TDS spectra were recorded after the indicated irradiation time.

N_2O formed scales directly with the NO coverage. Similar dependence is observed by comparing the N_2O PID signal to the NO and N_2O TDS intensities when the surface is preadsorbed with different amount of K. While the N_2O TDS intensity increases as the K precoverage increases, both the N_2O PID signal and the NO TDS intensity decrease and track each other.

On GaAs(110), N_2O is also formed readily from NO either thermally or via photon irradiation. The lower temperature NO peak (~ 110 K) is due partly to the cracking of N_2O in the mass spectrometer while the higher N_2O TDS peak (~ 180 K) arises mostly via photosynthesis from NO.

The formation of N_2O from NO occurs readily on different surfaces. The reaction is observed via exposures of NO at 80-90 K, thermal annealing of the substrate, and photon irradiation. Wavelength and power dependence measurements indicate that both the excitation of adsorbate/substrate complex and substrate mediated mechanisms play a role in the surface photochemistry of NO exposed surfaces. It is necessary to consider the photoelectronic excitation of the adsorbate-substrate complex as one entity since the electronic structure of molecules can be significantly modified upon adsorption.

6.3 Photochemistry Between Two Different Types of Adsorbates

The photosynthesis of N_2O is derived from surfaces exposed to a single type of molecule, NO. Surface photochemistry has been observed for the coadsorption of O_2 and CO on Pt(111) where selective photodissociation of O_2 leads to "hot" O atoms which collide and react with CO to form CO_2. The product CO_2 is desorbed (not trapped at 100 K) and detected by a mass spectrometer.

A schematic of the photo-induced reaction sequence $O_2 + CO + h\nu \rightarrow O + \{ O + CO\} \rightarrow O + CO_2$ is shown Fig. 6. The photons preferentially dissociate O_2 molecules which are adsorbed parallel to the surface with reduced bond order. The adsorbed O_2 resembles peroxides, with an absorption spectrum significantly different and red-shifted from gas phase O_2. Coadsorbed CO molecules are not affected directly by the irradiation.

The wavelength dependence of the CO_2 photoyield follows that of peroxides, suggesting that the initial step of the photoreaction involves direct electronic excitation of the adsorbed O_2/substrate complex. (The wavelength dependence of the chemisorbed atomic O was monitored by EELS and found to track the wavelength dependence of the CO_2 photoyield.) However, more detailed experimental investigations and analysis are needed in order to unambiguously determine the mechanism. In particular, it is desirable to determine directly the electronic structure of the adsorbed molecules. The contribution of substrate excitation to the observed photoyield is often a factor to be reckoned

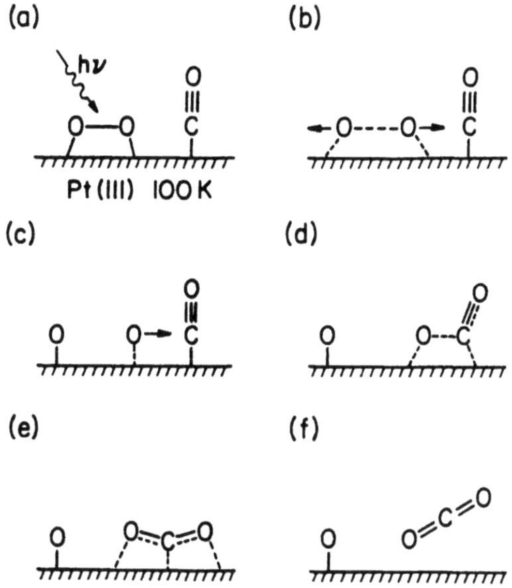

Fig. 6. Schematic of the photoreaction between coadsorbed O_2 and CO on Pt(111) at 100 K. Thermal desorption and decomposition of O_2 occur at ~ 150 K, where CO_2 is also observed to form thermally.

with, possibly by determining the dependence of the photoyield on the incident photon wavelength, power, and polarization.

The energetics of the initial step in the photosynthesis of CO_2 is illustrated in Fig. 7, where direct photoelectronic excitation of the adsorbed O_2/substrate complex is believed to be an initial step. Photoabsorption causes an optical transition from the ground state $O_{2,a}$ to an excited electronic state $O^*_{2,a}$. The excited state potential energy curve overlaps the potential energy curve for atomic O. Curve crossing can occur, which corresponds to the photodissociation process. The competing pathways are de-excitation of $O^*_{2,a}$ to $O_{2,g}$, leading to photodesorption, and energy transfer to the substrate.

When curve crossing occurs from the $O^*_{2,a}$ to the $2O_a$ potential energy curves, the O atoms have large excess energies. The "hot" O atoms can have high translational energy and/or be electronically excited. In the gas phase, the $O(^1D)$ lies 1.97 eV above the ground state $O(^3P)$[73]. The observed CO_2 photoyield has an onset of ~ 2.7 eV (~ 450 nm).

It is likely that a significant amount of energy is released in the photosynthesis, which appears as excess translational, rotational, and vibrational energies of the desorbing CO_2. It is not necessary for the photo-produced atomic O to translate a significant distance on the surface before reacting with coadsorbed CO. The coadsorbed O_2 and CO could be closely coupled and it is believed that close proximity would favor the cross section for photoreaction. It should be emphasized that the oxidation of CO could proceed directly via abstraction of an O atom from an excited O_2 molecule which is in the process of dissociating. In this case, the potential energy curves shown in Fig. 7 ought to be modified.

It should be noted that the adsorbed layer is continuously evolving during irradiation, which can influence the cross sections for various photo-induced processes. Deviations from simple behavior could be due to such coadsorbate effects.

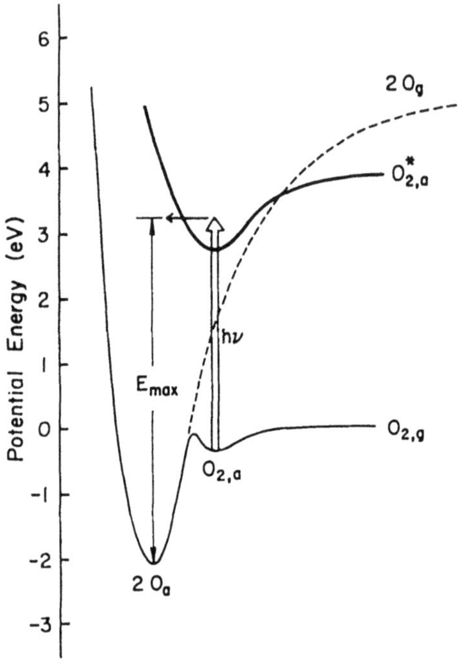

Fig. 7. Schematic of the potential energy curves for photoabsorption and dissociation of molecularly adsorbed O_2. The subscripts *g* and *a* denote gas and adsorbed phase, respectively. E_{max} is the maximum energy of the photolytically formed O atoms above the bottom of their chemisorption well. The energetics is relevant for the photoreaction of coadsorbed O_2 and CO on Pt(111) at 100 K. The desorption energy, or the well depth for $O_{2,a}$ as measured from $O_{2,g}$, is ~ 8.8 kcal mol^{-1} (0.38 eV) [70]. The barrier for dissociation, from $O_{2,a}$ to 2 O_a, is ~ 7 kcal mol^{-1} (0.3 eV) [70,71]. The dissociative adsorption energy, $O_{2,g} \rightarrow 2O_a$, is ~ 48 kcal mol^{-1} (2.1 eV) [70]. The dissociation energy of O_2 gas molecule, $O_{2,g} \rightarrow 2O_g$, is ~ 120 kcal mol^{-1} (5.2 eV) [72]. From these and the photon energy $h\nu$, the energy scale for the potential energy curves is determined.

6.4 The Effects of Coadsorbed Species on Photo-Induced Processes

It is natural to extend the study of CO_2 photo-production to other systems. The coadsorption of O_2 and NO on Pt(111) did not lead to the photosynthesis of NO_2. However, dramatic effects on the photodesorption and photodissociation are observed. These coadsorption-induced changes can explain the lack of occurrence of the photo-production of NO_2 or other molecules.

Vibrational spectroscopy indicates clearly the emergence of a new NO adsorption state when coadsorbed with O_2. In Fig. 8 are shown EELS spectra of NO adsorbed alone, coadsorbed with O_2, and the coadsorption system after UV irradiation (~ 330 nm). A new peak with high vibrational energy (~ 230 meV) appears when NO is coadsorbed with O_2. This peak is just slightly lower in energy than the gas phase value (236 meV) and disappears after UV irradiation. The shoulder at ~ 50 meV correlates with the 230 meV peak and is attributed to vibration against the surface of the 230 meV NO species. The lowest N-O stretch (~ 180 meV) is absent in the coadsorbed system and appears after irradiation. Coadsorption with O_2 has introduced a new photoactive NO state.

Molecular O_2 can be completely displaced by NO. The O – O stretch at ~ 95 meV is lowered in energy by the coadsorbed NO and is shifted towards higher energy after UV irradiation of the coadsorbed system. The distinct vibrational peak (~ 60 meV) associated with atomic O is missing after UV irradiation, which is in contrast to photon irradiation

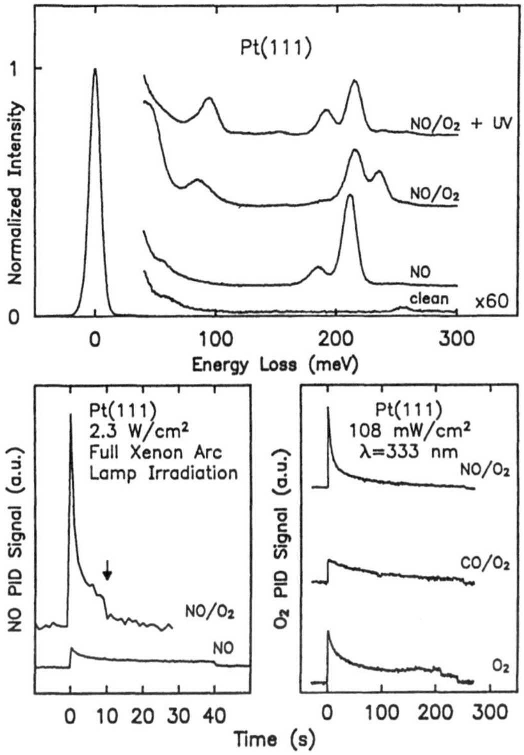

Fig. 8. EELS and PID spectra for coadsorption of NO and O_2 on Pt(111) at 100 K. Also included for comparison are the EELS spectrum of NO adsorbed alone and O_2 coadsorbed with CO after UV irradiation (333 ± 45 nm), and PID spectra of NO and O_2 adsorbed individually and O_2 coadsorbed with CO. The arrow in the NO PID signal for NO coadsorbed with O_2 indicates closure of the shutter for that spectrum.

of coadsorbed O_2 and CO. The absence of O_2 photodissociation is consistent with the lack of photoreaction for coadsorbed O_2 and NO.

The lack of photoreaction can be further understood from changes in the O_2 and NO photodesorption. The NO PID signal increases dramatically in the presence of coadsorbed O_2. Furthermore, NO photodesorption is observed even with near-IR radiation. The cross section for NO photodesorption of NO adsorbed alone on Pt(111) is extremely small (see Table I) and full arc irradiation is required for observing a photodesorption signal with reasonable intensity. For comparison, the NO PID signal for NO coadsorbed with O_2 is significantly higher, allowing a detailed wavelength dependence measurement.

The O_2 PID signal is also enhanced when O_2 is coadsorbed with NO. The cross section is increased relative to O_2 alone or O_2 coadsorbed with CO. The O_2 PID cross section for O_2 coadsorbed with CO is even lower than O_2 alone (the cross section is directly proportional to the rate of decay of the PID signal). Thus for the NO/O_2 system, photodesorption of both NO and O_2 is enhanced and photodissociation of O_2 is suppressed. Both changes lead to preferential photodesorption of the reactants over the photoreaction channel.

Coadsorbed species can alter the relative cross sections for photodesorption and photodissociation and influence the photoreaction yield.

7. Synopsis and Future Prospects

Photochemistry of oriented molecules on well-characterized single crystal solid surfaces is essentially a sparsely explored field at the present time. The present status of surface photochemistry is indeed primitive compared to that of gas phase photochemistry. However, progress in surface photochemistry is rapidly advancing and will benefit from the vast knowledge, technology, and experience gathered over the past years for gas phase photochemistry.

In addition to the rich, and often complicated, chemical and physical processes occurring on solid surfaces, the surface provides an energy transfer pathway and a medium for orienting the adsorbed molecules. The restricted degrees of freedom in the adsorbed state offers a natural medium for studying the dynamics of surface reactions. In addition to the angular orientation, the reaction occurs effectively in a two-dimensional space. Photon irradiation increases the energetics of the system, prepares the adsorbed reactants in well-defined states, and introduces the starting time of the reaction. Furthermore, the absorbed energy is both selective and significantly higher than is accessible thermally. Sizable reaction barriers can be overcome to affect dramatically the product distributions. Since photochemistry can occur even at low temperatures, the Boltzmann spread in the reactant state distributions can be effectively suppressed.

The results from the study of surface photochemistry are expected to extend to and advance our understanding of chemical reaction dynamics in general. In addition to identifying systematically the interesting surface photochemical systems, it is highly desirable to measure at least the translational energy distributions of the photoreaction products. However, translational energy only represents one part of energy disposal in a chemical reaction. In addition, the products can have non-equilibrium rotational and vibrational distributions and electronic excitation. Some of the excited products may decay via emission of near-UV, visible, or IR radiation. It is only by measuring these additional degrees of freedom that a direct analysis can be made for arriving at a basic, detailed understanding of the entire process of surface photochemistry.

Acknowledgments

We greatly appreciate over the years the generous support of this research in part by the Office of Naval Research Grant No. N00014-81-K-0505, the National Science Foundation Grant No. DMR-87-14616, and the Cornell Materials Science Center through the National Science Foundation Grant No. DMR-88-18558. The author is indebted to Charles Ying, Shu So, Walter Mieher, and Thomas Germer for their ingenuity in laboratory skills and scientific insights, and is enlightened and educated by them through numerous discussions. The results presented from our laboratory in this paper are extracted in parts from their thesis work. In addition, the author has benefited from close collaborations with René Franchy and Bengt Kasemo, whose insights into surface photochemistry and science in general have been refreshing.

References

[1] *Photo-effects in Adsorbed Species,* Faraday Discussions of the Chemical Society, No. 58 (The Faraday Division Chemical Society, London, 1974).

[2] D. Lichtman and Y. Shapira, CRC Critical Rev. Solid State Mater. Sci. **8**, 93 (1978).

[3] B.E. Koel, J.M. White, J.L. Erskine,and P.R. Antoniewicz, in *Interfacial Photoprocesses: Energy Conversion and Synthesis,* edited by M.S. Wrighton, Advan. Chem. Ser. 184 (Amercian Chemical Society, Washington, DC, 1980) p. 27.

[4] T.J. Chuang, Surf. Sci. Rep. **3**, 1 (1983).

[5] G.S. Selwyn and M.C.Lin, in *Lasers as Reactants and Probes in Chemistry.,* edited by W.M. Jackson and A.B. Harvey (Howard University Press, Washington, DC, 1985) p. 269.

[6] M.C. Lin and G. Ertl, Ann. Rev. Phys. Chem. **37**, 587 (1986).

[7] W. Ho, Comments Cond. Mat. Phys. **13**, 293 (1988).

[8] P. Avouris and R.E. Walkup, Annu. Rev. Phys. Chem. **40**, 173 (1989).
[9] A. Many, CRC Critical Rev. Solid State Sci., May 1974, p. 515.
[10] J.P. Cowin, D.J. Auerbach, C. Becker, and L. Wharton, Surf. Sci. **78**, 545 (1978).
[11] D. Burgess, Jr., R. Viswanathan, I. Hussla, P.C. Stair, and E. Weitz, J. Chem. Phys. **79**, 5200 (1983).
[12] R.B. Hall, J. Phys. Chem. **91**, 1007 (1987).
[13] C.H. Mak, J.L. Brand, B.G. Koehler, and S.M. George, Surf. Sci. **191**, 108 (1987).
[14] T.J. Chuang and I. Hussla, Phys. Rev. Lett. **52**, 2045 (1984); I. Hassla, H. Seki, T.J. Chuang, Z.W. Gortel, H.J. Kreuzer, and P. Piercy, Phys. Rev. B **32**, 3489 (1985).
[15] P.M. Ferm, S.R. Kurtz, K.A. Pearlstine, and G.M. McClelland, Phys. Rev. Lett. **58**, 2602 (1987).
[16] W. Hoheisel, K. Jungmann, M. Vollmer, R. Weidenauer, and F. Träger, Phys. Rev. Lett. **60**, 1649 (1988).
[17] Y. Shapira, R.B. McQuistan, and D. Lichtman, Phys. Rev. B **15**, 2163 (1977).
[18] F.A. Houle, J. Chem. Phys. **79**, 4237 (1983); Phys. Rev. Lett. **61**, 1871 (1988).
[19] E. Ekwelundu and A. Ignatiev, Surf. Sci. **179**, 119 (1987).
[20] C.F. Yu, M.T. Schmidt, D.V. Podlesnik, and R.M. Osgood, Jr., J. Vac. Sci. Technol. B **5**, 1087 (1987).
[21] Z. Ying and W. Ho, Phys. Rev. Lett. **60**, 57 (1988).
[22] S.A. Buntin, L.J. Richter, R.R. Cavanagh, and D.S. King, Phys. Rev. Lett. **61**, 1321 (1988); S.A. Buntin, L.J. Richter, D.S. King, and R.R. Cavanagh, J. Chem. Phys., submitted.
[23] E.P. Marsh, T.L. Gilton, W. Meier, M.R. Schneider, and J.P. Cowin, Phys. Rev. Lett. **61**, 2725 (1988); T.L. Gilton, C.P. Dehnbostel, and J.P. Cowin, J. Chem. Phys. **91**, 1937 (1989).
[24] E.B.D. Bourdon, P. Das, I. Harrison, J.C. Polanyi, J. Segner, C.D. Stanners, R.J. Williams, and P.A. Young, Faraday Discuss. Chem. Soc. **82**, 343 (1986).
[25] K. Domen and T.J. Chuang, Phys. Rev. Lett. **59**, 1484 (1987).
[26] S.A. Costello, B. Roop, Z.-M. Liu, and J.M. White, J. Phys. Chem. **92**, 1019 (1988); B. Roop, S.A. Costello, Z.-M. Liu, and J.M. White, *Solvay Conference on Surface Science*, edited by F.W. De Wette, Springer Series in Surface Science, Vol. 14 (Springer-Verlag, Berlin Heidelberg, 1988) p. 343.
[27] J.R. Swanson, C.M. Friend, and Y.J. Chabal, J. Chem. Phys. **87**, 5028 (1987).
[28] C.E. Bartosch, N.S. Gluck, W. Ho, and Z. Ying, Phys. Rev. Lett. **57**, 1425 (1986); N.S. Gluck, Z. Ying, C.E. Bartosch, and W. Ho, J. Chem. Phys. **86**, 4957 (1987).
[29] N. Djeghri, M. Formenti, F. Juillet, and S.J. Teichner, in *Photo-effects in Adsorbed Species*, Faraday Disscussions of the Chemical Society, No. 58 (The Faraday Division Chemical Society, London, 1974) p. 185.
[30] J. Cunningham, E. Finn, and N. Samman, ibid., p. 160; R.I. Bickley and R.K.M. Jayanty, ibid., p. 194.
[31] F. Steinbach and R. Harbroth, ibid., p. 143.
[32] D.H. Maylotte and A.N. Wright, ibid., p. 292.
[33] H. Gerischer, ibid., p. 219.
[34] S.J. Teichner and M. Formenti, in *Photoelectrochemistry, Photocatalysis and Photoreactors*, edited by M. Schiavello (Reidel, Dordrecht, 1985) p. 457.
[35] G.N. Schrauzer and T.D. Guth, J. Am. Chem. Soc. **99**, 7189 (1977).
[36] C.T. Lin, T.D.Z. Atvars, and F.B.T. Pessine, J. Appl. Phys. **48**, 1720 (1977); C.T. Lin and T.D.Z. Atvars, J. Chem. Phys. **68**, 4233 (1978).
[37] L.P. Levine, J.F. Ready, and E.G. Bernal, J. Appl. Phys. **38**, 331 (1967).
[38] J.M. Chen and C.C. Chang, J. Appl. Phys. **43**, 3884 (1972).
[39] K. Christmann, O. Schober, G. Ertl, and M. Neumann, J. Chem. Phys. **60**, 4528 (1974).
[40] R.O. Adams and E.E. Donaldson, J. Chem. Phys. **42**, 770 (1965); W.J. Lange, J. Vac. Sci. Technol. **2**, 74 (1965).
[41] P. Kronauer and D. Menzel, in *Adsorption-Desorption Phenomena*, edited by F. Ricca (Academic, New York, 1972) p. 313.
[42] F. Budde, A.V. Hamza, P.M. Ferm, G. Ertl, D. Weide, P. Andresen, and H.-J. Freund, Phys. Rev. Lett. **60**, 1518 (1988); P.M. Ferm, F. Budde, A.V. Hamza, S. Jakubith, G. Ertl, D. Weide, P. Andresen, and H.-J. Freund, Surf. Sci, in print.
[43] Z.C. Ying and W. Ho, to be published.
[44] S.K. So and W. Ho, Appl. Phys. A **47**, 213 (1988).

[45] R. Franchy, S.K. So, and W. Ho, to be published.
[46] S.K. So, R. Franchy, and W. Ho, to be published.
[47] T.A. Germer and W. Ho, J. Chem. Phys., submitted.
[48] K. Mase, S. Mizuno, M. Yamada, I. Doi, T. Katsumi, S. Watanabe, Y. Achiba, and Y. Murata, J. Chem. Phys. **91**, 590 (1989).
[49] J.A. Prybyla, T.F. Heinz, J.A. Misewich, M.M.T. Loy, and J.H. Glownia, to be published.
[50] X.-Y. Zhu, S.R. Hatch, A. Compion, and J.M. White, J. Chem. Phys., submitted.
[51] X. Guo, L. Hanley, and J.T. Yates, Jr., J. Chem. Phys. **90**, 5200 (1989).
[52] L. Hanley, X. Guo, and J.T. Yates, Jr., J. Chem. Phys., submitted.
[53] X. Guo, J. Yoshinobu, and J.T. Yates, Jr., J. Chem. Phys., submitted.
[54] St. J. Dixon-Warren, I. Harrison, K. Leggett, M.S. Matyjaszczyk, J.C. Polanyi, and P.A. Young, J. Chem. Phys. **88**, 4092 (1988).
[55] E.P. Marsh, M.R. Schneider, T.L. Gilton, F.L. Tabares, W. Meier, and J.P. Cowin, Phys. Rev. Lett. **60**, 2551 (1988).
[56] Z.-M. Liu, S. Akhter, B. Roop, and J.M. White, J. Am. Chem. Soc. **110**, 8708 (1988).
[57] K. Domen and T.J. Chuang, J. Chem. Phys. **90**, 3318, (1989).
[58] E. Hasselbrink, S. Jakubith, S. Nettesheim, M. Wolf, A. Cassuto, and G. Ertl, J. Chem. Phys., submitted.
[59] J.R. Creighton, J. Appl. Phys. **59**, 410 (1986).
[60] Z.C. Ying and W. Ho, to be published.
[61] S.K. So and W. Ho, to be published.
[62] Z.C. Ying and W. Ho, to be published.
[63] T.A. Germer and W. Ho, J. Chem. Phys. **89**, 562 (1988).
[64] F.G. Celii, P.M. Whitmore, and K.C. Janda, J. Phys. Chem. **92**, 1604 (1988).
[65] C.-C. Cho, J.C. Polanyi, and C.D. Stanners, J. Chem. Phys. **90**, 598 (1988).
[66] A. Mödl, K. Domen, and T.J. Chuang, J. Vac. Sci. Technol. A **7**, 1764 (1989).
[67] J.Y. Tsao and D.J. Ehrlich, J. Chem. Phys. **81**, 4620 (1984).
[68] W.D. Mieher and W. Ho, J. Chem. Phys. **91**, 2755 (1989).
[69] G.L. Geoffroy and M.S. Wrighton, *Organometallic Photochemistry* (Academic, New York, 1979).
[70] J.L. Gland, B.A. Sexton, and G.B. Fisher, Surf. Sci. **95**, 587 (1980).
[71] J.L. Gland, Surf. Sci. **93**, 487 (1980).
[72] R.T. Sanderson, Chemical Bonds and Bond Energy (Academic, New York, 1971), p. 28.
[73] J.G. Calvert and J.N. Pitts, Jr., *Photochemistry* (Wiley, New York, 1966); H. Okabe, *Photochemistry of Small Molecules* (Wiley, New York, 1978).

Photon Stimulated Desorption of Fluorine from Silicon Etched by XeF_2

J.A. Yarmoff[1,*], *S.A. Joyce*[1,**], *C.W. Lo*[2], *and J. Song*[2]

[1]Surface Science Division, National Institute of Standards and Technology (formerly the National Bureau of Standards), Gaithersburg, MD 20899, USA
[2]Department of Physics, University of California, Riverside, CA 92521, USA
*Present address: Department of Physics, University of California, Riverside, CA 92521, USA
**Present address: Surface Science Division, Sandia National Laboratory, Albuquerque, NM 87185, USA

Abstract. We have measured soft x-ray photoemission and photon stimulated desorption (PSD) from silicon surfaces which were etched by XeF_2. These surfaces are covered with a layer approximately 20Å thick composed of a mixture of silicon fluoride reaction intermediates. The relative intensities of both photoemission and PSD spectral features arising from each of the various SiF_x species depends strongly on the structure of the overlayer. Since photoemission measurements probe the entire reaction layer and PSD is sensitive to only the very outermost atomic layer, a comparison of these data can reveal structural details. The relative proportion of fluorides measured with photoemission varied slightly from run to run, but always showed a mixture of SiF, SiF_2 and SiF_3, along with a small amount of trapped SiF_4. The species observed with PSD, however, varied considerably more than the corresponding photoemission spectra. Some surfaces showed only SiF_3 spectral features while others showed features indicative of all three of the subfluorides. The implications of these results on the structure of the reaction layer are discussed.

1. Introduction

Many studies of the reaction of XeF_2 with clean silicon surfaces have been undertaken in an effort to model the basic mechanisms operative during the plasma-assisted etching of silicon. This system provides a convenient method for studying the most fundamental chemical interaction in fluorocarbon-based processes, the reaction between atomic fluorine and silicon surfaces. The observed gas-phase reaction products are primarily (~85%) SiF_4, with small amounts of other products such as SiF, SiF_2, SiF_3 and Si_2F_6 [1, 2, 3]. Soft x-ray photoemission has shown that an approximately 20Å thick reaction layer is formed on these surfaces after etching [4]. This layer is composed of a mixture of SiF, SiF_2 and SiF_3 intermediate species, as well as some trapped SiF_4. Photoemission alone, however, cannot be used to determine the structure of this layer.

The quantity and distribution of fluorides found on the surfaces is independent of the pressure used for the reaction, provided that a sufficient flux of XeF_2 has been used so that the steady-state reaction regime has been reached [5]. Despite this, however, reproducible photoemission spectra are obtained only under the most careful conditions. The efficiency of the reaction is dependent on the degree of passivation of the walls of the vacuum chamber and on the quantity of impurities which may be present.

Springer Series in Surface Sciences, Vol. 19
Desorption Induced by Electronic Transitions, DIET IV
Editors: G. Betz · P. Varga

Therefore we often see variations in the fluoride distributions for samples which were etched with the same nominal exposure to XeF_2, but at different times or in different chambers.

In contrast to photoemission, studies of adsorbates with desorption induced by electronic transition (DIET) techniques are considerably more surface sensitive. We present a photon stimulated desorption (PSD) study of silicon surfaces which were subjected to steady state etching by XeF_2, and compare the photoemission results, which probe the entire reaction layer, to PSD data which are sensitive only to the outermost component of the reaction layer. We find that the PSD spectra can vary considerably from sample to sample in a much more dramatic manner than photoemission.

There has been previous speculation that there is a structure to the etching reaction layer in which the higher fluorides are stacked above the lower [6, 7]. In that case, it would be expected that PSD would show emission only from SiF_3 groups. For some of the samples studied this result was found, confirming the model of a structured reaction layer. However, the PSD spectra from other samples showed ions originating from each of the SiF_x groups. This indicates a lateral inhomogeneity in which some portions of the surface contribute to the PSD signal from SiF groups and some areas contribute to the signal from SiF_3. We believe that this inhomogeneity is correlated with damage induced in samples exposed to excessive etching.

2. Experimental

These experiments were performed on beamline UV-8b at the National Synchrotron Light Source at Brookhaven National Laboratory. Silicon (111) wafers (n-type, 1.0 Ω-cm), cut within 0.25° of the low-index plane, were cleaned via standard methods in ultra-high vacuum (UHV) [8]. The fluorinated surfaces were prepared by transferring them under UHV into a separate reaction chamber (base pressure 1×10^{-9} torr), and then exposing them to XeF_2. Following exposure, the samples were transferred back into the spectrometer chamber (base pressure 1×10^{-10} torr). The composition of the surfaces was measured via Si 2p core level photoemission, employing an ellipsoidal mirror analyzer [9] with a typical energy resolution of 0.1 eV.

PSD was measured by reversing the polarity of the optics so as to be sensitive to ions instead of electrons, and then biasing the entrance to the multichannel plate array by -1000 volts to provide additional acceleration of the ions. In this manner it was possible to collect both kinetic energy distributions of the emitted ions and the dependence of the ion yield on the incident photon energy. The spectra were normalized by the current measured from the gold-plated final focussing mirror in order to account for beam drifts and for the monochromator transmission function. Note that the spot size of the synchrotron beam used in these experiments was about $1 mm^2$, and that these measurements were averaged over this area.

3. Results

As a reference, a first sample was prepared by exposure of a clean Si (111)-7x7 wafer to 50 L (1 L = 1×10^{-6} torr-sec) of XeF_2, which produces a surface covered by approximately one monolayer of fluorine. The distribution of silicon fluorides produced by a 50 L exposure is dependent

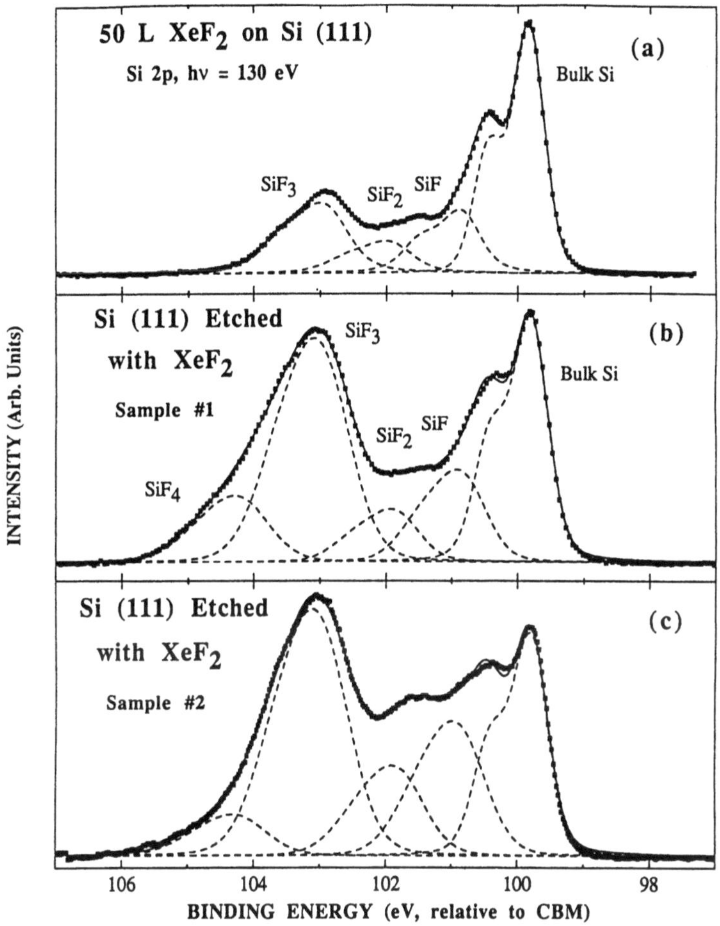

Figure 1. Photoemission spectra of Si (111) (a) after exposure to 50 L of XeF_2, (b) sample #1 after etching by XeF_2, and (c) sample #2 after etching by XeF_2. The symbols show the raw data after background subtraction, the solid line shows the result of a numerical fit to the data and the dashed lines show the individual components of the fit.

on the initial structure of the surface, and the 7x7 structure produces a surface which is primarily SiF and SiF_3 with only a small amount of SiF_2 [10]. The core-level soft x-ray photoemission spectrum collected from this surface is shown in Fig. 1a. The symbols in this figure show the raw data after removal of the background due to secondary electrons. The photoemission spectra were numerically fit to a sum of Gaussian broadened Lorentzian functions to represent the contribution from each Si oxidation state. Each of these functions is composed of a spin-orbit split (0.61 eV) doublet. The photoemission spectra are plotted in Fig. 1 with respect to the conduction band minimum (CBM) to illustrate the correspondence with transitions that lead to PSD.

The etching reaction layers were produced by exposing Si (111) wafers to ~5×10^{-6} torr of XeF_2 for 1 hour. This dose is sufficient to reach the steady-state reaction regime, and surfaces whose photoemission spectra are shown in Figs. 1b and 1c resulted. These surfaces are composed primarily of SiF, SiF_2 and SiF_3 species, but there is also some SiF_4 reaction product trapped within the layers. These layers are highly disordered solids whose composition is representative of the reaction intermediates in the formation of SiF_4. The fact that the photoemission intensity from SiF_3 dominates has been used to conclude that the bottleneck in the reaction is the addition of the final fluorine atom [4]

A rough estimate is that the average thickness of the layers is in the range of 6-7 monolayers, or about 20Å. Note that the photoemission peaks are considerably broadened in the spectra collected from the etched surfaces as compared to the spectrum from the monolayer surface, which is indicative of the high degree of disorder in the reaction layer. The data from samples #1 and #2 were collected at different times. Sample #1 was prepared during the same run as the sample used for the 50 L exposure of Fig. 1a, and sample #2 was prepared at a different time when the relative passivation and contamination level of the dosing chamber was much different. Sample #2 shows a larger amount of shifted silicon than sample #1, and a lot of this extra intensity is in the mono- and di-fluorides.

The photoemission spectra shown in Fig. 1 were measured prior to the PSD measurements. Additional photoemission spectra were recorded after completing the measurements in order to ascertain the effect of prolonged synchrotron radiation exposure on the composition of the reaction layers. A loss of fluorine had occurred during the course of the measurements, and the effect was to preferentially remove fluorine from the trifluoride species [11]. This makes it difficult to perform quantitative measurements with these experiments, but the level of the observed radiation damage was small enough so as not to alter any of the conclusions reported in this paper.

The monolayer surface formed the basis of our previous PSD work [12, 13], in which we found that the position of the edges in the PSD spectra coincided with the excitations from the bonding 2p level to the CBM and to any final state features associated with particular bonding sites. The upper curve in Fig. 2 shows the total PSD yield as a function of photon energy from the monolayer surface, and the arrows are used to illustrate the positions of transitions from Si 2p initial states to empty final states which lead to ion desorption. Two of the transitions illustrated are from the Si $2p_{3/2}$ of the SiF and SiF_3 surface species to the CBM. For clarity, only the features which arise from the $2p_{3/2}$ initial states are labeled, so it should be noted that the corresponding feature from the $2p_{1/2}$ initial state is located at a 0.61 eV higher photon energy, and can be clearly seen in the data. The PSD spectrum from the monolayer surface shows distinct edges for desorption from both SiF and SiF_3 surface sites. Note that there is also a small feature just prior to the 2p(SiF) $\rightarrow$ CBM edge peaked at ~100 eV photon energy. This feature is a result of electron stimulated desorption (ESD) via electrons produced by photon absorption in the bulk Si [13]. Additionally, the entire PSD spectrum is riding on a large background which is decreasing just prior to the edge. The scale on the abscissa shows that this background is about a factor of three larger than the edge.

The lower curve in upper panel of Fig. 2 shows the PSD spectrum measured from the first etched surface, sample #1. The two spectra in the upper panel of Fig. 2 are shown as raw data, without any normalization. It is thus seen that the yield from the monolayer surface is actually slightly higher that the yield from the etched surface. The spectrum from the etched surface shows only a small feature at the 2p(SiF) $\rightarrow$ CBM edge and a relatively much larger feature at the 2p(SiF_3) $\rightarrow$ CBM edge.

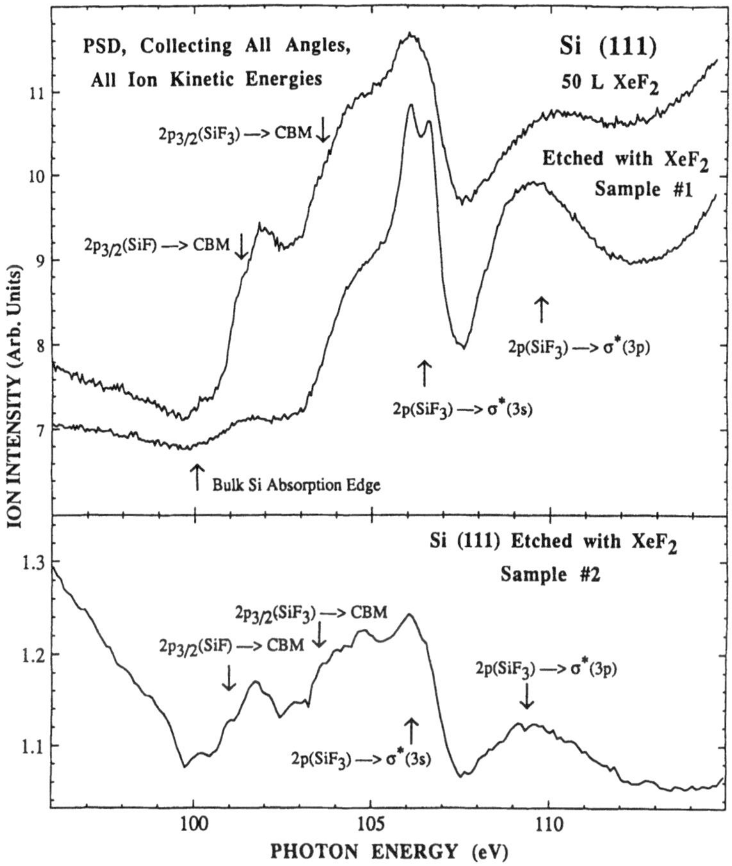

Figure 2. Yield of F^+ ions as a function of photon energy for a Si (111) wafer after exposure to XeF_2. These are raw data after normalization by the incident photon flux. The upper panel shows the data collected during the same run from a 50 L exposure and from the etched surface of sample #1. The bottom panel shows the results from sample #2.

The PSD spectrum from etched sample #2 is shown in the bottom panel of Fig. 2. Since this data was collected during a different run than the data from sample #1, there is no simple manner in which to normalize these spectra. The spectrum from sample #2 is riding on a background similar to the other spectra, and the scale on the abscissa shows that the background is about a factor of ten larger than the edge. This background is also decreasing prior to the edge, and it appears to decrease more steeply than the background in the spectra from the other surfaces. This is deceiving, however, because the total background intensity from this sample is larger. The spectrum from sample #2 shows both the 2p(SiF) → CBM and the 2p(SiF_3) → CBM edges as well as some structure at an intermediate photon energy, which is presumably the 2p(SiF_2) → CBM edge.

There are additional features in both the monolayer and etched spectra which are labeled in the figure as 2p(SiF_3) → σ^*(3s) and 2p(SiF_3) → σ^*(3p). These are transitions to empty σ^* states which have largely Si 3s

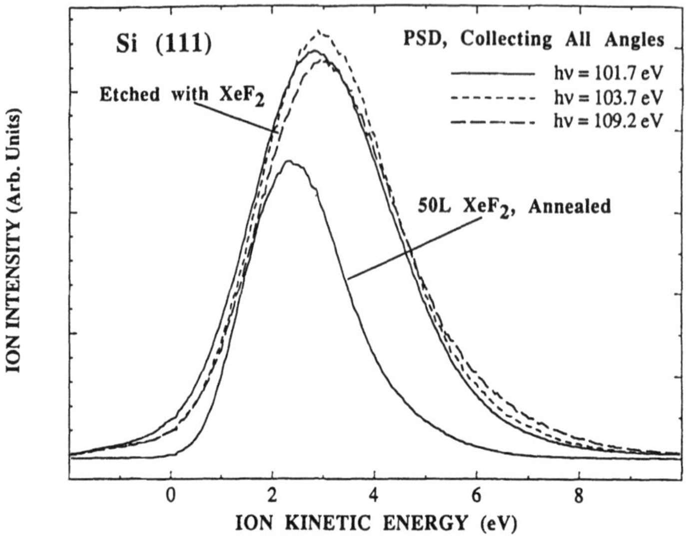

Figure 3. Kinetic energy distributions of ions from sample #2. These are the raw data which have been normalized only to the incident photon flux. For comparison, the kinetic energy distribution from an annealed 50 L exposure is shown (with an arbitrary normalization).

and 3p character, and have been previously identified in adsorbed SiF_3 [12, 13]. The trapped SiF_4 in the etching reaction layer may also contribute to these features, as the σ^*(3s) and σ^*(3p) final states have been observed in gas-phase absorption spectroscopy of SiF_4 [14] and in PSD of condensed SiF_4 [15]. Note that these features are much sharper in the spectrum collected from sample #1 than in either the PSD spectrum from the monolayer surface or from sample #2.

The kinetic energy distributions of the desorbed ions from sample #2 are shown in Fig. 3 at a variety of incident photon energies. The solid lines show data collected at hν = 101.7 eV for the etched surface and for a monolayer surface after annealing. The annealed monolayer surface is composed solely of SiF surface entities [12, 13], and this photon energy enhances the 2p(SiF) → CBM transition. Therefore, using hν = 101.7 eV with the annealed monolayer most closely shows the kinetic energy distribution of an ion originating via a direct process from an SiF surface site. The hν = 101.7 eV distribution from the etched surface is considerably broader than from the annealed monolayer, and the maximum is shifted to higher kinetic energy. The other two distributions shown are for sample #2 at hν = 103.7 and 109.2 eV. These photon energies enhance the 2p(SiF_3) → CBM and 2p(SiF_3) → σ^*(3p) transitions, respectively. The distributions obtained at these photon energies are shifted slightly higher in kinetic energy than the 101.7 eV distribution, and are also somewhat broader. The corresponding data collected from sample #1 showed the same qualitative behavior.

4. Discussion

We shall now discuss the implication of our data on the structure of the reaction layer. Because of the nature of such a highly disordered system, it is impossible to analyze this data quantitatively to the precision one might desire. However, a knowledge of the structural aspects of this reaction layer would be of considerable benefit to the etching community. In the following discussion we will outline the various factors that need to be considered in such an analysis. For the purposes of this discussion we must consider two types of spatial homogeneity, parallel to and perpendicular to the surface. With regard to the perpendicular homogeneity, we will consider two limiting cases: 1) a completely homogeneous reaction layer and 2) a reaction layer where the various silicon fluorides (SiF_x, x = 1-3) exist in spatially separate regions. For the latter case, it is physically more reasonable to place the SiF_3 species at the surface as they will act as chain terminators. The two cases will be most easily distinguished by the relative concentrations of SiF_3 and SiF in the surface region, and for this reason we will base our arguments on a comparison of the relative spectral signals of SiF_3 to SiF as determined by photoemission and PSD. Similar conclusions could in principle be made by comparing SiF_3 intensities with SiF_2.

In a previous publication we had proposed that the reaction layer has a perpendicular structure in which the higher fluorides are positioned above the lower fluorides [7]. The methodology used was to collect a series of photoemission spectra in the constant final state (CFS) mode, and to alter the sampling depth of the measurement by varying the kinetic energy of the measured photoelectrons. The variations in the proportions of the fluorides as a function of the electron attenuation length were then analyzed. This data was interpreted with the assumption that the variations of the photoemission cross section with photon energy were independent of oxidation state for electron kinetic energies greater than 16 eV. Further studies have indicated, however, that the variations of the cross sections are, in fact, strong functions of the oxidation state, possibly even for photoelectrons with as high as 100 eV kinetic energy. The magnitude of the cross section effects can be much larger than the electron attenuation length effects even if the layer were completely structured [6, 16]. Since the details of the photoelectron cross section variations are unknown, the results of the CFS measurements cannot be used to infer a structure for the reaction layer.

An additional effect of the cross section enhancements is that, since the photoemission cross sections are strong functions of the oxidation state, a simple analysis of the relative areas in a photoemission spectrum does not properly represent the actual relative abundances. Preliminary measurements of the relative cross sections indicate that at the 130 eV photon energy typically used for Si 2p photoemission, the SiF_3 cross section is about a factor of two larger than the SiF cross section [6]. (Also, as a confirmation, the cross section ratio of the 3+ to 1+ oxidation state at the SiO_2/Si interface is about 1.7 to 1 [17].) Therefore, although the data of Figs. 1b and 1c show SiF_3 to SiF photoelectron intensity ratios of 2.7 and 1.9, respectively, the actual proportion of detected photoemitters is closer to 1 to 1.

If we were to assume that the the reaction layer was laterally uniform, then the following conclusions can be drawn with respect to the two limiting cases mentioned above. In the case of a perpendicularly homogeneous reaction layer, the photoemission data would indicate that the proportion of SiF_3 species to SiF species is roughly equal throughout the layer. For the case of spatially separate phases with the SiF_3 groups at the surface, there would be an even higher concentration of SiF and SiF_2 species in the reaction layer than the raw photoemission intensity suggests, as the photoelectrons from these groups would be attenuated via inelastic scattering on transport to the surface. The photoemission data by itself cannot distinguish between these two possibilities, and the situation is further complicated by the possibility that the reaction layer may be laterally inhomogeneous. In this case, the photoemission data is simply indicative of the average distribution.

In contrast to photoemission, the PSD data is expected to be considerably more surface sensitive. This is because any ion created below the surface with only a few eV of kinetic energy would not be capable of traveling through the 20Å or so of reaction layer without suffering reneutralization. This would be particularly true for fluorine because of its high ionization potential. The data shown in Fig. 2 which was collected from sample #1 displays primarily features arising from SiF_3 surface species. Note that there is a small broad edge which overlaps the regions of the bulk Si absorption edge and the 2p(SiF) → CBM edge. Even if the layer were terminated absolutely by SiF_3, this feature would still be expected to have some intensity due to the indirect process wherein electrons created by absorption in the buried SiF moieties would induce ESD from the SiF_3 groups at the surface. The magnitude of this effect should be similar to the indirect effect seen at the bulk Si absorption edge in the monolayer surface. Thus, for sample #1, our data is consistent with a laterally homogeneous layer that is terminated by SiF_3 species.

Note that the relative intensity of the $2p(SiF_3) \to \sigma^*$ transition to the $2p(SiF_3) \to$ CBM transition has increased considerably in the spectrum collected from sample #1 when compared with the monolayer spectrum. In addition, the peaks are much sharper and the spin-orbit splitting in the $2p(SiF_3) \to \sigma^*(3s)$ transition is clearly resolved. This is a further indication that these SiF_3 moieties are positioned at the top of an amorphous SiF_x layer. Because of the spatial extent of these molecular orbitals, they are less likely to be quenched when they are further away fom the surface. These features are now very similar to the absorption and PSD data collected from SiF_4 [14, 15].

For sample #2, the situation is completely different. It is clear that the PSD spectrum shows features arising from each of the SiF_x species, thereby ruling out the case of complete perpendicular layering of the silicon fluorides in conjunction with lateral homogeneity. The increase in 2p(SiF) → CBM edge intensity cannot be accounted for simply by the increase in the total amount of SiF that was observed with photoemission. The radically different shape of the PSD spectrum collected from sample #2 indicates a radically different structure at the surface of the reaction layer.

The striking feature of the PSD spectrum from sample #2 is the strong 2p(SiF) → CBM edge. The only way to reconcile the shape of this feature is with a laterally inhomogeneous reaction layer in which there are areas of the surface at which the coverage is on the order of a monolayer. This feature is actually better resolved from the $2p(SiF_2) \to$ CBM and $2p(SiF_3) \to$ CBM edges in the data from sample #2 than in the 50 L data, which could indicate that the areas where the reaction layer is thinnest are composed primarily of silicon monofluoride. These thinner regions were produced during the etching of the substrate, and certainly have a different underlying structure than the (111)-7x7 structure which gives the monolayer fluoride distribution shown in Fig. 1a. The distributions of fluorides at monolayer coverages is a strong function of the structure at that surface [10], and an unreconstructed (111) surface would be expected to display exclusively silicon monofluoride. It is therefore not unreasonable that the thin regions would be terminated largely by SiF groups.

The large background signal associated with sample #2 in conjunction with the shape of the PSD spectrum at the Si 2p edge suggests that this surface is heavily damaged, and that our measurements are averaging over a surface that has considerably lateral inhomogeneity. We have previously noted that heavily etched samples tended to show extreme surface irregularities, even to the point of being visible to the eye. Further studies are planned in which we will study the evolution of the PSD spectra for a single sample over the entire range of exposures. On the basis of our current understanding, we would expect to see spectra such as that shown for sample #1 evolve towards the spectra shown for sample #2 as the exposures to XeF_2 increase.

5. Conclusions

We have measured soft x-ray photoemission and PSD from silicon surfaces which were etched by XeF_2. The photoemission data contains information dependent on the average distribution and thickness of the layer, but cannot be used for a structural determination. Structural information is available by combining photoemission with PSD, since the ions measured in PSD can only come from the outermost atomic layer of the sample. For some samples, the PSD data indicates that the surface of the reaction layer is terminated solely with SiF_3 species, and the reaction layer is laterally uniform. The reaction layer therefore has a structure in which the monofluoride is located at the interface to the Si substrate and the trifluoride is at the surface. Combining this result with the fact the cross section for photoemission is about a factor of two larger for SiF_3 than for SiF, leads us to conclude that the amount of SiF (and SiF_2) in the reaction layer is severly underestimated purely on the basis of photoemission intensities. For other samples, the PSD data is consistent with a reaction layer that is laterally inhomogeneous over the 1 mm^2 spot size. The thickness in these films varies from about a monolayer of fluorine coverage to many layers.

6. Acknowledgements

This research was carried out at the National Synchrotron Light Source, Brookhaven National Laboratory, which is supported by the U.S. Department of Energy (Division of Material Sciences and Division of Chemical Sciences) under Contract No. DE-AC02-76CH00016. The authors wish to acknowledge useful discussions with T.E. Madey and F.R. McFeely. SAJ wishes to acknowledge NAS/NRC for postdoctoral support.

References

1. F.A. Houle, J. Appl. Phys. **60**, 3018 (1986).
2. H.F. Winters and F.A. Houle, J. Appl. Phys. **54**, 1218 (1983).
3. M.J. Vasile and F.A. Stevie, J. Appl. Phys. **53**, 3799 (1982).
4. F.R. McFeely, J.F. Morar and F.J. Himpsel, Surf. Sci. **165**, 277 (1986).
5. J.A. Yarmoff and F.R. McFeely, Phys. Rev. B **38**, 2057 (1988).
6. F.R. McFeely, B.D. Silverman, J.A. Yarmoff and U.O. Karlsson in *Proceedings of the Eighth International Symposium on Plasma Chemistry*, edited by K. Akashi and A. Kinbara (International Union of Pure and Applied Chemistry, Tokyo, Japan, 1987), p. 927.

7. J.A. Yarmoff and F.R. McFeely in *Photon Beam and Plasma Stimulated Chemical Processes at Surfaces*, edited by V.M. Donnelly, I.P. Herman and M. Hirose (Material Research Society, Pittsburgh, PA, 1987), p. 451.
8. F.J. Himpsel, P. Heimann, T.-C. Chiang and D.E. Eastman, Phys. Rev. Lett. **45**, 1112 (1980).
9. D.E. Eastman, J.J. Donelon, N.C. Hein and F.J. Himpsel, Nucl. Instrum. Methods **172**, 327 (1980).
10. F.R. McFeely, J.F. Morar, N.D. Shinn, G. Landgren and F.J. Himpsel, Phys. Rev. B **30**, 764 (1984).
11. J.A. Yarmoff and S.A. Joyce in *Synchrotron Radiation in Materials Research*, edited by J.H. Weaver, J. Gland and R. Clarke (Materials Research Society Symposium Proceedings, Pittsburgh, PA, 1989), p.
12. J.A. Yarmoff and S.A. Joyce, Phys. Rev. B **40**, 3143 (1989).
13. J.A. Yarmoff, A. Taleb-Ibrahimi, F.R. McFeely and P. Avouris, Phys. Rev. Lett. **60**, 960 (1988).
14. H. Friedrich, B. Pittel, P. Rabe, W.H.E. Schwarz and B. Sonntag, J. Phys. B **13**, 25 (1980).
15. R.A. Rosenberg, J. Vac. Sci. Technol. A **2**, 1463 (1986).
16. F.R. McFeely, B. Robinson, J.A. Yarmoff and S.A. Joyce, to be published.
17. F.J. Himpsel, F.R. McFeely, A. Taleb-Ibrahimi, J.A. Yarmoff and G. Hollinger, Phys. Rev. B **38**, 6084 (1988).

Substrate Mediated Photodissociation of NO_2/N_2O_4 Adsorbed on Pd(111)

E. Hasselbrink, S. Nettesheim, M. Wolf, A. Cassuto, and G. Ertl*

Fritz-Haber-Institut der Max-Planck-Gesellschaft, D-1000 Berlin 33, Germany
*Permanent address: Laboratoire Maurice Letort, CNRS, B.P. 104, F-54600 Villers Les Nancy, France

Abstract. UV irradiation (hν = 3.5 to 6.4 eV) of NO_2/N_2O_4 adsorbed on top of a NO covered Pd(111) surface leads to photodissociation. State-resolved detection of the NO molecules released into the gas phase and determination of the yield as function of angle and polarization provide clear evidence that the dominant primary process consists in electronic excitation of the metallic substrate rather than in direct absorption by the adsorbate.

1. Introduction

UV photochemistry of adsorbed molecules is a rapidly growing field for which detailed understanding is still largely lacking [1]. In addition to the well-established mechanisms for photochemistry of the free molecules, the vicinity of a solid surface may open additional channels competing for primary excitation as well as quenching. Interpretation of the experimental observations existing so far [2-6] may be roughly classified into two categories, direct photoexcitation of the adsorbate or substrate mediated excitation. The latter process comprises electronic excitation of the solid as the primary step and subsequent conversion into valence level excitation of the adsorbate leading to bond breaking. Photon stimulated desorption of NO adsorbed on Si surfaces is, for example, believed to occur through the creation of 'hot' carriers by band-gap excitation in the solid [6], and photodissociation of CH_3Cl adsorbed on Ni(111) was likewise interpreted in terms of an electron-transfer mechanism [5].

2. Experimental

The experiments were performed with NO_2 adsorbed on a Pd(111) sample precovered with NO. This system was chosen for several reasons: UV absorption of NO_2 and its dimer, N_2O_4, had been studied in detail [7], as well as photodissociation of gaseous NO_2 and state-resolved detection of the product molecule NO [8-10]. While the cross section for photochemistry of a particle bound to a metal surface is generally very small [11], it may be much higher for desorption of weakly held species [12]. For this reason the metallic surface was precovered with a saturated layer of NO, on top of which NO_2 was adsorbed at 105 K. Adsorbed NO_2 dimerizes rapidly, and thermal desorption peaks at 135 K. Post-irradiation TPD experiments showed that the chemisorbed spacer layer is not modified by the laser radiation.

Springer Series in Surface Sciences, Vol. 19
Desorption Induced by Electronic Transitions, DIET IV
Editors: G. Betz · P. Varga

Irradiation was performed by pulsed light from an excimer laser operated at photon energies of 3.5 eV (351 nm), 4.0 eV (308 nm), 5.0 eV (248 nm), or 6.4 eV (193 nm). Low laser fluences, typically ≤ 5 mJ/cm^2·pulse, were applied in order to prevent noticeable heating of the substrate. A rotatable quadrupole mass spectrometer (QMS) served for recording temperature programmed desorption (TPD) spectra, as well as 'integral' time-of-flight (TOF) spectra without resolution of the internal state distributions. Rotational and vibrational state populations of the NO molecules released into the gas phase were probed by laser-induced fluorescence (LIF). State-resolved TOF spectra were recorded by varying the time delay between the dissociation- and probe-laser pulses. The state-resolved data presented here has been obtained using 6.4 eV photon energy.

3. Results

Fig. 1 reproduces a series of 'integral' TOF spectra for NO leaving the surface covered with less than 1 monolayer of NO_2 under the influence of UV radiation with different photon energy. These distributions are bimodal, exhibiting 'fast' and 'slow' molecules whose relative portion varies with wavelength. Analysis reveals that the mean translational energy of the 'slow' NO molecules (130 K) (as well as their average rotational energy of 115 K) corresponds to values close to those expected for thermal accommodation with the substrate, although purely thermal dissociation of NO_2 can be ruled out under these conditions. The 'fast' molecules, on the other

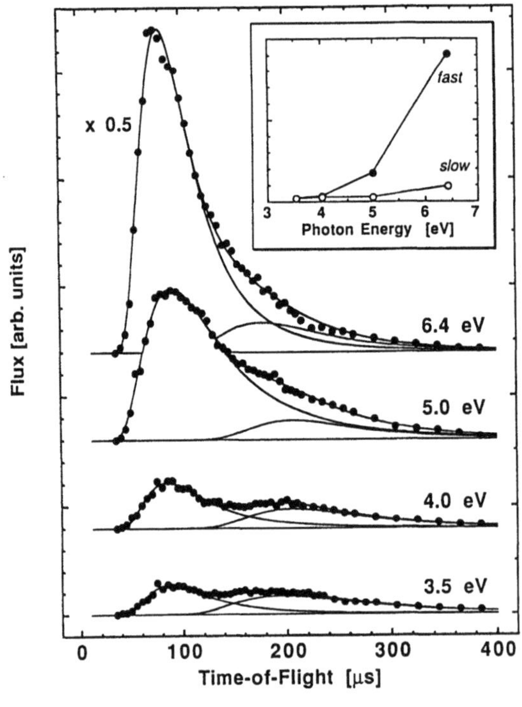

Fig.1: Time-of-flight spectra, obtained with the mass spectrometer, of NO resulting from photodissociation of adsorbed NO_2. The lines represent least square fits using two modified Maxwell-Boltzmann distributions. The inset shows the cross sections for the formation of both channels as function of excitation energy.

hand, are clearly of non-thermal origin. Their fraction increases with increasing photon energy and dominates completely at hν = 6.4 eV. Their mean translational energy, $<E_{trans}>$, of 800 K is practically independent of hν, which suggests that the formation of these molecules starts from the same level, regardless of the initial excitation. We find that $<E_{trans}>/2k$ increases linearly with internal energy, E_{int}, from about 700 K to around 2000 K, a feature which seems to be characteristic for molecules escaping from a surface having a limited lifetime on a repulsive potential [11,12]. The rotational energy distribution of the 'fast' molecules was found to correspond to a rotational temperature of 900 K.

The cross sections for photodecomposition were determined by monitoring through TPD the depletion of the NO_2-adlayer as a function of the number of laser shots. An exponential decrease over a wide range of coverages (1 – 0.05 ML) is found yielding absolute cross sections for the various photon energies. At 6.4 eV excitation a value of $3 \cdot 10^{-18}$ cm^2 is found, which is more than an order of magnitude higher than the corresponding gas phase value [7,13]. This can also not be attributed to the formation of dimers, because photolysis of these is known to yield only the monomers as fragments [14].

Discrimination between excitation within the adsorbed layer from excitation of the substrate electrons could be achieved by recording the desorption yield as function of the angle of incidence and polarization - switching between s- and p-polarization - of the UV-laser.

Absorption in the metal is described by the Fresnel equations [15]

$$A_s = 1 - r_s \cdot r_s^*; \qquad r_s = \frac{\cos\phi_1 - \sqrt{n_2^2 - \sin^2\phi_1}}{\cos\phi_1 + \sqrt{n_2^2 - \sin^2\phi_1}} \tag{1}$$

$$A_p = 1 - r_p \cdot r_p^*; \qquad r_p = \frac{n_2^2\cos\phi_1 - \sqrt{n_2^2 - \sin^2\phi_1}}{n_2^2\cos\phi_1 + \sqrt{n_2^2 - \sin^2\phi_1}}, \tag{2}$$

where ϕ_1 is the angle of incidence and n_2 the index of refraction of the metal [16].

The adsorbed layer, on the other hand, would respond to the intensity of the electric field at the surface which can be calculated [17] - neglecting the dielectric properties of the adsorbed layer - as

$$\langle E_{s,y}^2 \rangle = (1 + R_s) + 2\sqrt{R_s}\cos\delta_s \tag{3}$$

$$\langle E_{p,x}^2 \rangle = \cos^2\phi_1 \left\{ (1 + R_p) - 2\sqrt{R_p}\cos\delta_p \right\} \tag{4}$$

$$\langle E_{p,z}^2 \rangle = \sin^2\phi_1 \left\{ (1 + R_p) + 2\sqrt{R_p}\cos\delta_p \right\}, \tag{5}$$

with $\delta_i = \tan^{-1}[\mathrm{Im}(r_i) / \mathrm{Re}(r_i)]$ and $R_i = r_i \cdot r_i^*$. The coordinate system has been defined with z parallel to the surface normal and y perpendicular to the plane of incidence.

As can be seen from Fig. 2a, good agreement is obtained with the model based on optical absorption in the Pd metal as dominant process, in particular also with

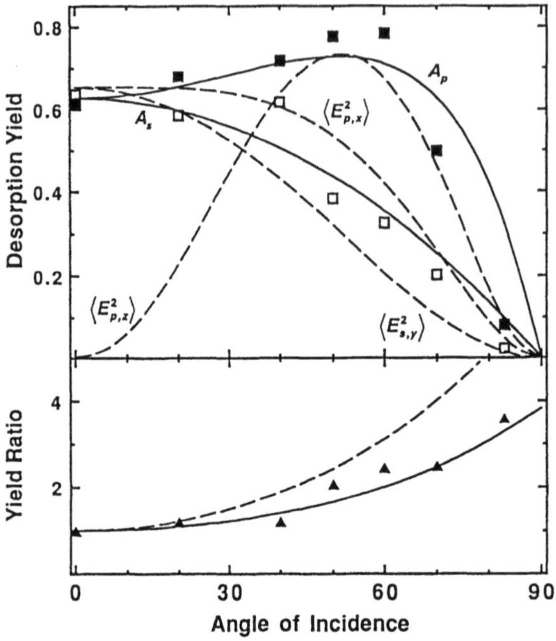

Fig. 2: (a) Desorption yield vs angle of incidence for excitation (hν = 6.4 eV) with p- (■)and s-polarized (□) light. The curves give the calculated absorption in the metallic substrate (——) and the field intensities (– – –) at the interface. (b) Ratio of the observed yields as function of the angle of incidence (▲). The solid line gives the prediction when assuming that absorption in the metal is the initial step in the photodissociation/desorption process. The dashed curve results from a model of direct absorption by the adsorbate.

respect to the occurrence of a maximum in the data for p-polarization. However, elimination of the direct absorption model is more complicated, since the orientation of the specific transition dipole moment of the adsorbate with respect to the surface is unknown. If it were exactly perpendicular to the surface, no signal could occur at normal incidence - contrary to the experimental evidence. If the transition dipole moment were parallel to the surface, the experimental yield for p-polarization should match with $\langle E^2_{p,x}\rangle$. This is also not the case; in particular $\langle E^2_{p,x}\rangle$ does not exhibit a maximum around 60°. A reasonable fit could, however, be reached if one assumes contributions from both, parallel and perpendicular, orientations with 1:1/3 weighting ratio. This possibility may, on the other hand, clearly be ruled out if one considers the *ratios* of the yields obtained with s- and p-polarization at each angle as shown in Fig. 2b. Again good agreement with the model of substrate mediated excitation is obtained, while the curve predicted by the model just sketched is outside the experimental error range.

In conclusion, we therefore believe that there is convincing evidence that the dominant primary process for photodissociation in the present system consists in electronic excitation of the substrate metal. Creation of 'hot' electrons or photoelectrons will be the consequence and these will interact with the adsorbate and cause dissociative electron attachment. The gas phase process $e^- + NO_2 \rightarrow O^- + NO$ has a threshold at 1.6 eV [18], while the reaction $e^- + N_2O_4 \rightarrow NO + NO_3^-$ is even exothermic by 2.6 eV [19]. The intermediate formation of NO_3 which might have a lifetime longer than a few μsec would also explain why no release of O atoms into the gas phase could be detected - their density would be too much spread out in time. An electron attachment process of the quoted type could further be reconciled with the observation that the energy distributions of the NO molecules are independent of the primary photon energies and different from those of gas phase photolysis of NO_2. The small amount of excess energy carried away by the ejected NO molecules indicates that a strong quenching interaction with the substrate is present despite the spacer layer.

The authors are indebted to Prof J. M. White for stimulating discussions. A. C. thanks the Max-Planck-Gesellschaft for a visiting scientist fellowship.

References:

1 T. J. Chuang: Surf. Sci. Rep. **3**, 1 (1983); P. Avouris and R. E. Walkup: Ann. Rev. Phys. Chem. **40**, in press

2 E. B. D. Bourdon, J. P. Cowin, I. Harrison, J. C. Polanyi, J. Segner, C. D. Stanners, and P. A. Young: J. Phys. Chem. **88**, 6100 (1984); I. Harrison, J. C. Polanyi, and P. A. Young: J. Chem. Phys. **89**, 1475 (1988)

3 E. P. Marsh, M. R. Schneider, T. L. Gilton, F. L. Tabares, W. Meier, and J. P. Cowin: Phys. Rev. Lett. **60**, 2551 (1988)

4 S. A. Costello, B. Roop, Z. M. Liu, and J. M. White: J. Phys. Chem. **92**, 1019 (1988)

5 E. P. Marsh, T. L. Gilton, W. Meier, M. R. Schneider, and J. P. Cowin: Phys. Rev. Lett. **61**, 2725 (1988)

6 Z. Ying and W. Ho: Phys. Rev. Lett. **60**, 57 (1988)

7 A. M. Bass, A. E. L. Ledford, Jr., and A. H. Laufer: J. Res. Natl. Bur. Std. **80A**, 143 (1976)

8 H. Zacharias, M. Geilhaupt, K. Meier, and K. H. Welge: J. Chem. Phys. **74**, 218 (1981)

9 T. G. Slanger, W. K. Bischel, and M. J. Dyer: J. Chem. Phys. **79**, 2231 (1983)

10 R. J. S. Morrison and E. R. Grant: J. Chem. Phys. **77**, 5994 (1982)

11 S. A. Buntin, L. J. Richter, R. R. Cavanagh, and D. S. King: Phys. Rev. Lett. **61**, 1321 (1988)

12 F. Budde, A. V. Hamza, P. M. Ferm, G. Ertl, D. Weide, P. Andresen, and H.-J. Freund: Phys. Rev. Lett. **60**, 1518 (1988)

13 H. Okabe: *Photochemistry of small molecules*, (Wiley, New York, 1978)

14 M. Kawasaki, K. Kasatani, H. Sato, H. Shinohara, and N. Nishi: Chem. Phys. **78**, 65 (1983)

15 M. Born and E. Wolf: *Principles of Optics*, (Pergamon, Oxford, 1980)

16 *Physik Daten*, Vol. 18-1, edited by H. Behrens and G. Ebel, (Karlsruhe, 1981)

17 J. D. E. McIntyre: in *Adv. in Electrochemistry and Electrochemical Engineering*, Vol. **9**, pp 68 (1973)

18 R. Abouaf, R. Paineau, and F. Fiquet-Fayard: J. Phys. **B9**, 303 (1976)

19 The heat of formation of NO_3^- is referenced as -3.49 eV in: *Inorganic Chemistry*, S. J. Ashcroft and G. Beech, p. 20, (VanNostrand, London, 1973)

Photochemistry of Halogenated Molecules Adsorbed on Ag(111)

X.-L. Zhou and J.M. White

Department of Chemistry, University of Texas, Austin, TX78712, USA

Abstract. The UV photodissociation of several halogen-containing molecules, adsorbed on Ag{111}, is described. Attention is focused on Cl_2CO.

Overview. The photochemically driven C-X (X = Cl or Br) bond dissociation in halogenated molecules, adsorbed on Ag(111), has been investigated using X-ray photoelectron spectroscopy (XPS), ultraviolet photoelectron spectroscopy (UPS), work function measurements ($\Delta\Phi$) and desorption mass spectrometry - both isothermal and temperature-programmed (TPD). Adsorbates include CH_3Br, CH_3Cl, C_2H_5Cl, ClC_2H_4Br, C_6H_5Cl and Cl_2CO. The light source is a filtered 100W high pressure Hg arc lamp. In the absence of illumination and with the Ag(111) held at 100 K, these adsorbates are all reversibly and weakly chemisorbed; there are no thermal decomposition products detected by TPD, XPS or $\Delta\Phi$.[1]

With illumination there is clear evidence for C-X bond cleavage for any adsorbate coverage in the range investigated, 0.2 monolayer (ML) to multilayers. XPS spectra of the halogen core levels show two peaks, the first corresponding to the parent molecule and the second, shifted to lower binding energy, corresponding to a halogen atom strongly chemisorbed to Ag(111).[2] For all the adsorbates containing a single halogen atom and a hydrocarbon fragment, R, in R-X, the TPD spectra taken after illumination show two well-defined product desorption peaks, R_2 and AgX.[3] We interpret these products in terms of a surface photochemical process that, as in the gas phase [4], breaks the C-X bond to form R and X. Based on XPS peak areas, all of the X and a large fraction of the R groups are retained at the surface. Depending on the adsorbate, particularly when R = CH_3, some of the R groups are lost to the gas phase.

For all the alkyl halides, the yields, but not the distribution or desorption characteristics, of the products are strongly wavelength dependent and molecule specific as indicated by the following facts. (1) Illumination with wavelengths longer than 420 nm leads to no detectable products for any of the adsorbates. (2) At constant incident absorbed energy, the yield increases with photon energy between 3.5 and 5 eV. (3) The cleavage of the R-Cl bond requires a higher photon energy than the R-Br bond.

Photodissociation of ClC_2H_4Br is different; at photon energies sufficient to break the C-Br bond in C_2H_5Br, but not the C-Cl bond in C_2H_5Cl, both the C-Br and C-Cl bonds are broken in ClC_2H_4Br, and C_2H_4 appears in TPD.[5]

Photochemistry of Phosgene. The photodissociation of Cl_2CO, which we focus on here, is also different; its dissociation extends to longer wavelengths and all of the CO product desorbs during illumination and can be easily detected as a time-

Springer Series in Surface Sciences, Vol. 19
Desorption Induced by Electronic Transitions, DIET IV
Editors: G. Betz · P. Varga

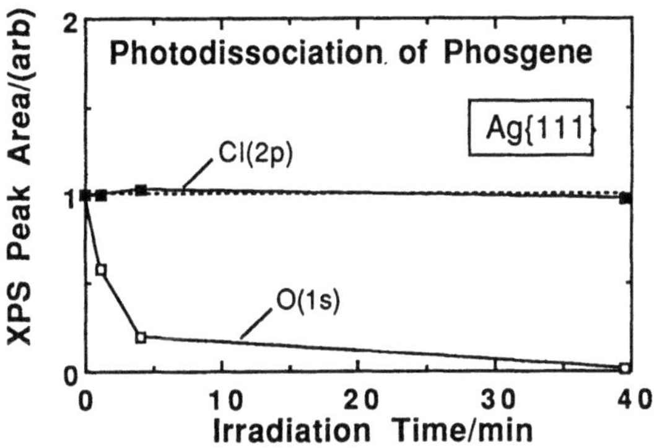

Figure 1. XPS peak areas measured after photolysis of 1 ML phosgene on Ag{111}.

dependent isothermal signal at the mass spectrometer.[6] The XPS spectra before and after illumination show that a full monolayer of Cl_2CO is easily dissociated. Figure 1 shows the XPS core level peak areas for O(1s) and Cl(2p) as a function of irradiation time using a 100 W Hg arc with no cut-off filter (the incident power is 75 mW/cm^2). To within experimental error, the Cl(2p) signal remains constant while the binding energy (BE) shifts from 201 to 197.3 eV. The O(1s) peak area decays rapidly and monotonically while its BE remains nearly constant at 533 eV. The C(1s) also decays with time but is much noisier than the O(1s). These XPS data confirm the retention of all Cl, i.e. no photodesorbed Cl_2CO or Cl, and loss of all O and C. The only product desorbed during illumination is CO and the only products desorbed after illumination for short periods are parent Cl_2CO and AgCl, the latter desorbing between 700 and 950 K.

The initial rise of the gas phase CO signal, upon opening an optical stop between the light source and the surface, can be used to evaluate the initial dissociation cross section. The results are summarized in Fig. 2 over the first monolayer. The dissociation cross section is remarkably high (10^{-18} cm^2 for very low coverages) exceeding the gas phase optical absorption cross section, which is around 10^{-20} cm^2 in the 250 nm range [4]. The cross section drops by a factor of two as the coverage increases from 0.2 to 1 ML, a strong dependence compared to alkyl halides on either Ag{111}[3] or Pt{111} [7]. As discussed below, this coverage dependence extends to the multilayer regime.

The wavelength dependence of the average photodissociation cross section is plotted in Fig. 3. In each case the initial coverage was 1 ML and the energy distribution of the incident light was controlled with cut-off filters. Clearly, there is a strong wavelength dependence; there is an apparent threshold near 2.5 eV, a steady rise up to 3.8 eV, a sharp rise between 3.8 and 4.2 eV, and, finally, a slow rise to 5 eV. Figure 3 also shows the work function of clean Ag{111} and 1 ML Cl_2CO on Ag{111}.

Several points arise from this figure. First, the optical extinction coefficient drops to negligible values for photon energies below 4 eV. [4] but the photochemistry extends to 2.5 eV. Second, there is a sharp rise of the cross section near, but always slightly below, the work function threshold of both the clean and monolayer-covered surfaces. Third, between 4.2 and 5 eV the cross section increases

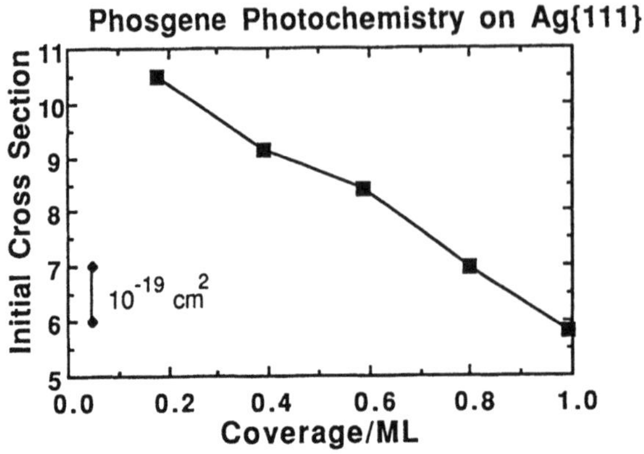

Figure 2. Photodissociation cross section as a function of coverage.

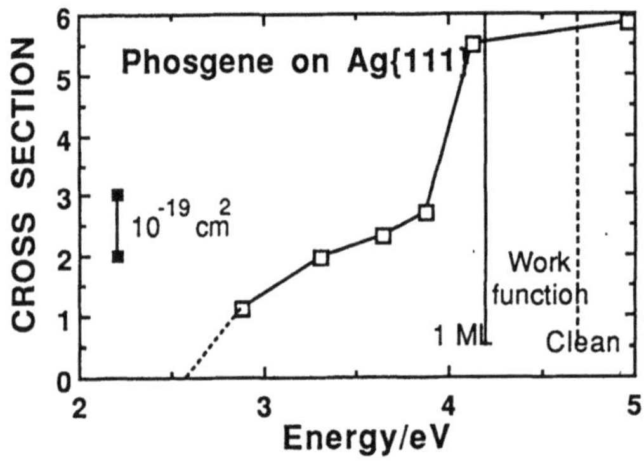

Figure 3. Cross section as a function of photon energy for first monolayer.

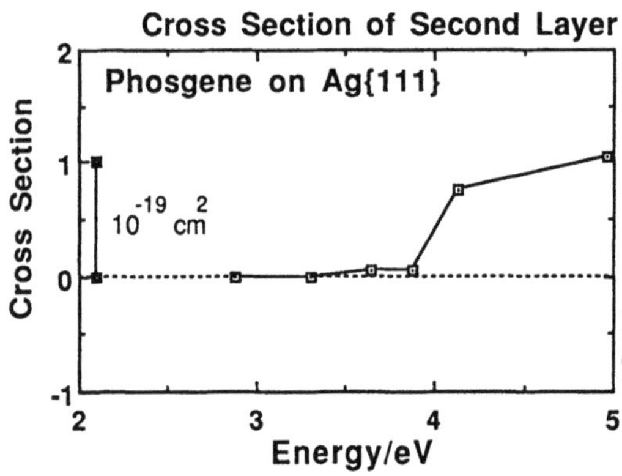

Figure 4. Cross section as a function of photon energy for the second monolayer.

by less than 10% while the optical extinction coefficient of gas phase phosgene increases by more than a factor of 10.

Figure 4 shows the cross section attributable to the second layer of phosgene. As measured either by the production of CO or the removal of Cl_2CO, the rate for a 1 ML photolysis was subtracted from that for a 2 ML photolysis. The resulting difference was converted to a cross section. The second layer has no measurable cross section between 2.8 and 3.8 eV. It then rises like the 1 ML cross section, quickly up to 4.2 eV and then slowly to 5 eV. At 4.2 and 5 eV the 1 ML cross section is 6 - 8 times the 2 ML cross section. Interestingly, all of the Cl is still retained when the second layer is photolyzed. It accumulates and, when the total coverage exceeds 1 ML, it desorbs with the same characteristics as bulk AgCl. [3]

These results, which require non-thermal excitation of the adsorbate may involve three excitation paths, depending on the wavelength and the nature of the adsorbate-substrate complex -- (1) direct photon absorption of the adsorbate-substrate complex to form a dissociative electronically excited state, (2) photon absorption in the bulk substrate to form chemically active photoelectrons, and (3) photon absorption in the bulk substrate to form chemically active hot electrons. The active photodissociation paths occur in competition with excited state quenching and thermalization by the substrate.

It is a simple matter to exclude photoelectron processes at some wavelengths; above 4.2 eV there are very few photoelectrons and there is readily detected photochemistry, particularly for the 1 ML case. The sharp increase in yield around 4 eV for both 1 and 2 ML may be ascribed to photoelectrons or hot electrons very near, but slightly under, the vacuum level. This is consistent with the very large cross section for electron-stimulated decomposition of adsorbed phosgene at low energies (10^{-15} cm^2) [8], a process analogous to dissociative electron attachment process in the gas phase [9]. Even hot electrons below the vacuum level will be able to reach the second layer if the local potential is low enough to allow passage, including tunneling. It would be very interesting to examine the dependence of the yield on multilayer thickness.

As noted above, gas phase absorption cross sections, even shifted ones will not describe these results. Direct excitation processes, if they contribute, must either have very different wavelength dependent absorption cross sections or very different variations with excitation energy of the subsequent quenching cross sections. These processes may contribute at all wavelengths but make a major contribution below 4 eV. While they are not separable from hot electron processes, it is worth noting that, as viewed by the first layer of molecules, the number and distribution of electrons near the vacuum level will change smoothly with photon energy; there is not a strong change in passing from below to above the work function threshold. Thus, the sharp rise in both the 1 and 2 ML results around 4 eV and the negligible cross section for the second monolayer below 4 eV suggests at least two pathways to products. There are at least two possibilities: (1) two distinct photoelectron/hot electron excitation processes, and (2) one direct excitation and one photoelectron/hot electron process.

The coverage dependence in the 1 ML regime indicates additional complexity. That the cross section drops for a given wavelength, may be related to one or more of the following: (1) Decomposition of an excited molecule may be a function of the number of vacant Ag sites in a given locality. (2) The quenching cross section may increase with coverage. (3) If there is a resonance in the electron excitation process, the excitation cross section may drop as the coverage increases and the macroscopic work function drops.

Acknowledgement. This work supported in part by the Army Research Office.

References.

1. X.-L. Zhou, F. Solymosi, P. M. Blass, K. C. Cannon and J. M. White, Surface Sci. **219**, 294 (1989).
2. Z.-M. Liu, S. Akhter, B. Roop, and J. M. White, J. Am. Chem. Soc. **110**, 8708 (1988).
3. X.-L. Zhou and J. M. White, (to be published).
4. H. Okabe, *Photochemistry of Small Molecules*, (Wiley, New York, 1978) p. 282.
5. X.-L. Zhou and J. M. White, (manuscript in preparation).
6. X.-L. Zhou and J. M. White, J. Chem. Phys. (in press).
7. S. A. Costello, B. Roop, Z.-M. Liu, and J. M. White, J. Phys. Chem. **92**, 1019 (1988).
8. X.-L. Zhou, S. R. Coon and J. M. White, J. Chem. Phys. (in press).
9. Z. Karpas and F. S. Klein, Intern. J. Mass Spectrom Ion Phys. **22**, 189 (1976).

Photophysics and Photochemistry of NO on Ag(111), Cu(111), and Si(111)7 × 7

R. Franchy, S.K. So, Z.C. Ying, and W. Ho*

Laboratory of Atomic and Solid State Physics and Materials Science Center, Cornell University, Ithaca, NY 14853, USA
*Present address: IGV der KFA Jülich, Postfach 1913, D-5170 Jülich, Fed. Rep. of Germany

We have performed photon induced desorption (PID) from NO adsorbed on Ag(111), Cu(111), and Si(111)7×7 at 85 K in the wavelength region 250–800 nm by using Xe and Hg-Xe arc lamps in combination with a monochromator ($P \leq 30$ mWcm^{-2}). The adsorption of NO on these surfaces and subsequent changes after irradiation were characterized by high resolution electron energy loss spectroscopy (HREELS) and thermal desorption spectroscopy (TDS). On Ag(111), PID of NO and N_2O have been observed from the NO saturated surface. On Cu(111), PID of both NO and N_2O has also been observed. N_2O is photosynthesized from photoinduced reactions. On Si(111)7 × 7, photon irradiation induces both desorption and dissociation of NO and photosynthesis of N_2O. The N_2O products partially desorb intact and partially dissociate into N_2 and O, which leads to N_2 desorption. PID of NO on Ag(111) and Cu(111) is state selective and occurs primarily for NO adsorbed in atop states. On Si(111)7×7, both atop and bridge bonded NO are photoactive. From the wavelength dependences of the PID, we suggest that the photoexcitation of the adsorbate ($2\pi_{occ}$ level) is the more important mechanism for Ag and Cu for photon energy≥3.4 eV; the substrate mediated process, however, is more dominant for Si.

1. Introduction

Although visible and UV photophysics and photochemistry of gas phase molecules are well understood, their extension to adsorbed molecules on solid surfaces is relatively unexplored. Recently, there has been a rapid growth of interest in photo-induced processes on metal and semiconductor surfaces. The interaction of photons with adsorbed molecules can lead to desorption of the entire molecule or dissociation of the molecule followed by a rearrangement of the photodissociated products [1, 2]. The nonthermal photo-products can react with other species leading to formation (or desorption) of new molecules on the surface [3–5]. In this paper, we will concentrate on photon induced desorption (PID).

The most investigated adsorption system is NO on metal [6] and semiconductor surfaces [7, 8]. The unoccupied 2π level above the Fermi level (E_F) can be filled by photogenerated electrons (originating from the 2π occupied level of the NO molecule and/or the substrate levels below

Desorption Induced by Electronic Transitions, DIET IV
Editors: G. Betz · P. Varga

E_F), leading to photon induced desorption. For a better understanding of PID, we have investigated the NO adsorbed on several surfaces including Ag(111), Cu(111), and Si(111)7 × 7. The band structures of these substrates are markedly different from each other. Different wavelength dependences of the PID cross sections are expected for substrate induced mechanisms.

2. Experimental

The experiments were performed in two similar UHV systems, each of which contains a quadrupole mass spectrometer and a high resolution electron energy loss spectrometer (HREELS). The base pressure was about 2×10^{-10} Torr. Details of the apparatus and sample cleaning have been published elsewhere [8, 9]. All adsorption and PID measurements were performed at $\sim$ 85 K. The photon source consists of a 1000 W Xe or 1000 W Hg-Xe arc lamp and a monochromator ($250 \leq \lambda \leq 680$ nm; $P \leq 30$ mW cm^{-2}; 11 nm bandwidth). PID has been performed only from the saturated NO overlayers.

3. Results and Discussions

Figure 1 shows HREEL and TD spectra before and after irradiation for NO on Ag(111), Cu(111), and Si(111)7 × 7. Detailed analysis of the adsorption and reactions of NO on Ag(111) [9], Cu(111) [10], and Si(111)7 × 7 [8] are published elsewhere. At saturation, different NO and NO-derived species (the 3-fold bridge and atop NO states, N_2O, O and N) coexist on Ag(111). The irradiation only removes the atop NO as demonstrated by EEL spectra. After irradiation for 180 s (λ = 340 nm), the loss peaks characteristic of the atop NO species (228 and 212 meV) are almost depleted, while the loss peaks due to the 3-fold bridge states (156 and 138 meV) are not affected. The relative intensity of the N-N stretching mode of N_2O (271 meV) remains unperturbed after irradiation. This observation is consistent with TDS data which show that the N_2O desorption peak (at 110 K) is not affected by the photons. The TD spectrum of NO after irradiation also demonstrates that the atop NO at 90 K is strongly diminished.

On Cu(111), NO is adsorbed in an atop (222 meV) and a 2-fold bridge site (187 meV) at saturation. The N-N stretching mode characteristic of N_2O at $\sim$270 meV is not observed in the HREEL spectrum [Fig. 1(b)], showing that no N_2O is formed on this surface from adsorption. The EEL peak at 263 meV is the C-O stretching vibration due to adsorption of CO from the residual gas in the UHV chamber during NO adsorption. After irradiation, the intensity of the HREELS peak at 222 meV corresponding to NO adsorbed in the atop state is diminished. The intensity of the peak at 187 meV (2-fold bridge position) [11, 12] does not show significant changes. Simultaneously, a new peak at 151 meV arises. The corresponding TD spectra have three peaks at

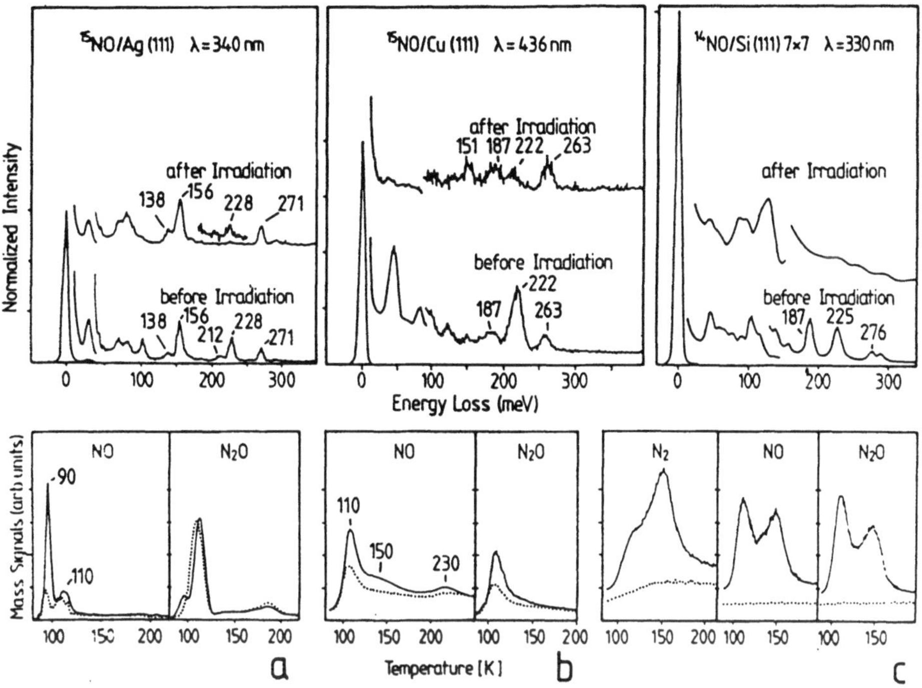

Fig. 1 HREEL and TD spectra before (solid line) and after (dotted line) irradiation for a saturated overlayer of NO on : (a) Ag(111), (b) Cu(111), and (c) Si(111)7 × 7.

$\sim$ 110, 150, and 230 K which all diminish in intensity after irradiation. Both PID and TDS [Fig. 1(b)] show strong N_2O signals. Since there is no direct evidence of N_2O on the surface, the observed N_2O in PID or TDS must be due respectively to photosynthesis and thermal synthesis from NO on the surface. In all cases, PID of CO cannot be observed.

On Si(111)7 × 7, NO is adsorbed molecularly in two different states: an atop state characterized by HREELS peak at 225 meV and a bridge state at 187 meV. In addition to the molecular adsorption, dissociative adsorption of NO and formation of N_2O are observed. The peaks at 155 and 276 meV are the N-O and N-N stretching modes of N_2O molecules, respectively. The intensities of both molecular NO stretches decrease under irradiation (as well as the two N_2O related peaks.) Under photon irradiation the Si-N and Si-O modes at 77 and 115 meV, respectively, increase in intensity and shift to higher energies. The surface after long irradiation time ($\geq$ 120 s) is depleted of all the molecular species and covered only with atomic N and O. Therefore, dissociation of molecularly adsorbed NO must occur under photon irradiation. Figure 1(c) shows a TD spectrum recorded for NO/Si(111)7 × 7 after irradiation for 120 s (λ = 330 nm). There is no observable TDS signal for NO and N_2O from the surface. A broad N_2 desorption feature is found, which is attributed to recombinative desorption from N atoms.

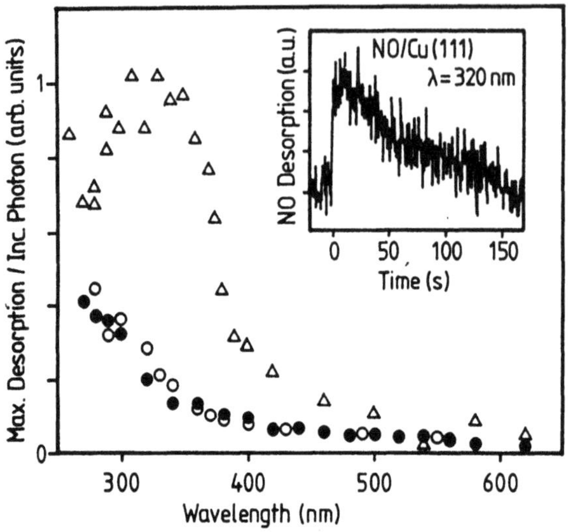

Fig. 2 The wavelength dependences of the PID signal of NO from saturated overlayer on Ag(111) (o), Cu(111) (•) and Si(111)7 × 7 (Δ). The inset shows a typical PID signal of NO from Cu(111). Irradiation began at t=0 s after the opening of a shutter.

Figure 2 shows the maximum PID signal of NO (normalized to the number of incident photons) versus wavelength. For comparison the curves for the different substrates are plotted relative to their absolute cross sections at $\sim$ 320 nm. The PID intensity increases with decreasing wavelength in the region between 280 and 600 nm. A more pronounced increase is observed below 360 nm (above 3.4 eV).

The PID cross sections are determined as follows. For NO/Cu(111), the NO PID signal can be fit by a single exponential decay. The cross section is evaluated from the inverse of the product of the incident photon flux and the time constant of the decay. On the other hand, the decays of the PID signals for NO/Ag(111) and NO/Si(111)7 × 7 are non-exponential. A single exponential is fit to the initial decay to derive a time constant from which the cross section for the initial decay is extracted. Following these procedures, the cross sections are found to be $\sigma_{NO} = 3.1 \times 10^{-18}\mathrm{cm}^2(\lambda = 320$ nm) for Ag(111), $\sigma_{NO} = 2.2\times10^{-18}\mathrm{cm}^2(\lambda = 320$ nm) for Cu(111), and $\sigma_{NO} = 1.1\times10^{-17}\mathrm{cm}^2(\lambda = 330$ nm) for Si(111)7 × 7. The corresponding cross sections for N_2O PID are $\sigma_{N_2O} = 1.7 \times 10^{-18}\mathrm{cm}^2(\lambda = 320$ nm) for Ag(111), $\sigma_{N_2O} = 1.9 \times 10^{-18}\mathrm{cm}^2(\lambda = 320$ nm) for Cu(111), and $\sigma_{N_2O} = 1.1 \times 10^{-17}\mathrm{cm}^2(\lambda = 330$ nm) for Si(111)7 × 7.

4. PID Mechanisms

In the following discussion, we will focus on the PID mechanisms of NO on the three substrates. Since the temperature rise of the samples during the entire irradiation period is small (less than 3 K), it is very unlikely that PID of NO is a thermal process. Instead, the process must involve excitation of the adsorbate and/or substrate electronic levels. There is no direct absorption for gas phase NO between 270 and 800 nm [13,14]. However, the electronic structure of an adsorbate may be changed by adsorption [15] and this change has to be considered in order to explain the PID phenomenon. In the following, PID mechanisms are discussed in terms of the $2\pi_{occ}$ and $2\pi_{unocc}$ levels of the NO molecule and the s and d levels of the Ag and Cu substrate and the p levels of Si. The densities of states (DOS) of these substrates are different: (i) for Ag(111) the 4d level has a width of 4 eV and the threshold for 4d-level excitation and a plasmon-like threshold apparently merge at about 4 eV [16]; (ii) for Cu (111) the threshold for 3d-level excitation is at about 2 eV [17]. A nearly constant DOS between E_F and the top of the d level is derived from the 5s electrons for Ag and 4s electrons for Cu as shown in Fig. 3.

In the adsorbate-substrate complex, NO–(S), two main PID excitation channels are possible:

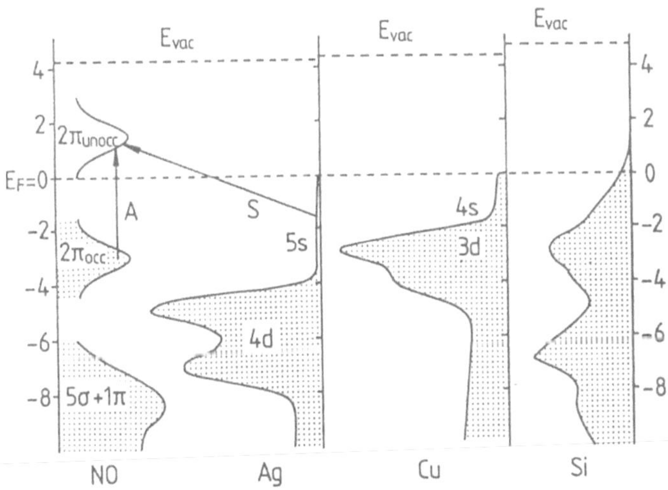

Fig. 3

Schematic representation of the first step of the PID mechanisms. Process A represents the adsorbate induced and S the substrate induced mechanisms. The 5s and 4d Ag bands [16], the 4s and 3d bands of Cu [17], and the 2p bands of Si [24] are obtained from photoemission data. The valence levels ($5\sigma+1\pi$, $2\pi_{occ}$) and the lowest unoccupied level ($2\pi_{unocc}$) of NO are derived from photoemission [21] and inverse photoemission [22, 23] data, respectively.

I. Direct Photoexcitation of the $2\pi_{occ}$-Level Below E_F

The incident photons can excite the $2\pi_{occ}$ level of NO and give rise to photogenerated electrons. If the energy of the incident photons is equal to $E_{2\pi_{unocc}} - E_{2\pi_{occ}}$, the electrons can be captured by the empty $2\pi_{unocc}$ level above E_F, resulting in a highly excited (NO)* molecule on the surface. The deexcitation can be followed by the desorption of NO from the surface. According to the Menzel-Gomer-Redhead mechanism [18, 19], the first step is the excitation from the ground state into a repulsive state. During the separation of the adsorbate from the surface, transition to the ground state can lead to recapture of the desorbing molecule. If, however, the transition occurs beyond a critical distance from the surface, the molecule has already acquired sufficient kinetic energy for desorption. A schematic representation of the first step of this PID mechanism is shown in Fig. 3. We denote the adsorbate induced mechanism by A.

II. Direct Photoexcitation of the Substrate Levels

The incident photons can also excite electrons from the substrate. If the energy of the incident photons is equal to $E_{2\pi_{unocc}} - E_F$, the electron can be captured by the $2\pi_{unocc}$ level, leading to a short-lived negative ion, $(NO)^-$, on the surface. The PID process can be described by the Antoniewicz model [20]. The first step is the excitation from the ground state into an excited ionic state by a Franck-Condon transition. Then the ion propagates on the excited state potential energy surface and subsequently decays via another Franck-Condon transition back to the ground state. Only those species projected on the ground state potential in the region of the repulsive wall with total energy greater than the binding energy desorb. We denote the substrate induced mechanism by S.

(i) NO/Ag(111) For NO chemisorbed on Ag(111) at 150 K, two peaks in the photoemission spectra at 2.5 and 3.3 eV below E_F are observed and assigned to the $2\pi_{occ}$ level [21]. In general the $2\pi_{unocc}$ level is observed as a broad level approximately 1.5 eV above E_F [22, 23]. The wavelength dependence of the PID signal indicates that both A and S mechanisms are valid. At low photon energies ($E_{ph} \leq$ 4 eV) photoexcitation of the adsorbate $2\pi_{occ}$ level and the substrate 5s band below E_F are the driving mechanisms. However, the increase of the PID cross section for photon energies above 3.4 eV can only be explained by mechanism A since for $3.4 \leq E_{ph} \leq 4.1$ eV (see Fig. 3) no contribution from the 4d level excitation is possible, and the contribution of the 5s level excitation is relatively small and structureless. At higher photon energies (above the threshold for Ag 4d level excitation) the d-level induced mechanism should also be possible.

(ii) NO/Cu(111) The positions of the $2\pi_{unocc}$ and $2\pi_{occ}$ levels are unknown for this system. Therefore, we can only speculate on the PID mechanism. In general both mechanisms A and S can be valid.

However, by assuming that the position of the $2\pi_{unocc}$ is not very different from NO/Ag(111), one expects that the threshold for substrate d-level induced mechanism is shifted by about 2 eV to lower photon energies (the threshold for d-band excitation for Cu is 2 eV lower than for Ag). As shown in Fig. 2, the wavelength dependence of the PID signal is almost the same for Cu(111) and Ag(111). There are two possible conclusions: (i) if mechanism A is valid, then the separation between the $2\pi_{unocc}$ and $2\pi_{occ}$-levels should be quite similar for NO on both Ag and Cu; (ii) if mechanism S is valid, then the position of the $2\pi_{unocc}$ above E_F has to be shifted by 2 eV in order to explain the increase of PID cross section above 3.4 eV. However, the position of the $2\pi_{unocc}$ level for many NO systems is located at about 1.5 eV above E_F [22, 23], and therefore, mechanism A is more likely.

(iii) NO/Si(111)7 × 7 The positions of the adsorbate 2π levels are also unknown. As shown in Fig. 2, an enhancement in PID is observed for photon energies above the transition energy of the first direct band gap of Si at 3.4 eV (365 nm). When the Si sample is irradiated with photons, electron-hole pairs are created. PID can be induced by hot carries which have not completely lost their kinetic energies by collisions with other carries and optical phonons. The data for NO/Si(111)7 × 7 strongly suggest that the photogenerated carriers from the substrate are responsible for the PID of NO [7].

5. Summary

(i) On metals [Ag(111) and Cu(111)] PID is state selective, only NO from atop states is photodesorbed.

(ii) On Si(111)7 × 7 PID is not state selective. NO in both atop and bridge states are photoreactive. N_2O is formed under irradiation and then desorbs from the surface and dissociates into N_2 and O.

(iii) The wavelength dependence of the PID signal is similar for Ag(111) and Cu(111), suggesting a common origin for the photoinduced desorption. For NO on Si(111)7 × 7, the photogenarated carriers are the dominant force for the observed PID.

In order to discriminate between adsorbate and substrate induced desorption, the position and distribution of the adsorbate and substrate levels should be measured by inverse photoemission, ultraviolet photoemission, and other techniques. In addition, it would be desirable to study PID over a larger photon energy range (1–10 eV). In this way, one can correlate the resonant transition of molecularly adsorbed NO or the absorption of the substrate with the wavelength dependence of the PID signals.

6. Acknowledgements

Support of this research by the Office of Naval Research Grant No. N00014-81-K-0505 and the Cornell Materials Science Center through the National Science Foundation Grant No.DMR-88-18558 is gratefully acknowledged. One of us (R.F.) would like to thank the School of Chemical Engineering, the MSC, and LASSP of Cornell University for financial support and Wilson Ho for hospitality.

References

[1] T.J. Chuang, Surf. Sci. Rep. 3 (1983).
[2] W.Ho, Comments Cond. Mat. Phys. 13, 293 (1988); W.Ho, this volume.
[3] W.D.Mieher and W.Ho, J. Chem. Phys. 91, 2755 (1989).
[4] K.G.Lloyd, A.Campion and J.M.White, Catalysis Lett. 2, 105 (1989).
[5] X.-Y.Zhu, S.R.Hatch, A.Campion and J.M.White, J. Chem. Phys. 91, 5011 (1989).
[6] See, for example, F.Budde, A.V.Hamza, P.M.Ferm, G.Ertl, A.Weide, P.Andresen, H.J-.Freund, Phys. Rev. Lett. 60, 1518, (1988); S.A. Buntin, L.J.Richter, R.R.Cavanagh and D.S.King, Phys. Rev. Lett. 61, 1321, (1988).
[7] See, for example, Z.Ying and W.Ho, Phys. Rev. Lett. 60, 57 (1988); S.K.So and W.Ho, Appl. Phys. A 47, 213, (1988).
[8] Z.C.Ying and W.Ho, J. Chem. Phys. 91, 2689 (1989).
[9] S.K.So, R.Franchy and W.Ho, J. Chem. Phys. 91, 5701 (1989).
[10] S.K.So, R. Franchy, and W.Ho, to be published.
[11] J.K.Wendelken, Appl. Surf. Sci. 11/12, 172 (1982).
[12] H.Ibach and D.L.Mills, *Electron Energy Loss Spectroscopy and Surface Vibrations* (Academic, New York, 1982).
[13] G.Herzberg, *Molecular Spectra and Molecular Structure I,* Spectra of Diatomic Molecules, 2nd ed. (Van Nostrand Reinhold, New York, 1950).
[14] J.Mason, J. Chem. Soc.(Dalton) 19, (1975) and references therein.
[15] W.Bach and H.D.Breuer, Faraday Disc. Chem. Soc. 58, 237 (1974).
[16] G.G.Tibbets and J.M.Burkstrand, Phys. Rev. B 16, 1536 (1977).
[17] J.Küppers, F.Nitschké, K.Wandelt and G.Ertl J. Chem. Soc. Faraday. Trans. I, 75, 984 (1979).
[18] D.Menzel and R.Gomer, J. Chem. Phys. 41, 3311 (1964).
[19] R.A.Redhead, Can. J. Phys. 42, 886 (1964).
[20] P.R.Antoniewicz, Phys. Rev. B 21 3811 (1980).
[21] K.Edamoto, S.Maehara, E.Miyazaki, T.Miyahara and H.Kato, Surf. Sci. 204, L739, (1988).
[22] P.D.Johnson and S.L.Hubert, Phys. Rev. B 35, 9427 (1987).
[23] W.Reimer, T.Fink and J.Küppers, Surf. Sci. 139, 259 (1988).
[24] J.Chelikowsky, D.J.Chadi and M.L.Cohen, Phys. Rev. B 8, 2786 (1973).

Dissociative Desorption of CH_3I by UV-Laser Radiation: Kinetic and Internal Energy of the Neutral Fragments

D. Feldmann, J. Kutzner, and K.H. Welge

Fakultät für Physik, Universität Bielefeld,
D-4800 Bielefeld 1, Fed. Rep. of Germany

Abstract. Short wavelength laser radiation is used to initiate photodissociation of condensed molecules and to detect the emitted particles species-, energy- and state-selectively. A case study with condensed methyliodine shows that direct dissociative photodesorption can be identified. A comparison with the gas phase process reveals several differences.

1. Introduction

The gas phase photodissociation of CH_3I at a UV wavelength of 266 nm has been studied to some extent /1/. A second reason for selecting this molecule is the fact that not only can the internal energy of CH_3 be probed, but also the two final states of the iodine atom $I(^2P_{3/2})$ and $I(^2P_{1/2})$ can easily be distinguished experimentally. In our experiments we have observed several modifications of the photodissociation of CH_3I when it is condensed in thick layers.

2. Experimental

Figure 1 shows the experimental arrangement schematically. The molecules are condensed in thick layers on a surface cooled by LN_2. The sample is irradiated perpendicularly by a UV-laser pulse. Neutral molecules from the surface can drift a distance of 38 mm into the ion source region of a

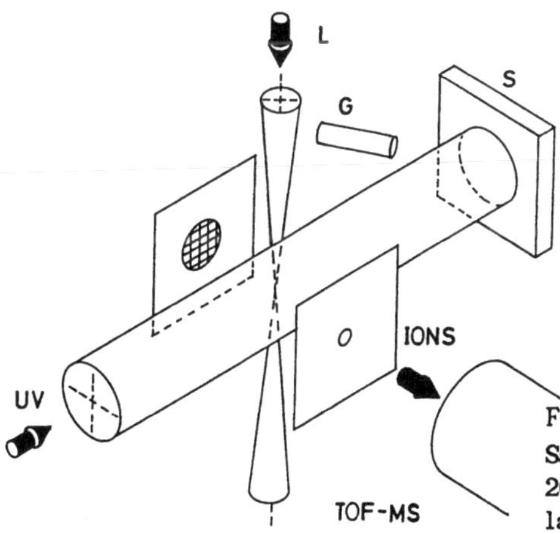

Fig. 1 Experimental arrangement. Sample S; UV-laser beam, wavelength 266 nm UV; gas doser G; detection laser L.

Springer Series in Surface Sciences, Vol. 19
Desorption Induced by Electronic Transitions, DIET IV
Editors: G. Betz · P. Varga

time-of-flight mass spectrometer. They are ionized by a second laser pulse either by resonantly enhanced multiphoton ionization (REMPI) using UV wavelengths around 335 nm or by single photon ionization with VUV-laser photons of 10.5 eV at 118 nm.

By variation of the delay time between desorption laser and detection pulse the velocity distribution of the neutrals from the surface can be determined. REMPI allows CH_3I, CH_3 and I to be detected state selectively whereas 10.5 eV single photon ionization allows us to detect these molecules simultaneously and in addition most larger molecules, however, not selectively with respect to internal energy. One advantage of these ionization methods should be mentioned: they provide almost background-free ion mass spectra, because they are insensitive to most background gases in vacuum systems like air or water. In addition, ionization by 10.5 eV photons is a "soft" ionization process with little or no fragment ion creation.

3. Results

We shall concentrate here on some results for UV desorption at 266 nm (the fourth harmonic from a Nd:YAG laser) combined with REMPI detection of CH_3 in the ground vibrational states. More details and results may be found elsewhere /2/. Figure 2 shows velocity distributions of CH_3 molecules in the vibrational ground state for different irradiation fluences F of the 266 nm UV-laser pulse. For $F \lesssim 1.5\ mJ/cm^2$ a low velocity contribution and two distinct non-thermal peaks around $v_1 \approx 3500$ m/s and $v_2 \approx 4600$ m/s are observed. Part of the low velocity component may be approximated by the thermal distributions shown in the figure with temperatures close to the substrate temperature. These "thermal" CH_3 molecules may be attributed to dissociation processes in deeper layers of the condensate. On their diffusion to the surface the molecules are thermalized. The two high velocity peaks can be attributed to direct dissociative desorption processes at the surface corresponding to the two gas phase dissociation channels

$$CH_3I + h\nu\ (266\ nm) \rightarrow CH_3(v{=}0) + I^*(^2P_{1/2}) \text{ and} \quad (1)$$

$$\rightarrow CH_3(v{=}0) + I\ (^2P_{3/2}). \quad (2)$$

The velocity at the peak maxima is 15 % lower than expected from gas phase values. The peak widths are larger than the experimental velocity resolution of $\Delta v \approx 100$ m/s at $v = 4000$ m/s.

These findings indicate that the signature of the UV-photon initiated dissociation process into the two decay channels (1) and (2) is preserved when the molecules stick together on the surface as ice or snow, however, a somewhat larger fraction of the photon energy is necessary to release $CH_3(\nu{=}0)$. One reason for this could be a modification of the CH_3I bond structure at the surface.

We observe two more indications for changes of the electronic ground or (and) excited state-potential(s) of the molecules at the surface:

First, it is known from gas phase /1/ experiments that UV photo-excitation at 266 nm is a "parallel" transition. One consequence of this is that CH_3 and I are emitted preferentially parallel to the polarization vector of the light. If this angular distribution were preserved for surface molecules we would not observe any fragments in our experimental geometry perpendicular to the light polarization. Our preliminary experiments with different geometries indicate that a large fraction of dissociation products is always emitted perpendicularly to the surface.

Secondly, from integrated signals for the two channels we obtain a branching ratio of 2 : 3 in favour of the iodine ground state decay (2). In

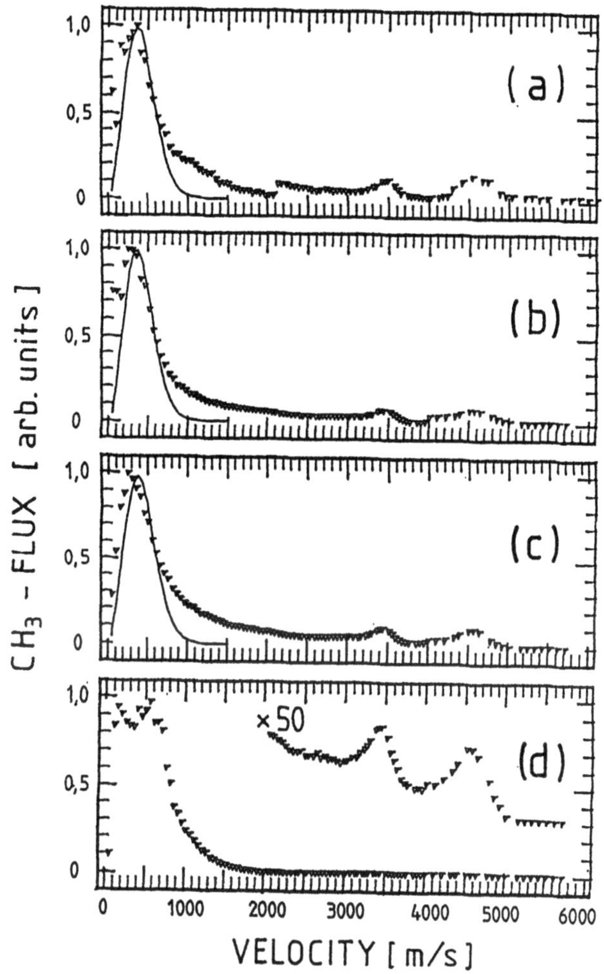

Fig. 2 Velocity distributions of $CH_3(v=0)$ for different UV-laser fluences F; a) F = 0,002 $mJcm^{-2}$, b) F = 0,3 $mJcm^{-2}$ c) F = 1 $mJcm^{-2}$ and d) 3 $mJcm^{-2}$. The full lines are thermal velocity distributions for a temperature T ≈ 125 K.

the gas phase it has been observed /3/ that 95 % of the $CH_3(v=0)$ molecules are correlated with excited I^*-atoms (channel 2). We have also measured the velocity distribution of I^*-atoms, it has a broad maximum at 400 m/s and a high velocity tail up to 1000 m/s. Velocities above 500 m/s can only be explained when one assumes that I^* has a recoil partner which has a larger mass than CH_3.

In the low UV-fluence range between 2µ J/cm² (Fig. 2a) and about 1 mJ/cm² (Fig. 2a-c) the total CH_3 signal increases approximately linearly with fluence. However, when the fluence is increased above about 1.5 mJ/cm², an increase of the total amount of desorbed particles by a factor of ten or more is observed (Fig. 2d), the direct dissociation products from the surface are only a minor fraction. Using VUV single-photon ionization, we observe in the mass spectrum a dominant contribution of parent molecules CH_3I and also cluster ions like $CH_3I_2^+$.

Under these strong irradiation conditions the signature of direct dissociative desorption is lost and the "explosive" removal of large amounts of material reminds one of the UV photoablation process of polymers.

Our experiments show the great potential of laser applications for the study of neutral desorption products. More detailed experiments with better defined surfaces are in progress.

References:

/1/ S.J. Riley and K. Wilson, J. Chem. Soc. Farad. Disc. 53, (1972) 132
G.N.A. van Veen, T. Baller, A.E. de Vries and N.J.A. van Veen, Chem. Phys. 87 (1984) 405
H.W. Hermann and S.R. Leone, J. Chem. Phys. 76 (1982) 4759

/2/ J. Kutzner, Dissertation 1989, University Bielefeld

/3/ D.W. Chandler and P.L. Houston, J. Chem. Phys. 87 (1987) 1445

Resonantly Enhanced Dissociation of Molecules into Excited Fragments Following Core-Level Excitation

R.A. Rosenberg [1], *C.-R. Wen* [2;*], *K. Tan* [3], *and J.-M. Chen* [4]

[1]Synchrotron Radiation Center
[2]Synchrotron Radiation Center and Department of Physics, University of Wisconsin-Madison
[3]Canadian Synchrotron Radiation Facility, 3731 Schneider Drive, Stoughton, WI53589, USA
[4]Synchrotron Radiation Research Center, Tapei, Taiwan, Rep. of China
*Present address: Department of Electronic Engineering, National Taiwan Ocean University, Keelung, Taiwan, Rep. of China

Abstract. We have performed the first experiments to examine the neutral fragmentation paths following direct core-level excitation in a molecule. Using monochromatized synchrotron radiation in the range 100 to 140 eV we have monitored the dispersed UV/optical fluorescence resulting from excitation of a Si 2p electron in SiF_4 and $SiCl_4$. The main features in the SiF_4 fluorescence spectrum have been identified as emission from the SiF_4^+ D state and from excited SiF, Si, F, and Si^+. For $SiCl_4$ fluorescence the fluoresence is assigned to emission from the $SiCl_4^+$ C state and Si^*. Features in the fluorescence excitation spectra are assigned to excitation of a Si 2p electron to unoccupied valence orbitals, Rydberg orbitals, and shape resonances. There is a large enhancement in the yield of excited-state fragments following core-to- Rydberg excitation, which is due to the greater probability of the core-excited Rydberg state decaying, via a resonant Auger process, to highly excited, unbound states of SiF_4^+ and $SiCl_4^+$.

1. Introduction

A great deal of recent experimental and theoretical effort has been devoted to understanding the decay processes following core-level excitation in molecules. Synchrotron radiation photoabsorption [1-4] has provided insight into the nature of the resonances, while electron spectroscopic techniques [5-9] have had great success in monitoring the resonant enhancement of various electronic states of the ion produced by decay of the core-excited neutral or ion. Studies of the ion fragments produced following core-level excitation have yielded insight [10-14] into dissociation pathways. Many of these studies have concluded that 2-hole, 1-electron (2h1e) states, in which 2 holes are produces in valence levels and 1 electron is excited to an antibonding level, are often involved in ionic dissociation. These 2h1e states have also been found to be important precursor states in DIET from surfaces [14]. However, until now, the fate of the remaining electron-hole pair has never been determined.

In this paper we present results of experiments which monitored the monochromatized UV/optical fluorescence resulting from excitation of a Si 2p electron in SiF_4 and $SiCl_4$. A primary motivation for studying both $SiCl_4$ and SiF_4 is that their preedge Si 2p spectrum is rich in structure which has been assigned by most researchers to excitation to valence (V^*) and Rydberg (R) orbitals [1-3], although other interpretations have been postulated [4]. Electron spectroscopic techniques have been utilized to monitor the decay channels of the $(2p)^{-1}V^*$ and $(2p)^{-1}R$ excited states which have been found to decay primarily by a resonant Auger process to form 2h1e states [5-9]. These complementary results have enabled us to gain unique insight into the mechanisms of excited neutral production following core-level excitation.

2. Experimental

Experimental details are given elsewhere [15], so only pertinent details will be given here. The experiments were performed on the Canadian Synchrotron Radiation Facility "grasshopper" beamline located at the Synchrotron Radiation Center, University of Wisconsin, Madison. Fluorescence was detected with a 0.2 m spectrometer using f/2.0 fused silica extraction optics located normal to and in the plane of polarization of the synchrotron radiation (SR). Due to the low signal levels during dispersed fluorescence measurements, the beamline was operated with a 100 μm entrance slit and a 200 μm exit slit (resolution = 1.4 eV at 100 eV), while the slits for the spectrometer were set to their maximum value of 2 mm (160 Å resolution). For measurements in which the beamline or the spectrometer were operated at zero order (white light), slits were set to 50 μm (resolution = 0.34 eV for the beamline and 4 Å for the spectrometer).

3. Results and Discussion

Fig. 1(a) shows a fluorescence spectrum of SiF_4 taken with the beamline operating at zero order. Also shown in Fig. 1 are assignments for the main features in the spectra. These assignments are based on previous fluorescence studies of SiF_4 using electron, ion, and photon impact [16-21] and on atomic reference data [22]. The feature marked SiF is due to α band system of SiF [20,21]. The intense broad feature peaked at 3050 Å has been observed previously and assigned to emission from SiF_4^+ D→A [17]. Most of the structure between 5200 and 6200 Å is due to the D→C transition in SiF_4^+ [19], although there is some second-order light contribution from the 3050 Å peak. The other structure in Fig. 1(a) is identified as arising from excited states of Si, F, and Si^+ [22].

In Fig. 1(b) is shown a fluorescence spectrum obtained with 117.5 eV incident photons. All the main features of 1(a) are seen, but with different relative intensities. In particular the emissions due to the various fragments have greatly increased; if it were possible to obtain this data at high resolution as in Fig. 1(a), the fragment peaks would be much more dominant.

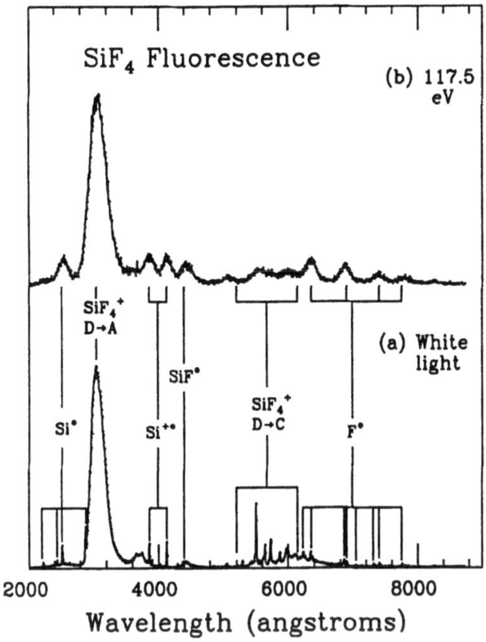

Figure 1. Dispersed fluorescence spectra of SiF_4 following excitation with white light (a) and 117.5 eV photons (b). The spectra in (a) and (b) were obtained with 4 Å and 160 Å resolution respectively.

To gain detailed information on the decay channels we monitored the relative intensities of the most intense features observed in Fig. 1(b) as a function of photon energy. The data for the fragments is shown in Fig. 2. Excitation curves for the molecular ion SiF_4^+ D→A and the total-electron yield (TEY) has also been obtained and is presented elsewhere [23]. All the spectra show the same gross features; however, there are dramatic differences in the relative intensities of the various peaks labelled A, B, C, D using the notation of Dehmer [24]. The assignments and characterization of these 4 peaks has been the subject of debate in the recent literature. High resolution photoabsorption data [1-3] (see Fig. 3) reveal the presence of 2 peaks in feature A and 6 peaks in feature B. Most of the papers have assigned the 2 peaks in feature A to excitation to antibonding valence orbitals and the higher-energy peaks in feature B to excitation to Rydberg states [1]. The intermediate peaks are due to a combination of valence and Rydberg transitions, although an alternative assignment of the structure in peaks A and B has been offered recently [4]. The assignment of the features C and D is less controversial. It is generally agreed that these peaks arise from Si 2p excitation to continuum shape resonances [1-4].

The main objective of these studies is to understand how the various fragmentation channels vary as a result of Si 2p excitation. The excitation spectra resulting from production of various fragments are shown in Fig. 2.

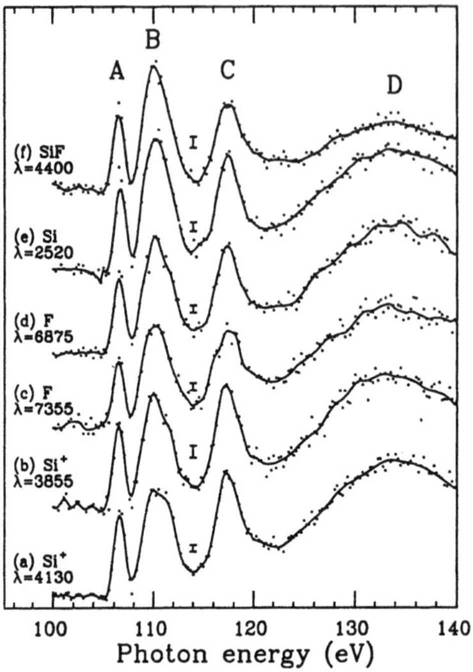

Figure 2. Yield of the fragment light-emitting species observed in Fig. 1 as a function of photon energy. All spectra were taken with a photon energy resolution of ~1.4 eV. The "I" represents a change of 10% of the signal level at 100 eV.

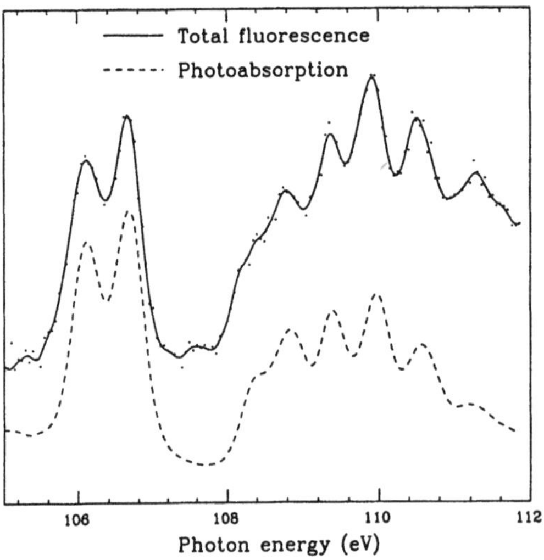

Figure 3. Total fluorescence yield (solid line) compared to the photoabsorption [1] (dashed line) of SiF_4 as a function of photon energy. The photon energy resolution was ~0.36 eV for the total fluorescence yield and ~0.16 eV for the photoabsorption data.

There is a dramatic enhancement in the yield of all the fragments at peak B (Rydberg excitation). In order to understand the possible causes for this one must examine the decay channels of the core-excited state. Following excitation of a Si 2p electron to a bound valence (V*) or Rydberg (R) level (Eq. 1), the excited state may autoionize (Eq. 2), or decay by a resonant Auger process, whereby the excited electron remains a spectator (Eqs. 3,4):

$$SiX_4 + h\nu \rightarrow SiX_4(2p)^{-1}(V^*, R) \tag{1}$$

$$SiX_4(2p)^{-1}(V^*, R) \rightarrow SiX_4^+(V_a)^{-1} \tag{2}$$

$$SiX_4(2p)^{-1}V^* \rightarrow SiX_4^+(V_a)^{-1}(V_b)^{-1}V^* \tag{3}$$

$$SiX_4(2p)^{-1}R \rightarrow SiX_4^+(V_a)^{-1}(V_b)^{-1}R \tag{4}$$

where V_a and V_b represent occupied valence levels, and X = F or Cl.

The direct, autoionization (non-spectator) decay process leads to enhancement of electronic states of the molecular ion. However, the resonant Auger or spectator decay route has been shown to dominate [5,7-9]. Therefore, following core-valence excitation (peak A), the molecule will decay primarily to a 2h1e state of type $V_a^{-1}V_b^{-1}V^*$, while a core-Rydberg excitation (peak B) will result primarily in a 2h1e state of type $V_a^{-1}V_b^{-1}R$. The dissociation dynamics and potential energy surfaces of a large molecule like SiF_4 are quite complex and not amenable to present-day theoretical treatments. It seems clear, however, that of the two types of 2h1e states produced by a resonant Auger process (Eqs. 3 and 4), the type most likely to result in an excited-state fragment, is one which contains an excited Rydberg electron as opposed to an excited valence electron. The reason is that the wave function of a loosely-bound, diffuse Rydberg electron, as opposed to an excited valence electron, has less overlap with the molecular-ion core. The excited valence electron is more likely to relax during dissociation, thereby resulting in an ion and a ground-state neutral.

If the preceding explanation is valid then the higher-lying, more diffuse Rydberg states, which have less overlap with the ion core, should have a greater probability to produce an excited photofragment. To test this hypothesis requires high resolution fragmentation excitation spectra in order to monitor the fragmentation process in a state selective manner. Due to the low signal levels it was not possible to do this for the individual fragments. However, it was possible to measure the total fluorescence yield in high resolution. This data is shown in Fig. 3 and compared with the high resolution photoabsorption data of Bozek *et al.*[1]. Since most of the increase in fluorescence signal above the Si 2p edge is due to the fragments (the SiF_4^+ D state emission shows only a 7% enhancement [23]), the spectrum in Fig. 3 is a weighted average of the overall excited fragment yield. It is clear that, relative to the photoabsorption spectrum, the higher lying Rydberg excitations result in a greater yield of excited fragments.

To completely characterize the decay channels which lead to the excited fragments requires knowledge of the 2h1e states formed prior to the dissociation and of the other fragments produced by the reaction. Such information is best obtained through coincidence studies which are presently beyond our capabilities. However, some insight may be gained by examining the relative partial cross sections for production of various 2h1e states and the ion yields as a result of Si 2p excitation. Recently, de Souza, Morin, and Nenner reported the relative partial cross sections of various satellite lines as a function of Si 2p excitation energy [5]. One level, in particular, with a binding energy of 60.4 eV, showed a strong enhancement at photon energies corresponding to peak B excitation. Although the electronic configuration of this level is not certain, this coincidence suggests that the 60.4 eV satelliite level may be an important precursor state for production of excited fragments. In a companion paper, Lablanquie *et al.* monitored the yield of the fragment ions produced following Si 2p excitation [12]. Some of the ion fragments have spectral features that are similar to the total fluorescence yield data of Fig. 3. In particular, Si^{++}, Si^{+}, and F^{+} have pronounced structure in the higher energy, Rydberg excitations, and a greater overall relative intensity in peak B relative to peak A. This suggests that these ions may be formed by the same decay process that produces the excited fragments.

Enhanced production of excited fragments following core-to-Rydberg excitation is very evident in the fluorescence excitation spectra of $SiCl_4$ [15]. This data is shown in Fig. 4(b-e) and compared with the TEY from $SiCl_4$ in Fig. 4(a). All the spectra show the same gross features; however, there are dramatic differences in the relative intensities of peaks due to excitation to valence levels (peak A) and that due to excitation to Rydberg levels (peak B). As compared to the TEY spectrum, the relative intensity of peak B is much lower in the molecular ion fluorescence spectra, Figs. 4(b,c), and much higher in the Si fragment fluorescence spectra, Figs. 4(d,e).

The $SiCl_4^+$ C state emission results from ionization of an electron from the $7t_2$ orbital of $SiCl_4$. Resonant enhancement of this level has been studied previously [5] using photoelectron spectroscopy (PES) and explained in terms of an autoionization process (Eq. 2). The PES studies also revealed a strong enhancement for the $SiCl_4(7t_2)^{-1}$ state as a result of excitation to valence levels (peak A, Fig. 4), but no enhancement following excitation to Rydberg levels (peak B) [5]. The reason for this difference is that the valence state wave function has a significant overlap with the core vacancy, while there is a much lower overlap between the Rydberg state wave function and the core hole [5].

The excitation spectra resulting from excited Si atoms are shown in Fig. 4(d) and 4(e). There is a dramatic enhancement in the yield at peak B (Rydberg excitation), similar to that observed in the case of SiF_4 (Fig. 2). This observation supports the model developed to explain the enhanced production of excited fragments following Rydberg excitation for SiF_4 (Fig. 2), and suggests that similar processes may be observable in other molecules.

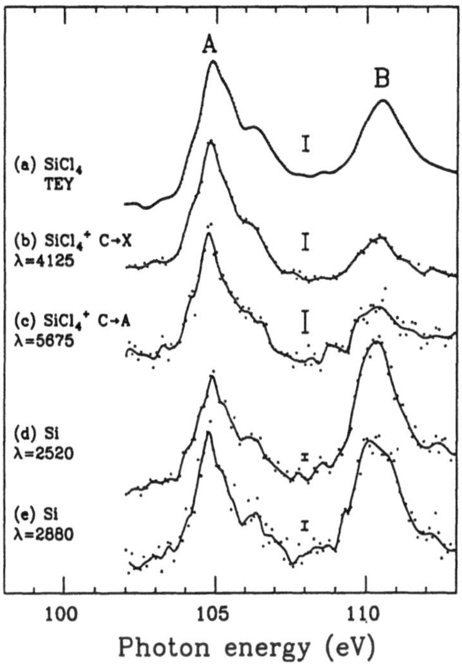

Figure 4. Yield of the various light-emitting species observed in Fig. 1 (b-e) compared to the total-electron yield (TEY) (a) of $SiCl_4$ as a function of photon energy. All spectra were taken with a photon energy resolution of ~1.4 eV. The "I" represents a change of 10% of the signal level at 102 eV.

4. Conclusion

We have observed enhanced yields of excited-state fragments following core-to-Rydberg excitation in both $SiCl_4$ and SiF_4, which suggests that this process may be a general one. If so, studies such as these may be used to help determine the identity of an emitting species, by using the excitation spectrum as a "fingerprint". Our findings also suggest that the 2h1c states formed following Si 2p to Rydberg excitation are likely precursors for excited-state fragments as well as ion production. In the future, by using more intense light sources and more efficient detection methods we hope to be able to perform high resolution and coincidence studies, which will enable us to determine which Rydberg states are involved in the various dissociation processes and to gain more detailed information on the exact manner in which the fragments are produced.

5. Acknowledgements

We would like to thank the entire staff of the SRC for their assistance. We would also like to thank Long Lee, Masako Suto, Erwin Poliakoff and Mike Bancroft for useful discussion and John Bozek for use of the high resolution SiF_4

photoabsorption data. These experiments were performed at the Synchrotron Radiation Center, which is supported by the National Science Foundation.

References

1. J.D. Bozek, G.M. Bancroft, and K.H. Tan, to be published.
2. W. Hayes and F.C. Brown, Phys. Rev. A 6, 21 (1972).
3. H. Friedrich, B. Pitte, P. Rabe, W.H.E. Schwarz, and B. Sonntag, J. Phys. B13, 25 (1980).
4. S. Bodeur, I. Nenner, and P. Millie, Phys. Rev. A34, 2986 (1986).
5. S. Aksela, O.-P. Sairanen, H. Aksela, G.M. Bancroft, and K.H. Tan, Phys. Rev. A37, 2934 (1988).
6. T.A. Carlson, D.R. Mullins, C.E. Beall, B.W. Yates, J.W. Taylor and F.A. Grimm, J. Chem. Phys. 89, 4490 (1988).
7. G.G.B. de Souza, P. Morin, and I. Nenner, J. Chem. Phys.90, 7071 (1989).
8. G.M. Bancroft, S. Aksela, H. Aksela, K.H. Tan, B.W. Yates, L.L. Coatsworth, and J.S. Tse, J. Chem. Phys. 84, 5 (1986).
9. T.A. Ferrett, M.N. Piancastelli, D.W. Lindle, P.A. Heimann, and D.A. Shirley, Phys. Rev. A38, 701 (1988).
10. P. Morin and I. Nenner, Phys. Rev. Lett. 56, 1913 (1986).
11. D.A. Lapiano-Smith, C.I. Ma, K.T. Wu, and D.M. Hanson, J. Chem. Phys. 90, 2162 (1989).
12. P. Lablanquie, A.C.A. Souza, G.G.B. de Souza, P. Morin, and I. Nenner, J. Chem. Phys. 90, 7078 (1989).
13. R.A. Rosenberg, J. Vac. Sci. Technol. 4, 1463 (1986).
14. D.E. Ramaker, J. Vac. Sci. Technol. 2, 1137 (1983).
15. R.A. Rosenberg, C.-R. Wen, K. Tan, and J.-M. Chen, Physica Scripta, to be published.
16. J.A. Hesser and K. Dressler, J. Chem. Phys. 47, 3443 (1967).
17. J.F.M. Aarts, Chem. Phys. 101, 105 (1986).
18. M. Suto, X. Wang, L.C. Lee, and T.J. Chuang, J. Chem. Phys. 86, 1152 (1987).
19. S.M. Mason and R.P. Tuckett, Mol. Phys. 60, 771 (1987).
20. E.H. Eyster, Phys. Rev. 51. 1078 (1937).
21. R.C. Johnson and H.G. Jenkins, Proc. Roy. Soc. A116, 327 (1927).
22. Atomic Transition Probabilities: Volume II Soldium Through Calcium, W.L.Wiese, M.W. Smith, and B.M. Miles, National Stand. Ref. Ser., Nat. Bur. Stand., v. 22 (1969).
23. R.A. Rosenberg, C.-R. Wen, K. Tan, and J.-M. Chen, submitted to Phys. Rev. Lett.
24. J.L. Dehmer, J. Chem. Phys. 56, 4496 (1972).

Resonances in Core-Level Shake-off Spectra of CO and N_2 Adsorbed on Ni(100): Implications for Photon Stimulated Desorption

A. Nilsson, H. Antonsson, and N. Mårtensson

Department of Physics, Uppsala University, Box 530, S-751 21 Uppsala, Sweden

In PSD (Photon Stimulated Desorption of ions) the ejection of ions is studied as a function of photon energy /1/. Excitations occur in the adsorbate-substrate complex leading to electronic states which are repulsive in terms of certain dissociation channels. There are many possible paths leading to desorption in the PSD process. The various fragments show different threshold energies and have different polarization properties. Especially it is seen that several ion thresholds are considerably delayed compared to the core ionization thresholds. For this reason it has been proposed that the fragmentation of molecular adsorbates are mainly related to multiple electron excitations such as shake-up and shake-off processes.

Using XPS (X-ray excited Photoelectron Spectroscopy) the various types of electronic excitations which are related to the emission of photoelectrons can be studied. The creation of a core hole disturbs the whole system which leads to additional valence electron excitations. A transition from a filled to an unfilled level is denoted shake-up while an additional ionization is denoted shake-off. It is generally found that a substantial fraction of the total intensity in a photoelectron spectrum is due to multielectron excitations. In this contribution XP spectra of CO and N_2 on Ni(100) are presented and discussed with main emphasis on the multielectron excitations. The XPS satellite spectra for CO/Ni(100) are compared to previously reported PSD features measured for CO/Ru(001) /2/.

The XP spectra were recorded using monochromatized X-rays. The X-ray source incorporates a rotating anode and a wide angle crystal monochromator /3/. The present satellite spectra have been measured with a resolution of about 0.5 eV. A major advantage of monochromatization in the study of satellite spectra is the improvement of the signal-to-background ratio.

The N1s spectra for N_2/Ni(100) are shown in Fig. 1. The spectra from the two nitrogen atoms have been separated using the forward focusing effect /4,5/. The N_2 molecules are adsorbed perpendicular to the surface. Due to directed elastic scattering the photoemission intensity from the inner nitrogen atom is enhanced in the direction along the intramolecular axis, i.e. normal to the surface. By performing measurements at different angles it is thereby possible to vary the relative contributions from the inner and outer nitrogen atoms. By taking appropriate linear combinations of these the pure spectra connected to each of the two nitrogen atoms can be constructed, see Fig. 1. An interesting observation which can be made directly from the spectra is that most of the intensity is contained in the satellite regions. This immediately shows that a proper understanding of core hole induced desorption

Desorption Induced by Electronic Transitions, DIET IV
Editors: G. Betz · P. Varga

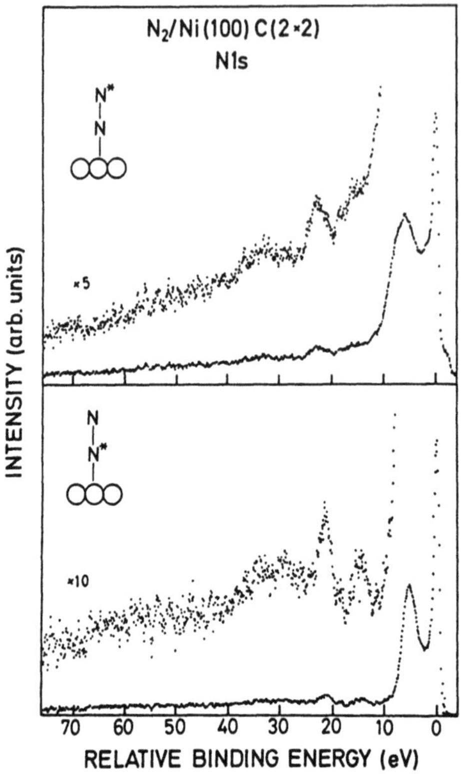

Fig. 1 The N1s shake-up/shake-off spectra of N_2/Ni(100) adsorbed in the c(2x2) phase. A measured background has been subtracted. The subspectra from the two nitrogen atoms have been separated using photoelectron diffraction effects, see text. N* is used to denote which nitrogen atom is core ionized.

processes requires not only considerations of the core hole states but also of all the satellite states. It can also be noted that both spectra contain several characteristic features at high satellite energies and furthermore that the two spectra are significantly different.

Fig. 2 shows extended parts of the shake-up and shake-off regions of adsorbed CO. The satellite spectra of CO/Ni(100) have been reported recently with main emphasis on the low energy shake-up region /3/. The same labelling of the peaks is used as in that paper. It should also be mentioned that the satellite spectra for the adsorbate lie on top of a large background of inelastically scattered electrons and weak Auger transitions. For the analysis of the satellite spectra it is essential that this background is carefully subtracted, see Ref 6.

The PSD yields of CO^+, O^+, C^+ and O^{++} for CO/Ru(001) have been reported by Treichler et al./2/. The polarization of the synchrotron light was used to determine the symmetry of the dissociating states (σ or π). Particularly pronounced structures were seen above the O1s threshold (the threshold energy is in the following de-

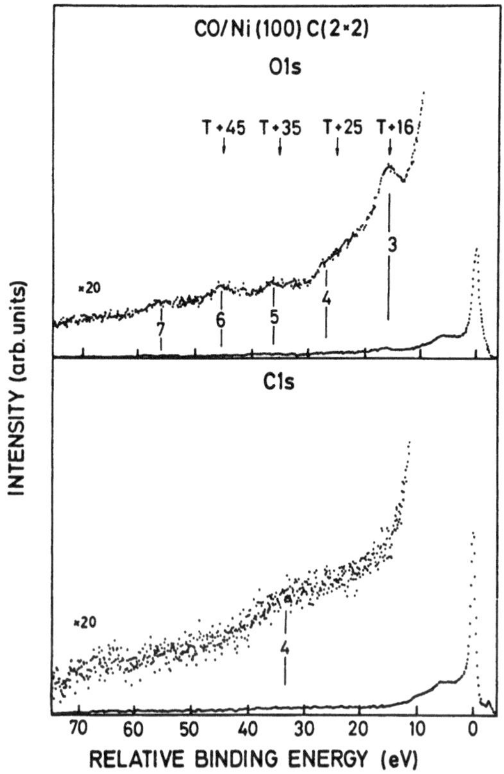

Fig. 2 The C1s and O1s shake-up/shake-off spectra of CO/Ni(100) adsorbed in the c(2x2) phase. Measured background spectra have been subtracted. The characteristic PSD threshold energies from Ref. 2 are marked in the O1s spectrum.

noted T). The CO^+ yield shows mainly resonances at T and T+16 and it seems to follow the O(KVV) Auger electron yield. In the photoabsorption spectra the first and second resonances are assigned to transitions to the 2π and the 6σ states. However, the polarization dependence of the T+16 feature shows that it has considerable π-character as well. It was therefore proposed that it is due to a multielectron process. The O^+ yield shows strong resonances of mainly π-character at T+16 and T+35. A small intensity of σ-character was also identified at T+25. The O^{++} and C^+ yield curves have delayed thresholds at T+35 and they increase up to T+45.

The characteristic O1s PSD features have been inserted in Fig. 1 (The threshold energy has for simplicity been set equal to the XPS main line position. This may lead to a small discrepancy which, however, for the present comparison can be neglected). It is immediately seen that the PSD structures coincide well with the various XPS satellite features. This strongly suggests a direct connection between the PSD process and various multielectron effects. In particular it can be concluded that a large fragmentation requires additional valence electron excitations, e.g. from bonding to antibonding orbitals.

In shake-up spectra it is usually found that the main excitation processes are of monopole type, i.e. between orbitals of the same symmetry. We interpret peak 3 as due to a 1π-2π shake-up process, which is consistent with the proposed assignment of the corresponding PSD feature /2/ and a previously observed XPS satellite at this energy /7/. Peak 4 we identify with transitions from the 4σ or 5σ valence states to the 6σ shape resonance. For symmetry reasons this cannot be the interpretation of the PSD-resonance at T+25 since this would lead to π-symmetry. A possible final state for the latter may instead be $1\underline{\sigma}1\underline{\pi}2\pi6\sigma$ (hole states underlined). Shake-up feature 5 we identify with the first excitations involving the 3σ level. Probably the first excitations involve final state orbitals of molecular Rydberg-like character. A tentative interpretation of peak 6 is then that the 3σ shake-off intensity is enhanced due to the presence of the 6σ shape-resonance. Resonances 6 and 7 (as well as some of the lower excitations) could also have strong contributions from multiple shake-up processes. However, for the detailed assignment of the structures at higher energies one has to be somewhat careful since this part of the spectrum is more sensitive to the background subtraction procedure due to the appearance of weak Ni Auger transitions.

The comparison between the XPS satellites and the PSD spectra indicates that there is a strong enhancement of the CO^+ and O^+ yields due to 1π-2π shake-up excitations. Furthermore, it can be concluded that the production of O^{++} and C^+ ions requires excitations involving the 3σ level. This is, however, not sufficient for determining the desorption mechanism. One possibility is that O^{++} ions are emitted due to multiple holes in bonding valence levels. Alternatively the dissociation may occur via 3σ excited O^+ species which autoionize to O^{++}.

In the high energy C1s shake-up spectrum only little structure is resolved. The broad structure 4 in the C1s shake-up spectrum probably contains a number of unresolved states. In the CO molecule, σ-$n\sigma^*$ transitions and double excitations appear in this energy range /9/. It can be estimated that shake-up excitations involving the 3σ level in the adsorbate should have energies in the range 30-40 eV, i.e. around structure 4. Note, however, that the 3σ orbital has mainly oxygen character;why these contributions to the C1s satellite spectrum may be rather weak. The PSD yield curves above the C1s threshold show only CO^+ ions. This yield curve follows essentially the photoabsorption cross section without extra structures /2/.

In conclusion we have shown that XPS core level spectra for adsorbates may be dominated by multielectron effects such as shake-up and shake-off. A comparison between the energy positions for characteristic shake-up/shake-off features and PSD features for adsorbed CO suggests that the desorption and fragmentation of adsorbed CO are to a large extent determined by multiply excited states.

References:

1. "Desorption Induced by Electronic Transitions, DIET III", Eds. R.H. Stulen and M.L. Knotek, Springer Series of Surface Sciences 13 (Springer-Verlag, Heidelberg, 1988)
2. R. Treichler, W. Riedl, W. Wurth, P. Feulner and D. Menzel, Phys. Rev. Lett. 54, 462 (1985).
3. A. Nilsson and N. Mårtensson, Phys. Rev. B40 (1989), in press.
4. S. Kono, S.M. Goldberg, N.F.T. Hall and C.S. Fadley, Phys. Rev. B 22, 6085 (1980)
5. W.F. Egelhoff, Jr., Surf. Sci. 141, L324 (1984).
6. A. Nilsson and N. Mårtensson, to be published.
7. E. Umbach, Surf. Sci. 117, 482 (1982).
8. J. Schirmer, G. Angonoa, S. Svensson, D. Nordfors and U. Gelius, J. Phys. B 20, 6031 (1987).

Part III

Laser-Induced Desorption

Laser-Induced Desorption

E. Matthias[1] *and T.A. Green*[2]

[1]Fachbereich Physik, Freie Universität Berlin, D-1000 Berlin 33, Germany
[2]Sandia National Laboratories, Albuquerque, NM 87185, USA

Abstract. The present understanding of laser-induced emission of neutrals and ions from surfaces of transparent dielectric crystals at prebreakdown laser intensities is reviewed. Differences in the desorption characteristics following bulk or surface absorption of the light are pointed out, and the influence of temperature and crystalline structure on the desorption yield is examined. Emphasis is placed on the resonance enhancement of multiphoton transitions by electronic states in the band gap, which could originate from either structural defects and impurities, or as the result of electronic processes following upon the absorption of the laser light.

1. Introduction

Laser-induced desorption (LID) is desorption induced by electronic transitions (DIET) involving **valence photoexcitation.** It is distinguished from typical DIET by a substantially smaller energy input per incident quantum and the absence of inner-shell Auger processes, ruling out the Knotek-Feibelman mechanism for desorption [1]. In wide band gap insulators the laser photons currently available can only excite the valence band via multiphoton processes, and this in turn requires the intense source which the laser provides. LID has been investigated mostly in connection with adsorbates on well-defined substrate surfaces [2-4]. Also LID from compound semiconductors has been studied [5] and a possible desorption mechanism has been proposed [6,7]. In this contribution we focus our attention on the emission of particles from **clean surfaces of transparent dielectric materials** during laser irradiation. The motivation for this is threefold: (1) to more readily identify the possible **electronic transitions** when $h\nu < E_g$, where E_g is the band gap energy; (2) to study the **desorption mechanisms** for neutrals and ions in order to learn to what extent the ones known from higher energy ESD/PSD apply here as well, or whether LID has its own specific desorption mechanism; (3) to gain a better understanding of the **onset of optical damage,** for which LID is considered to be the precursor.

For transparent materials absorption of laser light has to be assisted by defects or impurities which provide electronic states in the band gap. Since the electronic structure of color centers in alkali and alkaline-earth halides has been studied thoroughly [8-12], we choose these crystals as model substances and assume that the electronic states of color centers and impurities are responsible for both single and

Springer Series in Surface Sciences, Vol. 19
Desorption Induced by Electronic Transitions, DIET IV
Editors: G. Betz · P. Varga

multiphoton absorption. In addition it will be supposed for this discussion that the primary electronic process leading to laser-induced **neutral atom desorption** is the same as that already established in the alkali and alkaline-earth halides for excitation with particles and quanta having energies greater than the band gap energy. This is the production of electron-(valence) hole pairs, leading to the formation of self-trapped holes (V_k centers) and then self-trapped excitons (STE). When the STE decay nonradiatively, a fraction of the decays result in the formation of separated F center-H center pairs [7-12]. The F center is an electron occupying a regular halogen vacancy site; an H center is an interstitial neutral halogen atom associated primarily with one halogen ion and sharing that ion's lattice site. The details of STE formation and decay are the subject of ongoing research [10]. Besides being the source of F-H defect pairs, the STE provide a transient set of occupied and empty localized electronic states in the band gap, which lead to single photon absorption and therefore contribute to the heating of the crystal by a laser beam [13-15]. A typical level scheme of the STE is shown in Fig.1, adapted from the work of Refs. [8-13].

The interstitial H center is very mobile even below room temperature, while the F center is only mobile somewhat above room temperature, depending on the particular crystal under consideration. When it diffuses to the crystal surface the H center dissociates to release a desorbed halogen atom; when enough F centers diffuse to the surface they can neutralize surface metal ions and form surface metal atoms. If the crystal is sufficiently hot, the metal atoms can evaporate [16,17]. Thus, insofar as neutral atom desorption is concerned, the problem of laser induced desorption will be reduced to that of laser induced electron-hole (e-h) pair

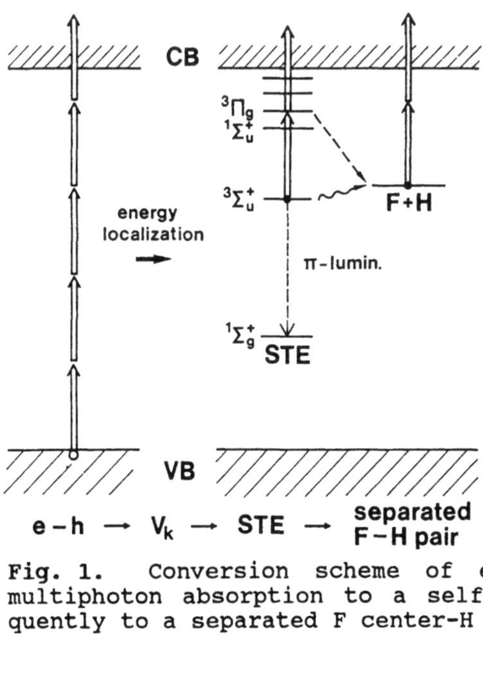

Fig. 1. Conversion scheme of electron-hole excitation by multiphoton absorption to a self-trapped exciton and subsequently to a separated F center-H center pair [8-13].

focused into the bulk, little desorption should be expected. We shall therefore assume that the laser is focused on the surface and direct our attention to the portion of the crystal within a micron or so of the surface. This is also the region where impurities and thermally and impurity-induced cation and anion vacancies tend to segregate [23]. Finally, and most importantly, cleavage-induced defects and absorbate-induced states are localized at the surface [18,19]. The shallow energy deposition depths of low energy electrons and synchrotron photons provide a perhaps more controlled way to create surface defects for LID studies. The work of Karlsson et al. [24] in which the surface halogen layer was removed from a CaF_2(111) surface is an interesting example. These authors also showed that the surface damage could be removed and the stochiometry restored by annealing. Apart from epitaxial layers [24,25] the reproducible preparation of dielectric crystal surfaces and their analysis is still an unsolved problem.

In the past, prebreakdown LID from transparent ionic crystals has received little attention. For example, up to now there seems to be no work, in which the electronic structure of color centers was purposely exploited to stimulate particle desorption from the surface, although proof of multiphoton induced neutral halogen emission oriented along certain crystal directions in alkali halides was reported as early as 1975 by Schmid et al. [26,27]. We will return to this example in section 3.3.

Apart from the low photon energy, requiring specific absorption processes, LID is characterized by the fact that it is only one out of several **competing energy channels** in laser surface interactions. Others are heat production, photoemission, fluorescence, collective excitations, morphological changes, and photochemistry. The relative importance of the LID channel depends on the material, crystal temperature, and on the radiation parameters intensity I, wavelength λ, and laser pulse width τ. For a satisfying understanding of the LID mechanism, as much complementary information as possible must be collected about the simultaneously active energy channels.

2. Single Photon Absorption

Much of what is known about the electronic states of the metal halides comes from single photon absorption and emission studies of the stable F, F_n(n=2-4), colloid, H, and H_2 centers, and transient $F^{\pm}$, $F_2^{\pm}$, $F_3^{\pm}$, V_k, and STE centers. These are complemented by the stable and transient impurity and impurity-induced centers, such as the F_A center. Any of these centers and impurities can absorb laser photons. Because of the strong electron lattice interaction, the absorption bands of these species for a given electronic transition are usually very broad. The F center provides a good example. The colloid bands in the alkaline-earth halides provide some of the narrowest absorption features other than the narrow lines associated with impurity f-electrons [28]. Only in the tranformation. The origin of **positive ion desorption** by visible laser photons at intensities below the threshold for optical breakdown is still an unsolved problem [18,19]. It will be discussed in Section 3.5.

It is important to recognize that the consequences of laser light absorption in the bulk and on the surface can be quite different. First of all the mean free paths for the diffusion of electrons, holes, V_k centers, STE, H centers and F centers are likely to be less than 1000 or perhaps 10000 Å, because of the strong electron lattice coupling and because of the unavoidable presence of impurities and dislocations which are known traps for all these species. In particular, F and H centers can form bulk aggregates when a crystal is irradiated uniformly throughout its volume in the correct temperature range [20,21]. Holes, V_k centers and electrons generate luminescence at impurity centers [22]. Thus, if the laser is sient STE can single photon absorption produce or enhance F center-H center pair production and, thus, desorption [11,12,29,30]. Single photon absorption in all the defects can contribute to energy absorption, however. Currently, the experiments of Braunlich and his collaborators suggest that the contribution of transient defects to the multiphoton induced heating of an alkali-halide crystal is small [13].

3. Desorption Following Multiple Photon Absorption

Bradford et al. [31] were the first to generate F-centers in KCl by two-photon absorption in the bulk at 266 nm, with an efficiency of about 15 % and a production time constant of 11 ps at 25 K. Schmid et al. [26] demonstrated e-h excitation via 4-photon absorption by a 4th power intensity dependence and by anisotropic emission of neutral halogen (cf. Section 3.3). Braunlich and his collaborators showed that multiphoton transitions across the band gap play a significant role in the development of bulk optical damage by free carrier heating [13-15,32]. However, they pointed out [14] that multiphoton absorption of order 4 to 5 requires such high electric fields that it can be dominant over avalanche breakdown only in crystals of highest purity. In these works the laser power was assumed to be sufficient for non-resonant multiphoton creation of electron-hole (e-h) pairs and possible evidence for resonance enhancement of the multiphoton excitation cross section was not sought, for example by using a tunable laser.

In contrast to these works, we take the view [18,19] that for customary crystals of optical quality there will always exist a finite density of defects with electronic states in the band gap and that these can assist multiphoton absorption across the band gap. This will be discussed in the next Sections. In fact, the higher concentration of defects near the surface makes multiphoton absorption at the surface more likely than in the bulk. This is compatible with the observation that optical damage mostly starts at the surface, which under some conditions also carries adsorbates that play a similar role as defects.

3.1 Resonance Enhancement of Multiphoton Absorption

The extremely small cross sections of 10^{-112} cm^8s^3 [33] to 10^{-114} cm^8s^3 [14] for 4-photon absorption and of 10^{-140} $cm^{10}s^4$ [34] for 5-photon absorption require enormous photon fluxes to achieve reasonable transiton rates. In addition, highly puri-

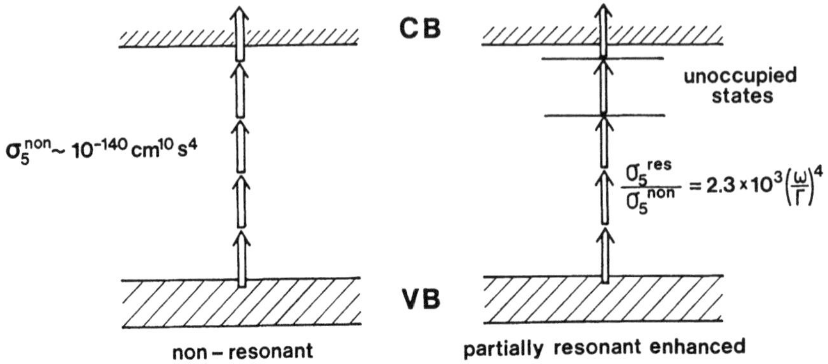

Fig. 2. Partial resonance enhancement of 5-photon absorption by two intermediate states.

fied samples are needed to avoid avalanche breakdown at these intensities. It was proposed [18,19,35] that the multiphoton transition probability can be greatly amplified by electronic states in the band gap which serve as intermediate states. Based on the theory of Bepp and Gold [36] for multiphoton ionization, Reif [35] has given an estimate for the expected enhancement when a 5-photon transition has two resonant intermediate states, as illustrated in Fig. 2. The cross section for an m-photon absorption is proportional to the square of the m'th order matrix element (see Eq.(28) of Ref. [36])

$$\sigma_m \propto \left| \frac{d^m}{\prod_{\mu=1}^{m-1} (\Omega_\mu - \mu\omega + i\Gamma_\mu/2)} \right|^2 .$$

Here, d is the generalized transition dipole moment, ω the photon frequency, and $\hbar\Omega_\mu$ the energy of the μ'th intermediate state having a width of Γ_μ. From this formula one can derive the ratio of cross sections for a partially resonant to non-resonant m-photon absorption [35]. Neglecting the width of the final state one obtains for example for a two-photon transition the ratio

$$\frac{\sigma_2^{res}}{\sigma_2^{non}} = 4\left(\frac{\omega}{\Gamma}\right)^2$$

and, assuming a width of $\Gamma = 10^{-2}\omega$ for the intermediate state,

$$\frac{\sigma_2^{res}}{\sigma_2^{non}} = 4 \times 10^4 .$$

The enhancement is even larger when two intermediate states are resonant in a 5-photon transition as indicated in Fig. 2. For this level structure one finds [35]

$$\frac{\sigma_5^{res}}{\sigma_5^{non}} = 2304\left(\frac{\omega}{\Gamma}\right)^4 .$$

Again assuming a width of $\Gamma = 10^{-2}\omega$, as evidenced by the observed spectral resonances in 5-photon ionization from BaF_2(111) shown in Fig. 3 [18,19], this ratio leads to an enhancement factor of

$$\frac{\sigma_5^{res}}{\sigma_5^{non}} = 2.3 \times 10^{11}$$

for only two resonant intermediate states. Since the m'th order transition rates are proportional to F^m [36], where F is the photon flux per cm^2 per second, one resonant intermediate state is roughly equivalent to a flux increase by two orders of magnitude. Therefore it is evident that resonant enhancement will enable multiphoton transitions to proceed at intensities where no absorption would take place in the absence of intermediate states. By the same argument, the damage threshold of the material is significantly lowered whenever there exists a certain density of states in the band gap assisting multiphoton absorption [37]. It should be noted that the above considerations apply only to single mode laser light. Multimode laser light will increase the transition rate still further, in the extreme of Gaussian light by a factor m! [36].

A characteristic consequence of resonance enhancement is that it introduces a **wavelength dependence** of the cross section. This provides a way of testing whether resonance enhancement takes place. In principle it also offers the opportunity of gaining information about the electronic structure in the band gap. Examples of wavelength dependent photoemission and positive ion desorption from BaF_2 were reported for the green (displayed in Fig. 3) and blue spectral ranges [18,19,38]. However, the observed spectral dependences of the yield still remain unidentified.

3.2 Accumulation of Defects and the Resonant Enhancement of the Multiphoton Cross Section for e-h Pair Production

At present, the main evidence for the resonant enhancement of multiphoton photoemission and ion desorption in the metal fluorides comes from Refs. [18,19, and 38] (see Fig. 3 and Section 3.5). In addition, surface-state resonant enhancement in compound semiconductors has been reported in Ref. [5]. Currently, the nature of the electronic defect or impurity states in the band gap which cause the multiphoton resonant enhancement are unknown, although in the fluorites surface F center aggregates appear to be closely related to the resonance enhancement [18,19,24,39,40]. The natural starting point for further investigation is to look for resonant enhancement by the known stable and transient defects mentioned earlier.

Two defects which could conceivably play a role in the kinetics of e-h pair formation are the F and F^+ centers. The ionization of F centers by the laser leads to a population of electrons and F^+ centers, i.e. halogen ion vacancies. These vacancies provide unoccupied states for absorption in the α band of the alkali-halides [41]. The species which result from the two excitations are the electron from the first ionization, an F center from the F^+ center, and a valence hole. The α band lies slightly below the exciton bands. In this example the preexisting F center allows an e-h pair and subsequent additional F center-H center pair to be created at an energy

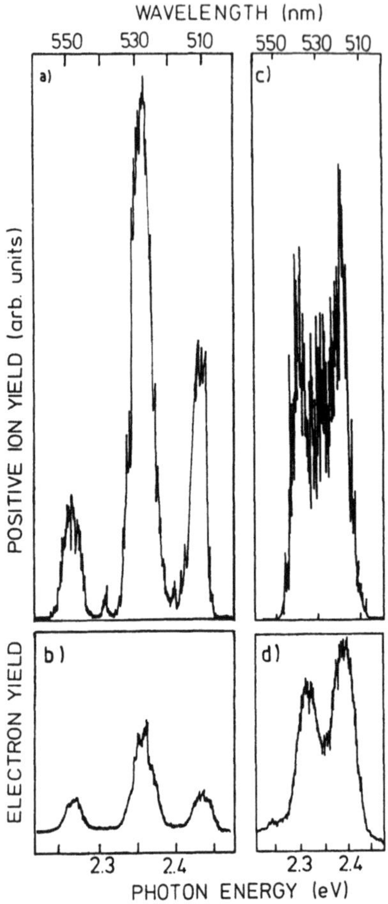

Fig. 3. Spectral and polarization dependence of the desorption yield of positive ions (a,c) and of multiphoton photoemission (b,d) from BaF_2 (111), induced by laser pulses of about 8 ns width at an intensity of about 5 MW/cm^2 [18,19]. The measurements were carried out under UHV conditions with a non-mass selective charge detector. The angle of incidence was about 70°. With the plane of incidence parallel to the <$\bar{1}\bar{1}$2> direction, the yield spectra in (a,b) are for p-polarized and the ones in (c,d) for s-polarized light. When the plane of incidence is oriented along the <$\bar{1}$10> direction the polarization dependence will be reversed.

below the band gap and exciton energies. Arguably, the rate for this mode of F center formation can be greater than that for F center formation by non-resonant multiphoton e-h pair formation in the perfect crystal. Moreover, the rate of F center production in this channel is proportional to the existing number of F centers and hence would lead to an exponential buildup with time of F center-H center pairs during single and multiple pulse irradiation. Near room temperature the F centers will not be mobile. Hence the buildup of F centers would be localized on sites adjacent to those of the first-formed or preexisting F centers. Such a laser-induced F center clustering could lead to colloid formation and a new set of adsorption bands and possible mechanisms for desorption.

While the above is merely an interesting speculation at present, it underscores the importance of further studies of the role played by stable and transient defects in the resonant enhancement of multiphoton absorption in transparent insulators. The reality of exponentially increasing defect concentrations has to be demonstrated experimentally. In this case the increase in absorbing centers is proportional to the total number of existing ones. As long as the growth rate is small one would still observe an intensity dependence of the desorption yield corresponding to F^m. Only at larger growth rates would yield versus intensity rise more steeply. Although measurements of damage thresholds as a function of number of laser shots at constant intensity have been done for years [see e.g. 13,42] the arduous problem of growth rate of absorbing centers has not been solved to any satisfaction. For example, except in the STE [11], the influence of controlled defect accumulation on LID or the damage threshold has not yet been studied by tuning the laser to the absorption bands of color centers.

Up to now the accumulation of defects during multishot irradiation has been treated phenomenologically as **incubation**, which denotes the decrease of damage threshold with increasing number of laser shots at the same spot [13, 42-44]. It can be described by a time- and intensity-dependent absorption coefficient [43]

$$\alpha(t,I) = \alpha(t=0) \cdot [1 + \beta \cdot I(r,t)] ,$$

where β is an empirical parameter. The dependence on t describes either the temporal shape of one laser pulse or, for multishot mode, the time of irradiation. A more detailed physical model for the incubation based on the build-up of local stress due to the volume increase by the production of F-H pairs has been presented by Jones et al. [13].

3.3 Emission of Neutral Halogen

To our knowledge there has been only one report on LID of neutral halogen from alkali halide crystals, definitely correlated to multiphoton excitation of e-h pairs. Using ruby laser pulses, Schmid et al. [26] observed the 4-photon induced emission of neutral Cl atoms from KCl and NaCl, as well as of neutral Br atoms from KBr. The highly oriented emission of Br from KBr(100) and Cl from KCl(100), taken from Ref. [27], are shown in Fig. 4. The halogen desorption is consistent with the sequence of e-h pair production by multiphoton absorption, followed by a STE which in turn transforms into a separated F-H pair. The desorption patterns clearly demonstrate that the H-center (interstitial X_2^- molecular ion) only can move along specified directions through the perfect lattice. If, by chance, it arrives at the surface, it emits the neutral halogen atom along the direction of its molecular axis, thereby restoring the surface stoichiometry [6,7,26,45].

The efficiency of the thermal conversion STE → separated F-H pair, and hence the desorption yield of neutral halogen

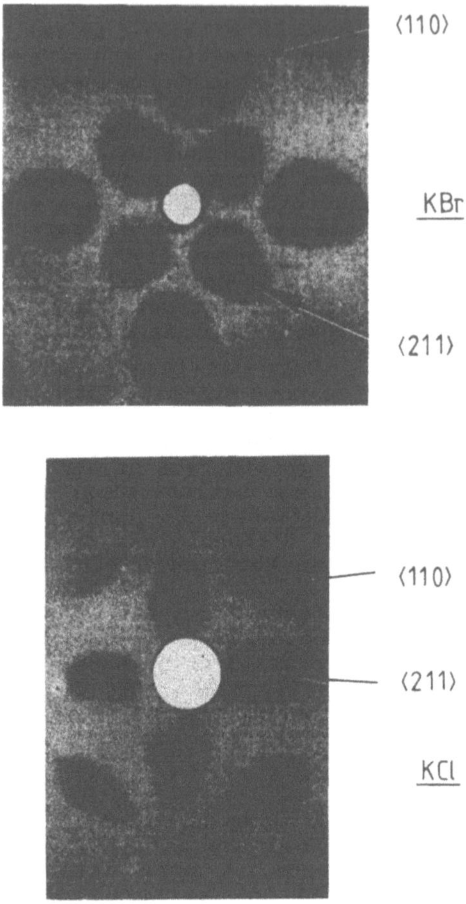

Fig. 4. Directional emission pattern for neutral Br- and Cl-atoms from KBr (100) and KCl (100), respectively, following multiphoton excitation by a Q-switched ruby laser [26,27]. The emission pattern was recorded by a silver nitrate-coated glass plate in 10^{-6} Torr vacuum. (From Ref. [27], courtesy of P. Braunlich)

depends on temperature and lattice structure. In alkali halides, on the one hand, the X_2^- molecular ion is aligned along the <110> axis and moves chiefly along this direction [6-8,26,45]. In the fluorite lattice, on the other hand, the X_2^- is pointing along the <111> direction [11,45], and the larger distance of the X^- neighbors makes the existence of an oriented collision chain along this axis problematic. No measurements of the directional emission of halogen from CaF_2-type crystals are known to us. The results of such measurements would be very enlightening.

The question must be raised whether the desorption of neutral halogen atoms is the only possible channel for the halogen to leave the crystal. An alternative is the emission of molecules. There exists an extensive literature on mass spectroscopic studies of molecules emitted by laser sputtering (see e.g. [46-48]) but little has been reported about possible LID of molecules at low laser intensities. DIET of halogen molecules from polycrystalline alkali halides has been observed and interpreted by Szymonski and his collaborators [49]. Cronberg et al. [50] report in these Proceedings the emission of neutrals from CaF_2 irradiated with 532 nm pulses below the plasma onset. With the final identification still pending there is some evidence that CaF is part of the neutral signal. It appears to us that a more thorough search for laser stimulated emission of neutral molecules from alkali and alkaline-earth halide crystals is a worthwhile task for the future.

3.4. Emission of Neutral Metal Atoms

This topic has been at the center of classical ESD/PSD [51] and sputtering work [42,52]. In contrast, little has been done in the way of metal desorption caused by laser irradiation acting on ionic crystals. Motivated by studying the damage onset, a few measurements were carried out near or above the damage threshold. Examples are Al emitted from Al_2O_3 (sapphire) by 248 nm excimer laser irradiation [46,47,53] and Na from NaF stimulated by 266 nm [54]. The data for the latter, taken for consecutive shots on the same spot, are shown in Fig. 5. The onset of Na emission at about 2 J/cm^2 lies more than a factor of three below the damage threshold. Time-of-flight measurements at 3.8 J/cm^2 revealed two velocity components for the Na atoms. A small fraction with high kinetic energy (10-100 eV) was suggested to stem from microscopic cracks of the surface, while the majority of the Na atoms had at this fluence a particle temperature of about 300 K close to the sample temperature. Even if the temperature varies with fluence, the emission appears to be a cold process and evaporation or sublimation can be ruled out. Chase and Smith [54] suggested electronic excitation of defect states as a likely cause for the Na desorption. Rather than speculating on the basis of the data shown in Fig. 5 we should regard them as a challenge requiring a more comprehensive investigation of neutral metal emission.

We should mention that the LID work on cleaved $BaF_2(111)$ in UHV described in Refs. [18,19,38] never revealed any trace of neutral Ba, while Chase and Smith [54] did observe Ba from polished and air-cleaved surfaces of BaF_2. These contrasting results may indicate the importance of surface preparation and environment; but in comparison to NaF they may also emphasize the significance of vapor pressure and lattice structure in these types of experiments.

Braunlich and collaborators proved in a series of experiments that bulk damage occurs by local melting due to free carrier heating [13-15,32]. Close to the surface, free carrier heating is expected to be less effective since electrons in the conduction band will be easily ionized and escape. Hence surface melting and subsequent evaporation of atoms or molecules seems unlikely for wide band gap crystals. The results of Chase and Smith [54] shown in Fig. 5 are consistent with this conjecture. Still the question of possible microscopic surface melting and evaporation of metal atoms deserves further attention and can be tested, among others, by measuring the photothermal deformation, as demonstrated by Dreyfus et al. [55] for sapphire surfaces.

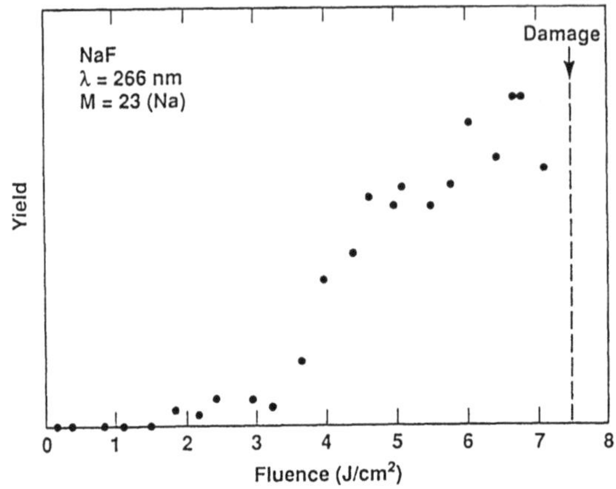

Fig. 5. Laser-induced desorption yield of neutral Na-atoms from NaF as a function of fluence below the damage threshold [54]. The yield was obtained from successive shots of increasing fluence on the same spot. The laser pulsewidth was 5-8 ns. The crystal was mounted in UHV, and Na-atoms were registered by a quadrupole mass spectrometer. (From Ref. [54], courtesy of L.L. Chase)

3.5 Multiphoton Photoemission and Desorption of Ions

In the previous Sections the desorption of neutral atoms has been associated with the production of e-h pairs. Thus, photoemission is expected to accompany LID of neutral atoms from the alkali and alkaline-earth halides, just as it does in DIET, even though in LID we are usually dealing with multiphoton photoemission at prebreakdown laser intensities. Siekhaus et al. [56] demonstrated multiphoton-induced photoemission from insulator and semiconductor surfaces at 1.06 μm and 355 nm. In Ref. [18] a tunable laser was used in the green spectral range to measure 2- and 5-photon photoemission rates in BaF_2. As shown in the lower part of Fig.3 (b and d), surprisingly sharp spectral resonances were observed. At present it has not been possible to conclusively establish the origin of these conspicuous resonant features, although progress has been made in showing how surface F centers create states of the proper character in the band gap [18,39,40]. The intensity dependence of the electron yield, measured at the peak near 518 nm for a polarization leading to the spectrum in Fig. 3d, indicated 2-photon ionization from occupied states in the band gap at intensities around 1 MW/cm^2 (see Fig. 6). It was suggested that this reflects photoemission from F-centers, as indicated in Fig. 1 [39], although the observed width of the resonances seems to be too narrow for that. The data in Fig. 6 show that at only slightly higher intensities above 1.6 MW/cm^2 the 5-photon photoemission dominates. This was attributed to ionization of valence electrons. The rapid transition from 2-photon to 5-photon absorption despite the vastly different cross sections cannot be explained by the higher density of states in the valence band but testifies to an enhancement mechanism as discussed in Section 3.1. The fact that the spectral dependence of the photoemission displayed in Fig. 3b and d does not change with intensity [18] might indicate that the

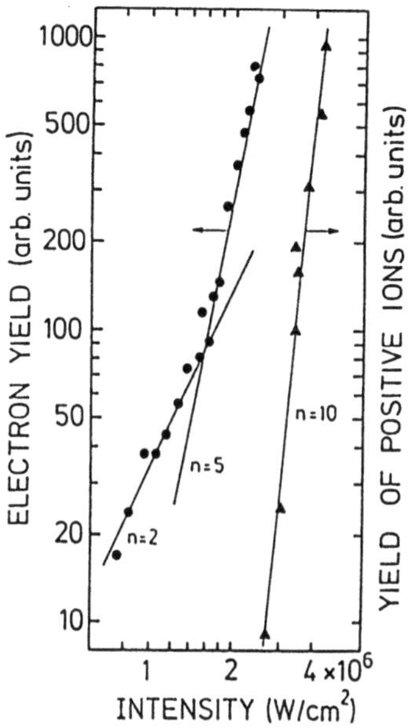

Fig. 6. Yield dependence of electrons (dots) and positive ions (triangles) on laser intensity for BaF_2 (111), irradiated at 518 nm using a polarization direction corresponding to the yield spectra (c,d) in Fig. 3 [18,19]. The solid lines represent the slopes indicated in the figure.

same states are active in the resonance enhancement. A program is currently underway in Berlin to repeat these types of measurements and extend them to CaF_2 [50]. It is worth noting that in this type of work the reproducibility of the crystal surface is actually the most crucial and difficult problem.

While multiphoton photoemission from alkali and alkaline-earth halides seems natural, the observation of resonantly enhanced LID of positive ions, as illustrated in the upper part of Fig. 3 (a and c) is surprising. We note that the spectral yield of the ions exhibits the same resonance features as observed for electrons. This implies that the positive ion emission is initiated by preceding photoemission. While data in Fig. 3a and b were obtained without mass selection, the mass analysis of the positive ion yield shows that the three ions BaF^+, Ba^+, and F^+, are desorbed with comparable yields [38]. Coupled with the possibility [24,39,40] that irradiation of BaF_2 can produce a layer of surface F centers, i.e. an outer surface layer of BaF replacing the normal (111) layer of BaF_2, the similar magnitude of the three ion yields suggests that one might seek a common source of the yields in the creation of such a BaF layer by the laser. This appears to be an interesting focus for further study.

The non-mass selected ion yield, measured again at 518 nm, exhibits an approximate 10th power dependence on the laser intensity, as shown in Fig. 6. The reason for this is presently as little understood as the origin of the positive ions itself. The reader will note that the ion signal turns on at higher powers. If the steep ion-yield slope is the result of electron avalanching and dielectric breakdown, the emission of positive ions would not constitute a conceptual problem. However, all experimental evidence speaks against dielectric breakdown at these low intensities. Even in air, the plasma onset on cleaved BaF_2 surfaces lies well above 100 MW/cm^2 [37]. Also the quenching of the mass-selected positive ion emission after a few thousand shots on one spot at intensities around 50 MW/cm^2 testifies against plasma formation [38]. Finally the striking spectral dependence of the positive ion yield in Fig. 3, together with its dramatic change when the laser polarization is turned by 90° with regard to the crystal orientation, contradicts any notion that these ions may be generated by dielectric breakdown. A different explanation is based on gas phase ionization [57] of desorbing neutral atoms or molecules, either by secondary electrons or the laser light itself. Objections against this mechanism are based on the small density of photoelectrons from transparent materials at prebreakdown intensities and their low kinetic energy, but it will be necessary to quantify these objections with direct measurements of the photoelectrons. The dependence of the spectral emission pattern on the relative orientation of the crystal axes with respect to the light polarization is important. It implies that ions are indeed ejected from the surface. This requirement could be met by assuming the existence of neutral BaF molecules in the upper surface layer, equivalent to surface F-centers, which are ionized (forming BaF^+) and dissociated by multiphoton absorption. While the dissociation channel $BaF^+ \rightarrow Ba^+ + F^{(0)}$ is well within the available energy range, the production of F^+ meets difficulties energetically. Nevertheless, it appears to us that this concept is well worth pursuing when trying to explain positive ion emission. Another possibility would be to speculate that the negative-U, hole-pair coalescence mechanism proposed for neutral oxygen desorption from ZnO by Itoh and Nakayama [58] might apply to the desorption of F^+ ions from BaF_2. The observed approximate slope of 5 for electrons and 10 for ions in Fig. 6 may be interpreted as to suggest that two holes are required for the emission of one positive ion.

Presently we conclude that the precise ion desorption mechanism from transparent ionic crystals still remains to be determined. However, Figs. 3 and 6 illustrate two important facts which must be remembered: (1) the resonance structure for ions is nearly identical to that found for photoemission and (2) the ion yield rises very steeply as the laser fluence is increased. At present the most pressing need is for additional measurements to confirm the earlier work, to better establish the conditions required for reproducibility of the ion desorption yields, and to make comparative studies on another alkaline-earth halide such as CaF_2 and an alkali halide such as NaCl or LiF. In order to make contact with theoretical models, electron and ion energy distributions and quantitative measurements of the per photon yields are required, similar to those obtained for single quantum excitation via DIET on alkali halides. These measurements of the charged particle yields need to be complemented by measurements of the neutral particle fluxes, about which almost nothing is known for the fluorites. And, of course, the observation of positive ion desorption below the laser fluence threshold for optical breakdown must be confirmed.

4. Summary and Outlook

Desorption from ionic crystals stimulated by photons with sub-band gap energy raises two problems; one is the light absorption and the other the desorption mechanism. With regard to absorption we have presented arguments and experimental evidence that, close to the surface, valence electron excitation across the band gap will proceed by multiphoton transitions, resonantly enhanced by electronic states in the band gap, belonging to defect states. Consequently, surface defects and adsorbates play a key role for the light absorption. Concerning the desorption mechanism we have to distinguish between neutral halogen atoms, neutral metal atoms, and ions. For the desorption of neutrals the concepts developed in ESD and high-energy PSD apply, except for the initial e-h pair excitation. In contrast, the desorption of ions is presently not understood.

LID is still in its infancy, however promising a great richness of various phenomena, ranging from defect physics to lattice and surface dynamics to nonlinear optical processes. Despite a beautiful beginning in alkali halide crystals [26], and the demonstration of multiphoton-generated emission of electrons and ions from the (111) surface of BaF_2 [18,19], much remains to be elucidated concerning the physics of LID. The published results have raised interesting questions about the desorption mechanisms and demonstrated the importance of surface F center production in connection with the resonance enhancement of the multiphoton excitation of (e-h) pairs [39,40]. They have also shown what directions should be fruitful for future work. The following points seem most significant to us:

1. In order to take advantage of the considerable amount of information about the alkali-halides from DIET, from studies in the bulk, and from theory, LID experiments with tunable lasers and quantitative yield determinations of both neutrals and ions should be carried out on alkali-halides.
2. In order to provide needed points of reference for current LID studies on the fluorites, DIET studies should be carried out on BaF_2 and CaF_2. Such studies are only beginning [59,60], and it is important to understand desorption mechanisms experimentally without the added difficulties connected with multiphoton excitation below the breakdown threshold.
3. Because the surface is considerably more complex and more difficult to monitor than the bulk, it is essential that more complete experiments should aim for simultaneous quantitative information about the various energy dissipation channels: desorption of electrons, neutrals, and ions, luminescence, acoustic waves, heating, and surface morphology changes.
4. In the case of DIET from alkali-halides, the introduction of time-resolved experiments has clarified the dynamics of the desorption process [17,61,62]. Similar LID studies have the potential of distinguishing desorption which results from surface energy deposition from that which results from energy deposition in the bulk.
5. Because of the apparently important role played by impurities and defects in the multiphoton absorption process, systematic absorption and desorption studies based on doped and colored crystals of known defect concentration should be carried out. The shallow penetration depth of electron and synchrotron beams suggest that they may be good vehicles for the production of controlled near-surface damage.

5. Acknowledgement

This work was supported by the Deutsche Forschungsgemeinschaft, Sfb 6, and one of us (TAG) acknowledges financial help from Sfb 6 during his stay in Berlin. This work was also supported by Sandia National Laboratories under contract No. DC-AC-0476DP00789. We are grateful for fruitful discussions with, among others, R.W. Dreyfus, H.B. Nielsen, S. Petzoldt, M. Reichling, J. Reif, A. Rosén, N.H. Tolk, and E. Westin.

6. References

1. M.L. Knotek and P.J. Feibelman: Phys. Rev. Lett. **40**, 964 (1978)
2. T.J. Chuang: J. Vac. Sci. Technol. **B3**, 1408 (1985); Surf. Sci. Rep. **3**, 1 (1983)
3. J. Heidberg: p. 230 in "DIET II", editors; W. Brenig and. D. Menzel, Springer Ser. Surf. Sci., Vol. **4** (Springer, Berlin, Heidelberg 1985)
4. W. Ho: these Proceedings
5. N. Itoh: these Proceedings
6. T. Nakayama and N. Itoh: p. 237 in "DIET II", editors: W. Brenig and D. Menzel, Springer Ser. Surf. Sci., Vol. **4** (Springer, Berlin, Heidelberg 1985)
7. N. Itoh: Nucl. Instrum. Methods **B27**, 155 (1987)
8. M.N. Kabler and R.T. Williams: Phys. Rev. **B18**, 1948 (1978)
9. N. Itoh: Advances in Physics **31**, 491 (1982)
10. R.T. Williams: Opt. Engineering **28**, 1024 (1989)
11. N. Itoh and K. Tanimura: ibid. **28**, 1034 (1989)
12. R.T. Williams, K.S. Song. W.L. Faust, C.H. Leung: Phys. Rev. **B33**, 7232 (1986)
13. S.C. Jones, P. Braunlich, R.T. Casper, X.A. Shen, P. Kelly: Opt. Engineering **28**, 1039 (1989)
14. S.C. Jones, A.H. Fischer, P. Bräunlich, P. Kelly: Phys. Rev. **B37**, 755 (1988)
15. X.A. Shen, P. Braunlich, S.C. Jones, P.Kelly: Phys. Rev. **38**, 3494 (1988)
16. M. Szymonski: Rad. Eff. **52**, 9 (1980)
17. T.A. Green, G.M. Loubriel, P.M. Richards, N.H. Tolk, R.F. Haglund, Jr.: Phys. Rev. **B35**, 781 (1987)
18. E. Matthias, H.B. Nielsen, J. Reif, A. Rosén, E. Westin: J. Vac. Sci. Technol. **B5**, 1415 (1987)
19. H.B. Nielsen, J. Reif, E. Matthias, E. Westin, A. Rosén: p. 266 in "DIET III", editors: R.H. Stulen and M.L. Knotek, Springer Ser. Surf. Sci., Vol. **13** (Springer, Berlin, Heidelberg 1988)
20. U. Jain and A.B. Lidiard: Phil. Mag. **35**, 245 (1977)
21. L.W. Hobbs, A.E. Hughes, D. Pooley: Proc. Roy. Soc. **A332**, 167 (1973)
22. H.B. Dietrich, A.E. Purdy, R.B. Murray, R.T. Williams: Phys. Rev. **B8**, 5894 (1973)
23. K.L. Kliewer and J.S. Koehler: Phys. Rev. **140**, A 1226 (1965)
24. U.O. Karlsson, F.J. Himpsel, J.F. Morar, F.R. McFeely, D. Rieger, J.A. Yarmoff: Phys. Rev. Lett. **57**, 1247 (1986)
25. U. Barjenbruch, S. Fölsch, M. Henzler: Surf. Sci. **211/212**, 749 (1989)
26. A. Schmid, P. Bräunlich, P.K. Rol: Phys. Rev. Lett. **35**, 1382 (1975)
27. P. Braunlich, S.C. Jones, X.A. Shen, R.T. Casper, P. Kelly: NIST Spec. Publ. **756**, 476 (1988)
28. R. Alcala and V.M. Orera: J. de Physique **37**, C7-520 (1976)
29. K. Tanimura, T. Katoh, N. Itoh: Phys. Rev. **B40**, 1282(1989)

30. X. Suzuki, H. Abe, M. Hirai: p. 274 in "Ultrafast Phenomena VI", editors: T. Yajima, K. Yoshihara, C.B. Harris, S. Shinoya, Springer Ser. Chem. Phys. Vol. **48** (Springer, Berlin, Heidelberg 1988)
31. J.N. Bradford, R.T. Williams, W.L. Faust: Phys. Rev. Lett. **35**, 300 (1975)
32. X.A. Shen, P. Braunlich, S.C. Jones, P. Kelly: Phys. Rev. Lett. **59**, 1605 (1987)
33. X.A. Shen, S.C. Jones, P. Bräunlich, P. Kelly: Phys. Rev. **B36**, 2831 (1987)
34. P. Bräunlich, A. Schmid, P. Kelly: Appl. Phys. Lett. **26**, 150 (1975)
35. J. Reif: Opt. Engineering **28**, 1122 (1989)
36. H.B. Bepp and A. Gold: Phys. Rev. **143**, 1 (1966)
37. J. Reif, S. Petzoldt, A.P. Elg, E. Matthias: Appl. Phys. **A49**, 199 (1989)
38. J. Reif, H. Fallgren, H.B. Nielsen, E. Matthias: Appl. Phys. Lett. **49**, 930 (1986)
39. A. Rosén, E. Westin, E. Matthias, H.B. Nielsen, J. Reif: Phys. Scripta **T23**, 184 (1988)
40. E. Westin, A. Rosén, E. Matthias: these Proceedings
41. F. Luty, in "Physics of Color Centers", editor: W. Beall Fowler (Academic Press, New York 1968), Figs. 3-8
42. Y. Jee, M.F. Becker, R.M. Walser: J. Opt. Soc. Am. B5, 648 (1988)
43. S. Küper and M. Stuke: Appl. Phys. **B44**, 199 (1987)
44. S. Küper and M. Stuke: Appl. Phys. **A49**, 211 (1989)
45. R.T. Williams: Rad. Effects and Defects in Solids **109**, 175 (1989)
46. R.W. Dreyfus, R. Kelly, R.E. Walkup, R. Srinivasan: Excimer Lasers and Optics, SPIE Vol. **710**, 46 (1986)
47. R.W. Dreyfus: these Proceedings
48. H.J. Heinen: Int. J. Mass Spectrom. Ion Phys. **38**, 309 (1981)
49. M. Szymonski, H. Overeijnder, A. de Vries: Surf. Sci. **90**, 274 (1979)
50. H. Cronberg, V. Muyderman, H.B. Nielsen, E. Matthias, N.H. Tolk: these Proceedings
51. see Proceedings of the preceeding conferences DIET I - III
52 R. Kelly: Surf. Sci. **90**, 280 (1979)
53. R.W. Dreyfus, R.E. Walkup, R.Kelly: Rad. Effects **99**, 683 (1986)
54. L.L. Chase and L.K. Smith: NIST Spec. Publ. **756**, 165 (1988)
55. R.W. Dreyfus, F.A. McDonald, R.J. von Gutfeld: Appl. Phys. Lett. **50**, 1491 (1987)
56. W.J. Siekhaus, J.H. Kinney, D. Milam, L.L. Chase: Appl. Phys. **A39**, 163 (1986)
57. R.E. Walkup, Ph. Avouris, A.P. Ghosh: Phys. Rev. **B36**, 4577 (1987)
58. N. Itoh and T. Nakayama: Phys. Lett. **92A**, 471 (1982)
59. C.L. Strecker, W.E. Moddeman, J.T. Grant: J. Appl. Phys. **52**, 6921 (1981)
60. K. Saiki, Y. Sato, K. Ando, A. Koma: Surf. Sci. **192**, 1 (1987)
61. H. Overeijnder, R.R. Tol, A.E. de Vries: Surf. Sci. **90**, 265 (1979)
62. T.A. Green, G.M. Loubriel, P.M. Richards, L.T. Hudson, P.M. Savundararaj, R.G. Albridge, A.V. Barnes, N.H. Tolk: these Proceedings

Desorption Stimulated by Collective Electron Excitation

W. Hoheisel, U. Schulte, M. Vollmer, and F. Träger

Physikalisches Institut der Universität Heidelberg, Philosophenweg 12, D-6900 Heidelberg 1, Fed. Rep. of Germany

In recent years laser-induced desorption processes have found a continuously growing interest [1-3]. Desorption stimulated with laser light can take place by different mechanisms. Most frequently it is simply promoted by thermal heating. Light is absorbed at the substrate surface, the temperature rises and the particles desorb. Bond breaking at the surface can also be stimulated by vibrational excitation of the adsorbate molecules with infrared laser light. Coupling of the vibrational excitation to the phonon bath of the substrate is the dominant channel for energy relaxation. As a result, the surface temperature rises causing desorption. This means that the energy is absorbed by a resonant process initially, but the desorption is thermal, a phenomenon denoted as "resonant heating". It is responsible for the lack of molecular or isotopic selectivity in such experiments. In addition to vibrational excitation laser induced photodesorption can also be accomplished by *electronic* excitation. Even though the study of such processes is a new field and the number of experiments is still small compared to that of classical DIET processes [4-6] stimulated by electron bombardment or synchrotron radiation, different types of excitation have been observed and different mechanisms established. Electronic excitation may either be accomplished in the adsorbate, in the substrate or in the adsorbate-substrate complex. The majority of experiments have been performed with high intensity pulsed lasers, where the available energy of the photons usually does not exceed about 6 eV. Therefore, valence excitation, ionization or fragmentation are possible but processes where inner shell excitation is involved are out of reach. Major difficulties of such desorption experiments are to identify the initial electronic excitation and to understand the bond breaking mechanism in detail.

The present article summarizes recent experiments on laser-induced desorption stimulated by electronic excitation in small metal particles [7-10]. These studies are substantially different from earlier work for several reasons. First, the electronic excitation preceding the rupture of a surface chemical bond has been identified unambiguously and is a collective rather than a single electron excitation. Second, desorption can be stimulated with low intensity continuous-wave laser light in the visible spectral range rather than with high power pulsed lasers. Third, desorption of metal atoms from the surface of small metal particles is observed. Thus, the experiments illustrate that even on a metal surface the rupture of the bond by electronic excitation can compete with quenching and generation of heat in the substrate. Finally, the desorption process outlined below has a very interesting application for the generation of small metal particles with a very narrow size distribution. This offers new prospects for technological applications, particularly for catalysis.

The principle of the experiment can be seen from Figure 1. Small sodium particles with radii ranging from about 50 Å to 1,000 Å are adsorbed on a LiF(100) single crystal surface in ultrahigh vacuum. If the particles are electronically excited with light of an argon or krypton ion laser, metal atoms desorb from the surface of the particles. These atoms are detected with a quadrupole mass spectrometer. The particles can be characterized by recording optical transmission spectra with a Xe arc lamp and a monochromator. Details of the experimental arrangement and of the results have been described elsewhere [7,8]. In short, the most important features of the desorption process can be summarized as follows.

Springer Series in Surface Sciences, Vol. 19
Desorption Induced by Electronic Transitions, DIET IV
Editors: G. Betz · P. Varga

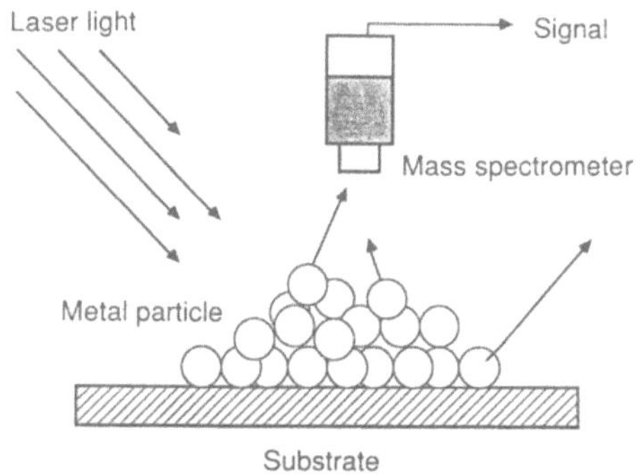

Fig. 1. Schematic view of atom desorption from the surface of small metal particles by excitation with visible laser light

The desorption rate under continuous-wave laser illumination is relatively large. Even with light intensities as low as 40 mW/cm^2 a signal can still be observed. The desorption rate depends linearly on the light intensity over a range of almost four orders of magnitude. No threshold for the desorption signal is observed. Time-of-flight measurements indicate that the translational energy of the desorbed atoms is 0.4 eV, i.e. much higher than one would expect for a thermal process. In addition, the desorption yield is quite high. Depending on the laser wavelength and on the particle size, as much as 80% of the total coverage can be desorbed with the laser light. The quantum efficiency of the desorption process is on the order of 10^{-5}. Most importantly, the desorption rate depends on the particle size and on the laser frequency.

The LiF substrate being transparent, the metal particles must be responsible for the absorption of light. A theoretical treatment of the interaction of small metal particles with electromagnetic waves was proposed by *Mie* as early as 1908 [11]. Today, the Mie resonances are mostly denoted as surface plasma resonances, i.e., collective oscillations of the conduction electrons (see e.g. [12,13]). The absorption cross section σ for such collective oscillations calculated with the Mie theory is displayed in Fig. 2. The three dimensional plot illustrates that σ depends on the particle radius as well as on the excitation wavelength of the light. Specifically, a resonant dependence on the wavelength occurs. The different maxima in the picture are the dipole and quadrupole excitation, respectively. The latter one comes into play for particles with radii larger than approximately 30 nm. Comparison of the literature data with the results obtained in our experiments shows that the absorption is in the same spectral range and has a similar width. The spectra are also redshifted with increasing particle size and become broader. Therefore, desorption in our experiment is attributed to excitation of a surface plasma resonance. This is also supported by comparison of the desorption spectra with the optical spectra of small sodium particles in beams and in matrices.

The high kinetic energy of the desorbed atoms, the absence of a threshold, the linear dependence of the rate on the light intensity and only a very moderate temperature rise of the surface during laser irradiation indicate that the desorption process is *nonthermal*. Obviously atom desorption takes place as a result of a single excitation/deexcitation cycle of a surface plasma resonance. The mechanism of the desorption process can therefore be visualized with the diagram depicted in Figure 3. Absorption of laser light excites either a collective dipole or quadrupole oscillation. The optical spectra of the particles being very broad, the excitation is damped rapidly on a time scale of around 10^{-14} s. Coupling of electronic excitation energy to the phonon bath of the substrate seems to be the dominant relaxation channel. However, collective electron oscillations can also decay into a *localized* single electron excitation [14]. If such a transition populates a repulsive potential energy surface, a sodium atom can gain sufficient kinetic energy to escape from the surface potential. Again, quenching comes into play as a competing process. An interesting observation is that only excitation of the dipole oscillation seems to cause desorption. So far, coupling of the quadrupole

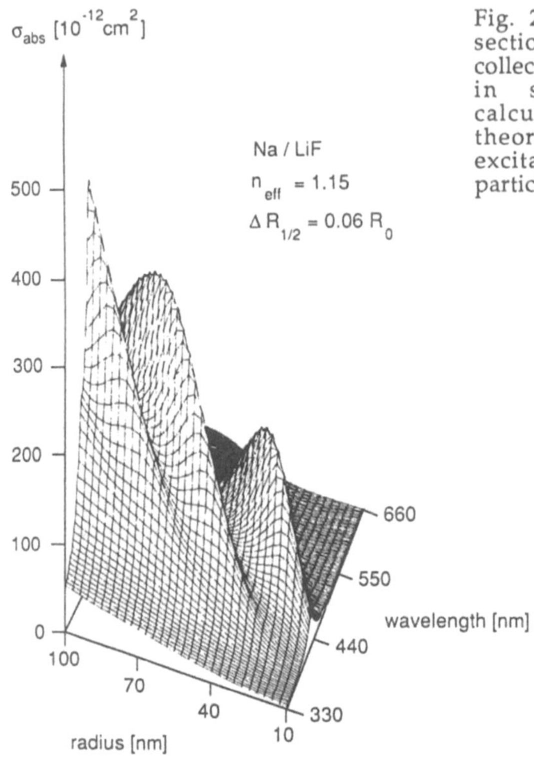

Fig. 2. Absorption cross section σ for excitation of collective electron oscillations in small Na particles calculated with the Mie theory as a function of excitation wavelength and particle size

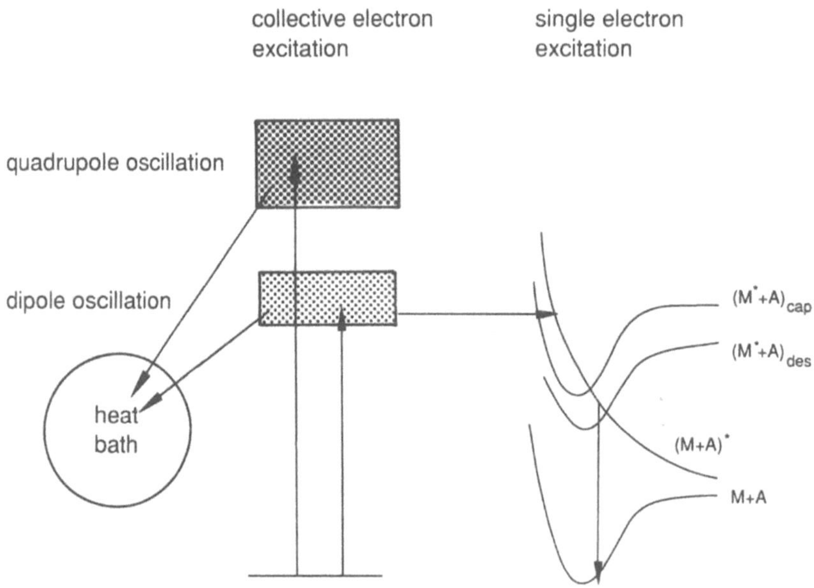

Fig. 3. Energy level diagram and mechanism of desorption stimulated by collective electron excitation

excitation to the desorption channel has not been observed, presumably because of even more rapid quenching than for the dipole oscillation. Furthermore, the scenario in Fig. 3 indicates that desorption finally takes place along the lines of the well-known Menzel-Gomer-Redhead model [15,16] with a collective excitation preceding the rupture of the bond. As has been demonstrated before for thermal desorption of the same system, the Na atoms desorb preferably from the perimeters or edges of the particles [17]. These sites can also serve as defects where the electronic excitation is localized. Similarly, it was pointed out [18] for desorption of molecules from ices like NH_3 or H_2O that edge molecules play a special role since they have particularly low coordination numbers and lower binding energies than other surface molecules. They desorb preferably and neighboring particles become the new edge molecules. It is therefore expected that the total photodecomposition rate is proportional to the number of edge molecules, an assumption also made in the model outlined below. Consequently, rough surfaces should exhibit efficient detachment. This is in agreement with the experimental observations for ices as well as for metal substrates [18,19]. Finally, our experiments constitute an important example of desorption from metal surfaces where quenching was often assumed to be so effective that bond breaking by electronic excitation is prohibited. However, even on a metal surface there is not necessarily complete relaxation of the electronic excitation on a time scale in which photodesorption can occur. Another example reported recently [20] is the UV-induced photochemical decomposition of $Fe(CO)_5$ and the desorption of CO from a Ag(110) surface. Of course, details of the mechanism outlined above have to be investigated in future experimental and theoretical work. Particularly the repulsive single electron state from which the atoms finally desorb remains to be identified. Recent experiments on Al films on a glass substrate indicate that desorption of single metal atoms can also occur from smooth planar metal films in which surface plasmons are excited with visible laser light [21].

Another interesting point is that a collective dipole oscillation can also be induced in sodium particles of totally different size. *Knight* and coworkers have discovered in a gas phase experiment that a plasma resonance like this can be excited in clusters as small as the pentamer [22]. This means that even for sodium particles that contain as few as five atoms, the electrons are at least partially delocalized. An interesting question is if such small clusters also exhibit nonthermal desorption.

Clusters and small particles (see e.g. [23-25]) are systems intermediate between atoms and molecules on the one hand and condensed matter on the other. Therefore, their properties are size dependent. In order to fully utilize these interesting characteristics and synthesize new materials with adjustable properties, e.g. for catalysis or photography, it is highly desirable to make the cluster size distribution as narrow as possible. One of the key issues in cluster physics and chemistry is thus to generate particles that are monodisperse or have a very narrow distribution [26]. Attempts have been made to generate such systems on surfaces by the deposition of particles on a substrate from a monodisperse beam. However, an unsolved problem is that of how and whether fragmentation can be avoided when the particles hit the surface. An alternative way might be to change the given distribution on a surface in a well-defined way and make it as narrow as possible. The desorption mechanism described above offers new ways of manipulating the size distribution along these lines [9,10] for two reasons. First, the surface plasmon excitation exhibits a resonance as a function of laser frequency. Second, the transition frequency of this resonance depends on the particle size. Consequently, the desorption rate depends on the particle size and is different for different laser wavelengths. The result of this kind of measurement is shown in Fig. 4. Particles with mean radii of R_0 = 45 nm, for example, interact preferably with green laser light of λ = 514 nm whereas particles with R_0 = 30 nm resonantly absorb blue light of λ = 458 nm. Therefore, at a given laser wavelength, a change of the size distribution takes place. It is correlated to a variation of the desorption rate as a function of illumination time. This variation is the interplay of two effects. First, the number of atoms which can desorb decreases as a function of time due to the overall decrease of the surface coverage. Second, the desorption can result in a shift in or out of

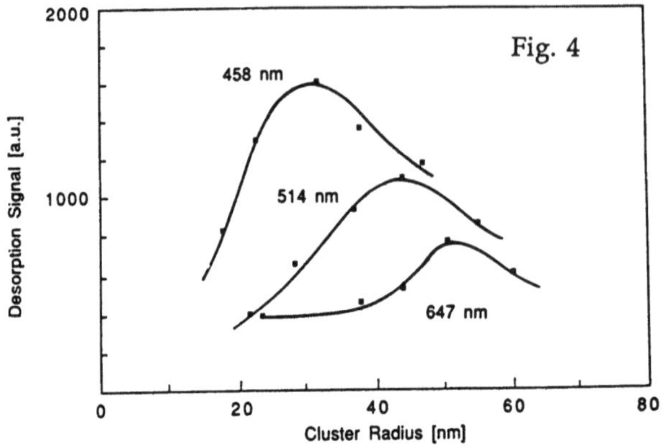

Fig. 4. Dependence of the laser-induced desorption signal on the mean particle radius for fixed laser wavelengths of λ = 458, 514 and 647 nm

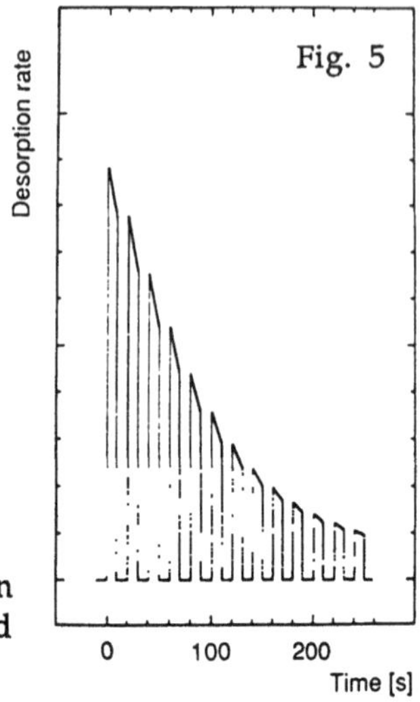

Fig. 5. Calculated change of the desorption rate during laser illumination with λ = 488 nm. The initial average cluster size was R_0 = 33 nm with a FWHM of 11.7 nm

resonance depending on the position of the laser wavelength with respect to the (size dependent) absorption profile. The combination of these two effects determines the change of the desorption rate and of the size distribution. The process naturally comes to an end if the size has decreased so much that the particles are out of resonance. This results in a considerable narrowing of the size distribution. A model has been developed to describe this process quantitatively, for details see [9,10]. For a fixed laser wavelength λ, the absorption resonance curve as a function of radius is approximated by a Gaussian centered at R_λ. The number of desorption sites on a cluster of radius R is written as aR, where a depends on the shape of the cluster [17]. The cluster size distribution f(R) is described by an asymmetric function. The desorption rate dN/dt originating from particles in the size interval [R,R+ΔR] at time t can then be calculated as a function of the light intensity, the cross section for absorption of light, the quantum efficiency and the number of illuminated clusters on the surface. The change of the cluster size distribution as a function of illumination time t is directly related to the rate dN/dt. To obtain the time dependence of f(R), the change of each cluster size during the desorption process was calculated, followed by an integration over the total size distribution.

The computation has been done for a number of cases with different particle sizes, excitation frequencies and laser intensities. One example will be described here. The wavelength of the incident light was λ = 488 nm. An absorption resonance curve which peaks at 36 nm and has a full width at half maximum (FWHM) of 15 nm was used. The mean cluster radius was R_0 = 33 nm with a FWHM of the distribution of 11.7 nm. Figure 5 displays the obtained desorption rate as a function of time. Since the laser is turned on and off during the experiment at 10 s intervals in order to distinguish between signal and background, the same procedure has been included in the calculation. Figure 6 shows the change of the cluster size distribution corresponding to the desorption rates of Fig. 5. The different traces (dashed lines) refer to the distribution assumed initially on the surface (0) and obtained 60 s (1), 120 s (2), 180 s (3) and 240 s (4) after starting the illumination with laser light, respectively. The assumed absorption profile (solid line) for light of λ = 488 nm is also included. The

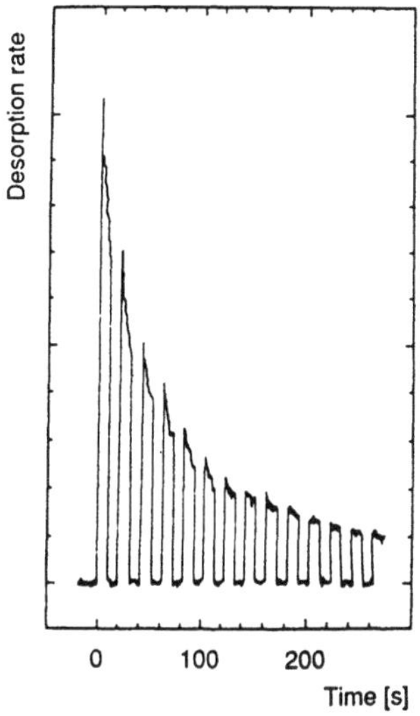

Fig. 6. Calculated change of the particle size distribution during laser illumination corresponding to the desorption rate of Fig. 5. The different traces (dashed lines) refer to the distribution assumed initially on the surface (0) and obtained 60 s (1), 120 s (2), 180 s (3) and 240 s (4) after starting the illumination, respectively. The absorption profile (solid line) for light of λ = 488 nm is also included

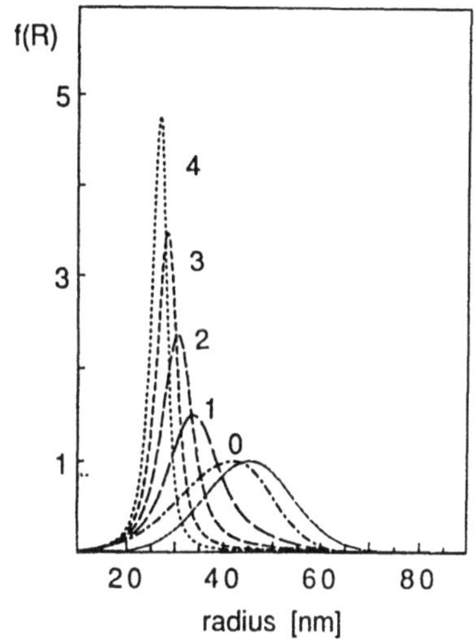

Fig. 7. Experimentally determined desorption rate as a function of illumination time with laser light. The initial average particle size was R_0 = 33 nm. The measurement was made with λ = 488 nm and I = 100 W/cm^2

desorption rate exhibits a monotonous decrease, see Fig. 5. As can be seen from Fig. 6, the change in the distribution is very pronounced. The larger particles in the distribution interact more strongly with the light than the smaller ones. They shrink and move out of resonance. Since the desorption comes to an end at about the same position of the absorption profile for all particles, a considerable narrowing of the distribution results. Figure 6 shows that this narrowing is as large as a factor of 6 after 240 s. It proceeds further if the illumination is continued and can amount to a factor of 10. The size manipulation can be repeated for different wavelengths to make the distribution even narrower.

In order to verify this model, the desorption rates for many different cases, particularly for the conditions used in the above example have been measured. The results are shown in Fig. 7. The average initial cluster size was R_0 = 33 nm. The clusters were illuminated by laser light of λ = 488 nm with an intensity of I = 100 W/cm^2. A decrease in the rate is observed starting at the very beginning of the illumination. A comparison with Fig. 5 shows that the signal shape is in very good agreement with the theoretical predictions. The good agreement of the theoretical predictions with experiment for different sets of parameters provides convincing evidence for the validity of the presented model and for the narrowing of the cluster size distribution under illumination with laser light. The model is also supported by the above mentioned measurements of optical transmission spectra of the clusters before and after irradiation with laser light [27]. Further experiments with the specific aim of verifying the predicted manipulation of the size distributions also using in situ electron microscopy are in progress.

Another interesting application of plasmon stimulated desorption might be chemistry on the surface of metal clusters. It seems possible that the localized electron excitation, into which the plasma oscillation decays, might also populate a repulsive state in an adsorbed molecule by which it is detached from the surface. Such an effect could even be molecularly selective. In preliminary experiments it has already been demonstrated that as little as one third of a molecular monolayer of ammonia on the surface of the sodium clusters has a dramatic influence on the desorption signal. In the future the desorption from other metal clusters will be investigated.

References

1. T.J. Chuang: Surf. Sci. Rep. **3**, 1 (1983)
2. W. Ho: Comments Cond. Mat. Phys. **13**, 293 (1988)
3. F. Träger: *In Photoacoustic, Photothermal and Photochemical Processes at Surfaces and in Thin Films*, ed. by P. Hess, Topics in Current Physics, Vol. 47 (Springer, Berlin, Heidelberg 1989)
4. N.H. Tolk, M.M. Traum, J.C. Tully, T.E. Madey (eds.): *Desorption Induced by Electronic Transitions, DIET I*, Springer Ser. Chem. Phys., Vol. 24 (Springer, Berlin, Heidelberg 1983)
5. W. Brenig, D. Menzel (eds.): *Desorption Induced by Electronic Transitions, DIET II*, Springer Ser. Surf. Sci., Vol. 4 (Springer, Berlin, Heidelberg 1985) p. 237
6. R.H. Stulen, M.L. Knotek (eds.): *Desorption Induced by Electronic Transitions, DIET III*, Springer Ser. Surf. Sci., Vol. 13 (Springer Berlin, Heidelberg 1988)
7. W. Hoheisel, K. Jungmann, M. Vollmer, R. Weidenauer, F. Träger: Phys. Rev. Lett. **60**, 1649 (1988)
8. W. Hoheisel, U. Schulte, M. Vollmer, R. Weidenauer, F. Träger: Appl. Surf. Sci. **36**, 664 (1989)
9. R. Weidenauer, M. Vollmer, W. Hoheisel, U. Schulte, F. Träger: J. Vac. Sci. Technol. **A7**, 1972 (1989)
10, M. Vollmer, R. Weidenauer, W. Hoheisel, U. Schulte, F. Träger: Phys. Rev. **B**, in press
11. G. Mie: Ann. Phys. **25**, 377 (1908)
12. C.F. Bohren, D.R. Huffman: *Absorption and Scattering of Light by Small Particles* (J. Wiley & Sons, 1983)
13. J.A.A.J. Perenboom, P. Wyder, F. Meier: Phys. Rep. **78**, 171 (1981)
14. T. Inagaki, K. Kagami, E.T. Arakawa: Appl. Opt. **21**, 949 (1982)
15. D. Menzel, R. Gomer: J. Chem. Phys. **41**, 3311 (1964)
16. P.A. Readhead: Can. J. Phys. **42**, 886 (1964)
17. M. Vollmer, F. Träger: Surf. Sci. **187**, 445 (1987)
18. N. Nishi, H. Shinohara, T. Okuyama: J. Chem. Phys. **80**, 3898 (1984)
19. K. Höning, W. Hoheisel, M. Vollmer, F. Träger: to be published
20. F.G. Celii, P.M. Whitmore, K.C. Janda: Chem. Phys. Lett. **138**, 257 (1987)
21. I. Lee, J.E. Parks II, T.A. Callcott, E.T. Arakawa: Phys. Rev. **B39**, 8012 (1989)
22. W.A. de Heer, K. Selby, V. Kresin, J. Masui, M. Vollmer, A. Châtelain, W. D. Knight: Phys. Rev. Lett. **59**, 1805 (1987)
23. F. Träger, G. zu Putlitz (eds.): *Metal Clusters* (Springer, Berlin, Heidelberg 1986)
24. P. Jena, B.K. Rao, S.N. Khanna (eds.): *Physics and Chemistry of Small Clusters* (Plenum, New York 1987)
25. C. Chapon, M.F. Gillet, C.R. Henry (eds.): Proc. 4th Int. Meeting on Small Particles and Inorganic Clusters, Z. Phys. D**12** (1989)
26, J. Kowalski, T. Stehlin, M. Vollmer, F. Träger: in Phase Transitions Journal (special issue on clusters), in press
27. W. Hoheisel, U. Schulte, M. Vollmer, F. Träger: to be published

On the Mechanism of High-Yield Desorption

R. Kelly

IBM Research Division, T.J. Watson Research Center,
Yorktown Heights, NY 10598, USA

Abstract. High-yield desorption, as with laser pulses, leads to near-surface collisions which can be so prominent that the system loses memory of the initial desorption step. Thus, this step could variously be thermal or electronic but the collisions lead to a primary distribution function having the form appropriate to a "Knudsen layer". We show how there are four phases due to the collisions: that of the Knudsen layer (the primary phase), that of unsteady adiabatic expansion, that of termination of the event, and that of free flight. The first two phases are well understood, but the third and fourth have never been adequately analyzed. We finally consider an effect which will be experimentally important: the tendency for desorbed species to return to the target surface and, moreover, to do so nonstoichiometrically.

1. Introduction

Desorption events are variously carried out under collision-free conditions and under conditions of more or less extensive collisions. The former is usual with individual incident particles, whether ions, electrons, or photons, and the latter with pulses, most commonly photon (laser) pulses. What we wish to do here is to summarize the results which occur when collisions are extensive [1-4]. We will, in so doing, give new information on termination of the event, on the passage to free flight, and on what may be expected for a binary target. In some respects what we will present may be termed the mechanism of high-yield desorption in the sense that, because of the dominant role of the collisions, the system tends to lose memory of the initial desorption step.

There are some surprises with binary targets, as it turns out that the collision regime may lead to non-stoichiometry. For example, starting with a target $YBa_2Cu_3O_{7-x}$ which is bombarded with laser pulses [5-7], one could expect the target surface to become enriched in Cu and the collector would therefore enrich to some extent in Ba.

A major exception to the rule that a collision regime occurs only in pulsed desorption occurs with condensed gases. Here there is good reason to believe that the impact of an individual heavy ion leads to an impact volume which is brought above the thermodynamic critical temperature, T_{tc}, and therefore behaves as a gas [3,8]. Extensive collisions would occur as the gas flowed from the impact volume.

2. The Phases of High-Yield Desorption

2.1 Initiation

As in virtually all treatments of near-surface collision regimes, we continue to restrict ourselves to an initial desorption step which is thermally activated and which is

Springer Series in Surface Sciences, Vol. 19
Desorption Induced by Electronic Transitions, DIET IV
Editors: G. Betz · P. Varga

therefore characterized by a velocity distribution function of the form:

$$f_s^{+}(v_x, v_y, v_z, E_I) = \frac{\rho_s}{(2\pi kT_s/m)^{3/2}} \frac{E_I^{j/2-1}}{\Gamma(j/2)(kT_s)^{j/2}} \times \exp\{-[2E_I + m(v_x^2 + v_y^2 + v_z^2)]/2kT_s\},$$

$$v_x \geq 0, \quad -\infty < v_y, v_z < \infty. \tag{1}$$

Here v_i ($i = x,y,z$) stands for a velocity component, E_I is the total internal energy, ρ_s is the gas number density next to the surface, and j is the number of internal degrees of freedom easily accessible at T_s. Eq. (1) is characterized by having only positive v_x and by the lack of a flow velocity, u. If the emission is non-thermal then eq. (1) is replaced by an appropriate form. One of the most useful is that for a monoenergetic flux with arbitrary angular distribution:

$$f_s^{+}(v, \theta) = (p\rho_s/2) \cos^{p-1}\theta \sin\theta. \tag{2}$$

Here v is the unique, positive velocity of the emitted particles, θ is the polar angle of emission, and p describes the arbitrary forward peaking. Eq. (2) is a possible description of some forms of electron or photon-stimulated desorption.

We finally note that the form of eq. (1) is appropriate to the evaporation coefficient being unity [9-11].

2.2 Knudsen-Layer (KL) Phase

If the desorption event is followed by a small number of collisions, typically as few as ~3 per atom [12,13], then a KL forms. We regard a sufficient definition of a KL to be the near-surface region in which the distribution function evolves from eq. (1) or (2) (characterized by $u = 0$) to a form with an explicit flow velocity, u_K. Furthermore, we suggest that this form will be largely independent of the initial desorption mechanism and will effectively constitute the primary distribution function:

$$f_K^{\pm}(v_x, v_y, v_z, E_I) = \frac{\rho_K}{(2\pi kT_K/m)^{3/2}} \frac{E_I^{j/2-1}}{\Gamma(j/2)(kT_K)^{j/2}} \times \exp\{-[2E_I + m((v_x - u_K)^2 + v_y^2 + v_z^2)]/2kT_K\},$$

$$-\infty < v_x, v_y, v_z < \infty. \tag{3}$$

Here ρ_K and T_K, with "K" for "Knudsen", differ from the values at the surface and it is, in fact, one of the prime objectives of KL theory [9-11] to evaluate these quantities (along with u_K and the recondensed fraction). There are thus 4 unknowns.

Table 1 summarizes the values of T_K/T_s, ρ_K/ρ_s, and F^{-} for the case that the initial desorption is thermally activated. F^{-} stands for the fraction of the emitted number flux which recondenses at the surface. The remaining unknown is u_K, and is found (again for thermal emission) to be similar to the speed of sound, a_K:

$$u_K \approx a_K = (\gamma kT_K/m)^{1/2},$$

where γ is the specific-heat ratio, given for a perfect gas by

Table 1
Parameters to describe a fully developed Knudsen layer [1].

Degrees of internal freedom, j	$\gamma = C_p/C_V$ (eq. (4))	T_K/T_s [1]	ρ_K/ρ_s [1]	F^- (eq. (10))
0	5/3	0.669	0.308	0.184
2	7/5	0.782	0.301	0.212
4	9/7	0.837	0.298	0.224
6	11/9	0.871	0.297	0.231
8	13/11	0.893	0.297	0.236
infinity	1	1	0.296	0.257

$$\gamma = C_p/C_v = (j + 5)/(j + 3). \tag{4}$$

Since the Mach number, M, is defined by $u = Ma$ it follows that a KL is characterized by $M = 1$.

Another way of looking at a KL is shown in two studies [14,15] in which the KL due to vaporization of I_2 was studied by laser-induced fluorescence. The work identified an on-going vaporization at $x = 0$, KL formation by roughly $x = \ell_K$, and a subsequent steady flow up to a condensing surface at $x = \ell$. The system is out-of-equilibrium from 0 to ℓ_K, with an ever increasing u and decreasing T, but in-equilibrium beyond ℓ_K.

Eq. (3) can be rationalized as follows. Eq. (1) easily leads to a value for $<v_x>$, a quantity proportional to the number flux:

$$<v_x> = \int f_s^+ v_x dv_x \,/\, \int f_s^+ dv_x = (2kT_s/\pi m)^{1/2} = 0.798(kT_s/m)^{1/2}. \tag{5}$$

If collisions occur it is reasonable to expect all values of v_x to be represented. This means that $<v_x>$ is now zero so that to conserve number flux a flow velocity (or some equivalent modification of the distribution function) is necessary. As stated above, the latter turns out to be that for $M = 1$, namely $u_K = Ma_K = a_K$ [9-11].

The significance of u_K can be seen by again evaluating $<v_x>$:

$$<v_x> = \int f_K^{\pm} v_x dv_x / \int f_K^{\pm} dv_x = u_K = 1.06(kT_s/m)^{1/2}. \tag{6}$$

Thus the flow velocity is identical with $<v_x>$. Also, as would be expected, eqs. (5) and (6) are similar. They are not identical, however, as the argument is oversimplified in that it should have taken into account the product $\rho <v_x>$ and should have allowed for the recondensing number flux.

A rough estimate of the thickness of a KL can be made starting from the observation that it is usual for a KL to exist in desorption experiments involving 10 ns laser pulses [5,6,12,13,16,17]. It follows that, in such experiments, the KL forms in a time given by $t \leq 10^{-8}$ s. We will see from eq. (8B) that an expansion front proceeding from a flowing source moves with a velocity similar to $\hat{u} = 4 <v_x>$. For atoms ($\gamma = 5/3$) of mass 40 at $T_s = 1000$ K, $\hat{u}$ is 2×10^5 cm/s, so that the KL thickness follows as ≤ 20 μm. If in one pulse precisely 10 times as many atoms were removed as needed for a KL, then the thickness would be 2 μm.

2.3 Unsteady Adiabatic Expansion (UAE) Phase

For more collisions than about 3 per desorbed atom, the system can be expected to undergo a UAE. This problem can be coupled explicitly to a KL provided the latter is assigned a definite width, a result supported by the work on I_2 vaporization [14,15]. In one dimension there are closed-form solutions [4,18]:

$$a_M = u_K\left(1 - \frac{\gamma - 1}{\gamma + 1}\frac{x}{u_K(t - t_K)}\right) = u_K\left(1 - \frac{x}{4u_K(t - t_K)}\right); \tag{7A}$$

$$u_M = u_K\left(1 + \frac{2}{\gamma + 1}\frac{x}{u_K(t - t_K)}\right) = u_K\left(1 + \frac{3x}{4u_K(t - t_K)}\right). \tag{7B}$$

Here and in what follows subscript "M" refers to results characterizing the UAE phase, since they are strong functions of M. Also, the second form of each equation applies to atoms, $\gamma = 5/3$. The KL boundary occurs at $x = 0$ (fig. 1) and t_K is the time after the beginning of the pulse when the KL is fully formed.

Eqs. (7A) and (7B) can be easily rearranged to give

$$(a_M/u_K)^2 = T_M/T_K = (\rho_M/\rho_K)^{\gamma - 1} = \left[\frac{\gamma + 1}{2 + (\gamma - 1)M}\right]^2 = \left(\frac{4}{3 + M}\right)^2, \tag{8A}$$

where we have made use of the relation appropriate to any adiabatic expansion of a perfect gas, $T \propto \rho^{\gamma - 1} = \rho^{2/3}$. As far as M is concerned it can be evaluated from eqs. (7A) and (7B) by forming $M = u_M/a_M$, and it follows that it increases rapidly with distance. Finally, we note that the value of x/t for which $a_M = 0$ defines the expansion front, the flow velocity of which is the maximum in the UAE:

$$\hat{u} = \frac{\gamma + 1}{\gamma - 1}u_K = 4u_K. \tag{8B}$$

The corresponding distribution function is like eq. (3) but with new values for ρ, T, and u. We note, however, in view of eq. (8A), that ρ_K and T_K can be retained in the pre-exponential part:

$$f_M^{\pm}(v_x, v_y, v_z, E_I) = \frac{\rho_K}{(2\pi kT_K/m)^{3/2}}\frac{E_I^{j/2 - 1}}{\Gamma(j/2)(kT_K)^{j/2}} \times \exp\{-[2E_I + m((v_x - u_M)^2 + v_y^2 + v_z^2)]/2kT_M\}. \tag{9}$$

2.4 Termination of the Event

For a semi-infinite reservoir of gas, eqs. (7) remain always valid and therefore accurately describe an unending pulse. Expressed in terms of density (eq. (8A)), they appear as in fig. 1. For sputtering of condensed gas by individual heavy ions one has an unusual situation in which the solid gas is brought to a temperature exceeding T_{tc}, and gas flow therefore occurs as from a finite reservoir. In this case a rarefaction wave moves into the reservoir and is finally reflected from the back, and the solutions seen in eqs. (7) are supplemented by "long-time" solutions of well-defined form [3]. Without going into detail, the result for sound speed, $\underline{a}$, is as in fig. 2. An important detail (see sect. 2.5) is that $\underline{a}$ is nearly flat from $x = -\ell$ to the line of contact (LOC), the maximum value lying at $x = -\ell$.

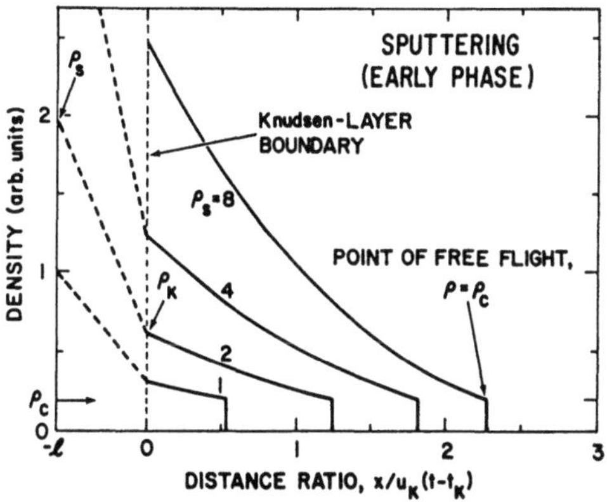

Fig. 1. Calculated values of ρ vs x appropriate to high-yield desorption starting at $x/u_K(t - t_K) = 0$ and proceeding to larger values of the same parameter [4]. Each curve corresponds to a different desorption yield, thence a different value of ρ_s and ρ_K. The KL lies between $x = -\ell$ and $x = 0$, where we note that the position of the target surface, $-\ell$, is mathematically ill-defined. The expansion space lies at $x > 0$. The calculations are based on the $\underline{a}$ values of Eq. (7A) as evaluated for atoms (i.e. $\gamma = 5/3$) and transformed into ρ values using Eq. (8A). ρ_K/ρ_s was taken as 0.308 (table 1) and ρ_c as 0.2, the latter being completely arbitrary. The point of free flight, assumed to be abrupt [4,20], is shown for each curve by a vertical segment.

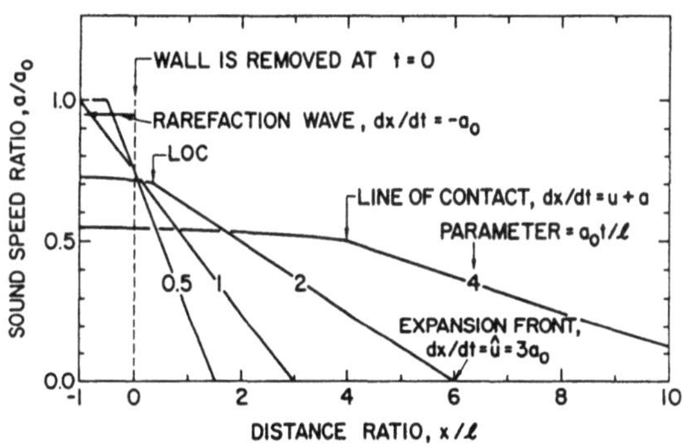

Fig. 2. Calculated values of $\underline{a}$ vs. x appropriate to the gas-flow model of condensed-gas sputtering [3]. The region of target brought to $T > T_{tc}$ lies between $x = -\ell$ and $x = 0$ and the expansion space lies at $x > 0$. The calculations are appropriate to atoms, i.e. $\gamma = 5/3$. The idea is that the "wall" is removed at $t = 0$, UAE immediately proceeds for $x > 0$, a rarefaction wave moves towards $x = -\ell$, the wave is reflected, and the wave then exists between $x = -\ell$ and the "line of contact".

For desorption with a short pulse, the termination of the pulse soon renders the solutions based on eqs. (7) and shown in fig. 1 invalid. We have recently obtained explicit long-time solutions to this problem and can report that they are similar (but not identical) to the long-time solutions of which that for $\underline{a}$ was seen in fig. 2 [19]. It is found that $\underline{a}$ is nearly flat from the $x = 0$ to the LOC but that the maximum value lies at an intermediate point.

2.5 Free-Flight Phase

The escaping gas has a density profile which decreases rapidly with distance (eq. (7A)). It follows that the density for large x will be low enough for free flight to occur and a useful approximation [4,20] is to assume that free flight sets in abruptly (and without invalidating eqs. (7)) for a critical density, ρ_c, and sound speed, a_c: see the vertical segments in fig. 1. Then if ρ_K/ρ_c is unity the system shows a KL without UAE and for $\rho_K/\rho_c > 1$ there are both a KL and UAE.

The idea of an abrupt passage to free flight for large x has a close gas-dynamic analog in a UAE where the expansion space instead of being semi-infinite ($x \geq 0$) is finite ($0 \leq x \leq \ell_0$). As pointed out by Stanyukovich (p. 169 of [18]), when the expanding gas reaches ℓ_0 it effectively disappears (as if in free flight) and, most important, this does not initiate a disturbance. This is because the velocity of propagation of a disturbance would be $v = u - a$ and this quantity is positive for $x > 0$ (eqs. (7)). Since there is no disturbance it follows that the passage to free flight does not invalidate eqs. (7).

Referring now to fig. 2, thence to condensed gas sputtering, it is clear that the passage to free flight, which first occurs at increasing distance from the surface, will finally approach the surface and will go to completion when $a = a_c$ at $x = -\ell$. Indeed, since $\underline{a}$ vs. x is nearly flat, one can approximate the entire entry to free flight as occurring abruptly when $a = a_c$ at $x = -\ell$ [3]. The problem is slightly different for desorption due to a short pulse as the long-time solution for $\underline{a}$ has a maximum between $x = 0$ and the LOC [19]. This means that as the position of the passage to free flight approaches the surface, an abrupt passage (involving about one-half the desorbed particles) occurs when $a = a_c$ at the position of maximum $\underline{a}$. In summary there are two stages in the passage to free flight in high-yield desorption: particles which throughout the pulse reach a low enough density plus a larger group of particles which go free after the pulse has terminated.

3. An Example Relevant to Experiment: Recondensation on the Target

In the theory of the KL one requires that certain fluxes (those of number, momentum, and energy) be conserved. Each flux is assumed to have 3 well-defined parts: out from the surface with distribution function f_s^+, returning to the surface with distribution function $a_{\bar{K}} f_K$ (i.e. eq. (3) with $v_x \leq 0$ and an unknown constant of proportionality, $a_{\bar{K}}$), and passing beyond the KL "boundary" with distribution function $f_K^{\pm}$. We are here interested in the number flux:

$$\iiiint dv_x dv_y dv_z dE_I v_x f_s^+ = \iiiint dv_x dv_y dv_z dE_I v_x a_{\bar{K}} f_K + \iiiint dv_x dv_y dv_z dE_I v_x f_K^{\pm},$$

i.e. $F_1 = F_2 + F_3$. It follows that the recondensing flux is just

$$F^- = F_2/F_1 = 1 - F_3/F_1 = 1 - (\rho_K/\rho_s)(2\pi\gamma T_K/T_s)^{1/2}. \tag{10}$$

We have already given values of F^- in table 1. Depending on the complexity of the desorbed particles (i.e. on the value of j) between 18 and 26% of the desorbed flux returns to the surface.

From another point of view, the existence of a non-zero value of F^- means that if, for example, an integrated circuit consisting of separate regions of Cu and Si were

Table 2
Parameters to describe recondensation when a two-component system shows a fully developed Knudsen layer. Component 1 is the heavier and 2 is the lighter. Due to Frezzotti [23].

Mass ratio, m_1/m_2	Atom fraction of component 1 at the surface	F_1^-	F_2^-
2	0.1	0.116	0.192
2	0.5	0.152	0.228
2	0.9	0.175	0.263
5	0.1	0.086	0.215
5	0.5	0.143	0.320
5	0.9	0.175	0.395

laser bombarded for any reason, then the outer surface would become a mixture of Cu and Si. In particular, if n monolayers of thickness λ were removed per pulse and 25% recondensed, then the thickness of alteration would be 0.25nλ, and this suggests that the contaminated surface could be restored if the laser treatment were terminated with a series of pulses sufficiently weak that a thickness 0.25nλ were removed under collisionless conditions. It is remarkable that recondensation has never been properly studied. Rather, its existence is merely hinted, as in work on desorption from Cl + Cu [21]. In another case recondensation was demonstrated to be significant when the desorption was made into an ambient gas, but the unwarranted conclusion was reached that it would be absent in vacuum [22].

Suppose now that the target were binary with masses m_1 and m_2. Then one of the basic results of KL theory is that the light component recondenses preferentially [23] (table 2). Thus recondensation is nonstoichiometric.

We would finally like to propose that another problem in which recondensation may play an explicit role concerns the clusters that form during high-yield desorption. For example, with $YBa_2Cu_3O_{7-x}$ cluster ions with masses up to ~30,000 u have been demonstrated by mass-spectrometry [6] and spheres with diameters of 150-350 nm have been demonstrated by scanning electron microscopy [7]. The spheres were confirmed only on the substrate but the target was not examined. We would like to propose that the clusters, whatever their origin, form a KL by colliding amongst themselves in the region beyond the normal KL. This layer would be independent of the normal layer, would be more extended in space than the ≤20 μm expected for the latter (sect. 2.2), and would in principle lead to the usual 18-26% recondensation. Indeed, because of the ease with which clusters are studied by scanning electron microscopy, it should be possible to make an explicit study of their repartition between the target and substrate as a test of whether they are governed by KL theory. We note that the form of eq. (10) is such that recondensation should occur for all masses, including, in principle, the enormous masses of clusters.

Unlike the recondensing of atoms and molecules, which can probably be counteracted by collisionless removal of a thickness ~0.25nλ, we see no way to avoid the problem of cluster recondensation. In this connection we would add that whether or not UAE follows the KL is irrelevant to the extent of recondensation, as the latter is confined wholly to the KL phase and is not affected by UAE.

REFERENCES

[1] R. Kelly and R.W. Dreyfus, Surf. Sci. 198 (1988) 263.

[2] R. Kelly and R.W. Dreyfus, Nucl. Instr. Meth. B32 (1988) 341.

[3] R. Kelly, Nucl. Instr. Meth. B (in press).

[4] R. Kelly, J. Chem. Phys. (in press).

[5] T. Venkatesan, X.D. Wu, A. Inam, and J.B. Wachtman, Appl. Phys. Lett. 52 (1988) 1193.

[6] C.H. Becker and J.B. Pallix, J. Appl. Phys. 64 (1988) 5152.

[7] R.B. Laibowitz (IBM, Yorktown Heights) (private communication, 1989).

[8] H.M. Urbassek and J. Michl, Nucl. Instr. Meth. B22 (1987) 480.

[9] T. Ytrehus, in: Rarefied Gas Dynamics, ed. by J.L. Potter (AIAA, New York, 1977) p. 1197.

[10] D.A. Labuntsov and A.P. Kryukov, Int. J. Heat Mass Transfer 22 (1979) 989.

[11] C. Cercignani, in: Rarefied Gas Dynamics, ed. by S.S. Fisher (AIAA, New York, 1981) p. 305.

[12] I. NoorBatcha, R.R. Lucchese, Y. Zeiri, J. Chem. Phys. 86 (1987) 5816.

[13] I. NoorBatcha, R.R. Lucchese, Y. Zeiri, J. Chem. Phys. 89 (1988) 5251.

[14] R. Schilder, G. Adomeit, and G. Wortberg, in: Rarefied Gas Dynamics, ed. by O.M. Belotserkovskii et al. (Plenum, New York, 1985) p. 577.

[15] R. Mager, G. Adomeit, and G. Wortberg, in: Rarefied Gas Dynamics (16th, Pasadena, CA, U.S.A., 1988) in press.

[16] R.W. Dreyfus, R. Kelly, R.E. Walkup, Appl. Phys. Lett. 49 (1986) 1478.

[17] R.W. Dreyfus, R. Kelly, R.E. Walkup, Nucl. Instr. Meth. B23 (1987) 557.

[18] K.P. Stanyukovich, "Unsteady Motion of Continuous Media" (Pergamon, London, 1960).

[19] R. Kelly, work in progress.

[20] K.L. Saenger, J. Chem. Phys. 79 (1983) 6043.

[21] J.H. Brannon and K.W. Brannon, J. Vac. Sci. Technol. (in press).

[22] G. Koren and U.P. Oppenheim, Appl. Phys. B42 (1987) 41.

[23] A. Frezzotti, in: Rarefied Gas Dynamics, ed. by V. Boffi and C. Cercignani (Teubner, Stuttgart, 1986) p. 313.

Studies of Laser Sputtering of GaP by Means of Resonance Ionization Spectroscopy: Response of Clean Solid Surfaces to Photo-excitation

Y. Nakai, K. Hattori, and N. Itoh

Department of Physics, Faculty of Science, Nagoya University, Fro-cho, Chikusa, Nagoya 464-01, Japan

We review the new experimental observation, using high-sensitivity measurements, of laser-induced sputtering of Ga atoms from GaP, as well as existing results of laser-induced sputtering of insulators and semiconductors. Emphasis is placed on clarification of the nature of the optical transitions which ultimately cause atomic ejection. It is pointed out that transitions involving surface states are responsible for laser sputtering from GaP and that high sensitivity measurements of laser-induced sputtering are useful for characterizing the surface transitions.

I. Introduction

It has become increasingly clear that irradiation of insulators and semiconductors with laser pulses causes sputtering [1,2]. One of the main features of laser sputtering is its superlinear dependence on the laser intensity; sputtering is induced only under intense excitation. The mechanism of the laser-induced sputtering may vary depending on the types of target materials. The purpose of the present paper is twofold: to survey several features of laser sputtering and to review our current high-sensitivity measurements of sub-band-gap-energy laser-induced sputtering from compound semiconductors.

II. A review on laser-induced sputtering

Experiments of laser sputtering from non-metallic solids having been carried out so far can be divided broadly into two types: those with photons above the band-gap energies and those with photons below the band-gap energies. (Laser sputtering from metals is analogous to sputtering from non-metals with photons above band-gap energies.) In the former type of experiments, particularly for photons above the direct band-gap energies, the bulk optical absorption coefficient is so large that the temperature rise of the surface layers has been regarded as the main cause of sputtering [3]. In the regime of desorption induced by electronic transitions, the laser-induced sputtering for photons above the direct band-gap energy is not the main concern here. We emphasize that the bulk optical absorption coefficient for lasers of sub-band-gap photon energies is extremely small and that most experiments of laser-induced effects in this regime have been carried out using laser pulses of nearly the same fluence as those for above band-gap energies. Therefore heating effects can be disregarded and hence the sputtering in this regime should be of electronic nature. Thus the primary concern for laser sputtering in this regime is to find the origin of the electronic transitions which result in sputtering and to find the process that converts electronic excitation energy to atomic energy responsible for sputtering. Since only atoms near the surface top layer are capable of being ejected, the electronic transitions responsible for sputtering should be those having the capability of transforming the electronic excitation energy to the kinetic energy of atoms near the surfaces.

Springer Series in Surface Sciences, Vol. 19
Desorption Induced by Electronic Transitions, DIET IV
Editors: G. Betz · P. Varga

Laser pulses of sub-band-gap energies will induce multi-photon band-to-band excitation and excitation of gap states originating either from surface dangling bonds or from surface defects. Therefore, in discussing laser sputtering, it is worthwhile to note that some non-metallic solids are subjected to sputtering, in proportion to the density of excitation by band-to-band excitation, but some are not [4]. We refer to the former type of non-metals as type II and the latter as type I. The most typical examples of type II are alkali halide crystals, for which extensive work on sputtering with low-intensity electron and photon beams has been carried out. In these materials, excitons and electron-hole pairs formed by electronic excitation are converted to self-trapped excitons, because of strong electron-lattice coupling [5], and the sputtering originates from the self-trapped excitons [6]. It has been shown that the sputtering dominates in the halogen sublattice, and hence band-to-band excitation of these materials leads to formation of non-stoichiometric surface layers. A few other insulators, alkaline earth fluorides [7] and silver halides [8] for example, belong to type II, while most elemental and compound semiconductors belong to type I. Under laser irradiation, sputtering is induced by multi-photon band-to-band transitions in these materials.

In type II materials, the surface is modified under laser irradiation because of band-to-band multi-photon transitions. The modified surface will be then the origin of single-photon surface transitions, which result in sputtering. Evidence for sputtering by a damaged surface has been obtained by Matthias et al. [9]. They measured the wavelength dependence of laser-induced sputtering of Ba^+ ions and found that the sputtering yield is largest at the F-band maximum, indicating that the electronic excitation at the defect sites causes sputtering. Since BaF_2 is a type II non-metal, the F centers are most likely generated by multi-photon band-to-band transitions. It follows that laser sputtering is a cumulative effect of multi-photon band-to-band transitions and of excitation of defect states. It has been found that neutral Ba atoms are also emitted by laser irradiation. Although the emission has been ascribed by Matthias et al. to thermal evaporation, we consider that it is due to thermal emission consequent to halogen emission by multi-photon band-to-band excitation. Defect excitation may also be a cause of Ba emission.

Laser-induced sputtering for photons below the band-gap energies has been studied extensively for Al_2O_3 [10], the categorization of which has not been definitely determined. The observations of ablation of surface layers and the velocity distribution of the ejected atoms and molecular species have shown that the sputtering is not a thermal effect. It has been suggested that the sputtering is due to excitation of the surface states.

It is of interest to study laser-induced sputtering of type I non-metals by sub-band-gap photons. Of particular interest are GaP(110) and ($\bar{1}\bar{1}\bar{1}$) surfaces, for which surface gap states have been observed by photoemission and inverse photoemission studies [11]. We describe here the work on laser-induced sputtering from GaP for photons of the indirect-band-gap energy [12]. Even though band-to-band transitions are induced in these experiments, the bulk optical absorption coefficient is small enough to exclude the heating effects. It has been demonstrated that sputtering and modification of the LEED pattern is induced by laser irradiation. Until now the sputtering experiments have been carried out using a conventional mass spectrometer, so that the surface is damaged substantially after a sputtering experiment is finished. In such a condition it has been found that the sputtering yield below the band-gap energy is lower than that at the indirect band-gap energy. In order to investigate the response of an undamaged surface to laser beams and to find the transitions that cause sputtering, it is necessary to investigate sputtering using a sensitive detection system, so that a measurement can be completed with destruction of only an extremely small part of the surface layer.

III. High-sensitivity measurements of laser-induced sputtering

We carried out high-sensitivity resonance-ionization-spectroscopy measurements of sputtering of Ga atoms from GaP($\bar{1}\bar{1}\bar{1}$) and (110) clean surfaces, which were analyzed by means of the LEED technique [13]. We found that a laser shot that ejects 10^{-5} of a monolayer of Ga atoms produces signals capable of being analyzed. Thus, in this experiment, the response of clean surfaces to successive excitation by laser pulses has been measured.

Fig. 1 shows the sputtering yield for the GaP($\bar{1}\bar{1}\bar{1}$) surface observed for 580 nm laser pulses as a function of the number of shots on the same spot on the specimen. It is clear that the yield is reduced by repeating the pulses; the number of sites which give rise to sputtering is gradually diminished on repeating the pulse. Thus the result can be best understood by considering that sputtering is caused by photoexcitation of the defect sites and that the optical absorption at the defect sites destroys them by emitting atoms. Similar fading emission of ions, which occurs at a lower intensity, was observed for the GaP (110) surface [14]. These experimental results indicate that sputtering due to defect-associated transitions below the band-gap energy occurs for GaP. It has been shown also that the emission of Ba^+ from BaF_2 fades on increasing the number of shots [9].

In order to answer the question whether sputtering by way of an "intrinsic" surface transition is induced in GaP($\bar{1}\bar{1}\bar{1}$), a similar experiment was carried out with a higher laser intensity. The result is shown in Fig. 2. It is clear that the decrease in the yield at the beginning of irradiation ceases at a low repetition and then the yield becomes constant. The constancy of the yield indicates that the number of sites for the optical transition inducing sputtering is not altered by repeating the pulse. LEED measurements were carried out before laser irradiation and after laser irradiation of several numbers of shots and it is found that the reconstructed (17x17) structure is not altered by repeating the irradiation within the range where the yield is constant. Similar constant yield was observed also for the GaP(110) surface. It follows that the sites responsible for optical transition are not related to the defects at surfaces in this range but probably to the surface states. In fact photoemission and in-

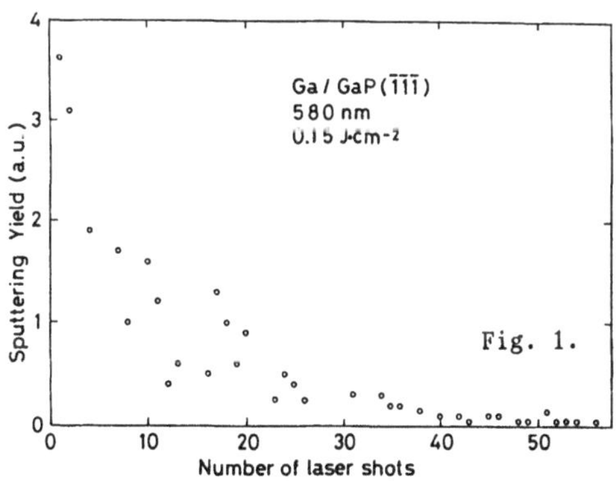

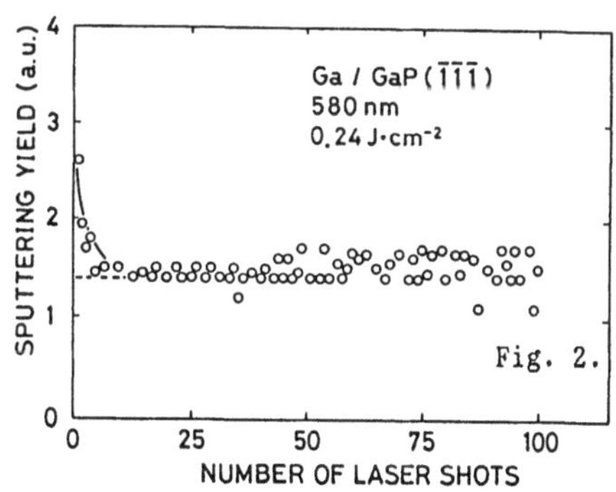

Fig. 1. Dependence of the sputtering yield of Ga atoms from the GaP($\bar{1}\bar{1}\bar{1}$) surface on the number of laser shots for a low laser intensity.

Fig. 2. Dependence of the sputtering yield of Ga atoms from the GaP($\bar{1}\bar{1}\bar{1}$) surface on the number of laser shots for a high intensity.

verse photoemission studies [11] indicate the existence of occupied surface states for the GaP($\bar{1}\bar{1}\bar{1}$) surface and of unoccupied surface states for the GaP (110) surface. Extended irradiation of the GaP($\bar{1}\bar{1}\bar{1}$) surface was found to lead to destruction of the reconstructed (17x17) structure [15] and later of the base (1x1) structure. Similarly the (1x1) structure of GaP(110) surface was destroyed by extended irradiation.

The results of the present investigation indicate that laser-induced sputtering by way of an "intrinsic" surface transition is indeed induced in GaP. Since sputtering is induced only when transitions at surfaces occur, measurements of the wavelength dependence of laser sputtering will yield information on the surface transitions. Even though laser sputtering at high intensities may be a complex process because of cumulative effects of the surface damage and of sputtering from damaged surfaces, the initial stage of laser damage appears to be purely electronic and to reflect optical transitions at the surfaces.

References

1. N. Itoh, Nucl. Instrum. Methods B 27, 155 (1987).
2. E. Matthias, H. B. Nielsen, J. Reif, A. Rosen and E. Westin, J. Vac. Sci. Technol. B 5, 1415 (1987).
3. B. Stritzker, A. Pospiezcky and J. A. Tagle, Phys. Rev. Lett. 47, 356 (1981).
4. N. Itoh, Crystal Lattice Defects and Amorphous Materials 12, 103 (1985).
5. Y. Toyozawa, Vacuum Ultraviolet Radiation Physics, eds. E. E. Koch, R. Hensel and C. Knuz (Pergamon, Braunschweig, 1974) p. 317, and Physica 116B, 7 (1983).
6. N. Itoh and T. Nakayama, Nucl. Instrum. Methods B 13, 550 (1986).
7. R. Souda and M. Aono, The Structure of Surfaces II, eds. J. F. van der Veen and M. A. Van Hove (Springer-Verlag, Berlin, 1987) p. 581.
8. H. Kanzaki and T. Mori, Phys. Rev. B 29, 3573 (1984).
9. J. Reif, H. Fallgren, H. B. Nielsen and E. Matthias, Appl. Phys. Lett. 49, 930 (1986), J. Reif, H. Fallgren, W. E. Cooke and E. Matthias, Appl. Phys. Lett. 49, 770 R (1986), A. Rosen, E. Westin, E. Matthias, H. B. Nielsen and J. Reif, Phys. Scripta 23, 184 (1988), and H. B. Nielsen, J. Reif, E. Matthias, E. Westin and A. Rosen, Desorption Induced by Electronic Transitions III, eds. R. H. Stulen and M. L. Knotek (Springer-Verlag, Berlin, 1988) p. 266.
10. R. Kelly, J. J. Cuomo, P. A. Laery, J. E. Rothenberg, R. B. Braren and C. F. Alliotta, Nucl. Instrum. Methods B 9, 329 (1985) and J. E. Rothenberg and R. Kelly, Nucl. Instrum. Methods B 1, 291 (1984), R. Kelly and J. E. Rothenberg, Nucl. Instrum. Methods B 7/8, 755 (1985) and R. W. Dreyfus, R. E. Walker and R. Kelly, Radiation Effects 99, 199 (1986), R. W. Dreyfus, F. A. MacDonald and R. J. van Gudfeld J. Vac. Sci. Technol. B 5, 1521 (1987).
11. F. Cerrina, Phys. Rev. B 31, 8314 (1985), D. Straub, M. Skibowski and F. J. Himpsel, J. Vac. Sci. Technol. A 3, 1484 (1985), D. Straub, V, Dose and W. Altman, Surface Sci. 133, 9, (1983), and K. Jacobi, Surf. Sci. 51, 29 (1975).
12. Y. Kumazaki, Y. Nakai and N. Itoh, Phys. Rev. Lett. 59, 2883 (1987), T. Nakayama, H. Ichikawa and N. Itoh, Surf. Sci. 123, L693 (1982), and T. Nakayama, Surf. Sci. 133, 101 (1983).
13. Y. Nakai, K. Hattori, T. Yamamoto, N. Itoh and M. Szymonski, Radiat. Effects Defects Solids 109, 213 (1989), and Y. Hattori, Y. Nakai and N. Itoh, Surf. Sci. to be published.
14. Y. Nakai, K. Hattori and N. Itoh, Appl. Phys. Lett. to be published.
15. J. van Bommel and J. E. Grombeen, Surf. Sci. 93, 383 (1980).

Do Electronic Transitions Play a Role in Laser Etching? The Case of Cl_2 on Si(100)

B. Bourguignon[1,2], *J. Boulmer, J.-P. Budin, and D. Débarre*

[1]Institut d'Electronique Fondamentale, Bât. 220, Université Paris-Sud, F-91405 Orsay, France

[2]Laboratoire de Photophysique Moléculaire, Bât. 213, Université Paris-Sud, F-91405 Orsay, France

Abstract. The possible role of desorption induced by electronic transition (DIET) in laser etching of silicon by chlorine is discussed, both in the case of the reaction of gas phase radicals at the Si surface ("photoreaction") and in the case of the reaction of chemisorbed chlorine ("photodesorption"). While DIET seems to be excluded in the first case, the data are still inconclusive in the second case. Laser heating which cannot be avoided in the experiments performed to date blurs the signature of DIET. However, purely thermal models seem to fail to account for all the experimental data.

1. Introduction

Although laser etching has attracted the interest of many researchers, the actual detailed reaction mechanism(s) remain to be elucidated. One reason is that the laser is responsible for three effects, namely: (a) electronic excitation of the gas phase reactant; in general photodissociation in the gas phase produces more reactive radical species at the vicinity of the surface; (b) electronic excitation of the solid; the laser excites electrons of the bulk (non resonant excitation) or electrons in particular states (surface states, surface plasmons, etc.; for these states, the excitation is resonant with respect to the laser wavelength); (c) heating of the substrate: heating occurs through phonon excitation at a time scale of the order of one picosecond. With excimer lasers, this results typically in a surface temperature rise of 1400 K at a fluence of 500 mJ cm^{-2} at 308 nm; this temperature rise is fast (the maximum temperature is reached before the end of the laser pulse) and the temperature decreases back to room temperature within a short time (typically one microsecond) after the end of the laser pulse. Hence, activated processes which do not occur at room temperature are induced by laser heating alone, provided that their activation energies

Desorption Induced by Electronic Transitions, DIET IV
Editors: G. Betz · P. Varga

and their time scales are not too large. As a result, the experimental measurements must result from both electronic and thermal effects, except where very low laser fluences are used; unfortunately, to date, etching has been observed at very low laser fluences only under experimental conditions (chamber pressure of several mbar) that preclude any detailed observation of the reaction products (identification, kinetic energy).

2. The possible role of DIET in the reaction of Cl radicals at Si surfaces ("photoreaction")

The group of Bäuerle [1-3] has measured etch rates under "high" pressure conditions (namely, above 1 mbar) using a two laser scheme in order to discriminate the role of gas-phase photodissociation against the role of electron-hole pair excitation in the substrate. Modelling of the measured etch rate versus laser power, chlorine pressure and laser wavelength (Ar^+ and Kr^+ laser lines) has provided conclusive evidence that under "high" pressure and low laser fluence (e.g. below ~ 300 $mJ.cm^{-2}$ under pulsed irradiation) the etch rate is controlled by the collision rate of Cl radicals at the surface. Nevertheless, etching does require laser induced electronic excitation of the surface; however, only very low fluences are necessary. Thus, thermal effects seem to be excluded (or at least unnecessary) in the reaction of gas phase Cl radicals at the silicon surface; the reaction probability per collision (10^{-2}) and the activation energy (0.06 eV) are derived from the fit of the experimental data to the model. The very low activation energy explains why laser heating is unnecessary.

The actual role of excited electrons of the bulk and/or the surface in collisions of gas-phase Cl radicals has been studied in more detail by Horiike et al.[4]. They have measured the etch rate as a function of the Si doping and orientation. They conclude that the reaction of Cl radicals requires the presence of electrons in the conduction band; these electrons "harpoon" the incoming Cl radicals; this cannot occur from the valence band in the case of Cl, by contrast with the case of F atoms which react spontaneously. In order to substantiate this interpretation, heavily doped n^+ type Si has been used, allowing electrons to be present in the conduction band ; in this case, spontaneous etching by Cl has indeed been observed. Hence, the reaction of Cl radicals at the Si surface appears as resulting from an initial electron transfer from Si to Cl. The subsequent reaction of Cl^- with the Si surface seems to involve the migration of Cl^- below the surface, as suggested by the influence of the orientation on the etch rate. Thus, there is no indication that a DIET is involved in the process. The reaction would simply proceed by evaporation of stable $SiCl_4$ as it forms by chlorine migration.

As pointed out above, the reaction products as well as their energy cannot be studied with the experimental apparatus used for these studies. Clearly, the identification of the reaction product and energy would be very relevant to confirm the above conclusions.

3. The possible role of DIET in the reaction of chemisorbed chlorine ("photodesorption")

Chemisorption on Si is known to be dissociative. However, detailed measurements of surface structures, desorption energies and thermal stability of chlorine overlayers have been reported for the Si(111) face only. It has been shown that a very slow reaction takes place, with $SiCl_4$ being the desorption product [5].

Chemisorbed chlorine has been shown to play a role in Bäuerle's group's experiments only as the excimer laser fluence is increased up to $\simeq 400$ mJ cm^{-2}. At such high fluences, the etch rate becomes almost independent of the chlorine pressure, which is the signature of chemisorbed chlorine under experimental conditions where the surface is saturated with chlorine. The large dependence of the etch rate on laser fluence demonstrates that, by contrast with the reaction of gas phase Cl radicals, the reaction of chemisorbed chlorine is highly activated. Similar results have been recently reported by Sesselmann et al. [6] who have also carefully studied the Si surface after exposure to Cl_2 and to the laser by means of XPS and AES.

The reaction of chemisorbed chlorine has been studied in more detail by the group of Baller, by time of flight (TOF) analysis of the mass spectrum and kinetic energy of the reaction products [7-9]. The experiments are conducted under relatively "low" (up to a few 10^{-5} mbar) Cl_2 pressure, and this allows a TOF analysis of the energy and mass spectrum of reaction products. On the other hand, under pulsed laser excitation, the number of molecules impinging on the surface during the laser pulse is exceedingly low, which precludes the use of the technique for the study of gas phase radicals colliding with the surface. Baller et al. have identified the main reaction product which is SiCl. $SiCl_x$ (x = 0, 2-4) and Cl are minor products. The TOF distributions are fitted to the Maxwell law. The fitted temperature is much higher than the laser-induced surface temperature rise. The total number of molecules desorbed per laser pulse exhibits an Arrhenius behavior with respect to the fitted temperature (the variable experimental parameter being the Cl_2 pressure, or equivalently the Cl coverage at the beginning of the laser pulse).

These results seem to suggest that etching occurs through thermal evaporation, with an activation energy of 0.8 eV for SiCl at 308 nm. However, one has to bear in mind two important points: (a) the fitted temperature (up to 4000 K) largely exceeds the surface temperature derived from models describing the laser heating of the substrate; (b) the analysis of TOF spectra, as well as the physical significance of Maxwell distributions, are not straightforward. The dynamical parameters of desorbing molecules have been discussed extensively by Comsa and David [10]. They pointed out that even at equilibrium, desorbing molecules need not follow the Knudsen (angular) and Maxwell (velocity) laws. In addition, the dynamical information is contained in both the TOF spectra and angular distributions. The results discussed in Comsa and David's paper are mainly associative desorption and catalytic oxidation occuring under thermal activation; by comparison with these phenomena, laser etching has an additional level of complexity because the substrate is involved in the chemistry; however, the basic feature is the same, namely a chemical event that ultimately leads to desorption; consequently, the remarks of Comsa and David are also relevant for laser etching. In this respect, Baller et al. have investigated recently [8,9] the angular dependence of the TOF distributions, and they did find a strong dependence of the fitted temperature with polar angle of desorption with respect to the surface normal. However, the results are still inconclusive with respect to the point adressed here, namely, does DIET play a role in laser etching? Besides the fact that very high temperatures are needed to fit the data to the Maxwell law, it is worthwhile to note that another feature in Baller et al's results may not be easily explained in the framework of a purely thermal model: the mass spectrum is different at 308 and 248 nm, with a significant increase of Si and $SiCl_2$ yields with respect to SiCl at 248 nm. It has been indeed concluded by Sesselmann et al.[6] that direct DIET occurs at 248 nm.

4. Results

We have undertaken the study of the Cl_2/Si(100) system with an additional feature with respect to the Baller et al. experiment, namely the use of a pulsed Cl_2 molecular beam [11,12]. This allows us to supply large Cl_2 doses to the substrate at very low chamber pressure, the ratio of beam to chamber pressures being in the range 10^3 at 30 Hz. The control of the timing of the experiment (pulsed Cl_2 beam, pulsed excimer laser, pulsed electron impact ionizer) enables us to study selectively the reaction of chemisorbed chlorine, as a function of both the initial chlorine coverage (i. e. at the beginning of the laser pulse) and the laser fluence. The SiCl signal has been calibrated using a mechanical profilometer, allowing for the direct measurement of etch rates (Å/pulse) with the TOF spectrometer. Experimental details are given elsewhere [11].

The results obtained so far can be summarized as follows:

(a) The etch rate has an Arrhenius behavior [12], with respect to the laser induced surface temperature as calculated using the available models [13,14] for laser heating (note that this temperature is not the one used by Baller et al. in their Arrhenius plot). The validity of such models is now experimentally established [15]. Not unexpectedly if one keeps in mind our remark of section 1, thermal activation must play some role in etching, and it shows up in the etch rate dependence on laser fluence (figure 1). The fitted activation energy is $E_d = 1.5 \pm 0.3$ eV [12]. It differs from the one obtained by Baller et al., and also from the activation energy of 2.4 eV obtained in temperature programmed desorption (TPD) experiments by Jackman et al. for a chemisorbed chlorine phase desorbing as $SiCl_2$ and $SiCl_4$ [16]. On the Si(100) surface each Si atom is bonded to two Si atoms of the underlying Si plane. A direct DIET would require the simultaneous excitation of two electrons belonging to the same Si atom. It may be concluded that 308 nm photons cannot produce such a double excitation; hence, the reaction rate may be limited by the thermal production of singly bonded Si atoms on the surface.

(b) In addition, the etch rate dependence on laser fluence exhibits a plateau, at least at laser fluences up to 1000 mJ cm^{-2}. This plateau corresponds to an etch rate of 0.6 Å/pulse, or approximately 0.5 monolayer (ML)/pulse, which is in good agreement with the measurements of

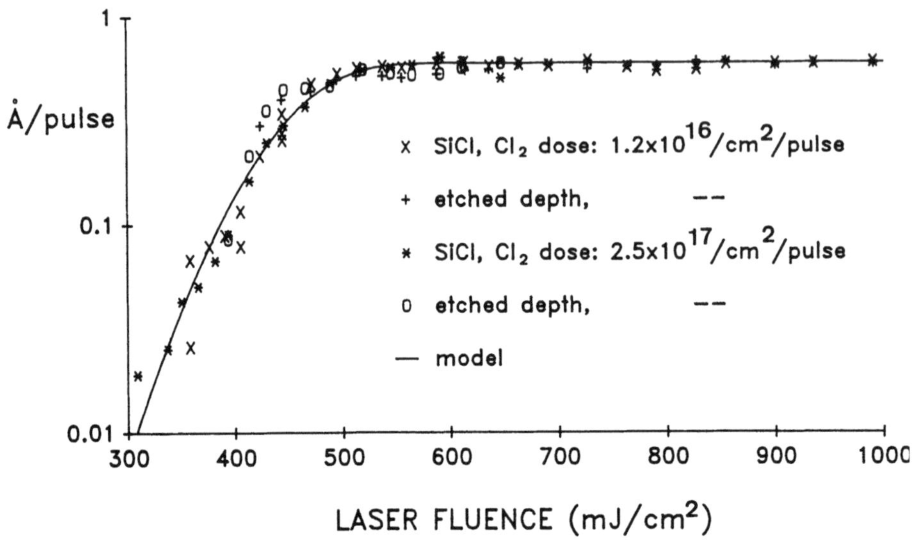

Figure 1. SiCl TOFMS signal and etch rate (profilometer measurement) versus laser fluence. The estimated surface melting fluence threshold is 500 mJ cm^{-2} .The solid line is calculated with the assumption of thermal desorption with 1.5 eV desorption energy .

Kullmer and Bäuerle [1] and Sesselmann et al.[6]. The asymptotic value of 0.5 ML/pulse is intriguing. Possibly, it reflects a structural property of the Si(100):Cl surface, which unfortunately has not been studied.

(c) The etch rate dependence on chlorine dose can be fitted assuming that the sticking probability of molecular chlorine on Si(100) is 0.08, and that the observed plateau corresponds to the desorption of all the chlorine present on the surface at the beginning of the laser pulse (figure 2). For comparison, the sticking probability of chlorine on Si(111) is ≃ 0.1 [5].

(d) The reaction products are SiCl, $SiCl_2$ and Cl, in agreement with Baller et al's findings [7]. Provided that the electron impact ionization yields of SiCl and $SiCl_2$ are similar, SiCl is the main reaction product. As pointed out above, the desorption products in TPD experiments are $SiCl_2$ and $SiCl_4$ [16], suggesting that the discrepancy between the desorption energies obtained in the two types of experiment is due to the fact that the photochemistry is different in laser desorption and thermal desorption experiments.

(e) The TOF distributions cannot be fitted to the Maxwell law (figure 3). This is in contrast with Baller et

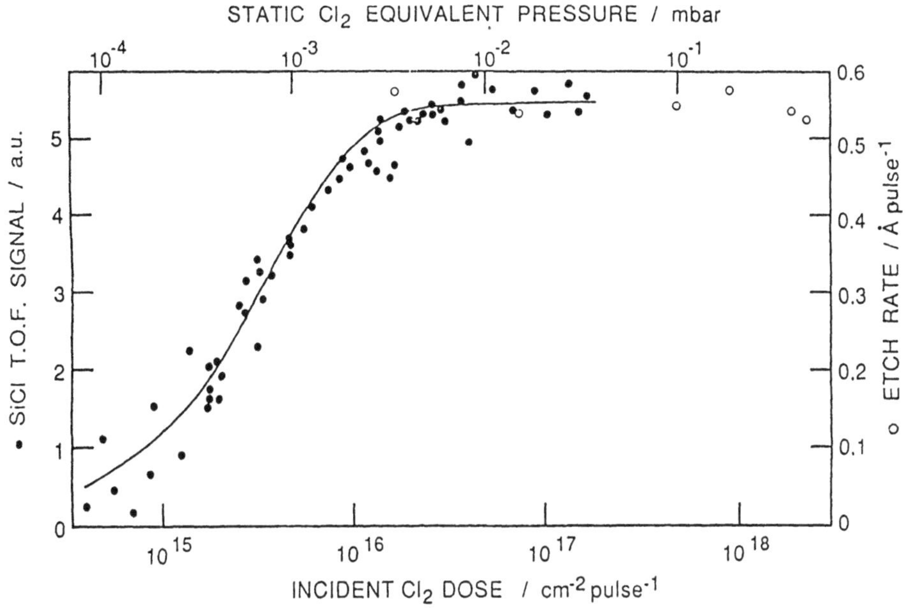

Figure 2 . SiCl TOFMS signal (full dots) versus Cl_2 dose. Also shown on top axis is the equivalent static chlorine pressure. Laser fluence was 500 mJ cm^{-2} . The full line is calculated with the assumption that all chemisorbed chlorine reacts at each pulse. It is assumed that one monolayer of chlorine corresponds to a Si : Cl ratio of 1. The fitted sticking coefficient is 0.08.

al's findings, and we do not explain this discrepancy. The fast part of the velocity distribution obtained at a laser fluence of 500 mJ cm^{-2} (surface temperature 1680 K) corresponds to a Maxwell distribution at 2000 K for both SiCl and $SiCl_2$. The width of the experimental spectrum is much larger than the Maxwell distribution, i. e. there is an additionnal slow component. The Maxwell distribution corresponds in principle to the desorption of species in equilibrium with the surface; by fitting to the Maxwell law, we implicitly assume that the adsorbate remains in equilibrium until very late in the desorption process. We may assume instead that the adsorbate is "equilibrated" with the surface in the well of the ground state potential energy surface (PES) along the surface-adsorbate coordinate, and is then suddenly desorbed (note that the most probable velocity of desorbing SiCl is 10 Å/ps; at such a velocity, the equilibration of the translational

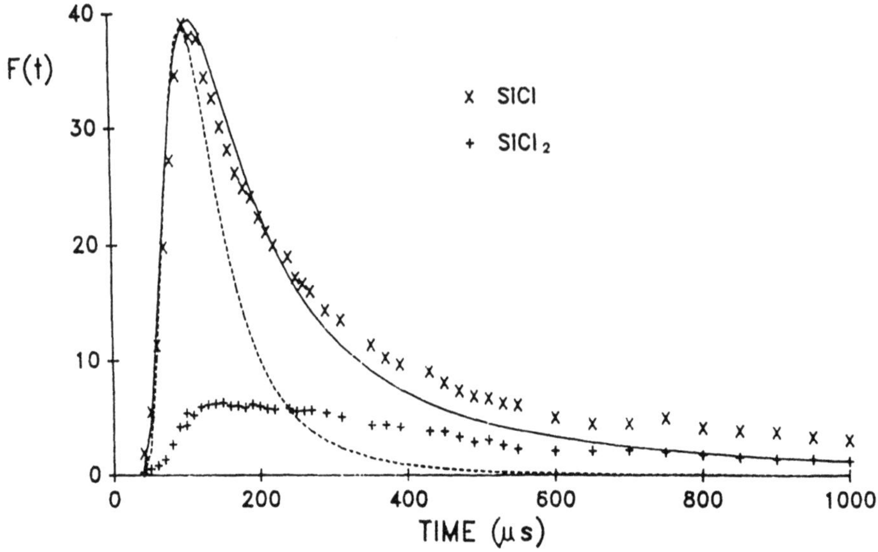

Figure 3 - TOF spectra of SiCl (× × ×) and $SiCl_2$ (+ + +). The flight path is 10 cm. Solid line : Maxwell-Boltzmann (MB) fit of the SiCl spectrum with a temperature of 2000 K. Broken line : fit of the SiCl spectrum corresponding to the tail of a MB distribution ($E \geqslant E_d$) at a temperature of 2800 K with $E_d = 1.5$ eV.

temperature with the surface temperature seems difficult to explain). The Maxwell distribution has the form [10]

$$f(v) \propto v^3 \exp\left(-\frac{mv^2}{2kT}\right)$$

(where v, m, k, T have their usual classical meanings). We make use of a truncated Maxwell distribution with $E \geqslant E_d$ to describe the distribution resulting from "sudden" desorption after initial equilibration [12] :

$$f(v) \propto v^3 \left(1 + \frac{2E_d}{mv^2}\right) \exp\left(- \frac{mv^2}{2kT}\right) .$$

The fit of the experimental data to the truncated Maxwell distribution is reasonably good (figure 3). E_d was taken to be 1.5 eV (see (a) above) and the fitted temperature is 2800 K, which is again much higher than the estimated surface temperature. The most significant feature is not the actual law that was used for the fit, but rather the fact that a broader and more excited population distribution than the thermal distribution at the surface temperature is created in the ground state PES of the surface-adsorbate system. The mechanism for building such a population distribution is yet to be understood. It may be the result of repeated electronic excitation induced by photon absorption and followed by fast de-excitation due to the coupling with electrons of the bulk. At 308 nm, with a laser pulse duration of 20 ns and at a laser fluence of 500 mJ cm^{-2}, one Si atom of the bulk absorbs approximately 1 photon every nanosecond; the absorption by Si atoms of the surface should not be much different. The lifetime of excited PES's is unknown; it is most probably shorter than 1 ps; nevertheless, the surface-adsorbate system spends a fraction of the laser pulse duration evolving on the excited PES, and this will cause the average population distribution in the ground state PES to change. The actual shape of the population distribution will depend on many parameters, namely the PES's shapes, the lifetime of the excited PES, and the laser fluence.

5. Conclusion

The mechanism of laser etching of Si by Cl_2 is discussed in the two cases of "photoreaction" and "photodesorption". In the first case, the reaction of laser generated chlorine radicals at the Si surface involves excited electrons in the conduction band that harpoon the incoming Cl atoms. The photoreaction seems to proceed by migration of Cl^- species on the surface and evaporation of stable $SiCl_4$.

In the second case (photodesorption of chemisorbed chlorine), the experimental results obtained so far are inconclusive with respect to the reaction mechanism. While a number of data are suggestive of non-thermal, purely electronic effects, the thermal activation of a number of processes during the short time of the laser pulse cannot be avoided and this blurs the signature of electronic excitation. It is suggested that 308 nm photons cannot produce a direct DIET, and that the exceedingly large translational temperatures measured in the reaction products by TOF techniques may be accounted for by photon induced electronic excitation that allows the surface-adsorbate system to evolve during a fraction of the laser pulse on an excited PES, resulting in the broadening of the population distribution in the ground state PES. More experiments are needed, such as the laser wavelength dependence of the etch rate, and the improvement of the sensitivity allowing for measurements of angular and velocity distributions at low induced surface temperatures.

References

[1] R. Kullmer and D. Bäuerle, Appl. Phys. A **43**, 227 (1987).
[2] P. Mogyorosi, K. Piglmayer, R. Kullmer and D. Bäuerle, Appl. Phys. A **45**, 293 (1988).
[3] R. Kullmer and D. Bäuerle, Appl. Phys. A **47**, 377 (1988).
[4] Y. Horiike, N. Hayasaka, M. Sekine, T. Arikado, M. Nakase and H. Okano, Appl. Phys. A **44**, 313 (1987).
[5] J.V. Florio and W.D. Robertson, Surf. Sci **18**, 398 (1969).
[6] W. Sesselmann, E. Hudeczek and F. Bachmann, J. Vac. Sci. Technol. B, accepted.
[7] T. Baller, D. J. Oostra and A. E. de Vries, J. Appl. Phys. **60**, 2321 (1986).
[8] T. S. Baller, J. van Zwol, S. T. de Zwart, G. N. A. van Veen, H. Feil and J. Dieleman, MRS Symposium B, Boston (1988).
[9] T.S. Baller, and J. Dieleman, Appl. Surf. Sci, in press.
[10] G. Comsa and R. David, Surf. Sci. Rep. **5**, 145 (1985).
[11] J. Boulmer, B. Bourguignon, J.P. Budin and D. Débarre, Appl. Surf. Sci., in press.
[12] J. Boulmer, B. Bourguignon, J.P. Budin and D. Débarre, Chemtronics, in press.
[13] S. de Unamuno and E. Fogarassy, Appl. Surf. Sci. **36**, 1 (1989).
[14] J.T. Lin and T.F. George, J. Appl. Phys. **54**, 382 (1983).
[15] J.M. Hicks, L.E. Urbach, E.W. Plummer and H.L. Dai, Phys. Rev. Lett., **61**, 2588 (1988).
[16] R. B. Jackman, H. Ebert and J. S. Foord, Surf. Sci. **176**, 183 (1986).

Intensity Dependence of Laser-Induced Emission of Particles from $CaF_2(111)$ near the Plasma Threshold

H. Cronberg [1], *W. Muydermann* [1], *H.B. Nielsen* [1], *E. Matthias* [1], *and N.H. Tolk* [2]

[1]Fachbereich Physik, Freie Universität Berlin, D-1000 Berlin 33, Germany
[2]Deptartment of Physics and Astronomy, Vanderbilt University, Nashville, TN 37235, USA

Abstract. Laser stimulated emission of electrons, ions and neutrals from $CaF_2(111)$ surfaces in UHV was studied around the breakdown threshold by time-of-flight techniques in UHV for 532 nm and 355 nm. The experiments were carried out with crystals cleaved in vacuum and heated to 300 ^{0}C. It is shown that the laser light is absorbed by the crystal via inverse bremsstrahlung/avalanche ionization. The intensity dependence of the particle and photon emission reflects the gradual transition from desorption of individual particles to plasma formation.

1. Introduction

A review of laser induced desorption (LID) from large band gap materials is given elsewhere in these proceedings [1]. In this area, the main issues are: 1. How is laser light absorbed by a crystal when the photon energy is much smaller than the band gap? 2. What is the mechanism by which the absorbed energy is converted into kinetic energy of desorbing particles? At higher intensities near and above the breakdown threshold a third question arises, namely how does a plasma form on the surface of dielectric materials? This last topic is of great importance in connection with laser damage of optical materials, and in general LID is considered to be the precursor of optical damage.

We have started a program to investigate these issues by systematically studying the yield and the kinetic energy of the emitted electrons, ions, and neutrals as a function of laser intensity and wavelength for nanosecond pulse durations. In this contribution we show that a CaF_2 crystal can absorb energy from nano-second laser pulses in the visible and near UV via inverse bremsstrahlung/avalanche ionization. We also show that, with increasing intensity, there is a smooth transition from individual particle desorption to massive emission. The plasma onset was monitored by the emitted UV light and fast electrons directly irradiating the channel plate (MCP in Fig. 1).

2. Experimental

All measurements were carried out in a UHV-chamber with a base pressure of 4×10^{-10} mbar. The CaF_2 crystals were cleaved in UHV and heated to several hundred degrees centigrade with a resistive heater. Temperatures were measured by a thermocouple clamped to the side of the crystal, and the real surface temperature is expected to be somewhat higher than the measured one.

A mode-locked Nd-YAG laser system equipped with facilities for generating higher harmonics radiation provided about 8 ns pulses at

Springer Series in Surface Sciences, Vol. 19
Desorption Induced by Electronic Transitions, DIET IV
Editors: G. Betz · P. Varga

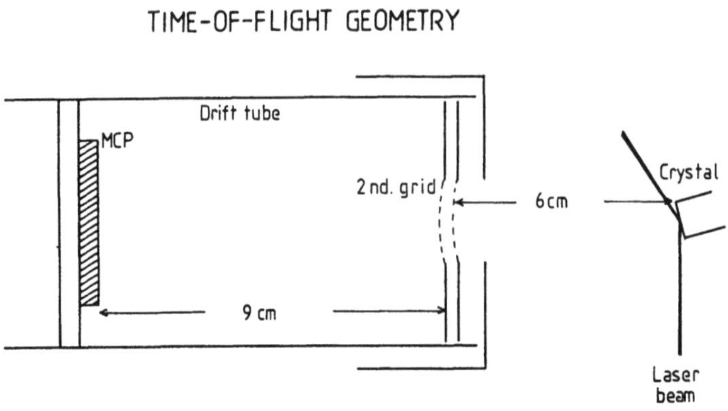

Figure 1. Schematic of the TOF spectrometer. The copper drift tube is electrically connected to the entrance of the microchannel plate and can be biased to ± 500 V. The second grid can be separately biased to ± 500 V. The first grid, the shield surrounding the TOF tube and the crystal are connected to ground.

532 nm and 6.5 ns pulses at 355 nm. The laser beam diameter at the crystal surface was approximately 0.3 mm. At 75^0 angle of incidence, used in the experiments, this corresponds to a spot size of about 0.2 mm^2. All data presented here were recorded with single pulses.

The emitted charged or neutral particles were detected by a microchannel plate placed in a 9 cm long drift tube (see Fig.1). The total distance from crystal to detector was 15 cm. Finely meshed grids placed at the entrance of the drift tube could be biased to several hundred volts in order to discriminate between negatively and positively charged particles and neutrals. These grids also allowed us to estimate the energy of the charged particles.

Reproducible surface conditions were prepared by heating the crystals to 300 ^{0}C in UHV for several days and subsequently cleaving them in situ. All experiments reported here were done at 300 ^{0}C.

3. Results and Discussion

Fig.2 shows the electron and ion yields as a function of laser intensity I for 532 nm and 355 nm in a double-logarthmic plot. For any multiphoton initiated process the yield would be $Y=I^n$, and therefore result in a straight line in this representation with a slope n indicating the order of the multiphoton absorption [2]. Obviously, the electron yield does not follow a straight line. It shows a threshold, and above it a rapidly increasing non-linear dependence on intensity corresponding to $n > 10$. For multiphoton ionization of 2p valence electrons from CaF_2 5-6 photons would have been required for 532 nm radiation, while for 355 nm 3-4 photons would have sufficed [3]. We therefore believe that the absorption of laser light in this intensity range is due to a process involving inverse bremsstrahlung and avalanche ionization [4]. The fact that the threshold for electron emission is significantly lower for 532 nm compared to 355 nm supports this view at least qualitatively since the threshold is expected to vary approximately as 1+const.ω^2, where ω is the frequency of the laser light [5].

For avalanche ionization to occur, it is necessary that the laser beam can excite "free" electrons in the conduction band to kinetic

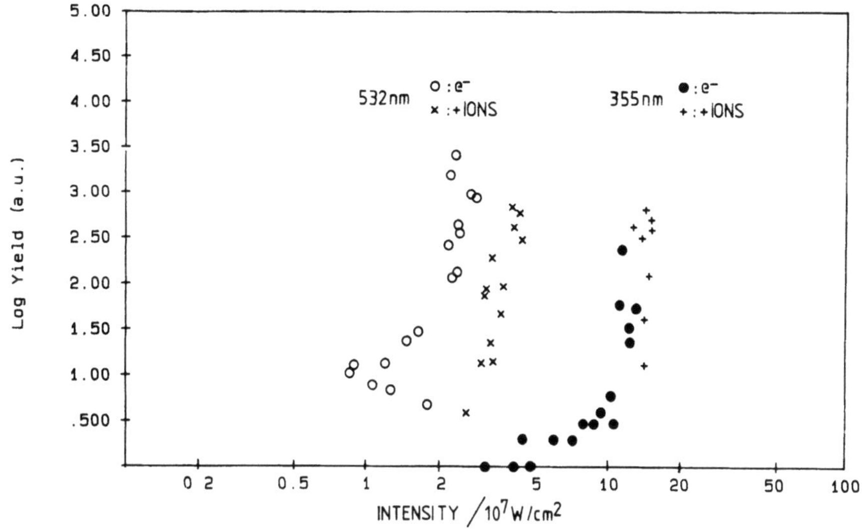

Figure 2. Electron and ion yields as a function of intensity for 532 nm and 355 nm radiation.

energies greater than the band gap, which for CaF_2 is approximately 12 eV [6]. We assume at this point that under intense laser irradiation there are always a few electrons, originating mostly from defect states, populating the conduction band. In the TOF spectra presented in Fig. 3 we see that electrons with such high kinetic energies indeed are observed at the threshold for electron emission. As the intensity is increased the weight of the energy distribution shifts very rapidly to higher energies. In Fig. 3 we also notice a signal due to UV photons, which at the threshold is about 10-20 % of the electron signal. Possible origins are recombination radiation and/or bremsstrahlung, but its exact nature needs further spectroscopic investigation. At this stage we take the UV peak as a marker for dielectric breakdown.

Returning to Fig. 2, it can be seen that positive ion emission starts at intensities 20-30 % higher than the threshold for avalanche ionization. This would be consistent with the model of Itoh [7] according to which a high density of electron-hole pairs leads to ion emission. Alternatively, and perhaps more likely, the kinetic energy of the electrons must be sufficiently high to achieve inner-shell excitation (for example of the F 2s level), the decay of which produces positive ions by the Knotek-Feibelman mechanism [8].

We have attempted to further elucidate the mechanisms leading to ion emission by trying to identify the masses and kinetic energies of the ions. The mass determination based on the TOF spectra was confirmed in a separate experiment using a quadrupole mass spectrometer. Figs. 4a and b show the TOF spectra of positive ions, recorded for increasing laser intensity with -150 V bias on the drift tube. Fig. 4a emphasizes more the plasma region at higher intensities, while Fig. 4b illustrates the onset of ion emission. For the lowest intensities the peak at 6-9 μsec flight time is due to CaF^+ ions with kinetic energies in the range 20-40 eV. At higher intensities the CaF^+ yield decreases and instead Ca^+ and F^+ ions are observed. Simultaneously the UV photon yield increases dramatically, which we take as an indication of a transition from

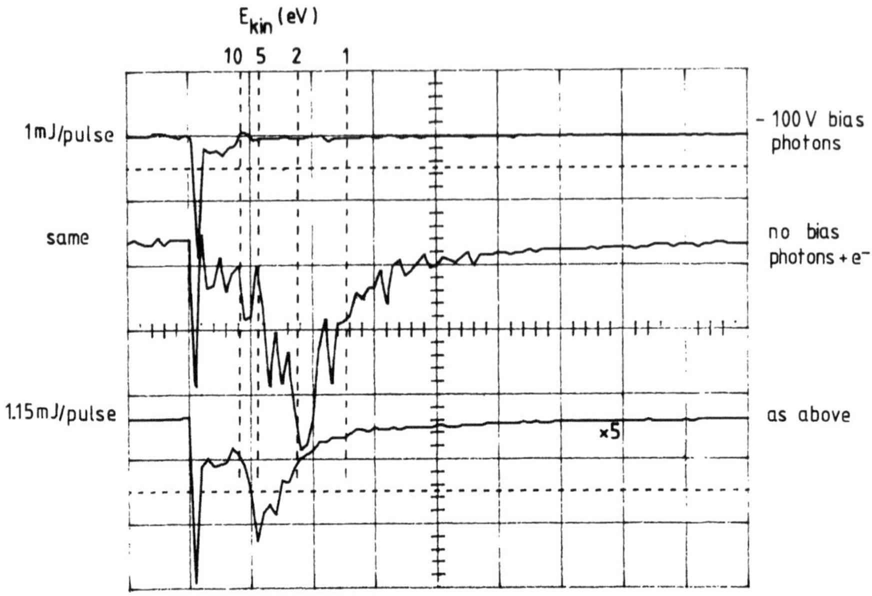

Figure 3. TOF spectra of electrons for 532 nm radiation. The upper trace is recorded with the 2nd grid biased to -100 V in order to suppress all but the photon signal. The lower two traces show both photons and electrons at different laser intensities.
(1 mJ ≃ 2.5×10^{7} W/cm^{2})

the situation where single ions are emitted to the one where a plasma is formed in front of the surface. The CaF^{+} ions are thus dissociated due to interactions in the plasma. A similar effect has been observed by Dreyfus [9], in the case of AlO molecules emitted from sapphire irradiated by 248 nm pulses.

The slower species arriving at flight times near 12 μs for lower intensities and near 6 μs for the highest intensities in Figs. 4a and b are as yet unidentified neutrals. As can be seen from Fig. 4d, there exists at all intensities a large contribution to the signal due to neutrals.

By reversing the bias on the drift tube we can observe negative ions as shown in Fig. 4c. The appearance of negative ions, which were identified as F^{-}, occurs at an intensity slightly higher than the onset of positive ion emission, but before the plasma becomes dense. These negative ions are surprisingly energetic (~ 100 eV) even at the onset.

After removing the bias between the drift tube and the 2nd grid we have used the flight times to determine the kinetic energy of the emitted particles. Fig. 5 shows the energy of F^{+} ions as a function of laser intensity. Energies of several hundred eV are reached. As the same peak in the TOF spectra also contains Ca^{+} ions, these must have energies close to 1 keV at the highest intensities used here.

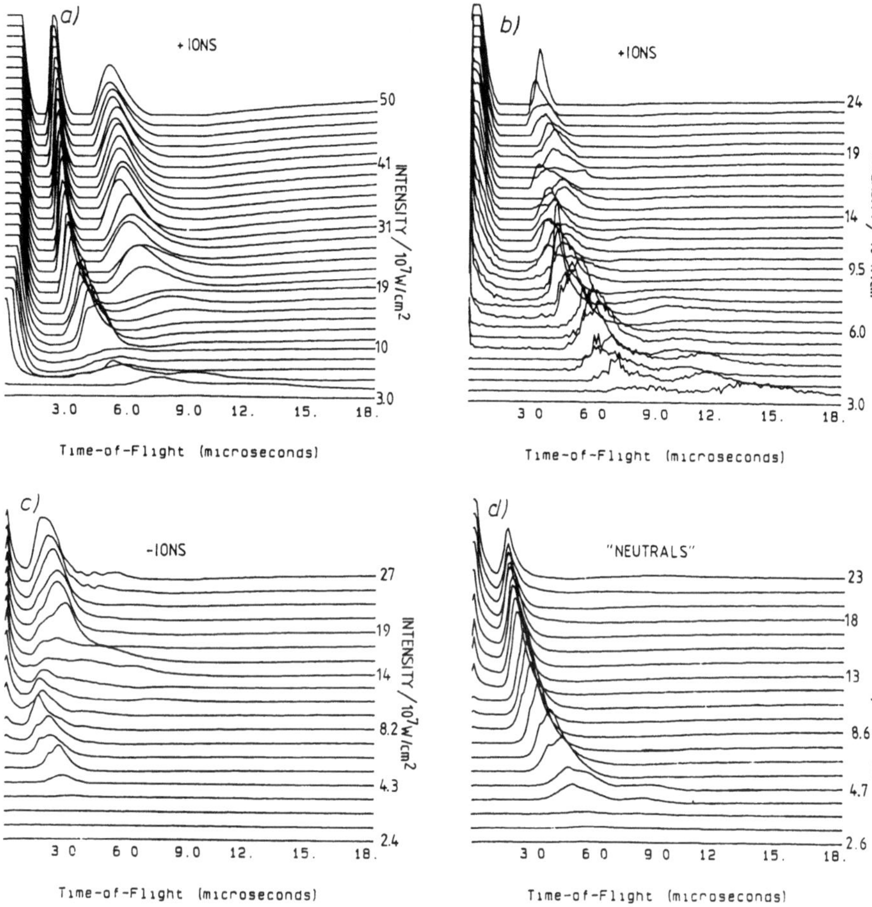

Figure 4. a and b: TOF spectra of positive ions for increasing laser intensity, measured with drift tube bias −150 V and 2nd grid bias −20 V.
c: TOF spectra of negative ions for various intensities, drift tube bias +150 V and 2nd grid bias +20 V.
d: TOF spectra of neutrals recorded with drift tube bias +500 V and 2nd grid bias −500 V.

4. Summary and Conclusion

We have studied the laser induced emission of electrons, ions and neutrals from CaF_2(111) in UHV near the breakdown threshold for the two wavelengths 532 nm and 355 nm. The intensity dependence of the electron yield shows that the initial absorption of the laser light proceeds via inverse bremsstrahlung and avalanche ionization. For positive ions the emission of CaF^+ ions sets in at intensities 20-30 % above the avalanche ionization threshold. At still higher intensities the density of emitted particles reaches the level where a plasma is formed, breaking up CaF^+ into Ca and F atoms and ions. The plasma manifests itself by intense emission of UV radiation.

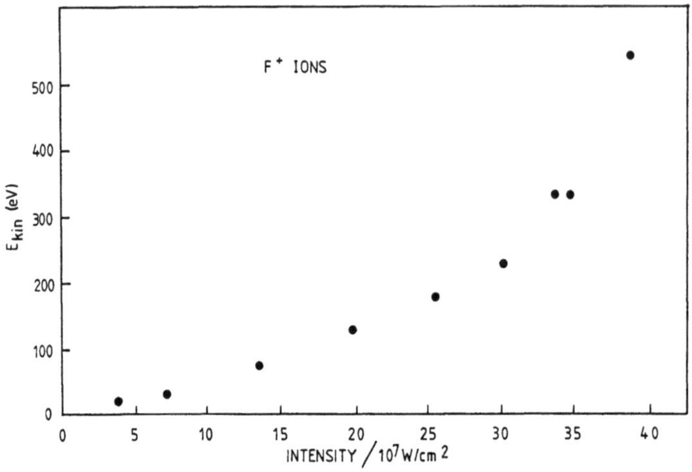

Figure 5. Kinetic energy of F^+ ions as a function of laser intensity as derived from the TOF spectra in Fig. 4a.

In order to understand the ion emission mechanism in more detail and gain more information about the development of the plasma, we intend next to accurately measure the electron energies at the onset of ion emission and to spectroscopically analyse the UV light.

5. Acknowledgement

This work was supported by the Deutsche Forschungsgemeinschaft, Sonderforschungsbereich 6. One of us (NHT) acknowledges the support of the Alexander von Humboldt Foundation.

6. References

1. E. Matthias and T. Green in these proceedings.
2. Vaidya Nathan, A.H. Guenther, and S.S. Mitra; J. Opt. Soc. Am. B2, 294 (1988).
3. R.T. Poole, J. Szajman, R.C.G. Leckey, J.G. Jenkin, and J. Liesegang; Phys. Rev. B12, 5872 (1975).
4. see e.g., A.S. Epifanov; IEEE J. Quant. Electr. QE 17, 2018 (1981); and
   M. Sparks, D.L. Mills, R. Warren, A.A. Maradudin, L.J. Sham, E. Loh, Jr., and D.F. King; Phys. Rev. B24, 3519 (1981).
5. N Bloembergen; IEEE J. Quantum Electron. QE-10, 375 (1974)
6. G.W. Rubloff; Phys. Rev. B5, 662 (1972).
7. N. Itoh, and T. Nakayama, Phys. Lett. 92A, 471 (1982).
8. M.L. Knotek; Rep. Progr. Phys. 47, 1499 (1984).
9. R.W. Dreyfus, R.E. Walkup, R. Kelly; Radiation Eff. 99, 683 (1986).

Laser-Induced Ion Emission from Si and Ge Surfaces

Z. Wu

Department of Physics, Rutgers University, Newark, NJ 07102, USA

Abstract. We discuss a possible mechanism for the recent observations of ion emission from Si and Ge surfaces under low flux UV laser irradiation. The mechanism is based on the Phillips' bond charge model and hole stimulated desorption.

Ion emission from solid surfaces under electron or photon bombardment has considerable current interest [1]. The majority of these processes can be understood using the Menzel-Gomer-Redhead model [2] or the Knotek-Feibleman mechanism [3]. The MGR model is a very general one, since all desorption results from antibonding. For each individual case one's task is to fill in the details of this bonding to antibonding transition. Thus in the KF mechanism the bonding to antibonding transition is effected by interatomic Auger decay.

Recently, Moison and Bensoussan [4] have reported the emission of Si^+ and Si when a Si sample was irradiated by 3.6 eV photons from an excimer-laser-pumped dye laser. There is a photon flux threshold of $160\,mJ/cm^2$ for Si^+ emission. Chen *et al.* [5] reported similar flux threshold behavior for higher photon energies: around $10\,mJ/cm^2$ for 6.4 eV photons and $40\,mJ/cm^2$ for 5.0 eV photons. The flux threshold for Ge is higher than that for Si (by a factor of 2 or 3) under similar surface conditions. Ion emission current is found to be larger for rough surfaces.

Itoh and Nakayama [6] proposed a two-hole localization mechanism to explain the threshold behavior for neutral particle emission from non-metal surfaces. In this paper we discuss a different mechanism [7] for ion emission from Si and Ge surfaces. Our basic assumption is that Si or Ge ions are emitted from the surface defect sites. Here surface defect sites are identified only as sites where the electronic excitation can be trapped for as long as $10^{-13} - 10^{-12}$ seconds, the time scale for nuclear motion. The exact nature of the defect sites is not otherwise known. The experimental observation that the Si^+ yield is enhanced for rough surfaces supports our assumption. The proposed mechanism for ion emission is as follows. The transition from bonding to antibonding is assumed to be effected by an energetic hole. Energetic holes can be produced in

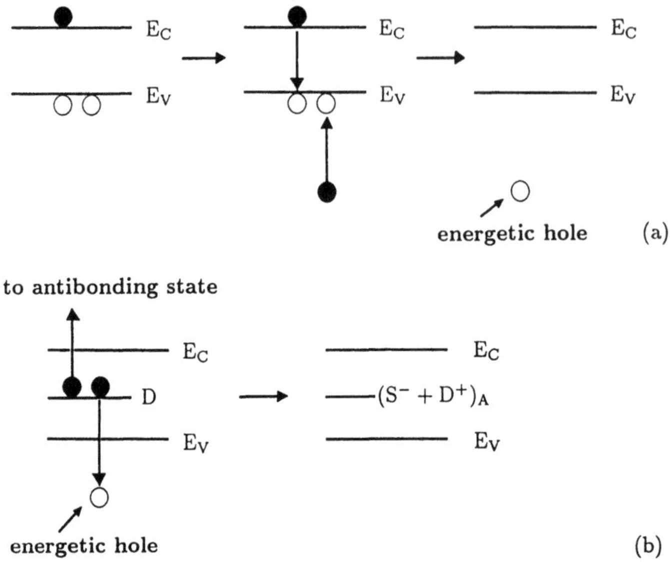

Fig. 1. (a) The e-h-h Auger process. When an electron in the conduction band recombines with a hole in the valence band, the energy released can be used to create a deep hole (an energetic hole) in the valence band. (b) Hole stimulated desorption. When an energetic hole happens to be near a surface defect site D, an electron from the Si atom at this defect site can recombine with this hole, converting the Si atom into Si^+ and, at the same time, releasing enough energy to excite another electron of this Si^+ into an antibonding state. This Si^+ can then desorb. Here S represents the substrate and $(S^- + D^+)_A$ the antibonding state of the Si^+.

several ways: surface enhanced e-h-h Auger process (Fig.1a) for example, or excitation of electrons from deep in the valence band to the conduction band by incident photons. Now, if an energetic hole happens to be near a defect site (Fig.1b), then an electron from the Si atom at this defect site can combine with the hole, converting the Si atom into Si^+ and, at the same time, releasing enough energy to excite another electron of this Si^+ to an antibonding state. The Si^+ can then desorb. Also the crossing of the Si^+ antibonding curve and the Si antibonding curve can lead to the desorption of Si neutrals. Now, the energy of the energetic holes depends upon the energy of the incident photons. If the photon energy is not high enough, then the holes may not be energetic enough to induce the bonding to antibonding transitions. It is here that the flux threshold comes into the picture. According to the Phillips' bond

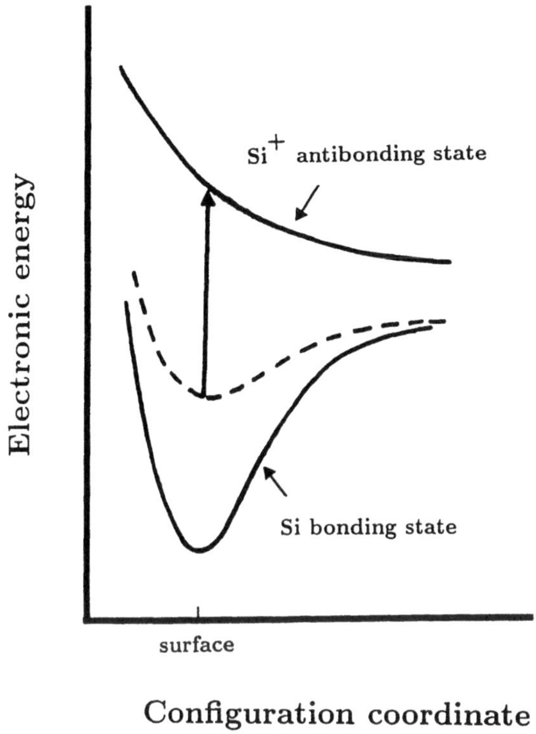

Fig. 2. Excitation to a Si$^+$ antibonding state from a weakened Si bonding potential well (as indicated by the dashed curve) is effected by an energetic hole. The reduction in the potential well is caused by the electron-hole plasma.

charge model [8], the electron-hole plasma produced by the laser beam near the surface weakens the binding of the surface atoms. When the bond weakening is sufficient for the energetic holes to induce bonding to antibonding transitions, Si ion desorption occurs. The laser flux threshold is interpreted as corresponding to the minimum e-h plasma density at the surface needed to weaken the potential well so that the excitation to the antibonding state (as indicated by the arrow in Fig.2) becomes feasible.

We would like to determine whether we can use the above mechanism to understand the observation that Ge has a higher flux threshold than Si. The flux threshold is affected by the following three factors: (1) surface reflectivity; (2) efficiency of bond weakening through the creation of an e-h pair and (3) the diffusion of the e-h plasma into the bulk, which will affect the e-h density at the surface. For unpolarized incident photons of 6.4 eV, the reflectivities for Ge and Si surfaces are $R_{Ge} = 52\%$

and $R_{Si} = 63\%$, respectively [9]. To estimate the efficiency of the bond weakening by an e-h plasma we use the Phillips' bond charge model. According to this model, the magnitude of each bond charge, which is responsible for the covalent bonding, is $-2e/\epsilon$, where ϵ is the dielectric constant. Now, the bond charge in the E(k) picture is located at the top of the valence band. Therefore, the creation of an e-h pair removes a bond charge $-\alpha e$ and, therefore, destroys $\alpha\epsilon/2$ bonds. Here α depends on the antibonding nature of the bottom of the conduction band and a rough estimate [7] yields $\alpha_{Ge} \approx 0.67$ and $\alpha_{Si} \approx 0.96$. Thus the ratio of the number of e-h pairs needed to destroy one bond in Ge to that in Si is $(\alpha_{Si}\epsilon_{Si})/(\alpha_{Ge}\epsilon_{Ge}) \approx 1.1$. Since the ambipolar diffusion coefficient D for an e-h plasma in Ge is about four times as large as that in Si [7] and the e-h plasma density n at the surface is inversely proportional to $\sqrt{D}$, we have, for the same laser flux, $n_{Ge}/n_{Si} \approx 0.5$. Finally, taking all three factors together, in order to produce the same amount of bond weakening, the ratio of laser flux needed in Ge to that in Si is given by

$$\frac{1 - R_{Si}}{1 - R_{Ge}} \frac{\alpha_{Si}}{\alpha_{Ge}} \frac{\epsilon_{Si}}{\epsilon_{Ge}} \frac{n_{Si}}{n_{Ge}} \approx 1.7 \ .$$

Surface reconstruction and surface buckling (as in the Si(111) 7×7) seems to be a more severe process on Si than on Ge surfaces. This means that although most of the Si surface atoms are more tightly bound, the ionized ones are more weakly bound. This would increase our estimated ratio and bring it closer to the experimentally observed value.

We would like to make three comments related to the above model. First, the time scale for these energetic carriers to transfer energy to the lattice via electron-phonon scattering is of the order of 1 ps [10]. However, since the role of the energetic holes in the model is to induce bonding to antibonding electronic transitions, the energetic holes near the defect sites are required to exist only for a time period of the order of 10^{-15} sec (the time scale for electronic transitions). Thus, despite the extremely short lifetime of the energetic holes, one would still expect a non-negligible fraction of these holes to contribute to the ion emission process.

Second, the exact nature of the antibonding potential assumed in this model is not known. There are no theoretical calculations available for these antibonding ion potentials on Si or Ge surfaces. The recent calculations of F on Al by Avouris *et al* [11] indicate that the ionization of the 2*s* level of F^- can initiate a Franck-Condon transition to the repulsive part of the F^+ potential curve, leading to the desorption of F^+. It is interesting to note that this process bears some resemblance to the hole-stimulated desorption postulated in our model, since the F^+ desorption can be regarded as a hole-induced Franck-Condon transition to a repulsive state, resulting in the desorption of F^+:

$$F^-(2s^2 2p^6) + \text{hole} \rightarrow F^*(2s2p^6) \xrightarrow{Coster-Kronig} F^+(2s^2 2p^4) + e^-$$

where $F^+(2s^2 2p^4)$ is in the repulsive part and therefore desorbs. Here the energy resulting from filling the $2s$ hole is so large that the intermediate state F^* is ionized in addition to being brought to an antibonding state.

Third, the only state in Fig.2 that is required to exist for as long as $10^{-13} - 10^{-12}$ sec is the Si^+ antibonding state. The lifetime of the weakened state, represented by the dashed curve, can be as short as $10^{-16} - 10^{-15}$ sec because in this state no nuclear motion is required. This point is important in terms of what trapping properties are required of the defect sites.

ACKNOWLEDGEMENT. I am grateful to Dr. J. C. Phillips for his comments. This work was supported by JESP.

References

[1] Recent work has been reviewed in *Desorption Induced by Electronic Transitions, DIET I*, edited by N. H. Tolk, M. M. Traum, J. C. Tully and T. E. Madey (Springer-Verlag, Berlin, 1983); *Desorption Induced by Electronic Transitions, DIET II*, edited by W. Brenig and D. Menzel (Springer-Verlag, Berlin, 1985); *Desorption Induced by Electronic Transitions, DIET III*, edited by R. H. Stulen and M. L. Knotek (Springer-Verlag, Berlin, 1988).

[2] D. Menzel and R. Gomer, J. Chem. Phys. 41, 3311 (1965); P. A. Redhead, Can. J. Phys. 42, 886 (1964).

[3] H. L. Knotek and P. Feibelman, Phys. Rev. Lett. 40, (1978).

[4] J. M. Moison and M. Bensoussan, J. Vac. Sci. Technol. 21, 315 (1982); J. M. Moison and M. Bensoussan, in: *Surface Studies with Lasers*, eds. F. R. Aussenegg, A. Leitner and M. E. Lippitsch (Springer-Verlag, 1983) p.202.

[5] L. Chen, V. Liberman, J. A. O'Neill, Z. Wu and R. Osgood, J. Vac. Sci. Technol. A6, 1426 (1988).

[6] N. Itoh and T. Nakayama, Phys. Lett. 92A, 471 (1982).

[7] Z. Wu, Phys. Lett. 131, 486 (1988).

[8] J. C. Phillips, Phys. Rev. 166, 832 (1968).

[9] H. R. Philipp and E. A. Taft, Phys. Rev. 113, 1002 (1959); 120, 37 (1960).

[10] N. Bloembergen, H. Kurz, J. M. Liu, and R. Yen, Mat. Res. Soc. Symp. Proc. 4, 3 (1982); D. H. Lowndes and G. E. Jellison in *Semiconductors and Semimetals*, Vol. 23, p.313, Academic Press 1984.

[11] Ph. Avouris, R. Kawai, N.D. Lang and D. M. Newns, Phys. Rev. Lett. 59, 2215 (1987).

Adsorption and UV-Laser Desorption of NO/O/Ni(100)

Th. Mull[1], *H. Kuhlenbeck*, *G. Odörfer*[1], *R. Jaeger*[1], *C. Xu*[1], *B. Baumeister*[1], *M. Menges*[1], *G. Illing*[1], *H.-J. Freund*[1], *D. Weide*[2], *and P. Andresen*[2]

[1]Physikalische Chemie 1, Ruhr-Universität Bochum, D-4630 Bochum, Fed. Rep. of Germany

[2]Max-Planck-Institut für Strömungsforschung, D-3400 Göttingen, Fed. Rep. of Germany

1. Abstract

We have studied adsorption as well as thermal and UV-laser (193 nm) desorption of NO adsorbed on Ni(100), on O(c2x2)/Ni(100) and on epitaxially grown oxides NiO(111) and NiO(100) using HREELS, XPS, NEXAFS, LEED and TPD. We find NO on NiO to be weakly chemisorbed. Laser desorption is only induced from the oxidic surfaces. Laser desorption cross sections are two orders of magnitude higher than gas phase photoabsorption cross sections for NO.

NO_2/NiO(100) has been studied via XPS, LEED and TPD. NO is probably the predominant desorption product under UV laser impact /1/. Reaction also takes place under XPS influence and under impact of secondary electrons from the X-ray gun.

Possible desorption processes are discussed.

2. Introduction

Recently, the photochemistry of small molecules on metal oxide surfaces has received some attention /2-5/. Briefly, a study of the dynamics of the desorbing molecules reveals non-thermal features such as bimodal velocity distributions and spin dependent rotational state populations, i.e. an underpopulation of the $^2\Pi_{3/2}$ state by a factor of 40 as compared to the $^2\Pi_{1/2}$ state for low rotational quanta.

Here we present new data concerning the NO desorption yields from different systems and discuss possible desorption mechanisms in light of a better chemically and structurally characterized oxide surface.

3. Experimental

A standard UHV System was used containing facilities for LEED, AES, XPS and TPD. The sample could be cooled down to 90 K by a liquid nitrogen reservoir and heated radiatively up to sufficiently high temperatures. The Ni-crystal was cleaned by argon ion bombardment and repeated redox cycles as described in the literature (e.g./6/). NiO(111) was prepared by exposure of Ni(100) to 500 L O_2 at room temperature (/7/). Repeated cycles of oxygen dosing (1000 L, $p=10^{-5}$mbar) at T=570 K and subsequent annealing to T=670K lead to pure NiO(100).

4. Results

4.1. NO adsorbed on clean and oxidized Ni(100)

We have investigated NO desorption from clean Ni(100), c(2x2)O/Ni(100) and epitaxially grown nickel oxides NiO(111) and NiO(100). We find measurable

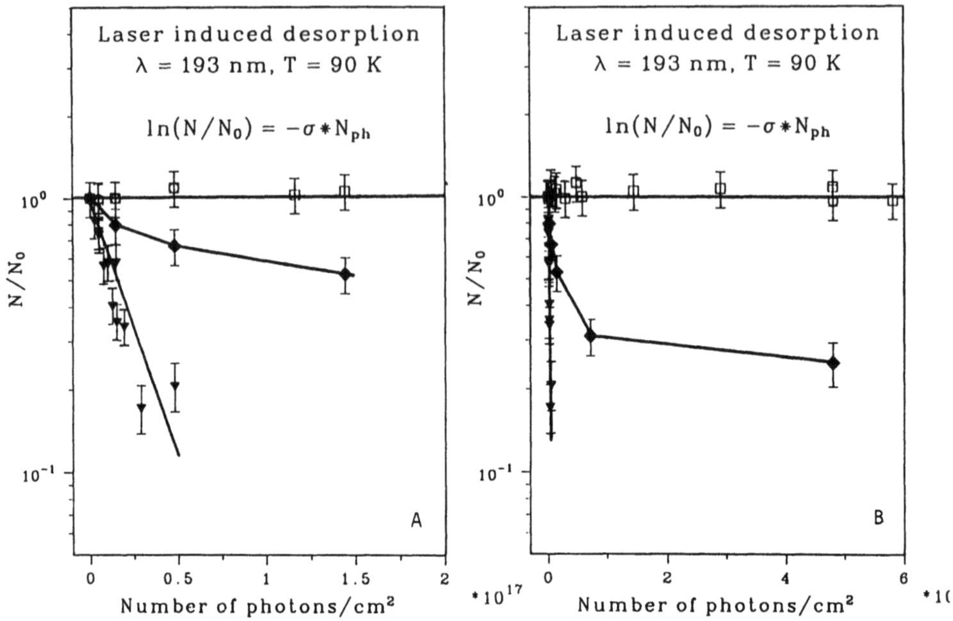

Fig. 1, a) and b): Laser desorption of NO from the different surfaces

▼ NO/NiO(100)/Ni(100), $\sigma \approx 7x10^{-17}$ cm^2
◆ NO/(NiO(111)/Ni(100),$\sigma \approx 1.4x10^{-17}$ cm^2
○ NO/c(2x2)O/Ni(100)
□ NO/Ni(100)

desorption of NO (as judged by N1s intensities) under UV laser irradiation only for the nickel oxides. See fig. 1. The highest desorption cross section is reached for NO/NiO(100) with $7x10^{-17}$ cm^2 /4/. This finding is compatible with EELS data and XPS N1s structures and energies /4/ showing that NO/NiO(100) is a weakly bound system.

It is known that the NO desorption from NiO(100) starts from only one adsorption site and that in this case the NO axis is probably tilted to the surface normal by about 45°. The desorption cross section is more than two orders of magnitude above the average NO gas phase photoabsorption cross section in the wavelenght region around 193 nm of $1.1x10^{-19}$ cm^2, /8/.

It was found that NO desorption takes place also under the influence of soft X-rays of the X-ray gun (Al K_α).

4.2. NO_2/NiO(100)

NO_2 is of interest because in NO_2 adsorbate systems NO is found to be the main desorption result under 193 nm laser irradiation (/1/), as well as under electron impact (/9/).

NO_2 adsorbs on NiO(100). TPD- and XPS-spectra show that there are several species on the surface, i.e. N_2O_4, growing in intensity with higher exposure and NO_2 already present at low coverage. Our binding energies are in excellent agreement with those reported by Fuggle and Menzel for the same molecular species on W(110) (/10/).

UV laser desorption with the NO_2 species - as judged by XPS N1s (see fig.2) and O1s intensities - takes place with cross sections from $3x10^{-19}$ cm^2

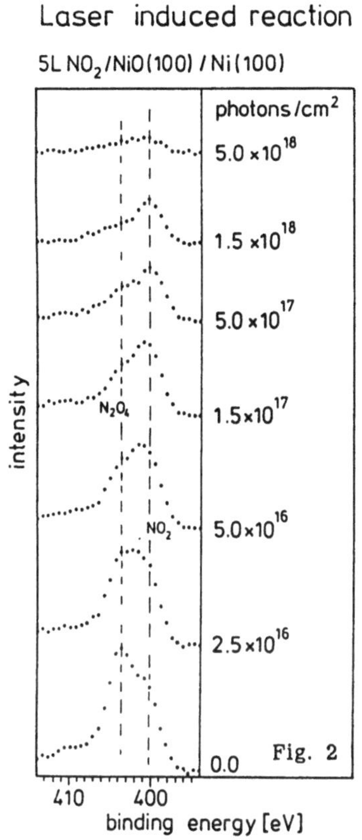

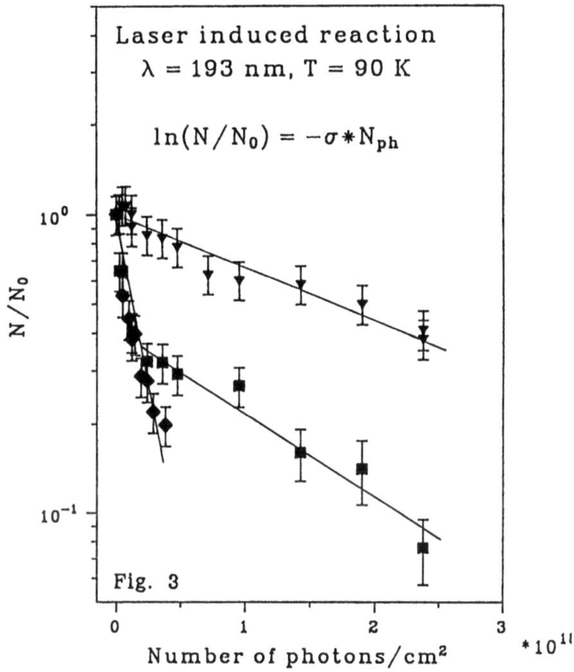

Fig. 2): XPS spectra of 5 L NO_2/NiO(100) under laser impact

Fig. 3): Decrease of XPS intensities
▼ NO_2/NiO(100), N1s, $\sigma \approx 3x10^{-19}$ cm^2
■ N_2O_4/NiO(100), N1s, $\sigma \approx 4x10^{-18}$ cm^2 ... $7x10^{-19}$ cm^2
◆ N_2O_4/NiO(100), O1s, $\sigma \approx 4x10^{-18}$ cm^2

for the monomer species up to $4x10^{-18}$ cm^2 for the dimer species (see fig. 3). The monomer value has to be compared with the NO_2 gas phase photoabsorption cross section ($6x10^{-19}$ cm^2, as derived from /11/).

The system NO_2/NiO(100) is unstable under XPS influence, see fig. 4. This destruction may partially occur via the same - unknown - reaction channels as the laser desorption at a much lower photon energy. As the photon flux of the X-ray gun is not yet calibrated, the cross sections cannot be compared directly. Reaction of the dimer species is also caused by secondary electrons emitted from the Al-foil of the X-ray window. This can easily be seen by the influence of a bias of −30 V that was applied to the sample.

However we want to state that electron- and XPS influences are irrelevant for the cross sections shown in fig. 3, because every point is subject to the same alteration and the intensities are related to each other.

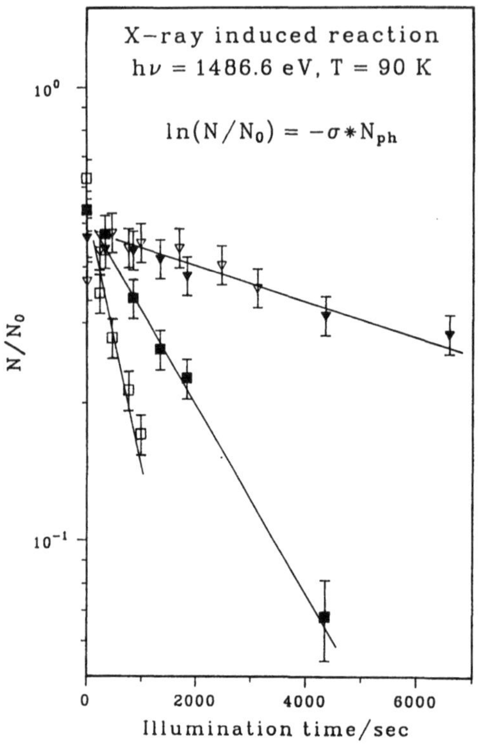

Fig. 4): Influences of X-rays and electrons on the NO_2 adsorbate
▽ NO_2/NiO(100)/Ni(100), unbiased
▼ NO_2/NiO(100)/Ni(100), biased
□ N_2O_4/NiO(100)/Ni(100), unbiased
■ N_2O_4/NiO(100)/Ni(100), biased

5. Discussion

Two observations indicate a non-thermal cause for the desorption of NO molecules from NO/NiO (/2,3/) as well as from NO_2 adsorbates on metal surfaces (/1/):

One of them is the observation of bimodal velocity distributions. Both the low and high velocity-channels seem to be of non-thermal origin for both are not well described by Maxwell velocity distributions, although this is particularly true for the fast channel.

The second point is a large difference in the populations of different spin states at low rotational quanta /3/. Such an underpopulation can be understood as a consequence of the desorption mechanism in different ways:

The desorption can be a direct process: According to quantum mechanical calculations of Corey et al. (/12/) a desorption process induced by a neutral-neutral transition could be described in terms of the second half of a scattering process. The populations of the final rotational states are dependent on the orientation of the unfilled π orbital of the NO molecule in the initial configuration. It can lie either in the plane spanned by the molecular axis and the surface normal (NO being tilted by approximately 45°) or perpendicular to this plane and is responsible for different potentials felt by the desorbing NO molecule. The calculated final state distributions for different potentials already show underpopulation of the $^2\Pi_{3/2}$ state. But the strength of the experimentally observed effect is not described by such calculations quantitatively.

On the other hand the desorption could be an indirect process as well. For example, the photon creates an electron-hole pair or a free electron and the electron is transferred to the molecule. Consequently a negative ion is formed which may lead to desorption. Some evidence is found by Hasselbrink et al. (/1/) who report the desorption yield as a function of laser energy, the light polarization and angle of incidence to be incompatible with direct excitation.

Thinking about the spin effect one may have to consider the antiferromagnetic properties of NiO (/13/). But in how far the transfer of a spin polarized electron may be involved in the desorption process and thus lead to the strong spin effect cannot be decided at present.

Acknowledgements

We thank the Deutsche Forschungsgemeinschaft and the Ministerium für Wissenschaft und Forschung des Landes Nordrhein-Westfalen for financial support.

References

/1/ E. Hasselbrink, S. Jakubith, S. Nettesheim, M. Wolf, A. Cassuto and G. Ertl, to be published

/2/ F. Budde, A. V. Hamza, P. M. Ferm, G. Ertl, D. Weide, P. Andresen and H.-J. Freund Phys.Rev.Lett. 60 (1988) 1518; see also: F. Budde, Thesis, Technische Universität Berlin, 1988

/3/ P. M. Ferm, F. Budde, A. V. Hamza, S. Jakubith, G. Ertl, D. Weide, P. Andresen and H.-J. Freund, Surf.Sci. 218 (1989) 467

/4/ H. Kuhlenbeck, G. Odörfer, R. Jaeger, C. Xu, T. Mull, B. Baumeister, G. Illing, M. Menges, H.-J. Freund, D. Weide, P. Andresen, G. Watson, E. W. Plummer, Vacuum, Proceedings of the 11th International Vacuum Congress and 7th International Congress on Solid Surfaces, Cologne, FRG (1989), to be published

/5/ Th. Mull, M. Menges, B. Baumeister, G. Odörfer, H. Geisler, H. Kuhlenbeck, H.-J. Freund, D. Weide, U. Schüller, P. Andresen, F.Budde, P. M. Ferm, A. V. Hamza and G.Ertl., Phys.Scripta, to be published

/6/ B. E. Koel, D. E. Peebles, J. M. White, Surf.Sci. 125 (1983) 709

/7/ G. Dalmai-Imelik, J. C. Bertolini and J. Rousseau, Surf.Sci 63 (1977) 67

/8/ S. F. Marmo, J.Opt.Soc.Am. 43 (1953) 1186

/9/ A. R. Burns, D. R. Jennison and E. B. Stechel, Phys.Rev.B, to be published

/10/ C. Fuggle and D. Menzel, Surf.Sci. 79 (1979) 1

/11/ T. Nakayama, M. Y. Kitamura, K. Watanabe, J.Chem.Phys. 30 (1959) 1180

/12/ G. C. Corey, J. E. Smedley, M. H. Alexander, W.-K. Lin, Surf.Sci. 191 (1987) 203

/13/ C. Kittel, Einführung in die Festkörperphysik, Oldenbourg Verlag, München - Wien, 6th ed. (1983) 515.

Part IV

Adsorbed Layers

Electron Stimulated Desorption Ion Angular Distributions (ESDIAD) of Negative Ions

T.É. Madey[1], *S.A. Joyce*[2,*], *A.L. Johnson*[2,**], *and J.A. Yarmoff*[2,***]

[1]Rutgers, The State University, Department of Physics and Laboratory for Surface Modification, P.O. Box 849, Piscataway, NJ 08855, USA
[2]National Institute of Standards and Technology, Gaithersburg, MD 20899, USA
* Present address: Sandia National Laboratory, Albuquerque, NM 87185, USA
** Present address: Department of Physical Chemistry, Cambridge University, Cambridge, UK
*** Present address: Department of Physics, University of California, Riverside, CA 92521 , USA

1. Introduction

Electron bombardment of molecules adsorbed on surfaces can lead to the desorption of positive ions as well as ground state and excited neutral species. In addition, in many systems desorption of negative ions has been observed [1]. The measurement of the angular distributions of the desorbing species (ESDIAD: electron stimulated desorption ion angular distributions) can yield important information about the bonding geometry of species adsorbed on surfaces because the initial direction of desorption has been found to be related to the bond vector directions [2].

Until recently, most ESDIAD measurements have focused on angular distributions of positive ions [2]. We have now developed methods for measuring ESDIAD of negative ions from surfaces, and have found strong anisotropies in their angular distributions [3,4]. Using time-of-flight methods and a video display system, we have measured ESDIAD of F^- for a series of fluorine-containing molecules (PF_3, NF_3, and $(CF_3)_2CO$) on Ru(0001); the influence of coadsorbed K on yields and angular distributions of negative ions from PF_3 has also been studied [5,6]. On the clean surface, negative ion yields appear to be higher from undissociated molecules than from dissociation fragments; positive ion yields are often higher from dissociation fragments. The variation of yields of positive and negative ions with coverage of the molecules and of coadsorbed K will be discussed, and comparisons will be made between angular distributions for positive ions, negative ions, and (in certain cases) metastable species. The results to date indicate that positive and negative ion ESDIAD are complementary.

Springer Series in Surface Sciences, Vol. 19
Desorption Induced by Electronic Transitions, DIET IV
Editors: G. Betz · P. Varga

2. Experimental Procedures

The primary difficulty associated with any electrostatic detection scheme for negative ions is the large background of electrons (primaries and secondaries) present (~10^6 larger than the ion signal). The large mass difference between electrons and ions, which results in large velocity differences for particles with comparable energies, suggests that time-of-flight (TOF) techniques can be used to separate them. We employ TOF methods in an apparatus [4] adapted from the existing digital display-type ESDIAD apparatus [2]. Typical electron bombardment energies are ~280 eV, with use of a pulsed electron beam (repetition rate ~10^4/s). Under typical conditions, the electron flight time is <30 ns and the flight time for an F^- ion is ~900 ns. This experimental apparatus can be used for mass-resolved positive-ion studies as well.

The procedures for measuring negative ion ESDIAD are described in detail in Ref. [4]. For these studies, we have chosen three molecules [PF_3, NF_3, and $(CF_3)_2CO$] known to yield F^- ions upon electron impact in the gas phase [7,8]. Adsorption studies are carried out on a Ru(0001) substrate, a surface on which each of the molecules has been studied previously [9-11].

3. Results

For all of the systems studied [PF_3, NF_3, and $(CF_3)_2CO$] both the positive- and negative-ion ESDIAD are dominated by fluorine ion emission. The observed yields for F^+ and F^- ESDIAD for the systems studies are summarized in Table 1 [3]. We observe that the total angle integrated yields from these fluorinated molecules are a significant fraction of the total (F^+ + F^-) ion current [from 3% for NF_3 to 40% for monolayer $(CF_3)_2CO$, to 80% for multilayers of $(CF_3)_2CO$].

TABLE 1. Ion yields (ions/electron) of saturation coverage adlayers on Ru(0001). Electron impact energy is 280 eV; $T_S \approx 110K$ Yields are angle integrated over the entire ESDIAD image [3].

	F^+ yields	F^- yields	$F^-/(F^-+F^+)$
PF_3	3×10^{-6}	2×10^{-7}	0.08
NF_3	4×10^{-7}	1×10^{-8}	0.03
$(CF_3)_2CO$	8×10^{-7}	6×10^{-7}	0.43
$(CF_3)_2CO$ (multilayer)	1×10^{-6}	6×10^{-6}	0.83

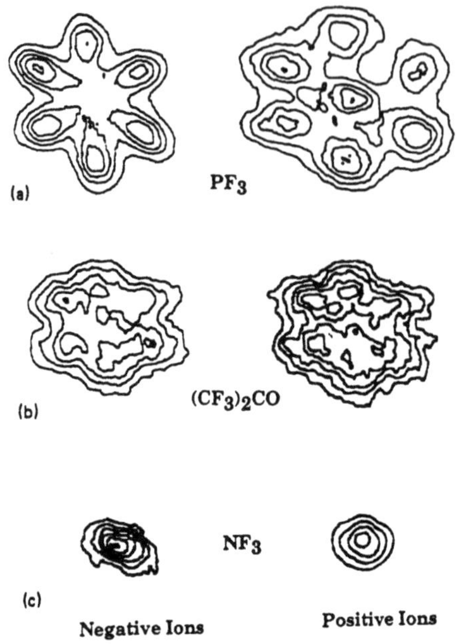

Fig. 1. Negative and positive ion ESDIAD images (contour plots) for a series of fluorinated molecules adsorbed on Ru(0001): (a) saturation coverage of PF_3, (b) monolayer coverage of hexafluoroacetone, and (c) saturation coverage of NF_3. All images acquired at a substrate temperature of 120K after annealing to 270K.

Strong angular anisotropies in the F^- emission are observed. Shown in Fig. 1 are contour plots of the negative ion images for PF_3, hexafluoroacetone, and NF_3 at monolayer coverages. The ESDIAD of both PF_3 and hexafluoroacetone show six distinct off-normal beams of ions along with some normal emission. Determination of the azimuthal direction of these beams by comparison with the position of the primary LEED beams indicates that the ion emission is along the azimuthal direction between the atom rows. The ESDIAD of NF_3 is characterized by a single beam sharply peaked about the surface normal. For comparison in Fig. 1 the corresponding positive ion images are also shown. While the positive and negative ion images for NF_3 and hexafluoroacetone are qualitatively similar, those for PF_3 are different. The positive ion ESDIAD has an intense beam along the surface normal and the six off-normal beams have a larger polar angle.

The behavior of the positive and negative ion signals during prolonged electron beam exposure are different. In all cases, the total negative ion signal decreases with increasing electron exposure, while the total positive ion signal always increases. Based on the rate of disappearance of the F^- signal during electron bombardment, we calculate a cross section for the dissociation of PF_3 of 2.5×10^{-16} cm^2,

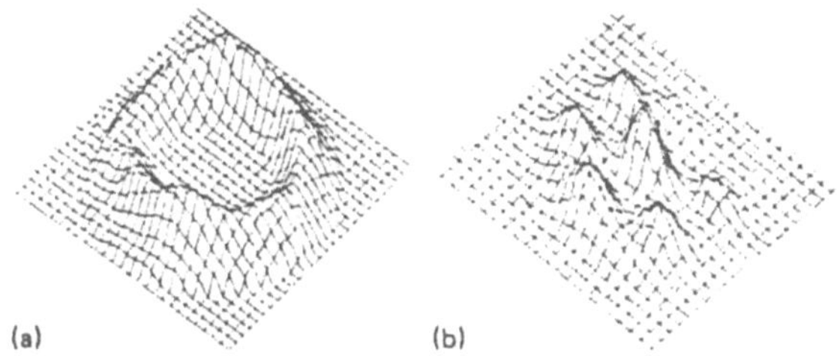

Fig. 2. F^+ ESDIAD images (perspective plots) from a half-saturation coverage of PF_3 on Ru(0001): (a) before and (b) after an electron beam exposure of 6 x 10^{-3} C/cm^2.

which compares favorably with gas phase values of the electron impact dissociation cross sections of fluorinated molecules [12]. In addition, the negative ion angular distributions do not change with electron beam exposure. In contrast, for PF_3, there is a significant change in the angular distributions of the positive ions with beam exposure.

Figure 2 shows the positive ion ESDIAD images of a half-saturation coverage of PF_3. At low beam exposures, there is a halo of ion emission due to free rotation or random azimuthal adsorption of the surface PF_3 [3,4]. Prolonged beam exposure produces six off-normal beams and an intense normal beam.

Based on ESDIAD studies combined with soft x-ray photoemission studies, we identify the F^+ and F^- hexagonal patterns (Fig. 1) as due to molecular PF_3, whereas the F^+ hexagon of Fig. 2b arises from adsorbed PF_2 produced by beam-induced dissociation. The normal F^+ beam (Fig. 2b, Fig. 1) appears to be due to PF. Similar results were found in a study of PF_3 on Ni(111) [13].

In all cases of molecules adsorbed on clean Ru(0001) studied so far, the ESD negative ion signal is dominated by undissociated molecular adsorbates, whereas the positive ion signal contains strong contributions from dissociation fragments (e.g., PF_2, PF).

Finally, we have studied the influence of coadsorbed K on ESD of PF_3 on Ru(0001) [5,6]. We find that K causes major changes in the ESD ion angular distributions, and that the desorption yields of three species (F^+, F^- and metastable F^*) vary strongly with K coverage. In the absence of K at low PF_3 coverages, no F^- or F^* are observed (although F^- appears at saturation coverage of PF_3). Negative ions (F^-) appear at $\theta(K) > 0.04$ ML and the F^- yield increases with K coverage. A metastable species identified as F^* (the $^4P_{5/2}$ state of atomic F) is observed for $\theta(K) > 0.06$ ML. The F^+ yield decreases with increasing $\theta(K)$. This is the first study in which angular distributions and yields of all three products (positive ions, negative ions and neutral species) have been measured simultaneously as a function of alkali coverage. The changes in angular distribution caused by the interaction of PF_3

with K (F^+, F^- and F^* are all centered about the surface normal) are due to a reorientation of the PF_3 and to chemical reaction to form other species with normally-oriented fluorine bonds, such as P-F, Ru-F or K-F. The variation in desorption yields (F^-, F^*) with K coverage are caused by reneutralization (charge-transfer) processes between the K-covered surface and the desorbing species [6].

4. Summary

In summary, we have used time-of-flight techniques to measure negative ion ESDIAD from PF_3, NF_3 and $(CF_3)_2CO$ on Ru(0001). For all of these systems, large ESD yields of F^+ and F^- are observed. The negative-ion yield is dominated by molecular adsorbates, whereas the positive-ion yield is higher for dissociation fragments. Negative-ion ESDIAD provides structural information that complements the information derived from positive-ion ESDIAD.

When K is coadsorbed with PF_3 on Ru(0001), changes in desorption yields and angular distributions of F^+, F^- and metastable F^* are observed. These measurements provide a new view of the role of resonant neutralization effects in charge transfer at surfaces. They illustrate how impurity-induced changes in the surface electronic structure can influence the desorption yields.

5. Acknowledgment:

This work has been supported in part by the U. S. Department of Energy, Office of Basic Energy Sciences.

References

[1] For a review of negative ion electron stimulated desorption, see L. Sanche, in Proceedings of DIET III, edited by M. Knotek and R. Stulen, Vol. 13 in Springer Series in Surface Science (Springer, Berlin, 1988), p. 78; E. Bauer, in Proceedings of DIET I, edited by N. H. Tolk, M. M. Traum, J. C.Tully, and T. E. Madey, Springer Series in Chemical Physics 24, (Springer, Berlin, 1988), p. 104; D. Lichtman, ibid, p. 117.

[2] T. E. Madey, Science 234, 316 (1986); J. J. Czyzewski, T. E. Madey, and J. T. Yates, Jr., Phys. Rev. Lett. 32, 77 (1974).

[3] A. L. Johnson, S. A. Joyce and T. E. Madey, Physical Review Letters 61, 2578 (1988).

[4] S. A. Joyce, A. L. Johnson, and T. E. Madey, J. Vac. Sci. Technol. A7, 2221 (1989).

[5] T. E. Madey, S. A. Joyce and C. Benndorf, in H. P. Bonzel, A. M. Bradshaw and G. Ertl, eds., Alkali Adsorption on Metals and Semiconductors (Elsevier, Amsterdam, 1989).
[6] S. A. Joyce, J. A. Yarmoff, T. E. Madey and P. Nordlander, Vacuum (1990), in press.
[7] J. G. Dillard, Chem. Rev. 73, 589 (1973).
[8] D. F. Torgeson and J. B. Westmore, Can. J. Chem. 53, 933 (1975).
[9] M. M. Walczak, A. L. Johnson, P. A. Thiel and T. E. Madey, J. Vac. Sci. Technol. A6, 675 (1987).
[10] M. M. Walczak, P. K. Leavitt and P. A. Thiel, J. Am. Chem. Soc. 109, 5621 (1987).
[11] F. Nitschke, J. Kuppers and G. Ertl, J. Chem. Phys. 74, 5911 (1981).
[12] E.g., see L. Christophorou, Electron-Molecule Interactions and Their Applications, (Academic Press, New York, 1984).
[13] M. D. Alvey, J. T. Yates, Jr., and K. J. Uram, J. Chem. Phys. 87, 7221 (1987); M. D. Alvey and J. T. Yates, Jr., J. Chem. Soc. 110, 1782 (1988).

Quantum-Resolved Stimulated Dissociation and Desorption of Chemisorbed Molecules

A.R. Burns, E.B. Stechel, and D.R. Jennison

Sandia National Laboratories, Albuquerque, NM 87185, USA

1. Introduction

The electronic excitations and subsequent dynamics responsible for the electron-stimulated dissociation of adsorbed molecules can be characterized in detail by examining the neutral products with quantum specific resolution[1]. We have examined the dissociation of chemisorbed NO_2 on Pt(111) through state-selective, time-of-flight laser resonance-ionization of the directly desorbed NO product[2]. The dissociation threshold and dynamics reflect the presence of two excitations: one hole in the $3b_2$ orbital and two-holes in the $1a_2$ orbital. The internal energy distributions of the NO product result from forces in the excited states that are largely determined by screening charge in the $6a_1$ orbital which is greater for the $1a_2^{-2}$ excitation than for the $3b_2^{-1}$ excitation, but still considerably less than the full electron in the analogous gas-phase process.

We also have examined the importance of substrate screening in the electron-stimulated desorption (ESD) of NO from Pt(111) by pre-covering the surface with atomic O. We find dramatic evidence that the desorption yield for the NO $5\sigma^{-1}$ excitation is increased 100-fold relative to the clean surface when the screening charge in the 2π orbital is reduced by the electrophilic coadsorbate. We also find that the translational and internal energies are clearly affected by the change in screening charge.

2. Experimental

All the experiments are conducted in a time-of-flight (TOF) mode whereby the pulse of neutral NO molecules produced by a 0.2-1.0 μsec electron beam traverse a 0.5 cm distance before they are resonantly-ionized at a specified delay time by the 4 nsec laser pulse. Details concerning the apparatus and its computer control have been published elsewhere[1-3].

Molecular adsorption of NO_2 on the clean Pt(111) crystal was confirmed by examining the thermal desorption spectra (TDS)[2,4]. In all the work described below, the crystal temperature was kept at 150 K during saturation dosing (0.5 ML) so that no N_2O_4 ice layer was allowed to form. In the NO+O coadsorption experiments, the Pt(111) surface was pre-covered with O atoms by exposing it to NO_2 at 400 K[5]. Coverages up to θ_O=0.75 ML are obtained, as verified by the 0.90 ratio of the Auger peaks O(510 eV)/Pt(238 eV)[6]. The sample was kept at 200 K while dosing with NO so that only well-defined chemisorbed states of NO are formed[5]. The saturated NO coverage was found[5] to be inversely proportional to θ_O with a slope of -1/2, such that at 0.75 ML O, the NO coverage is 0.15 ML. There is no evidence of bond formation between the NO and O at any temperature or coverage[5].

Springer Series in Surface Sciences, Vol. 19
Desorption Induced by Electronic Transitions, DIET IV
Editors: G. Betz · P. Varga

3. Results

Quantum-resolved translational energy (E_{trans}) distributions for the NO product resulting from 350 eV electron dissociation of the adsorbed NO_2 are shown in Figs. 1A and 1B. While the NO leaves the surface directly, the O atom remains[2]. Figure 1A exhibits slight shift to higher E_{trans} for the NO ν=1,2 populations relative to that for ν=0. Similarly, Fig. 1B shows a distinct shift to higher E_{trans} with increasing rotational energy of the dissociation product. The double-peaked distribution in Fig. 1B for an intermediate rotational level suggests that both the vibrational and rotational shifts are due to the presence of two dissociation channels, although the threshold data[2] indicate dissociation starts in the 10-15 eV range for all observed vibrational levels. The internal energy distributions for all the observed vibrational levels of the dissociation product are shown in Table I along with those obtained for the ESD of NO from clean and O-covered Pt(111).

The E_{trans} distributions for 350 eV ESD of NO (ν=0) as a function of θ_O are shown in Fig. 2. A remarkable observation is that the whole distribution shifts by +0.13 eV over the range from θ_O=0.4-0.75 ML. Meanwhile, the specific yield rises exponentially as a function of θ_O, as shown in Fig. 3. Since the NO ESD cross-section on the clean surface[3] is ~10^{-18} cm^2 for 350 eV excitation, the enhanced specific yield at high θ_O is clearly approaching, if not exceeding, gas-phase electron-impact ionization cross-sections[7]. It is important to note, however, that the yield includes con-

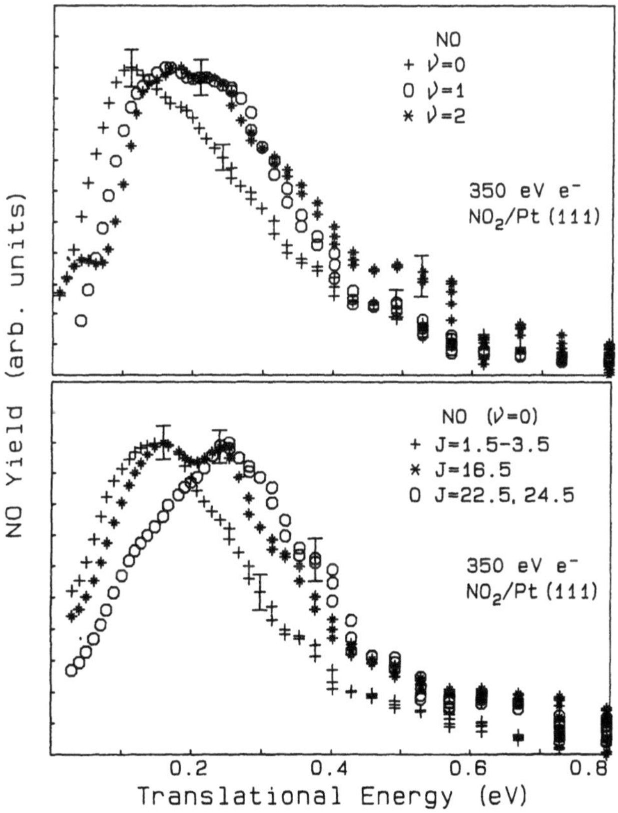

Fig. 1: On the top (A) is a comparison of the normalized TOF E_{trans} distributions for the three observed vibrational levels (ν=0,1,2) of the NO product of NO_2 dissociation on Pt(111). On the bottom (B), is a comparison of the normalized TOF E_{trans} distributions as a function of rotational energy for the ν=0 NO product of NO_2 dissociation on Pt(111). The low J distribution was obtained at the $P_{21}Q_{11}$ bandhead (J=1.5-3.5). The intermediate J distribution was obtained at the $R_{11}Q_{21}$ (J=16.5) line. The high J distribution was obtained at the $P_{21}Q_{11}$ (J=24.5) line which overlaps with the weaker R_{22} (J=22.5) line.

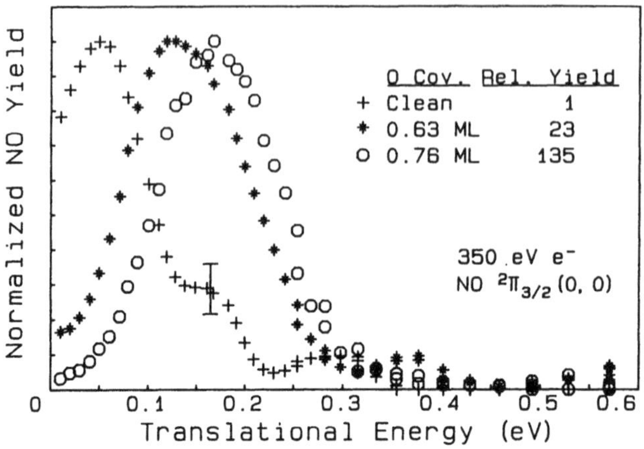

Fig. 2: Comparison of the normalized TOF E_{trans} distributions for the ESD of $NO(\nu=0)$ as a function of O-atom coverage (θ_O) on Pt(111).

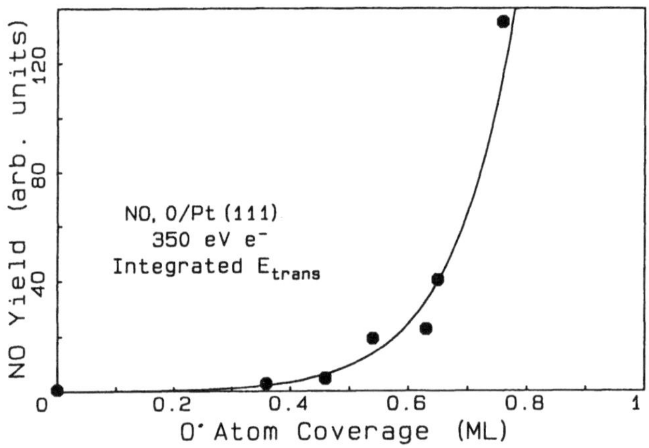

Fig. 3: Enhancement in the specific NO ESD yield as a function of O atom coverage (θ_O). The specific NO yield is obtained by integrating the TOF data as shown in Fig. 2 at each coverage and dividing the result by the NO coverage. The solid curve is a single exponential fit.

Table I. Energies of NO Products

ν	Population	E_{trans} Peak (eV)	E_{rot} (K)
	NO_2 dissociation on clean Pt(111) [2]		
0	1.00	0.15	862 ± 30
1	0.96 ± 0.01	0.20	821 ± 35
2	0.52 ± 0.13	0.20	743 ± 29
	NO ESD from clean Pt(111) [3]		
0	1.00	0.05	543 ± 10%
1	0.58 ± 5%	0.05	581 ± 10%
2	0.26 ± 5%	0.05	642 ± 10%
3	0.43 ± 5%	0.05	481 ± 10%
	NO ESD from 0.75 ML O-covered Pt(111)		
0	1.00	0.18	465 ± 15
1	0.44 ± 1%	0.18	565 ± 38
2	0.14 ± 5%	0.18	424 ± 18

tributions from secondary electrons. The thresholds for the enhanced ESD of NO coadsorbed with 0.75 ML O on Pt(111) are 9-10 eV for the ν=0,1, and 2 vibrational levels, which correspond nicely with the $5\sigma^{-1}$ excitation observed on the clean surface[1,3]. There is no sign of multiple-hole formation ($5\sigma^{-2}$) at 18-20 eV except a possible hint in the ν=2 data.

4. Discussion

A more complete discussion of electronic structure, excitation lifetimes, and thresholds for these chemisorbed systems is found elsewhere in this volume (D. R. Jennison, et al.), and is only briefly summarized here. Electronic excitations with thresholds < 7 eV are degenerate with Pt(111) valence levels and thus decay rapidly by direct resonant tunneling. Thus the metal effectively quenches the shallow 3.1 eV valence excitations which result in gas-phase NO_2 dissociation[8]. Excitations which are deeper than 7 eV below E_F decay predominantly via Auger processes involving electrons in the shallower molecular levels. Since the NO_2 chemisorption bond involves the $6a_1$ molecular orbital (MO), all one- and two-hole excitations will be screened by charge donated by the metal into this hybridized level. Similarly, the $5\sigma^{-1}$ excitation in chemisorbed NO is screened and decays by electrons in the 2π-metal hybrid orbital.

The 10-15 eV NO_2 dissociation threshold corresponds to the $3b_2^{-1}$ and $1a_2^{-2}$ excitations. The presence of screening charge in the antibonding $6a_1$ orbital will weaken the NO bonds and decrease the ONO bond angle[2]. The antibonding nature of the $6a_1$ orbital is evident in the gas-phase dissociation[8], where excitation to the $^2B_2$ state is associated with promotion of a $4b_2$ electron to the $6a_1$ level and results in considerable vibrational and rotational excitation. On the surface, however, there is a large reduction in internal excitation, particularly vibrational, relative to that observed in the gas-phase. The average rotational energy of ~ 750-860 K for all three observed vibrational levels is only 50% greater than the zero-point energy of 572 K for the NO_2 bend on Pt(111)[4]. Thus most of the observed rotational energy is not from the excited state geometry, but from ground-state zero-point motion of the parent molecule.

The low energy gas-phase excitations have a full electron promoted to the $6a_1$ level, whereas a screening charge from the metal substrate can be less than unity. If the $3b_1^{-1}$ excitation is only partially screened by the metal-$6a_1$ charge, then much smaller changes in NO bond length and ONO bond angle are expected and internal excitation would be dependent on the actual value of ζ in the $3b_2^{-1}6a_1^{+\zeta}$ electronic excitation. Evidence is found in the E_{trans} shifts to higher energy observed in Fig. 1A and Fig. 1B for increasing vibrational and rotational excitation. These shifts are most likely due to differences in screening charge for the one-hole $3b_2^{-1}6a_1^{+\zeta}$ excitation and the two-hole $1a_2^{-2}6a_1^{+\zeta'}$ excitation, although some of these effects may be due in part to extra forces in the $1a_2^{-2}6a_1^{+\zeta'}$ excitation due to hole-hole repulsion. Since $\zeta'>\zeta$, the two-hole excitation does result in more NO vibrational and rotational energy. The antibonding character of the $6a_1$ orbital is also evident in the slight increase in E_{trans} associated with the two-hole excitation.

The large increase in NO ESD yield and the shift to higher translational energies with increasing θ_O arises from the decreasing 2π screening charge. These effects are not due to the site change from bridge to on-top which occurs at θ_O=0.25 ML[5], because on the clean surface, where both bridge and on-top sites are occupied[5], no changes in the specific yield or translational energies are observed when the on-top NO is selectively desorbed at 300 K. It is thought [5] that the on-top NO is bent at θ_O=0.75 ML; it remains to be determined if this is an important qualitative in the ESD process.

Since the 2π orbital is antibonding, decreasing the metal screening charge in this level would reduce vibrational excitation in the desorbed NO. In Table I, one can see that the NO vibrational excitation is indeed significantly less for NO desorbing from the coadsorbed (θ_O=0.75 ML) surface than from the clean surface. The rotational energies are somewhat lower for the coadsorbed system than they are for the clean surface, which indicates that the retrapping of the low rotational states is reduced. The lower rotational energies and higher translational energies of the NO leaving the O-covered surface indicate that the excitation decays farther away from the surface relative to the clean system[3], consistent with the increased lifetime.

5. Conclusions

We have found that excitation screening, involving charge transfer from the metal through re-hybridized adsorbate-metal orbitals, directly influences the dynamics and lifetimes of stimulated processes. If the screening involves a strongly antibonding molecular orbital, excited-state <u>intra</u>molecular forces are then very sensitive to the amount of charge transfer, with observable consequences in desorbate internal energy distributions.

Acknowledgments

The authors are indebted to M. E. Bartram and B. E. Koel for allowing us to study their NO,O/Pt(111) experimental data work prior to publication. This work was supported by the U. S. Department of Energy under Contract DE-AC04-76DP00789.

References

1. A. R. Burns, E. B. Stechel, and D. R. Jennison, in <u>Desorption Induced by Electronic Transitions, DIET III</u>, edited by R. H. Stulen and M. L. Knotek (Springer-Verlag, New York, 1988), pp. 67-72, 136-143, 167-172.
2. A. R. Burns, D. R. Jennison, and E. B. Stechel, Phys. Rev. B, in press.
3. A. R. Burns, E. B. Stechel, and D. R. Jennison, Phys. Rev. Lett. <u>58</u>, 250 (1987).
4. M. E. Bartram, R. G. Windham, and B. E. Koel, Surf. Sci. <u>184</u>, 57 (1987).
5. M. E. Bartram, B. E. Koel, and E. A. Carter, Surf. Sci. <u>219</u>, 467 (1989).
6. J. L. Gland, B. A. Sexton, and G. B. Fischer, Surf. Sci. <u>95</u>, 587 (1980).
7. D. Rapp and P. Englander-Golden, J. Chem. Phys. <u>43</u>, 1464 (1965).
8. M.-R. Taherian, P. C. Crosby, and T. G. Slanger, J. Phys. Chem. <u>91</u>, 2304 (1987).

Velocity Distribution of Laser Photoionized Neutrals Ejected from Methanol-Dosed Aluminum(111) by Electron-Stimulated Desorption

C.E. Young[1], *J.E. Whitten*[1], *M.J. Pellin*[1], *D.M. Gruen*[1], *and P.L. Jones*[2]

[1]Materials Science/Chemistry Divisions, Argonne National Laboratory, Argonne, IL 60439, USA
[2]Department of Chemistry, The Ohio State University and Spectra Technology, Inc., 2755 Northup Way, Bellevue, WA 98004, USA

Abstract. Nonresonant multiphoton ionization at 193 nm wavelength was employed for efficient detection of electron-stimulated neutral desorption from Al(111) dosed with methanol to produce monolayer methoxide coverage. Velocity spectra were measured by the flight time from the crystal surface to the focal region of the laser beam with a pulsed primary electron beam of 3 keV and the sample at 300 K. Either the C^+ or HCO^+ photofragment indicated the same non-Boltzmann velocity spectrum for the neutral parent precursor with a peak kinetic energy of ~0.1 eV. Identical distributions were obtained when the cleaned crystal was pre-oxidized with O_2 prior to methanol dosing. As the crystal temperature was raised, HCO^+ photoion signal declined steadily, while C^+ increased until ~550 K. The total cross section for loss of parent signal with dose of 3 keV electrons was measured to be $(2\pm1)\times10^{-17}$ cm^2.

1. Introduction

Experimental velocity distributions provide a very discriminating probe of the dynamics of electron- or photon-stimulated desorption (ESD or PSD) of species from surfaces. In the case of the neutral channel, which typically dominates the total desorption flux by orders of magnitude, relatively few velocity distributions have thus far been measured [1]. For monolayer coverage on metal surfaces, the existing data on ESD/PSD of neutrals in the ground or low-lying states are characterized by peak kinetic energies in the ~0.1 eV region, and thus are translationally much "colder" than either neutrals in electronically-excited metastable states [2] or desorbed ions (for reviews, see [3, 4]). The present paper reports measurements by non-resonant laser ionization of the velocity spectrum of neutrals released by ESD from methanol-dosed Al(111). Nonresonant laser photoionization provides signal enhancement resulting from accessing simultaneously a large number of quantum states of the probed species, believed to be the CH_3O (methoxy) radical in this case. The well-characterized methoxy A-X electronic spectrum in the 316 nm region [5] provides the possibility of state-resolved neutral detection in future studies.

2. Apparatus and Procedures

The basic procedure employed in these experiments was nonresonant laser ionization of neutrals, following their desorption from a sample surface by a pulsed electron beam. The apparatus included a commercial concentric hemispherical analyzer (CHA), operated here for positive ion transmission. A detailed experimental description has been published [6]; also, a comprehensive paper on the present studies is in preparation [7]. For the acquisition of velocity spectra, the sample was bombarded with electron pulses of ~1 μs, followed after a variable delay time by ionization by a 20 ns pulse of laser radiation at 193 nm from an ArF excimer (Questek model 2660, unstable resonator optics, focused into the ionization volume by a 240 mm focal length lens, giving a beam waist ~0.15x0.41 mm). Velocity spectra were generated in this manner, using the known flight path of 2.5 mm. The energy analyzer provided a bandpass of 1-2 eV, allowing for complete discrimination against signal due to direct ion desorption from the sample in experiments studying neutral desorption by laser ionization. The potential drop between sample (biased at +92 volts) and the photoionization volume provided the necessary energy separation. Residual signal from background-gas ionization was subtracted by blanking the desorbing electron pulse on alternate laser shots. In independent experiments, kinetic energy distributions for desorbed ions were obtained by scanning the CHA energy window in the usual fashion. Time-of-flight mass spectrometry of either direct ions or photoionized neutrals was performed using the electron or laser pulse, respectively, as the initial timing mark. All of the ESD experiments reported here were performed with a primary electron energy of 3 keV (nominal gun voltage, referenced to ground). The aluminum crystal (99.999%) was oriented to (111) within 0.5° and cleaned by cycles of sputtering (1.5 keV Ar^+) followed by annealing at 700 K. Crystallinity was monitored by LEED analysis, and surface contamination by C and O was established at less than 1%, using the CHA in Auger mode. Dosing with degassed methanol followed standard procedures.

Although fragmentation occurs at the laser intensities employed for obtaining high photoionization efficiency, considerable dynamical information from the initial desorption step is retained. Mass spectrometry studies of the photoions from neutrals desorbed by long electron pulses (~10 μs duration, to avoid velocity discrimination effects) were performed as a function of laser intensity in the range 30 to 650 MW/cm^2. At low intensity, signal at mass 12 and 13 (C^+ and a more intense CH^+) and a mass group extending from 28 to 31 (CO^+, HCO^+, CH_2O^+, CH_3O^+) were observed. The C^+ and HCO^+ peaks showed comparable intensity at 60 MW/cm^2, but beyond ~300 MW/cm^2 at the onset of saturation in both ion signals, the C^+ was more intense by an order of magnitude.

3. Results and Discussion

3.1 Surface Initial Conditions

From various published evidence, it seems apparent that exposure of Al(111) to a saturation dose of methanol at 300 K, as used in the work reported here, results in monolayer coverage of the methoxide CH_3O with the O atom adjacent to the metal. Surface vibrational spectroscopy, via high-resolution electron energy loss spectroscopy (HREELS), led to this conclusion in a study by Yates and co-workers [8] with deposition at 90 K followed by warming to 200 K. The surface methoxide persisted to ~450 K where decomposition to surface oxides and carbides with release of methane occurred. Dosing at room temperature produced the same characteristic methoxide vibrational spectrum in a study on Al(110) by Waddill and Kesmodel [9]. Room temperature dosing was also employed in an XPS/UPS study on polycrystalline aluminum by Rogers et al. [10], in which splitting of the bonding orbitals indicated a bent C-O-Al bond. The decomposition path, as described above, was observed at ~500 K. At lower temperatures, the possible decomposition products, formic acid, formaldehyde, and CO were rejected on the basis of the UPS data. The secondary ions CH_3O^- and $AlCH_3O^+$ were directly observed after saturation dosing of polycrystalline Al with methanol at room temperature by Tindall and Vickerman [11]. Evidence for the initial oxidation of the surface (at a few L exposure) prior to methoxide formation was found. Heating to 373 K rendered the methoxy/methoxide ions undetectable. Quite similar behavior was observed with a heavily oxidized surface (1000 L methanol).

3.2 Velocity Distributions

Adequate signal for neutral velocity distribution determination required operation in the high laser intensity regime where C^+ was the most intense photoion. The upper panel of Fig. 1 shows data at 390 MW/cm^2 for the neutral desorbate, monitored via the C^+ fragment, for a saturation dose (55 L) of CH_3OH on Al(111). The detected photoion counts are taken to be proportional to the initial density of desorbed neutrals in the limit of arbitrarily short pulses of electrons and laser photons [6, 12]. These assumptions are very good for the exhibited data, except that convolution over the finite-duration electron pulse tends to shift the highest velocity data points to somewhat greater values [7]. In Fig. 1, fits to both a Boltzmann distribution (dashed curve) and an empirical "planar-barrier/power-law" form (solid curve) are displayed. The model distributions are $u^4 \exp(-u^2)$ and $u^4(1+u^2)^{-3}$, respectively, with $u = v/v_b$, v being the measured velocity and $v_b = (2k_B T/m)^{1/2}$ (Boltzmann) or $v_b = (2E_b/m)^{1/2}$ (planar barrier) [6]. The data displayed correspond to a binding parameter $E_b = 0.065$ eV. Within experimental statistics, identical behavior was observed irrespective of whether the C^+ or HCO^+ photofragment (Fig. 1, lower

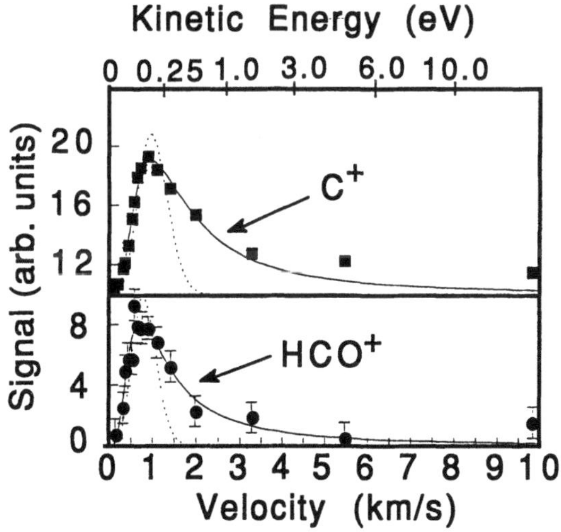

Fig. 1. Velocity distributions for neutral desorption from saturation dosing of methanol on clean Al(111), monitored by 193 nm photoionization and detected via the C^+ or HCO^+ photofragments. Dashed curve: Boltzmann fit; solid curve: empirical "planar-barrier" fit (see text).

panel) was used to monitor the TOF spectrum of the neutral desorbate. In a second set of experiments (not shown), the cleaned Al(111) surface was oxidized through several cycles of O_2 exposure (several hundred L) and heating to 700 K [13] prior to saturation dosing with methanol. This treatment is sufficient to produce at least an overlayer and underlayer pattern of oxygen in the aluminum surface layer [14]. This level of pre-oxidation gave no significant change in the neutral ESD velocity spectra. It is therefore likely that the two surface preparations produce similar adsorbed methoxy layers, with an equivalent oxidation occurring in the process of dosing clean Al(111) with methanol.

3.3 Electron Irradiation Effects

Experiments were performed monitoring the dose dependence of both the HCO^+ and C^+ signals from the deposited CH_3O/Al(111) monolayer films described above. Photoionization mass spectra were measured under long-pulse (10 μs duration) conditions for the 3 keV electron beam, interspersed with timed intervals of continuous bombardment. The data for the pre-oxidized sample, shown in Fig.2, indicate a continuous decrease of the HCO^+ signal with dose, while the C^+ signal levels off after the initial decrease. The former data can be interpreted in terms of an exp $(-J\sigma t)$ dependence on the bombardment time t, with electron current density J, and with σ being the total cross section for removal of the surface species which provides the observed HCO^+ signal. For the actual experimental parameters (2 μA in ~ 1 mm^2), a value of $\sigma \approx 2\times10^{-17}$ cm^2 is obtained from the initial slope; the subsequent slowing of the HCO^+ decay rate results from electron beam-profile effects. The results with the Al(111) surface were very similar. The corresponding desorption yield would be ~3×10^{-2} if desorption were to dominate the loss processes. Qualitatively, the neutral yield in the present experiments was comparable to that estimated in the earlier studies on an aluminum alloy surface, i.e., $\geq10^{-3}$ [6], so additional loss processes may be operative here. For comparison, Yates and co-workers [15] have employed HREELS to investigate the electron-induced destruction of the methoxy species adsorbed on Al(111) at lower temperatures. Their sample was held at 95 K and bombarded with 300 eV electrons, with the CH stretching frequency in the methyl group being monitored as an indicator of the concentration of surface methoxy. A total cross section of ~5×10^{-15} cm^2, ascribed to loss of CH_3O, was

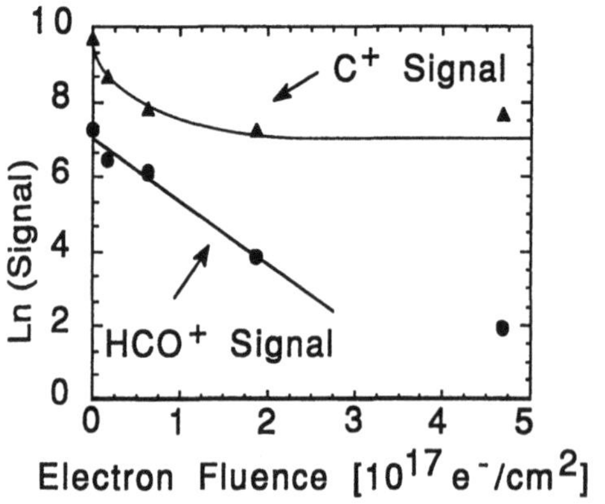

Fig. 2. Decrease of neutral desorption signal with total electron fluence. The HCO^+ signal indicates a cross section of 2×10^{-17} cm^2 for all processes resulting in loss of parent CH_3O adsorbate, from the pre-oxidized Al(111) surface.

measured at an electron dose of $\sim 10^{14}$ cm^{-2}, as well as a cross section of $\sim 4 \times 10^{-18}$ cm^2 at a dose of $\sim 5 \times 10^{17}$ cm^{-2}, ascribed to the loss of a much more desorption-resistant CH_x- species created by the bombardment. The lower cross section observed in the present work can be attributed to the higher primary electron energy employed (3000 vs. 300 eV), with possible influences from both lower excitation cross sections and lower secondary electron yields at the higher energy.

3.4 Surface Temperature Effects

Figure 3 shows laser ionization mass spectrometry data taken as the sample temperature was raised. Notable is the absence of any strong desorption effect. Methanol should give a large nonresonant ionization signal, for example [16]. The C^+ signal actually tends to increase in the high-temperature region (>500 K), where surface methoxide has been observed to decompose, possibly due to the desorption of new species with higher probability for total photofragmentation to C^+ ions and/or higher desorption yield. Additional isothermal data on the time-dependent decline in HCO^+ signal can be analyzed according to first-order kinetics to correspond to an activation energy of 1.9 eV at 560 K for an assumed pre-exponential factor of 10^{13}, indicating a species strongly bound to the surface. The empirical nature of the planar-barrier fits of Fig. 1 is emphasized ($E_b <<$ desorbate surface bond).

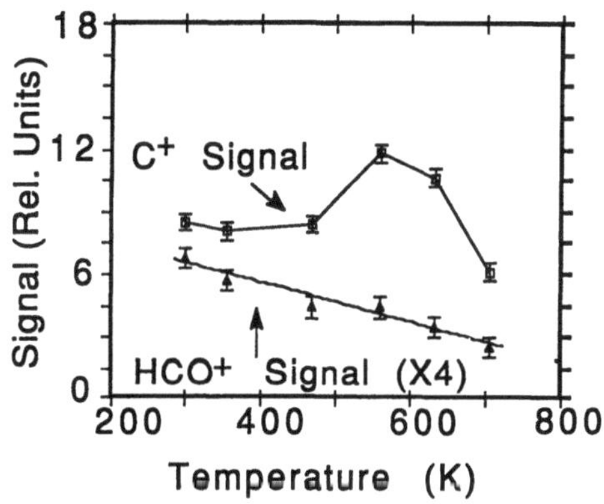

Fig. 3. Temperature dependence of the neutral desorption signal as monitored by the C^+ or HCO^+ photofragments from 193 nm laser ionization.

3.5 Conclusions

The desorption of methoxy from Al(111) is characterized by low velocities despite an initial strong bond to the surface. If the methoxy desorption proceeds through intermediate ionic states, reneutralization is apparently very efficient, no ions belonging to the parent group being detectable in these studies [7]. Relatively high desorption yield and low kinetic energy of desorption in the neutral channel are, in general, conflicting requirements in familiar mechanisms [17, 18]. Gortel et al. found rapid de-excitation from the upper state implied low kinetic energies and low yields [19], although recent calculations [20] suggest larger yields are possible for the case of neutral desorption of inert gases from a metal surface. Detailed calculations for systems of the type studied in the present work are unavailable, at present.

Acknowledgment

This work was supported by the U.S. Department of Energy, under Contract W-31-109-ENG-38.

References

1. For a review of recent work, see Desorption Induced by Electronic Transitions (DIET III), ed. by R. H. Stuhlen and M. L. Knotek, Part III. Springer-Verlag, Berlin (1988)
2. D. R. Sandstrom, M. J. Dresser and W. D. Dong: Phys. Rev. **B34**, 5125 (1986)
3. T. E. Madey and J. T. Yates Jr.: J. Vac. Sci. Technol. **8**, 525 (1971)
4. M. L. Knotek: Rep. Prog. Phys. **47**, 1499 (1984)
5. J. Kappert and F. Temps: Chem. Phys. **132**, 197-208 (1989)
6. C. E. Young, J. E. Whitten, M. J. Pellin, D. M. Gruen, R. L. Benbow and P. L. Jones: Surf. Interface Anal. **14**, 647 (1989)
7. J. E. Whitten et al.: to be published
8. J. G. Chen, P. Basu, L. Ng and J. T. Yates Jr.: Surf. Sci. **194**, 397 (1988)
9. G. D. Waddill and L. L. Kesmodel: Surf. Sci. **182**, L248-L252 (1987)
10. J. W. Rogers Jr., R. L. Hance and J. M. White: Surf. Sci. **100**, 388 (1980)
11. I. F. Tindall and J. C. Vickerman: Surf. Sci. **149**, 577 (1981)
12. H. F. Arlinghaus, W. F. Calaway, C. E. Young, M. J. Pellin, D. M. Gruen and L. L. Chase: J. Appl. Phys. **65**, 281 (1988)
13. J. E. Crowell, J. G. Chen and J. T. Yates Jr.: Surf. Sci. **165**, 37-64 (1986)
14. J. L. Erskine and R. L. Strong: Phys. Rev. **B25**, 5547 (1982)
15. P. Basu, J. G. Chen, L. Ng, M. L. Colaianni and J. T. Yates Jr.: J. Chem. Phys. **89**, 2406 (1988)
16. R. C. Sausa, A. J. Alfano and A. W. Miziolek: Appl. Optics. **26**, 3588 (1987)
17. D. Menzel and R. Gomer: J. Chem. Phys. **41**, 3311 (1964)
18. P. A. Redhead: Can. J. Phys. **42**, 886 (1964)
19. Z. W. Gortel, H. J. Kreuzer, P. Feulner and D. Menzel: Phys. Rev. **B35**, 8951 (1987)
20. R. E. Walkup, Ph. Avouris, N. D. Lang, and R. Kawai: Phys. Rev. Lett. **63**, 1972 (1989)

Desorption and Secondary Ion Production During Bombardment of CO/Ni(111) with Ne^+ and Ne^{++} Ions at Very Low Impact Energies

U. Diebold and P. Varga

Institut für Allgemeine Physik, Technische Universität Wien,
Wiedner Hauptstr. 8–10, A-1040 Wien, Austria

Abstract. We have studied ion - induced desorption at very low impact energies. Singly and doubly charged Ne ions with impact energies E_0 between 20 eV and 500 eV have been used. The desorbing species (neutrals and ions) have been analysed by means of a quadrupole mass spectrometer. The desorption cross section σ_{CO} for neutral CO which is desorbed predominantly was determined for different impact energies and impact angles. When using doubly charged Ne ions instead of singly charged ones, no difference in the desorption cross section for neutral CO could be found. For impact energies below 100 eV the secondary ion yield of CO^+ and Ni^+ was found to be higher for bombardment with Ne^{++} than for Ne^+. As an explanation, we propose an additional, new mechanism for secondary ion production; the particle emission is caused by momentum transfer during binary collisions, accompanied by ionization via electron capture by the primary ion from the sputtered particle.

1. Introduction

DIET processes (Desorption induced by electronic transitions) induced by ion bombardment have frequently been discussed in recent publications [1,2]. Experiments are reported where the dependence of the secondary ion yield on the charge state of ions impinging on low - conducting materials [3,4,5,6,7] as well as alkali halides[1,8,9] has been shown. For these processes, Bitenskii et al.[10,11] have proposed a process termed Coulomb explosion, i.e. electron transfer from the target into the projectile, followed by ion emission due to the repulsive force between the now positively charged target atoms.

To our knowledge, no studies of ion - induced DIET - processes on adsorbate - covered metal surfaces have been done, although most of the electron (ESD) and photon stimulated desorption (PSD) processes were studied with these systems. This may be caused by the high cross sections ($\sigma \sim 10^{-14} - 10^{-16}$ cm^2 [12,13,14]) for desorption induced by ion momentum transfer as compared to the low desorption cross section for ESD and PSD ($\sigma \sim 10^{-17} - 10^{-20}$ cm^2).

To study the latter processes, we have performed experiments on desorption due to impact of singly and doubly charged Ne ions with rather low kinetic energy (20 eV to 500 eV), thus keeping kinetic processes like sputtering improbable and increasing the possibility of observing electronic processes. As target system CO

Springer Series in Surface Sciences, Vol. 19
Desorption Induced by Electronic Transitions, DIET IV
Editors: G. Betz · P. Varga

adsorbed on Ni(111) has been chosen, which is an extensively studied system in surface science as well as in ESD and PSD [15,16,17].

2. Experiment

2.1. Experimental setup

The experiment consists of an UHV - chamber (base pressure 1.5 x 10^{-10} Torr) equipped with a sputter gun for target cleaning and an Auger electron spectrometer to monitor the target surface composition. The desorbed species (ions as well as neutrals) have been detected by means of a Quadrupole Mass Spectrometer (QMS, with pulse counting mode and ion transfer optics for external ions) in line of sight with the target.

The ion beam (impact energy of 20 to 500 eV) is mass - separated and involved a typical energy width of 3 eV FWHM. The amount of neutrals in the beam was estimated to be less than 0.1%. The beam could be scanned over an adjustable area and chopped with low frequency. Typical values for ion current densities as measured by means of a Faraday cup were 800 nA/cm^2 for Ne^+ and 30 nA/cm^2 for Ne^{++}. The angle between the ion beam and the QMS was 90°. Pulse counting and ion beam chopping proved to be crucial for measuring the desorbing neutral CO - molecules with the QMS because of a high background signal at mass 28.

The Ni(111) - single crystal was spot - welded on two 0.1 mm Mo - wires which could be resistively heated. The target was prepared by standard methods [18], involving several cycles of 700 eV Ar^+ - ion sputtering and subsequent annealing at up to 600°C. Remaining impurities of carbon could be removed by adsorbing 0.5 L O_2 with subsequent annealing. This procedure was regularly repeated after each desorption measurement to anneal the target from sputtering defects and to remove all carbon produced by ion - induced dissociation.

2.2.Measurements

For the desorption measurements, the target was loaded with 4L CO (112 s, p=4x10^{-8} Torr) at room temperature. This resulted in a c(4x2) CO layer [17] with a saturation coverage of 0.5 monolayers as proved by reproducible thermal desorption spectra (TDS).

Assuming a first-order desorption process and mainly neutral CO coming off the target, the measured desorption signal $dN_{des}(t)/dt$ [counts/s] at a time t for a short period Δt should follow the relation

$$dN_{des}(t)/dt = J\sigma ACN_0 e^{-J\sigma t} , \qquad (1)$$

where N_0 is the initial coverage [molecules/cm^2] at t = 0, J the ion flux [ions/cm^{2}s], A the bombarded target area [cm^2], σ the desorption cross section [cm^2] and C a proportionality factor covering the overall detection efficiency of the QMS.

σ appears in the pre-exponential factor as well as in the exponent of eq.1. From this, two possibilities for the determination of σ can be applied.

One possibility is to measure the decrease of the desorption signal during the bombarding time. A semilogarithmic plot of the desorption signal versus the ion fluency [ions/cm^2] yields a straight line, with its slope giving directly the absolute value of σ. This method (method 1) can only be applied provided the desorption signal decays at least by 20% during one measurement (bombardment time of a few minutes), i.e. if $J.\sigma$ is sufficiently large.

Second, with $J.\sigma$ being too low, the initial desorption signal (dN_{des}/dt (t=0)) can be utilized, since it is also proportional to σ. The proportionality factor can be determined by comparison with values of σ obtained by method 1.

With Ne^+ as projectile ion, σ has been determined by method 1, yielding the proportionality factor needed for method 2. For Ne^{++} bombardment, only the initial desorption signal could be used to evaluate the desorption cross section.

Measurements of secondary ions have been performed in the same way with the QMS tuned for an optimized secondary ion signal. Since the secondary ion yield was found to depend strongly on the CO - coverage of the surface, the CO - pressure has been elevated to $p_{CO}=5x10^{-7}$ Torr to keep the coverage constant during the measurement.

3. Results

Fig. 1a shows the desorption cross sections of neutral CO desorbed from Ni(111) by Ne^+ bombardment at different impact energies. The values are quite large, e.g. for impact energy of 300 eV, σ amounts to 3.9 x $10^{-15}cm^2$, corresponding to 3.6 desorbed molecules per incoming ion. This comparably high desorption yield can been explained by the low binding energy of CO in bridge sites (E_B = 0.7 eV[17]) combined with a high collisional cross section for the diatomic CO molecule [12] . The angular dependence of the desorption cross section at a fixed impact energy of 50 eV can be fitted with a $\cos\varphi^{-1}$ dependence as shown in fig.1b.

In fig.2a, the initial desorption yields (dN_{des}/dt (t=0) per incident ion) for bombardment with Ne^+ (crosses) are shown. Within the error bars they are proportional to the desorption cross section (small circles); Although the scatter of the data for bombardment with Ne^{++} - ions (open circles) is larger due to the lower ion flux, no systematic deviation from the Ne^+ impact data can be seen.

Fig.2b shows the ratio of the initial desorption yields, i.e. desorbed CO molecules per incident Ne^{++} ion over desorbed CO molecules per incident Ne^+ ion. As the ratio is unity in this energy range, one can conclude that either the charge state of the primary ion does not influence the desorption of neutral CO or such an influence can not be detected because desorption by momentum transfer is much more effective, even at these low impact energies.

This picture changes if one considers the desorbed CO - ions. In fig. 3a, CO^+ - yields for impact with Ne^+ and Ne^{++} respectively are shown. An estimated value of ~$2x10^{-4}$ desorbed CO^+ ion per incoming Ne^+ - ion at 300 eV can be given, if one assumes an ionization probability of CO of 10^{-3} in the ion source of the QMS. For

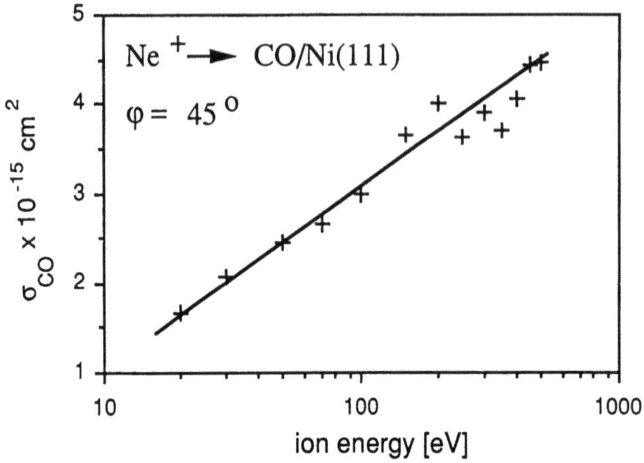

Figure 1a Desorption cross sections σ for neutral CO induced by bombardment of CO/Ni(111) with Ne^+. Initial coverage was 0.5 monolayers and impact angle was 45° with respect to the target normal. σ has been determined by the exponential decay of the desorption signal.

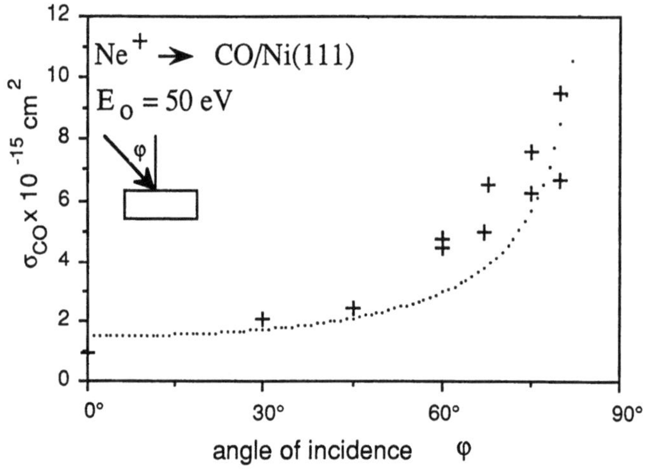

Figure 1b Desorption cross sections for neutral CO for bombardment with 50 eV Ne^+ at different impact angles φ.

bombardment with Ne^{++} at impact energies below 200 eV, the CO^+ signal is larger than that induced by Ne^+. The ratio of the CO^+ - secondary ion yields increases up to a factor of five for 50 eV impact energy as shown in fig.3b. The absolute amount of desorbed CO^+ ions decreases with decreasing energy for both projectiles.

In addition to CO^+, Ni^+, $NiCO^+$, Ni_2^+ and O^+ were found in the SIMS - spectrum of the CO covered Ni - surface. As compared to CO^+, their signals were ~1:3000 (Ni^+), ~1:30 ($NiCO^+$ and Ni_2^+) and ~1:0.3 (O^+). Fig.4a shows the secondary ion yield of Ni^+ when a fully CO - covered surface was bombarded by

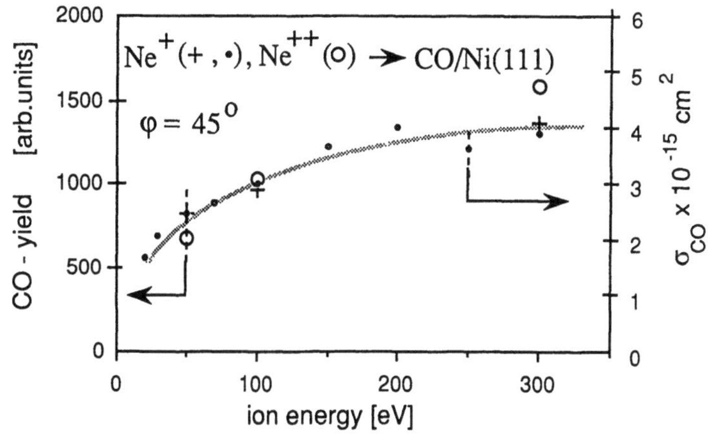

Figure 2a Initial desorption yield of neutral CO for bombardment with Ne^+ (crosses) and Ne^{++} (open circles). The yields follow qualitatively the Ne^+ - induced CO-desorption cross sections (small circles) from fig.1a which have been determined by the exponential decay of the desorption signal.

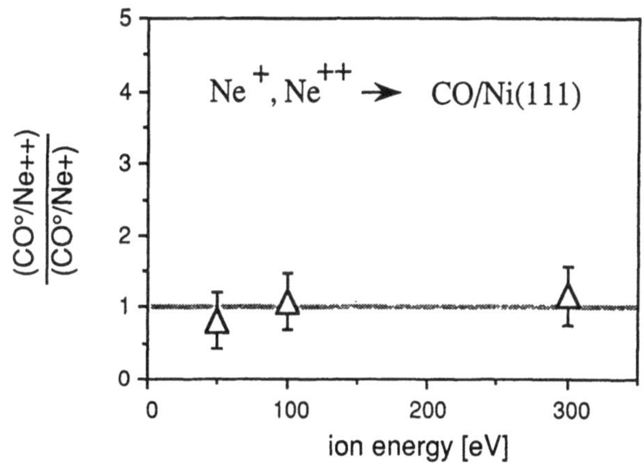

Figure 2b CO - yield per incoming Ne^{++} ion divided by CO - yield per incoming Ne^+ ion.

singly and doubly charged Ne ions at different impact energies. These measurements have been performed by using a very low ion flux of $6x10^{10}$ ions/cm^{2}sec for Ne^+ as well as for Ne^{++}. Again, a decrease of the secondary ion yield with decreasing impact energy of Ne^+ as well as Nc^{++} was found but an increase of the ratio of the secondary ion yields was apparent toward low impact energy (see fig.4b).

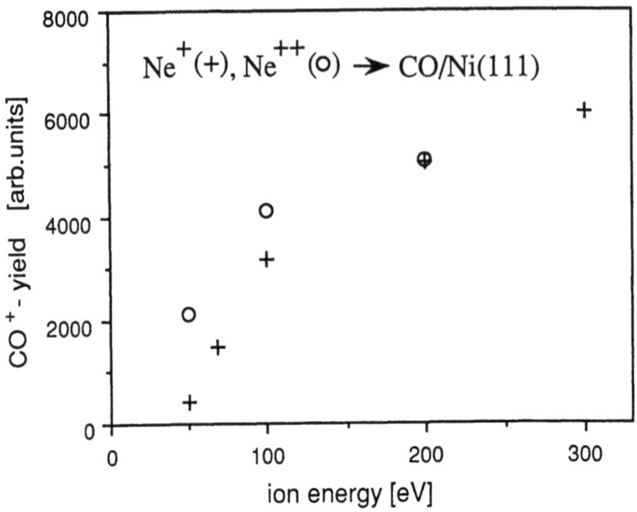

Figure 3a CO^+ desorption yields for bombardment with Ne^+ (crosses) and Ne^{++} (open circles). The ambient CO - pressure was 5 x 10^{-7} Torr during the measurements.

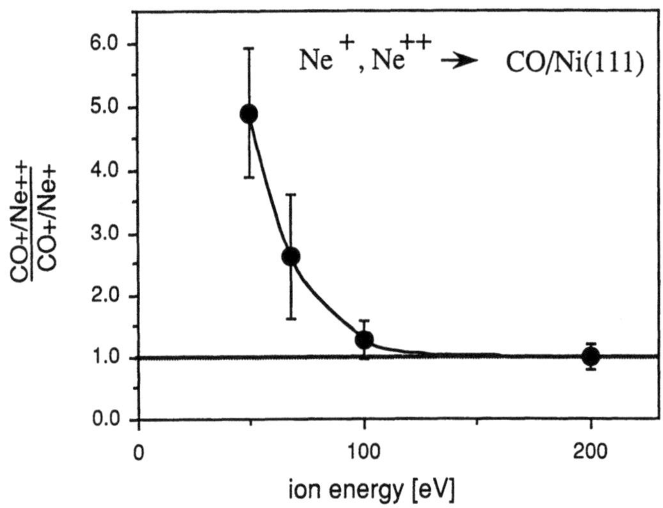

Figure 3b CO^+ - yield per incoming Ne^{++} ion divided by CO^+ - yield per incoming Ne^+ ion.

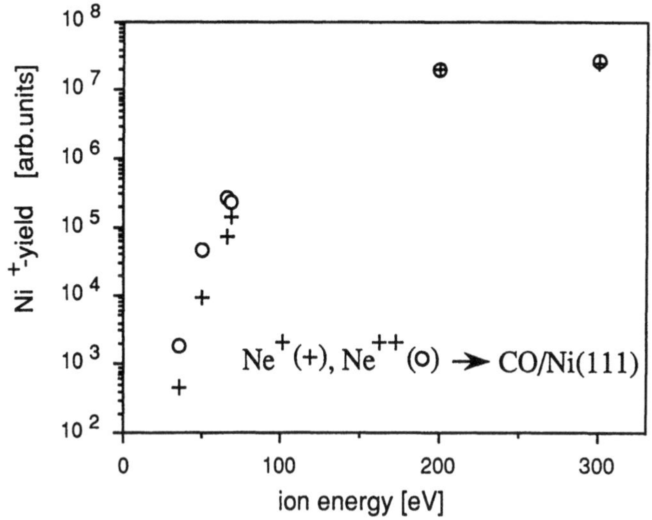

Figure 4a Ni^+ secondary ion yield for bombardment with Ne^+ (crosses) and Ne^{++} (open circles). The ambient CO - pressure was 5 x 10^{-7} Torr during the measurements.

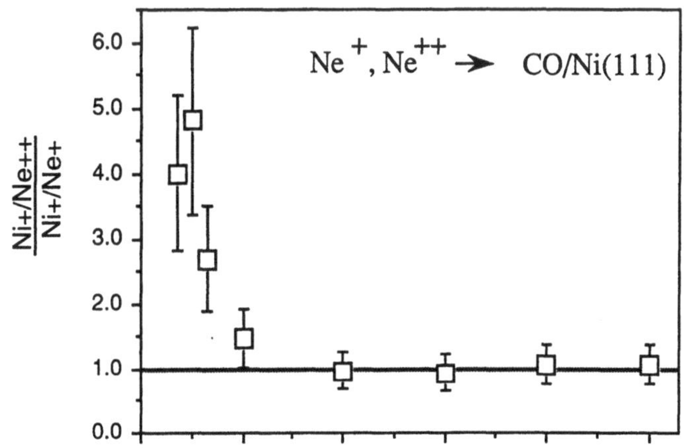

Figure 4b Ni^+ yield per incoming Ne^{++} ion divided by Ni^+ yield per incoming Ne^+ ion.

4. Discussion

4.1. Desorption of neutral particles (CO)

For these low ion impact energies there is strong evidence that direct knock - off caused by a collision between the primary ion and the adsorbed molecule is the dominating desorption process. For direct knock - off, Winters and Sigmund[13] give as a first approximation for the desorption cross section

$$\sigma \sim \frac{\ln E_0}{\cos\varphi} \quad . \tag{2}$$

To obtain eq.(2), a Born - Mayer potential and an isotropically bound adsorbate molecule, which is hit by the incoming ion with a primary energy E_0 and then reflected with zero energy loss from the substrate, has been assumed. Although these are very rough approximations, the observed energy dependence of the desorption cross sections is described quite satisfactorily as shown in fig.1a, where σ has been plotted versus E_0 in a semilogarithmic plot. Furthermore, good agreement is found by fitting the results for different impact angles to $\cos^{-1}\varphi$ dependence(fig.1b). Due to the almost equivalent masses of CO and Ne, a major part of kinetic energy can be transferred during the first collision. Accordingly, the effect of cascade sputtering is of minor importance.

4.2. Desorption of ionized particles (CO^+, Ni^+)

Benninghoven et al.[19] and Wittmaack [20] gave an overview of possible processes leading to secondary ion emission during particle bombardment. Neither of the proposed mechanisms involves a dependence on the primary ion charge state. Coulomb explosion, which has been proposed as a possible mechanism causing additional sputtering from insulators due the potential energy of the impinging particle [10,11], is unlikely for the CO/Ni(111) system. Because of the high mobility of the conduction electrons in the metal substrate, electron vacancies in the adsorbate created by charge exchange processes with the projectile will be recombined before the ejection mechanism is started. Production of ions due to secondary electrons can be excluded as well . The secondary electron yield due to potential emission decreases when a metal substrate is covered with an adsorbate [21] which is in contradiction to the large enhancement of the ion yield when adsorbing CO on the Ni - surface.

From our results of secondary ion emission depending on the charge state of the projectile with very low kinetic energy, two conclusions can directly be drawn. First, the main production process of desorbed (respectively sputtered) species is due to the collisional interaction between primary ion and adsorbate or substrate since the yield decreases very strongly with decreasing kinetic impact energy. This is also plausible in the light of results for desorption of neutral CO (fig.2a and fig.2b).

Second, the potential energy of the impinging ion influences only the charge state of the emitted particles, at least at very low kinetic impact energy (compare fig.2b and fig.3b).

Therefore one has to assume that the charge state of the projectile does not influence the amount of desorbed particles but only their charge state, and that the fraction of the ions of the sputtering yield increases with the potential energy of the projectile.

Considering the distance of closest approach (~ 1.5 Å for a head - on collision with 50 eV impact energy), the velocity of the particle and the distance where

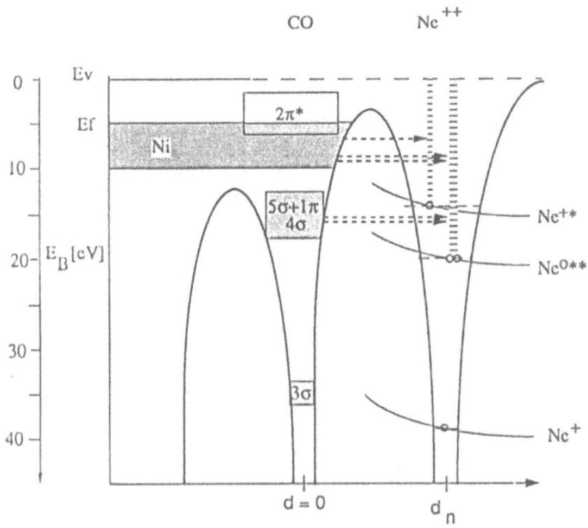

Figure 5a Electron energy diagram of a doubly charged Ne^{++} ion in front of a CO - covered Ni - surface. A shift to lower binding energy of the electronic states of Ne^+ due to the image force is drawn schematically. Possible resonance tunneling processes from the Ni - conduction band and the molecular states of the adsorbed CO into excited states of Ne^+ and autoionizing states of Ne^0 are indicated by arrows.

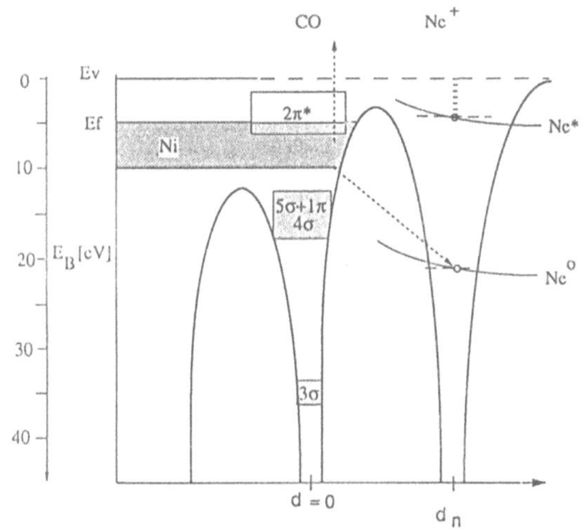

Figure 5b Electron energy diagram of a singly charged Ne^+ ion in front of a CO - covered Ni - surface. A shift to lower binding energy of the empty states of Ne due to the image force is drawn schematically. A possible Auger neutralisation process from the Ni - conduction band into the Ne^0 ground state is indicated.

electronic transitions start to become important (~ 3 - 5 Å), the collision time is in the order of 10^{-14} s. This is the typical time scale for Auger- and resonance-transitions.

Figures 5a and 5b show schematically the potential energy diagram of the Ni - surface, the adsorbed CO molecule and Ne^{++} and Ne^+ ions, respectively moving towards the surface. Ne^{++} provides many unoccupied excited and autoionizing states down to energies far below the Fermi level, which can be filled by resonance tunneling of either one or two electrons from the adsorbates $5\sigma+1\pi$ and 4σ states or from the Ni d band during the collision process.

For Ne^{+}, only Auger neutralization from the Ni d - band or the partly filled $2\pi^*$ level of CO into the Ne^{0} ground state is possible, since no unoccupied excited states of Ne^{0*} lie energetically within the occupied states of the substrate and the adsorbate.

Resonance tunneling is a much more probable process as compared to Auger transitions where two electrons are involved. Because of this and the fact that more final states are provided by Ne^{++} than Ne^{+}, the overall probability that a target particle is in an ionized state after the collision with a Ne^{++} ion is much higher than for interaction with Ne^{+}.

Sufficient energy has to be provided by a binary collision that the particle can leave the surface. So far, no decision can be made if this electron capture during the sputtering event takes place for higher impact energies as well. Only at very low impact energy, where ionization by kinetic processes strongly decreases, docs this effect gain enough importance to be detected.

5. Summary

The interaction of Ne ions with a CO - covered Ni(111) surface causes desorption of neutral CO - molecules with a high desorption cross section. Momentum transfer from the primary ion by binary collisions with the adsorbate causes the high desorption yield. The influence of impact energy and impact angle can be explained by applying conventional sputtering theory.

For very low impact energies the secondary ion yield is strongly influenced by the charge state of the primary ion. We propose a mechanism for the production of secondary ions by momentum transfer and simultaneous ionization due to electron transitions into the projectile which are more effective for Ne^{++} projectiles than for Ne^{+}.

6. Acknowledgement

This work was supported by Österreichischer Fonds zur Förderung der wissenschaftlichen Forschung, Projekt Nr.S 43/03 and by Jubiläumsstiftung der Stadt Wien.

7. References

[1] Desorption Induced by Electronic Transitions, DIET I, edited by N.H.Tolk, M.M.Traum, J.C.Tully and T.E.Madey, Springer-Verlag Berlin (1983)

[2] Desorption Induced by Electronic Transitions, DIET III, edited by R.Stulen and M.L.Knotek, Springer-Verlag Berlin (1988)
[3] S.P.Wolsky and E.J.Zdanuk, Phys.Rev. 121,**374** (1961)
[4] J.A.Van den Berg, Vacuum **36**, 981 (1986)
[5] P.Williams, Phys.Rev.B **23**, 6187 (1981)
[6] S.N.Morozov, D.D.Gurich and T.U.Arifov, Izv.Ak.Nauk SSSR, Ser.Fiz.**43**, 612 (1979)
[7] S.T.de Zwart, T.Fried, D.O.Boerma, R.Hoeckstra, A.G.Drentje, A.L.Boers, Surf. Sci. Lett.**177**, L939, (1986)
[8] P.Thompson and J.W.Taylor, Surf.Sci.**176**, 610 (1986)
[9] Sh.S.Radzhabov, R.R.Rakhimov and D.Absdulamov, Izv.Ak.Nauk SSSR, Ser.Fiz.**40**, 2543 (1976)
[10] I.S.Bitenskii, M.N.Murakhmetov and E.S.Parilis, Sov.Phys.Tech.Phys.**24**, 618 (1979)
[11] I.S.Bitensky and E.S.Parilis, Journal de Physique **50**, C2-227 (1989)
[12] E.Taglauer,W.Heiland and J.Oonsgard, Nucl.Instr.Meth.**168**, 571 (1980)
[13] H.F.Winters and P.Sigmund, J.Appl.Phys.**45**, 4760 (1974)
[14] E.Taglauer, W.Heiland and U.Beitat, Surf.Sci.**89**, 710 (1979)
[15] F.P.Netzer and T.E.Madey, J.Chem.Phys.**76**, 710 (1982)
[16] J.C.Campuzano, R.Dus and R.G.Greenler, Surf.Sci. **102**, 172 (1981)
[17] H.Froitzheim and U.Köhler, Surf.Sci.**188**,70 (1987)
[18] M.Grunze, H.Ruppender and O.Elshazly, J.Vac.Sci.Technol.A **6**, 1266 (1988)
[19] A.Benninghoven F.G.Rüdenauer and H.W.Werner, Secondary Ion Mass Spectrometry, John Wiley, New York (1987)
[20] K.Wittmaack in Inelastic Ion Surface Collisions, edited by N.H.Tolk, J.C.Tully, W.Heiland and C.W.White, Academic Press New York (1977)
[21] P.Varga, Comments At. Mol.Phys. **23**, 111 (1989)

Electron Stimulated Desorption of Hydrogen from Si(111) Surfaces Adsorbed with Hydrogen and Alkali Atoms

T. Yasue[1], *A. Ichimiya*[1], *and S. Ohtani*[2]

[1]Department of Applied Physics, School of Engineering, Nagoya University, Nagoya 464-01, Japan

[2]National Institute for Fusion Science, Nagoya 464-01, Japan

Electron stimulated desorption of hydrogen ions from hydrogen and alkali adsorbed on Si(111) surfaces is studied using a time-of-flight method. The Auger stimulated desorption is observed for hydrogen adsorption. For alkali adsorption, kinetic energies are explained using the potential energy diagram of alkali hydrides.

1.Introduction

Stimulated desorption studies for gas adsorbed metal surfaces and insulators such as alkali halides have been widely reported[1]. There are, however, few reports on stimulated desorption from semiconductor surfaces[2-6]. For the hydrogen adsorbed Si(111) surface, Madden *et al.*[2] showed that the desorption of H^+ ions was induced by the two-hole localization, which was created by the shake-up excitation of valence electrons in the threshold region. On the other hand, it is known that the surface structures of hydrogen and alkali adsorbed Si(111) surface are similar in RHEED analysis[7,8]. In the present work, we measured the desorption yield and the kinetic energy distribution of H^+ ions from hydrogen and alkali adsorbed Si(111) surfaces. The desorption processes are discussed using the potential energy diagram.

2. Experimental

The specimen was an n-type Si(111) surface. The surface was annealed by direct resistive heating. No ionic desorption was observed after annealing. Hydrogen was introduced through a palladium tube and dissociated by a hot tungsten filament. Lithium and sodium were evaporated using alkali metal dispensers. Desorbed ionic species were detected by a time-of-flight (TOF) mass spectrometer. In the present work, the desorption yield and the kinetic energy distributions were measured as a function of electron energy.

3. Results and Discussion

In a TOF mass spectrum obtained from the hydrogen adsorbed surface, a peak due to H^+ ions was clearly recognized but the other peaks due to impurities scarcely observed. For alkali adsorbed surfaces, alkali ions were scarcely observed, while H^+ ions were detected. The origin of hydrogen is considered to be outgas from the dispensers during evaporation. The bonding nature of alkalis to silicon is rather metallic[9]. Therefore the probability of alkali ion desorption is small as no localized repulsive states are produced with lifetimes long enough to enable desorption.

Figure 1(a) shows the desorption yield spectrum of H^+ ions for the hydrogen adsorbed surface. The threshold energy is determined to be 26 eV. In addition an enhancement of the desorption yield is observed at about 100 eV. For the threshold, the present results can be explained by the mechanism proposed by Madden *et al.*[2]. The form of

Springer Series in Surface Sciences, Vol. 19
Desorption Induced by Electronic Transitions, DIET IV
Editors: G. Betz · P. Varga

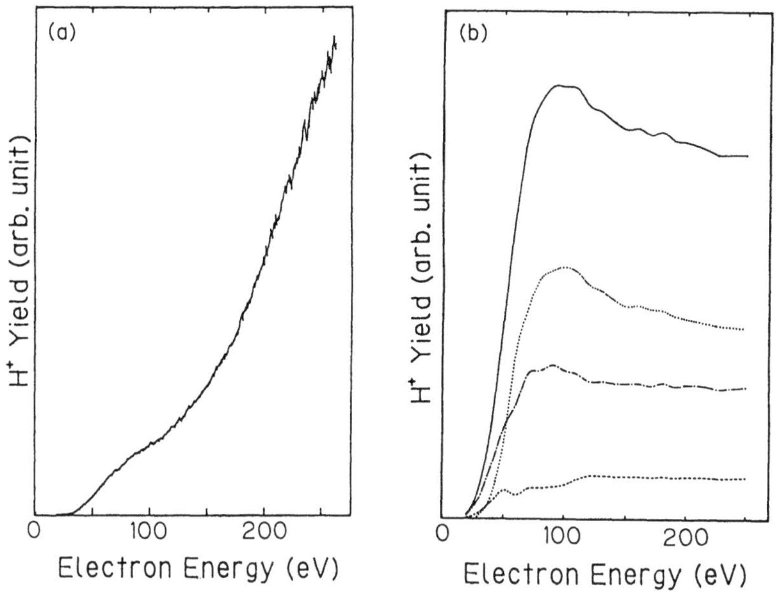

Fig. 1. The desorption yield spectrum of H^+ ions from (a) the hydrogen adsorbed surface and (b) the sodium adsorbed surface. In fig. 1(b) the solid line is the total H^+ yield spectrum, the dotted line the spectrum of the kinetic energy component of 6.7 eV, the dot-dashed line that of 4.4 eV and the broken line that of 2.9 eV. See also fig. 2(b).

the kinetic energy distributions of H^+ ions did not change over the range of electron energies used. This means that the final repulsive state did not change between below and above 100 eV. The binding energy of Si 2p electrons is 99 eV[10] and the satellite peak due to the localization was observed in the Si LVV Auger spectrum[2]. Therefore we assume that the enhancement of the desorption yield observed at about 100 eV is induced by the two-hole localization which is produced by the Si LVV Auger transition.

Figure 1(b) shows the desorption yield spectrum of H^+ ions for the Na adsorbed surface. The spectral profile is very different from that for the hydrogen adsorbed surface. The threshold energy is 13 eV, which is a fair amount lower than that for the hydrogen adsorbed surface. Figure 2(a) shows the dependences of the kinetic energy distributions of H^+ ions from the Na/Si surface on the incident electron energy. These distributions consist of three components of kinetic energies, whose centers are located at about 2.9, 4.4 and 6.7 eV as shown in fig. 2(b). As seen in the figure, intensities of each component show different dependences on the electron energy. In fig. 1(b), the dotted, the dot-dashed and broken lines are the decomposed desorption yield spectra, which correspond to the components whose kinetic energies are 6.7, 4.4 and 2.9 eV, respectively. For the highest kinetic energy component (the dotted line), the threshold energy is about 30 eV, which is slightly higher than those for the other components. For lithium adsorption, the threshold was 15 eV. The kinetic energy distribution consisted of two components, whose centers are located at about 3 and 4.5 eV.

Figure 3 shows the potential energy diagram of sodium hydride schematically. The cohesive energy of NaH is 8.22 eV[11]. For the transition from the ground state to the Coulombic repulsive state, the kinetic energy of desorbed H^+ ions is 6.5 eV, which is in good agreement with the measured highest kinetic energy of 6.7 eV. In this case the initial excitation is the excitation of Na 2p electron which is followed by the Auger transition. Then the excitation energy is about 30 eV, since the binding energy of Na 2p electrons is 33 eV[10]. This value is nearly the same as the threshold energy for the component with highest kinetic energy of 6.7 eV as shown in fig. 1(b). For the threshold behavior, the transition which requires the excitation energy of 13 eV is considered. If

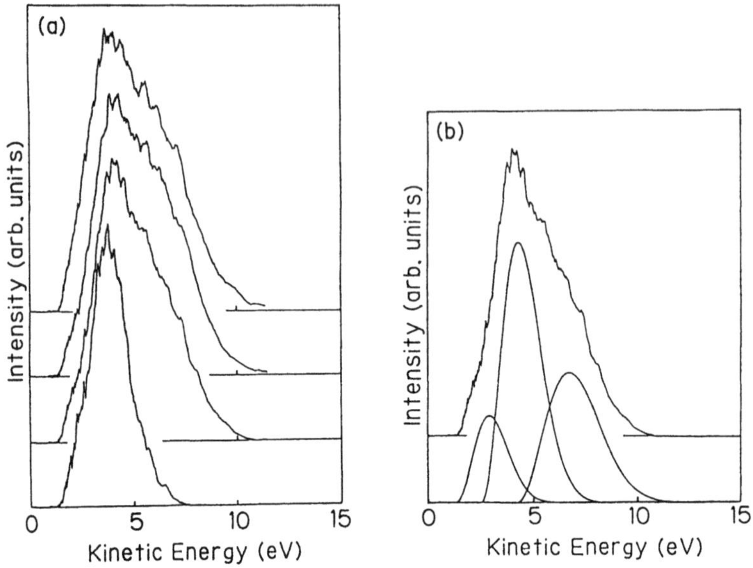

Fig. 2. (a) The kinetic energy distributions of H^+ ions from the sodium adsorbed surface. The incident electron energies are 150, 100, 60 and 20 eV from the top spectrum. (b) The kinetic energy distribution for the electron energy of 60 eV (top), and the three kinetic energy components (bottom).

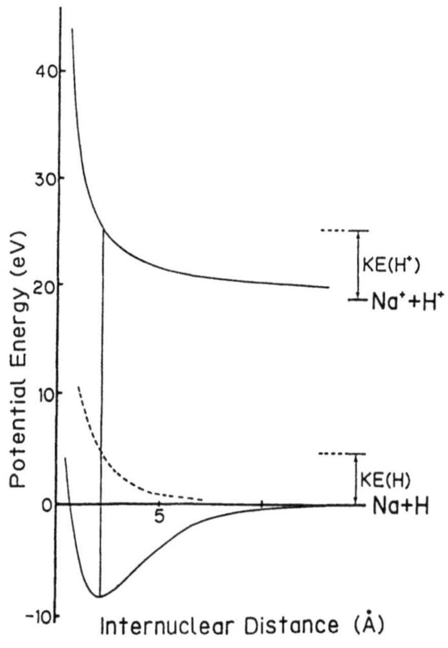

Fig. 3. The potential energy diagram of sodium hydride. The $(Na^+ + H^+)$ curve is given by the Coulombic form. The broken line represents a possible antibonding potential. KE's are the kinetic energies of outgoing species.

there is an antibonding state as indicated by the broken line, which is assumed to be the same as that of H_2 molecules because of the two-electron system, the kinetic energy of desorbed hydrogen is 4.8 eV. This kinetic energy is close to that measured at 4.4 eV in the present study. In the case of H^+ desorption from the Li adsorbed surface, the same mechanism is valid for the component of 4.5 eV by taking the cohesive energy of 9.38 eV [11] into account. The desorbed hydrogen, however, is neutral, so an ionization process is necessary in the outgoing path. Such a two-step mechanism is unlikely from the viewpoint of desorption probability. A similar gas-phase mechanism, however, has been reported in ESD of positive ions from alkali halides[12]. The desorption processes for the lowest kinetic energy component are unknown.

4. Conclusions

ESD of H^+ ions from hydrogen and alkali adsorbed Si(111) surfaces was observed using electron energies below 250 eV. The threshold behavior for the hydrogen adsorbed surface is explained in terms of the two-hole localization via the shake-up excitation. An enhancement of the desorption yield was first observed at an electron energy of about 100 eV. This enhancement is induced by the two-hole localization via the Si LVV Auger transition. For alkali adsorption, several kinetic energy components were observed, and each component showed a different dependence on the electron energy. It is considered that the different initial excitations result in the desorption of ions with different kinetic energies.

In ESD from alkali adsorbed Si(111) surfaces, the desorption of alkali ions was hardly observed. The reason is that the localization in the bonding orbital between alkalis and silicon scarcely takes place because of the metallic bonding character[9].

Acknowledgments

This work was carried out in collaboration with the Institute of Plasma Physics, Nagoya University. This work was supported by a Grant-in-Aid for Scientific Research from the Ministry of Education, Science and Culture.

References

[1] *Desorption Induced by Electronic Transitions DIET I*, ed. by N.H.Tolk, M.M. Traum, J.C.Tully and T.E.Madey (Springer, Berlin, 1983); *Desorption Induced by Electronic Transitions DIET II*, ed. by W.Brenig and D.Menzel (Springer, Berlin, 1985); *Desorption Induced by Electronic Transitions DIET III*, ed. by R.H.Stulen and M.L.Knotek (Springer, Berlin, 1988).

[2] H.H.Madden, D.R.Jennison, M.M.Traum, G.Margaritondo and N.G.Stoffel, Phys. Rev. **B26** (1982) 896.

[3] M.L. Knotek, G.M. Loubriel, R.H. Stulen, C.E. Parks, B.E. Koel and Z. Hussain, Phys. Rev. **B26** (1982) 2292.

[4] N.Matsunami, Y.Hasebe and N.Itoh, Surface Sci. **192** (1987) 27.

[5] C.U.S.Larsson, A.S.Flodström, R.Nyholm, L.Incoccia and F.Senf, J. Vac. Sci. Technol. **A5** (1987) 3321.

[6] J.A.Yarmoff, A.Taleb-Ibrahimi, F.R.McFeely and Ph.Avouris, Phys. Rev. Letters **60** (1988) 960.

[7] A.Ichimiya and S.Mizuno, Surface Sci. **191** (1987) L765.

[8] S.Mizuno and A.Ichimiya, Appl. Surface Sci. **33/34** (1988) 38.

[9] S.Nishigaki, M.Ohara, A.Murakami, S.Fukui and S.Matsuda, Appl. Surface Sci. **35** (1988-89) 121.

[10] K.D.Sevier, Atomic Data Nucl. Data Tables **24** (1979) 323.

[11] R.C.Bowman,Jr., J. Phys. Chem. **75** (1971) 1251.

[12] R.E.Walkup, Ph.Avouris and A.P.Ghosh, Phys. Rev. **B36** (1987) 4577.

Resonant Desorption of Negative Ions from Adsorbed Layers Bombarded by Very Low Energy Electrons

M. Bernheim and Ting Di Wu

Laboratoire de Physique des Solides, Bât. 510, Université Paris-Sud, F-91405 Orsay, France

1. Introduction

Positive or negative, atomic or molecular ions may be desorbed during an electron bombardment of layers adsorbed on solid samples. The charge and the nature of the desorbed species depend on the adsorbed layer composition as well as on the energy of the incident electron beam.

Whereas for positive ions the desorption generally starts above energy thresholds higher than 10eV, intense negative ion desorption yields can be measured during low energy bombardments of adsorbed layers[1, 2]. This low energy desorption has been attributed to resonant dissociative electron attachments similar to those in play during electron collisions with free molecules[3].

Up to now, the lack of data on the negative ion desorption might be attributed
(i) to the difficulty of using of intense low energy electron beams,
(ii) to the low initial energies of negative desorbed ions, most of which are not collected in field-free optical systems,
(iii) to a lesser degree to the elimination of all the backscattered and secondary electrons from the measuring cell.

Obviously, both ion desorptions might provide complementary information. The attractive features for the negative ion desorption could be :
(i) the high yields for atoms and polyatomic species having high electron affinities [4],
(ii) the very low energy process which might aid the understanding of heterogeneous catalytic reactions,
(iii) the possible selective desorptions on heterogeneous samples, according to the nature of the adsorbed species and to the energy of the incident electron beam ; thus, from ion images, analytical information on sample surface composition might be acquired without destruction of the sample itself in contrast to analysis by Secondary Ion Mass Spectrometry.

2. Experimental set-up

A cathode lens was used as an electrostatic "mirror" to decelerate an electron beam just in the last part of its path (Fig.1). Thus the set–up was similar to the one developed to control the surface potential of insulating samples during negative SIMS analyses[5]. Such a set-up was used for our first desorption experiments[2]:

The electrons reached the cathode, i. e. the investigated sample surface S, at normal mean incidence, with a high and nearly constant density for any incident energy. This energy was controlled by the differences in voltage (and in work function) between sample surface and the filament of the electron source. In such a configuration no photons left the filament to reach the sample surface.

The desorbed ions were collected over a well-defined imaged field. The size of the cross–over stop d_1[6] controlled the discrimination in initial energies ϕ_0 and desorption angle α_0. An image demagnified by a factor 2 of this cross–over was focussed by the transfer lens L_2 on the entry stop d_2 of the mass spectrometer. In this latter, a mass and energy analysis was achieved through successive magnetic and electrostatic prisms coupled in a high energy dispersion mode[7]. Then the filtered ion beam could be either

Springer Series in Surface Sciences, Vol. 19
Desorption Induced by Electronic Transitions, DIET IV
Editors: G. Betz · P. Varga

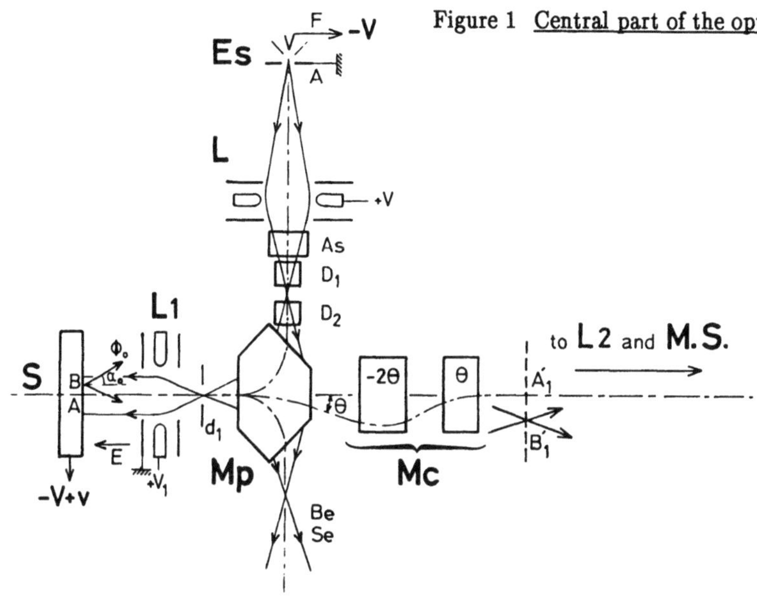

Figure 1 Central part of the optical system

The electrons emitted from the source E_s were focussed by lens L then deviated by a stigmatic prism Mp near to the focal point of the cathode lens L_1 axis. In this immersion lens the electrons were decelerated just as they reached the target surface S. The same lens ensured a very efficient collection of the desorbed ions over a well-defined field. A double correction prism Mc was added to re-align the ion beam in the transfer optics L_2 and mass spectrometer M.S. axis for each ion mass. On the symmetrical exit of the Mp magnetic prism, secondary and backscattered electrons could be recorded.

post–accelerated into an electron multiplier for direct intensity measurements, or focussed on a channel plate as an ion image of the investigated surface. Because of the ion path length only stable species could reach the detector (the transit time for 4000 eV H^- ions being $3.2 \cdot 10^{-8}$s).

A mass filtered Ar^+ ion beam was used to sputter off the adsorbed layers prior to introduction of the selected molecular species as a narrow molecular beam directed onto the sample surface. In order to prevent magnetic interferences in the electron beam optics, non-magnetic materials were selected for most elements near to this beam ; for the same purpose, the accelerating mode for lenses L, L_1, L_2 [8] and continuous soft magnetic material shields were used.

The whole set–up was very convenient for recording the variations of negative ion desorption yields with electron energy as well as the ion energy distributions for any incident energy. A more extensive description of the experimental set–up, the practical optical adjustments and the procedure used to control the mass spectrometer will be presented elsewhere [9].

3. Some of the first experimental results and discussion

This set-up was used to study the negative ionization of free molecules by low energy electron impact. Among the different gases investigated, we report in figure 2 the large variation of the H^- ion yields during electron collision on hydrogen molecules. A clear separation of several ionization processes was obtained in fair agreement with literature data[10] except for some small shifts in energy that might be attributed to the kinetic

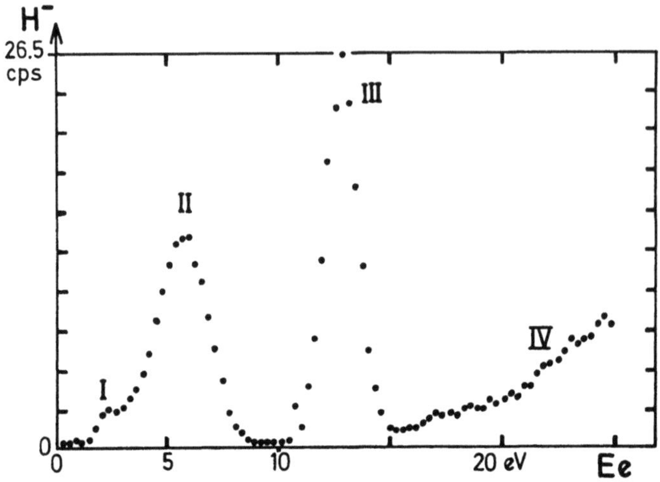

Figure 2 Ionization of free hydrogen molecules

The sample was moved backwards ; its voltage was increased to keep the same electrostatic field. The H— ions were produced in direct electron collisions with the H_2 molecules being in a region about 1 μm thickness, isolated by the narrow energy band pass (1eV) used for the mass spectrometer. The four ionization processes were identified clearly, though the kinetic energy transferred to the ion during the ionization process could not be isolated here.

energy acquired by the ion during the collision, rather than to differences in experimental conditions. The main features vary with the nature of the molecular beam interacting with the electron beam obviously.

Let us now consider the negative ion desorption from adsorbed hydrogen layers. The variation of desorption yield with electron energy presented a sharp peak at 6.5eV, followed by a low emission plateau and new increase of desorption yields above 15eV (Fig.3). The position of the desorption peak depends on the nature of the metallic sample only slightly. In addition, for any electron energy, as well as for any investigated sample, the hydrogen ions were desorbed with narrow energy distributions and low mean energies about 1.5eV (see Fig.3 insert). Such observations confirmed our first experiments [2] but not other literature data [11]. Nevertheless, using well-controlled molecular species adsorbed on different surfaces, we have been unable to reproduce surface conditions inducing a marked displacement in electron energy for the H^- resonant desorption peak up to now.

The practical experimentation was made difficult by the high efficiency of desorption itself, which induced rapid variations of surface composition (through neutral and negative ion desorption, the positive ions being retained by the electrostatic field). Therefore, the experiments could be performed either on adsorbed layers fabricated in different conditions beforehand or under dynamical equilibrium, keeping a weak continuous molecular flow directed onto the surface investigated.

It is noteworthy that even for the second conditions, the stabilisation of desorption yields and of shapes of desorption curves was attained after a time varying with the electron intensity and energy as well as with the molecular species used. Thus, apparently, the structure of the adsorbed species could be modified during the electron bombardment. Nevertheless an increase in electron density induced a proportional growth of the desorbed ion intensity ; a similar tendency occurred during modification of the exposure to the molecular beam until saturation was reached.

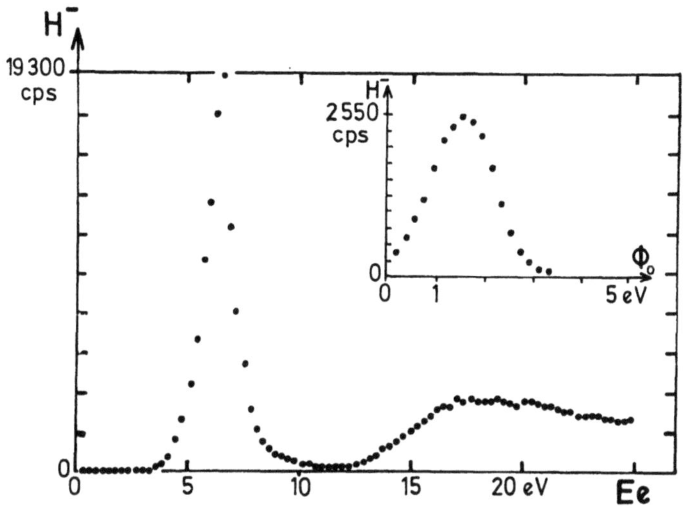

Figure 3 Incident electron energy influence on the H⁻ ion yield desorbed from an hydrogen layer adsorbed on a (100) aluminum crystal.

A pure continuous hydrogen beam was directed on the surface just after the end of the Ar^{+} ion sputtering. Due to the difference in work function between tungsten filament and aluminum crystal, the electron reached the surface with an energy about 1 eV higher than the energy reported on the x-axis. Thus the resonant process is centered at 7.5 eV. The corresponding energy distribution shown in the insert was recorded with a narrow energy band pass (1 eV).

Some features suggest that desorption takes place on quasi-physisorbed layers weakly bounded on top of chemisorbed layers, whereas some experiments established that desorption could occur also out of bare bulk material (silicon hydride for instance).

In their present state, our results leave many questions unanswered. Many new investigations are needed to reach a full comprehension of all the processes involved. New molecules and other ions species have to be studied. Some set–up adaptations are planned to extend the experimental possibilities, in order to record the angular distributions of desorbed negative ions [12] (for comparison with the angular effects for positive desorbed ions). Other information could be extracted from the variations of the secondary and backscattered electron intensities around the resonant desorption peak

This low energy desorption could influence many surface analyses. To end with, it seems that such a "mirror" like optical set-up could ease other surface investigations, inverse photo–emission studies for instance.

References

(1) Liu Zhen Xian and D. Lichtman, Surf. Science 114 (1982) 287 and Liu Zhen Xian and D. Lichtman, Surf. Science 125 (1983) 490.
(2) M. Bernheim, M. Chaintreau, R. Dennebouy and G. Slodzian, Surf. Science 126 (1983) 610.
(3) D. Rapp and D. D. Briglia, J. Chem. Phys. 43 (1965) 1480.
R. J. Van Brunt and L. J. Kieffer, J. Chem. Phys. 60 (1974) 3062.
(4) H. Hotop and W. C. Lineberger, J. Phys. Chem. Ref. Data 14, 3 (1985) 731.
A. Radzig and Smirnov Ref. Data, Springer Series in Chem. Phys. 31 (1985).

(5) G. Slodzian et al. SIMS V Conf. Springer Series in Chem. Phys. 9 (1985) 40.
G. Slodzian, Optik 77, 40 (1987) 148.
(6) R. Castaing and G. Slodzian, J. Phys. E 14 (1981) 1119.
(7) M. Bernheim and F. Le Bourse, Nucl. Instr. Meth. B 27 (1987) 94.
M. Bernheim, Rev. Phys. Appl. 24 (1989) 515.
(8) G. Slodzian and A. Figueras, J. Phys. (lettres) 39 (1978) 90.
(9) M. Bernheim and Ting Di Wu (to be published).
(10) H. S. W. Massey, Negative Ions,Cambridge Univ. Press (1976) 309.
(11) J. H. Craig, Surf. Science 127 (1983) 175.
(12) A.L. Johnson, S.A. Joyce, T.E. Madey, Phys. Rev. Lett. 61 (1988) 2578.

Note suggested by the referee.

The H^- desorption is attributed to a significant adsorption of hydrogen on metallic surfaces (Al, Mg, Ti especially). The ultra high vacuum conditions (8 10^{-11} torr, locally increased by a large 15 K cryogenic trap isolating the investigated surface) and the careful backing of the hydrogen introduction needle valve were supposed to eliminate any water vapour fixation on the investigated surface. The low level of water vapour on the investigated surface could be established either by electron stimulated desorption or by secondary ion emission of O^- ions. The intensities of O^- ions could be referred to the O^- intensities measured on water vapour adsorbed layers.

Resonant Desorption from Clean and Carbon-Covered Alkali Halide Surfaces: State-Specific Detection by REMPI

J. Heidberg, W. Claas, H.-C. Langowski, M. Suhren, and M. Warskulat*

Institut für Physikalische Chemie und Elektrochemie, Universität Hannover, Callinstr. 3–3A, D-3000 Hannover 1, Fed. Rep. of Germany

*Permanent address: Philips and DuPont Optical Deutschland GmbH, Dev/PM, Postfach 1409, D-3000 Hannover 1, Fed. Rep. of Germany

Desorption of NO from multilayer systems condensed on NaCl(100) as well as on C-films evaporated onto NaCl(100) was induced by resonant excitation of the symmetric stretch vibration ν_1 of the NO cis-dimer with the frequency doubled radiation of a CO_2 laser. On C-film-NaCl(100), additional non-resonant desorption was observed at lower condensate coverages (≈16 monolayers). Translational temperatures as derived from the time-of-flight distributions of the desorbed NO monomers are higher than or equal to the calculated maxima of the laser-induced surface temperatures. State-selective detection of desorbed NO molecules by resonance enhanced multi-photon ionization (REMPI) indicates rotational temperatures of 190 ± 20 K at corresponding desorbate translational temperatures of 120 K.

1. Introduction

Desorption induced by resonant excitation of adsorbed molecules with IR-laser radiation - reported for the first time by Heidberg and co-workers in 1978 [1] - has been studied by several groups on many adsorption systems [2-7] and multilayer condensate systems [8-12].

Characterization of desorbed NO molecules with regard to energy distribution among translational, rotational, in some cases also vibrational degrees of freedom has been performed for temperature programmed desorption, laser-induced thermal desorption or scattering of molecular beams. For an overview see [13, 14]. For UV-laser induced desorption, a resonant mechanism by electronic excitation of the adsorbate has been suggested [15].

In [12], resonant desorption of NO induced by excitation of the symmetric stretch vibration of the NO-cis-dimer in the multilayer condensate system NO/C-film-NaCl(100) with the frequency-doubled radiation of a CO_2 laser could be demonstrated by us.

Here a more detailed study concerning the influence of the substrate (NaCl-cleavage plane vs. C-film on NaCl(100)) and of condensate coverage on desorption yield and desorbate translational temperatures has been carried out. Additionally first REMPI results for this kind of desorption process are presented.

2. Experimental

Most of the experimental details have been described in [12]. Therefore, apart from a brief description, only additional features will be mentioned.

The experiments were performed in an ion-pumped UHV chamber, equipped with sample holder attached to a liquid He cryostat. The samples, NaCl single crystals cleaved in air, could be cooled down to 20 K. Carbon films were evaporated by resistive heating of graphite rods. NO, purified by vacuum distillation, was admitted onto the substrates via a capillary out of

Springer Series in Surface Sciences, Vol. 19
Desorption Induced by Electronic Transitions, DIET IV
Editors: G. Betz · P. Varga

a bakeable gas inlet chamber with known volume, allowing an estimation of the number of condensed molecules.

The radiation of a line-tunable CO_2 laser, frequency-doubled with a Te single crystal, was focussed with a NaCl-lens (f = 15 cm) onto a spot of 0.95 mm diameter in the sample surface, at an angle of incidence of 30°, yielding a fluence of 43 mJ cm^{-2}. Beam energy after a sapphire filter was 360 µJ.

Detection of desorbed molecules and recording of time-of-flight distributions was carried out either with a quadrupole mass spectrometer (QMS), or by resonance-enhanced two-photon ionization (REMPI). The QMS was situated at an angle of 30° relative to the surface normal and a distance of 30 cm from the sample surface. In between the QMS and the sample was a liquid-nitrogen-cooled diaphragm. The REMPI was achieved at wavelengths around 226 nm with the frequency-doubled radiation of a dye laser; ion detection was done with a micro channel plate (MCP). Signals coming from either QMS or MCP were amplified and stored in a signal memory recorder coupled with a PC.

IR spectra of NO on NaCl(100) and highly oriented pyrolytic graphite (HOPG) were taken in a separate UHV chamber - equipped with XPS - which could be connected to a FTIR spectrometer with optics for surface measurements according to Heidberg. With XPS no contaminations either on the surface nor in the adlayers could be detected.

3. Results

The resonant characteristic of the desorption process is shown in figs. 1a, b for both types of substrates, together with FTIR-spectra. The appearance of absorption bands around 1860 cm^{-1} (ν_1-) and at 1752 cm^{-1}/1780 cm^{-1} (ν_5-normal vibration) shows that the NO-condensate consists of dimers [16].

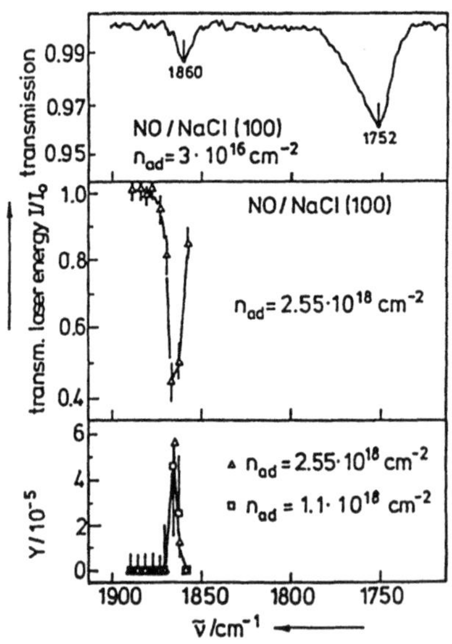

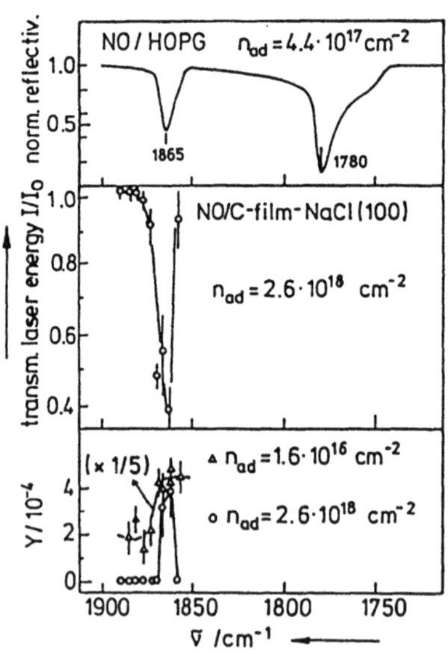

Figs. 1a(left), 1b(right): top: FTIR-transmission-absorption spectrum for NO/NaCl(100)(a), reflection-absorption spectrum for NO/HOPG (b), center: transmitted laser energy vs wavenumber for (a) NO/NaCl(100), (b) NO/C-film-NaCl(100), bottom: NO desorption yield induced by irradiation with frequency-doubled CO_2 laser radiation (43 mJ cm^{-2}) for different coverages.

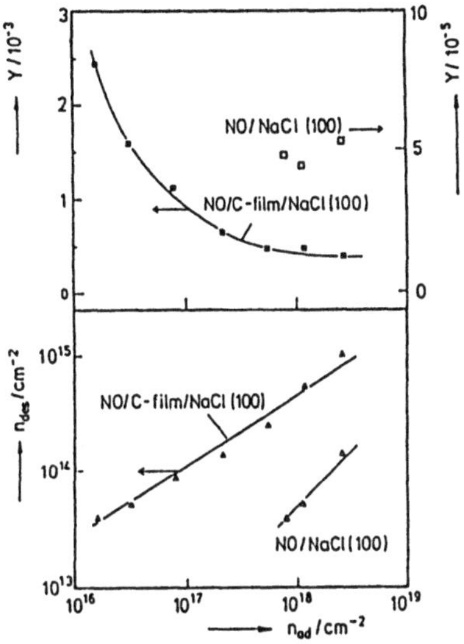

Fig. 2: Desorption yield Y (top) and number of desorbed molecules per unit surface area n_{des} (bottom) vs initial coverage n_{ad}. Desorption laser fluence: 43 mJ cm^{-2}.

The coincidence of the maximum desorption yield with maximum absorption of laser radiation at ν_1, the symmetric stretch vibration of the NO dimer, is evident.

An absorption cross section of the NO condensate of $4.5 \cdot 10^{-19}$ cm^2 at the wavenumber of absorption maxima (1862 cm^{-1} for NO/C-film-NaCl(100), 1865.9 cm^{-1} for NO/NaCl(100)) was derived from measurements at different condensate coverages. The absorption of the 4.5 nm-thick carbon film around 1860 cm^{-1} could be estimated to be 4.4 %. The energy of the laser photons at the resonance frequencies, equivalent to 22.3 kJ mol^{-1}, is larger than the condensation/desorption energy, which can be deduced from thermodynamic data [18] and from our own TPD-measurements [19] to be 15.5 ... 17.1 kJ mol^{-1}.

For C-film-NaCl(100) and lower condensate coverages ($N = 1.6 \cdot 10^{16}$ cm^{-1}), a substantial portion of the desorption yield appears to be independent of laser frequency. It should be noted that under these conditions only 0.7 % of the incident laser energy, which is by a factor of 6 smaller than the amount absorbed by the carbon film (see above), is absorbed by the condensate at resonance frequency.

The desorption yield **Y** is defined by $Y \equiv -\ln(1 - n_{des}/n_{ad})$ [1], with n_{des} number of desorbed, n_{ad} number of adsorbed/condensed molecules per unit surface area. Desorption yield dependence on n_{ad} is shown in fig. 2. Even at the highest condensate coverages investigated, the amount of desorbed molecules is not higher than the equivalent of one monolayer.

Translational temperatures as derived from time-of-flight distributions for the desorbed NO monomers are shown in fig. 3. No NO dimers could be detected.

Maxima T_{stheor} of the laser-induced surface temperature are also shown. They were calculated according to [17], thereby taking into account light absorption by the condensate as well as by the substrate and the heat capacity of condensate and substrate [18].

Relative populations of the rotational states with J = 3/2 up to J = 29/2 of the electronic ground state $X\ ^2\Pi_{1/2}$ as deduced from two-photon-ionization within the $(Q_{11} + P_{21})$ branch of the NO $A^2\Sigma^+ \leftarrow X^2\Pi$ transition are shown in fig. 4 for a reference measurement at 300 K and $2 \cdot 10^{-6}$ mbar NO as well as for a desorption experiment from a multilayer system of ≈ 2600 monolayers on NaCl(100). The slopes indicate a rotational temperature of 300 ± 15 K for the reference measurement and 190 ± 20 K for the desorption experiment.

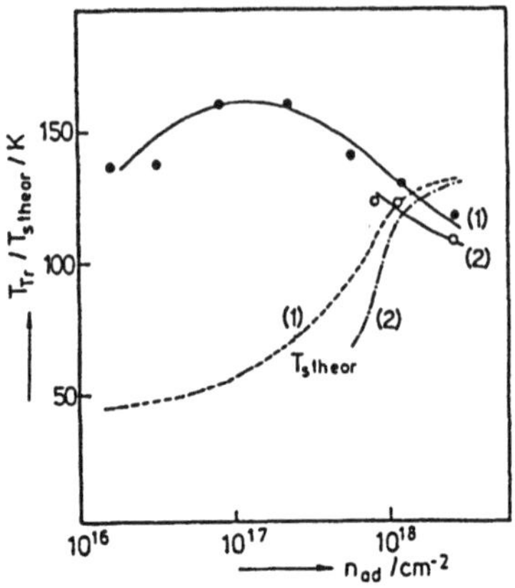

Fig. 3: Translational temperatures T_{Tr} and calculated maxima of laser-induced surface temperature T_{stheor} vs initial coverages n_{ad}.
(1): NO/C-film-NaCl(100)
(2): NO/NaCl(100)

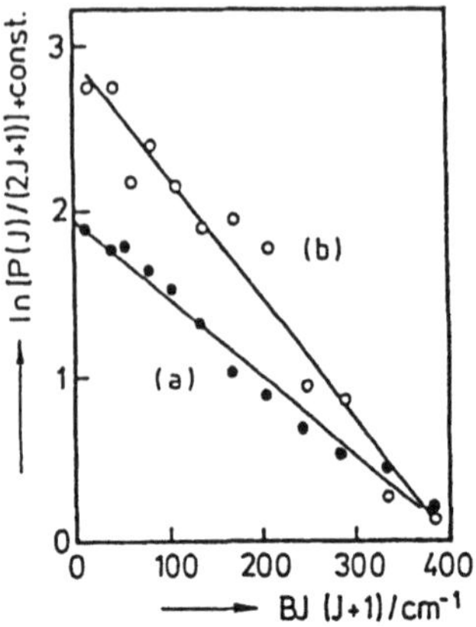

Fig. 4: "Boltzmann-plot" for the populations of the rotational states of NO molecules,
(a) gas phase, 300 K, $2 \cdot 10^{-6}$ mbar,
(b) desorbate from NO/NaCl(100), $n_{ad} = 2.55 \cdot 10^{18}$ cm^{-2}.
The slopes indicate:
(a) $T_{rot} = 300 \pm 15$ K,
(b) $T_{rot} = 190 \pm 20$ K.

4. Discussion

These results indicate that the desorption mechanisms for resonant excitation differ from that for non-resonant excitation: Compared with non-resonant excitation, an additional 16 percent of laser energy resonantly absorbed by the NO-layers at 1862 cm^{-1} (for NO/C-film-NaCL(100), $n_{ad} = 1.6 \cdot 10^{16}$ cm^{-2}) leads to an increase of desorption yield by 100 percent (see fig. 1 b).

This large additional effect - due to the small amount of additional absorbed laser energy - cannot be explained by "resonant heating", where the energy resonantly absorbed is assumed to be distributed within adsorbate and substrate, thus leading to thermal desorption.

The same argument holds for the observed translational temperatures, which are higher than the calculated maxima of the surface temperature for NO/C-film-NaCl(100) with coverages below $7 \cdot 10^{17}$ cm^{-2}. These temperatures indicate an incomplete energy redistribution within the condensate/substrate prior to desorption.

The rotational temperature of 190 K at $n_{ad} = 2.55 \cdot 10^{18}$ cm^{-2} (corresponding to a translational temperature of 120 K) indicates the absence of "rotational cooling" (e.g [14]). One possible explanation is given in a model proposed by Gadzuk et al. [20]. This model predicts for the desorbate molecules formal rotational temperatures, which depend on the angle of a cone in which the adsorbed molecules are assumed to rotate and are practically independent of surface temperature in the low range of desorbate rotational states.

In the frame of this model, 190 K of rotational desorbate temperature means for NO and the investigated desorbate rotational states of $J \leq 29/2$ that the adsorbed NO molecules are confined within a cone with an angle of about 20° relative to the surface normal.

Another model [21] suggests a librational motion of adsorbed molecules as a predesorption step. This should lead to desorbate molecules in high rotational states.

To verify either of these models, however, more information about the population of higher rotational states (up to J ≥ 35/2 c.f. [20]) and especially of vibrational desorbate states is necessary. Moreover, experiments with adsorbates in the monolayer region are required to give insight into the adsorbate-surface interaction in the course of the desorption event.

For appropriate experiments, an increase in detection sensitivity for desorbed NO molecules and - more important - in the efficiency of second harmonic generation for CO_2 laser radiation by more than one order of magnitude has to be accomplished.

5. References

[1] J. Heidberg, H. Stein, A. Nestmann, E. Hoefs, I. Hussla, Proc. Mat. Res. Soc. Symp. on Laser-Solid Interactions and Laser Processing, S.D. Ferris, H.J. Leamy, J.M. Poate (eds.), Am. Inst. of Physics, New York **49** (1979), J. Heidberg, H. Stein, E. Riehl, A. Nestmann, Z. Phys. Chem. NF **121**, 145 (1980), J. Heidberg, H. Stein, E. Riehl, Phys. Rev. Letters **49**, 666 (1982), Surface Science **126**, 183 (1983)

[2] J. Heidberg, D. Hoge, J. Opt. Soc. Am. B **4**, 242 (1987)

[3] J. Heidberg, H. Stein, H. Weiß, Surface Science **184**, L 431 (1987), J. Heidberg, K.W. Stahmer, H. Stein, H. Weiß, J. El. Spectr. Rel. Phen. **45**, 87 (1987)

[4] T.J. Chuang, J. Chem. Phys. **76**, 3828 (1982), Surface Science Reports **3**, 1 (1983)

[5] T.J. Chuang, I. Hussla, Phys. Rev. Letters **52**, 2045 (1984)

[6] I. Hussla, H. Seki, T.J. Chuang, Z.W. Gortel, H.J. Kreuzer, P. Piercy, Phys. Rev. B **32**, 3489 (1985)

[7] T.J. Chuang, H. Seki, I.Hussla, Surface Science **158**, 525 (1985)

[8] B. Schäfer, P. Hess, Chem. Phys. Letters **105**, 563 (1984), Appl. Phys. B **37**, 197 (1985)

[9] M. Buck, B. Schäfer, P. Hess, Surfac Science **161**,245 (1985)

[10] M. Buck, P. Hess, J. El. Spectr. Rel. Phen. **45**, 237 (1987)

[11] J. Heidberg, H.-C. Langowski, G. Neubauer, M. Folman, Surface Science **189/190**, 946 (1987)

[12] J. Heidberg, H.-C. Langowski, G. Neubauer, M. Folman, J. El. Spectr. Rel. Phen. **45**, 249 (1987)

[13] D.S. King, R.R. Cavanagh, Chemistry and Structures at Interfaces: New Laser and Optical Techniques, R.B. Hall, A.B. Ellis (eds.), VCH, Deerfield Beach (1986)

[14] H. Zacharias, Appl. Pho. A **47**, 37 (1988)

[15] D. Weide, P. Andresen, H.J. Freund, Chem. Phys. Letters **136**, 237 (1987), F. Budde, A.V. Hamza, P.M. Ferm, G. Ertl, D. Weide, P. Andresen, H.J. Freund, Phys. Rev. Letters **60**, 1518 (1988), W.C. Natzle, D. Padowitz, S.J. Sibener, J. Chem. Phys. **88**, 7975 (1988)

[16] J. Heidberg, E. Kampshoff, M. Suhren, to be published

[17] J.H. Bechtel, J. Appl. Phys. **46**, 1585 (1975)

[18] Y.S. Touloukian (ed.) Thermophysical Properties of Matter, Vol. 2, Vol. 6, IFI/Plenum, New York (1970), H.L. Johnston, W.F. Giauque, J. Am. Chem. Soc. **51**, 3194 (1929)

[19] H.-C. Langowski, Thesis, Univ. Hannover (1989)

[20] J.W. Gadzuk, U. Landman, E.J. Kuster, C.L. Cleveland, R.N. Barnett, Phys. Rev. Letters **49**, 426 (1982)

[21] A. Ben Ephraim, M. Folman, J. Heidberg, N. Moisejev, J. Chem. Phys. **89**, 3840 (1988)

Processes of Desorption and Desorption Cross Sections with Electrons at Silicon Surfaces

S.L. Bennett and E.M. Williams

Department of Electrical Engineering and Electronics, The University of Liverpool, P.O. Box 147, Liverpool L69 3BX, UK

Electron stimulated production of positive ions from water exposed silicon monocrystals (100) and (111) has been studied following adsorption, oxidation and electron irradiation. Measurements of desorption cross section are described for both H^+ ions as well as the range of minority OH^+ and O^+ ions observed in the study. The decay of adsorbate signal is further compared with the reduction of the oxygen (KLL) Auger signal.

1. Introduction

The interaction of low energy electrons (<1000 eV) with surfaces has been known for some time to have the capability of breaking surface and intramolecular bonds resulting in the ejection of ions, neutrals and metastables from the surface [1]. Removal of a bond follows a core or valence level excitation which evolves to a point where the surface group or fragment resides in a repulsive potential. The surface itself may influence this process to the extent that it provides pathways for the delocalisation of the initial excitation energy dependent upon the geometry and electronic structure of the surface. The nature of the desorption process makes the technique of ESD particularly sensitive to the topmost layer.

The mass separation of desorbed positive ions provides a route to the identification of surface groups, while the behaviour of the ion yield, energy distribution, threshold and angular distribution of ions reflect the electronic and geometric nature of the surface bonding. These variations in bonding between surface sites can often be more subtle than changes normally detected by techniques such as Auger electron spectroscopy. Indeed it has been noted that in some cases changes in the ESD yield are the only evidence of a multiplicity of bonding sites [2].

2. Experimental details

Desorption measurements were carried out in a UHV system equipped with a time of flight apparatus for electron stimulated desorption measurements, a cylindrical mirror analyser for Auger analysis, and facilities for Argon ion cleaning and sample heating. A pulsed electron gun provided a primary electron beam incident on the target. Desorbed positive ions were detected by a channel electron multiplier (CEM) positioned some 25 cms from the target. A

Springer Series in Surface Sciences, Vol. 19
Desorption Induced by Electronic Transitions, DIET IV
Editors: G. Betz · P. Varga

quadrupole filter in front of the CEM was used for the mass analysis of the desorbed species. The quadrupole unit could be biased to aid ion collection.

Si(100) and Si(111) targets were cleaned by argon ion bombardment and annealed at ~1000° C. Water dosing of the surfaces was carried out with a molecular beam effusion source. Previous experiments [3] have shown that in the absence of electron irradiation during exposure the resulting surface state is restricted to adsorbed surface groups. Simultaneous electron-water exposure, on the other hand, results in the oxidation of the surface. For sufficiently high electron doses, measurable depletion of surface groups takes place, from which the desorption cross sections may be calculated. Provision was made in the experiment for using an auxiliary filament to bombard the surfaces at high current density (up to 6.5 mA cm^{-2})

3. Results

The time of flight spectra obtained from the (100) and (111) surfaces under conditions where exposure results in adsorption without oxidation are presented in fig. 1a and 1b respectively. In the former case, ions identified as H^+ and OH^+ are observed, consistent with dissociative adsorption. Conversion of the flight time scale to one of energy, gives a value of 4.9 eV as the most probable energy. This is also close to values obtained for H^+ desorption from other substrates and is attributed to cleavage of the O-H bond in surface hydroxyl groups [3,4,5].

Desorption from the (111) surface following the same adsorption procedure yields the H^+ ion, with a much weaker OH^+ signal. The peak energy of the H^+ ion was measured as 5.5 eV.

Desorption from oxidised surfaces is markedly different from the case of pure adsorption. Figs. 2a, b and c show the detection of ions from both surfaces following the establishment of surface oxide (as seen in AES). The oxidised Si(100) surface (fig. 2a) yields H^+ and OH^+ as for electron-free water exposure, but is now further accompanied by a peak attributed to the O^+ ion. The growth of the OH^+ and O^+ peaks is concomitant.

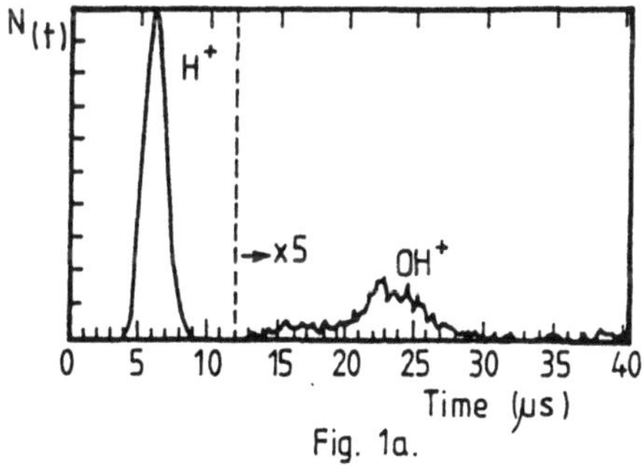

Fig. 1a.

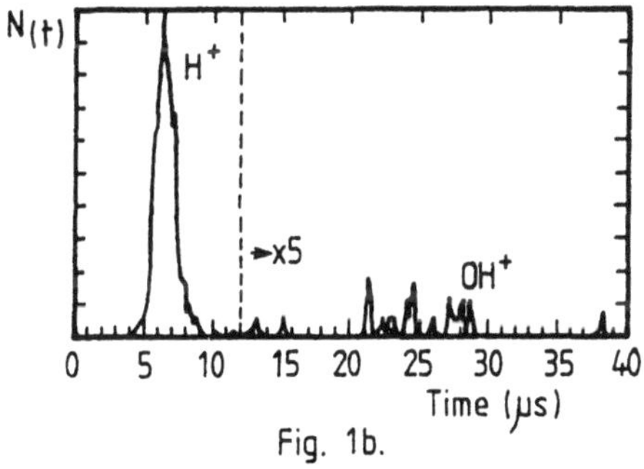

Fig. 1b.

Fig. 1a Time of flight spectrum of ions desorbed from Si(100) exposed to 10 L of water in an electron free environment.

Fig. 1b Time of flight spectrum of ions desorbed from Si(111) exposed to 200 L of water in an electron free environment.

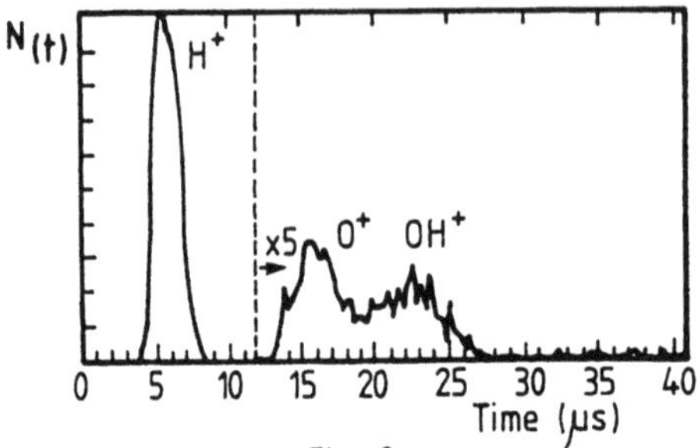

Fig. 2a.

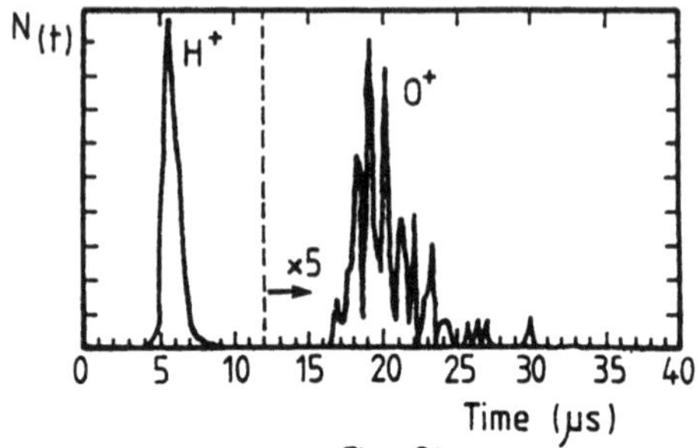

Fig. 2b.

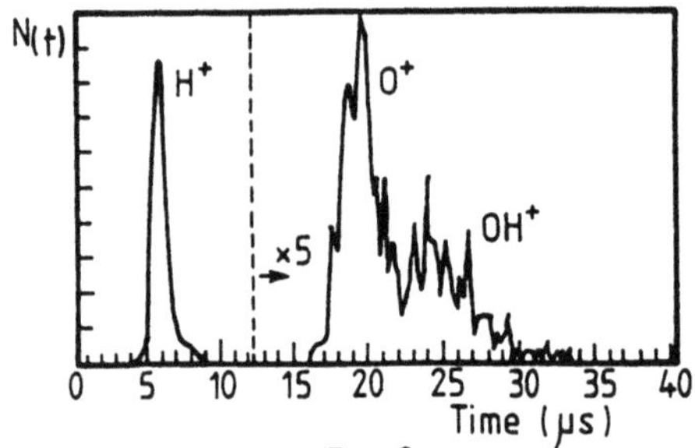

Fig. 2c.

Fig. 2a Time of flight spectrum of ions desorbed from Si(100) exposed to 10 L of water with simultaneous electron bombardment (note: Auger electron spectroscopy universally indicated the presence of oxidised silicon where the O^+ ion is detected).

Fig. 2b Time of flight spectrum of ions desorbed from Si(111) exposed to 120 L of water with simultaneous electron bombardment.

Fig. 2c Time of flight spectrum of ions desorbed from Si(111) exposed to 190 L of water with simultaneous electron bombardment.

Fig. 2b and 2c show two time of flight spectra taken at different points in the evolution of the Si(111) surface ions under this exposure procedure. Again the dominant H^+ signal is present, but in contrast with the response of the (100) surface, an O^+ peak is first detected without any significant signal attributable to OH^+ (fig. 2b). As the exposure progresses, formation of the OH^+ peak becomes evident (fig. 2c). The peak in the energy distribution for each ion is recorded in table 1. No shift in ion energy was noted in the course of experiments.

Irradiation of interfaces (both with and without oxidation) with 500 eV electrons at a current density of 95 μA cm^{-2} caused the exponential decay of the H^+ and OH^+ ion signals over a period of a few minutes, although significant changes in the O^+ ion signal were difficult to detect. The response of the H^+ and OH^+ signals are presented in fig. 3 in semi-logarithmic form, from which it can be seen that the plots of ion decay lie closely together with desorption cross sections in the range 4.4 to 6.0 x 10^{-18} cm^2 (see table 2). Results for oxidised Si(100) are not reported as readsorption of water from an

Table 1. Table of most probable ion energies (eV).

Surface Orientation	Exposure procedure	H^+	OH^+	O^+
Si(100)	Non-oxidised	4.9	4.9	—
	Oxidised	4.9	4.8	12.1
Si(111)	Non-oxidised	5.5	—	—
	Oxidised	5.4	4.0	7.0

Table 2. Table of derived total desorption cross sections (cm^2).

Surface Orientation	Exposure procedure	H^+	OH^+	
Si(100)	Non-oxidised	6.0×10^{-18}	4.9×10^{-18}	
Si(111)	Non-oxidised	4.9×10^{-18}	————	
	Oxidised	4.4×10^{-18}	5.3×10^{-18}	ESD
		O^+		
		1.9×10^{-18}	1.2 C to 2.4	
		6.4×10^{-19}	>2.4 C cm^{-2}	
		O(KLL)		
		1.0×10^{-21}	<60 C cm^{-2}	AES
		5.0×10^{-22}	>60	

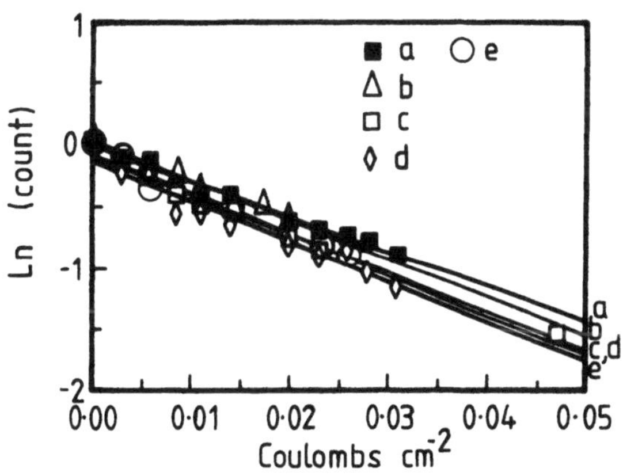

Fig. 3 Semilogarithmic plot of a) H^+ from oxidised Si(111); b) OH^+ from non-oxidised Si(100) c) H^+ from non-oxidised Si(111); d) OH^+ from oxidised Si(111); e) H^+ from non-oxidised Si(100).

abnormally high residual concentration in this case is believed to have affected the validity of the derived cross sections. The further results included in table 2 relate to the desorption cross section of O^+ measured from the oxidised Si(111) surface using an enhanced electron current density of 650 $\mu A\ cm^{-2}$, and the oxygen Auger response from the same surface, to irradiation at 6.5 mA cm^{-2}.

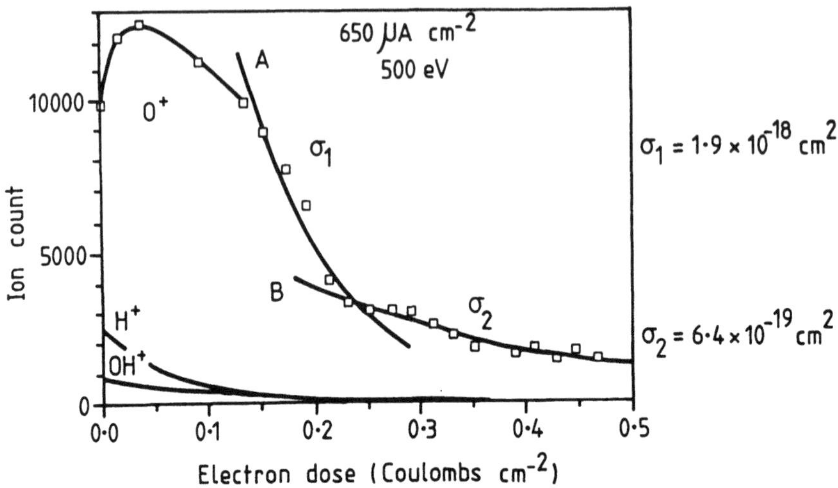

Fig. 4 Response of O^+, H^+ and OH^+ ion signals from oxidised Si(111) as a function of electron dose (650 μA cm^{-2} at 500 eV).

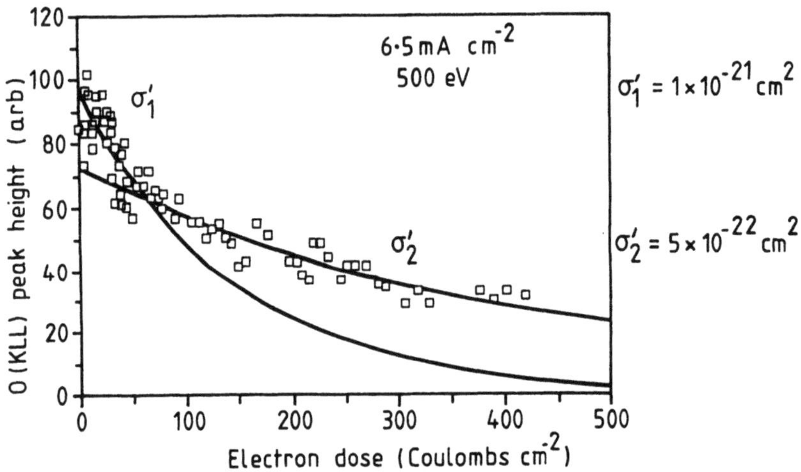

Fig. 5 Response of O(KLL) Auger signal from oxidised Si(111) as a function of electron dose (6.5 mA cm^{-2} at 500 eV).

From fig. 4 it may be seen that the O^+ ion signal increases over the exposure range up to 0.04 C cm^{-2}, after which desorption dominates. The decay above 0.14 C cm^{-2} does not conform to a unique desorption cross section and in the figure we have identified two exponential decay forms, curves A and B, which respectively fit the measurement points over the ranges 0.14 C cm^{-2} to 0.24 C cm^{-2}, and 0.24 C cm^{-2} to 0.5 C cm^{-2}. These correspond to cross sections of 1.9 x 10^{-18} cm^2 and 6.4 x 10^{-19} cm^2.

The response of the O(KLL) Auger signal to electron irradiation is shown in fig. 5. The relative insensitivity of this response required the use of a higher current density (6.5 mA cm^{-2}) to induce changes over a reasonable time scale. The measurements with AES again need to be fitted with two exponential curves. However these are not apparently related to the ESD case, being fitted between 0 C cm^{-2} to 60 C cm^{-2} and 60 C cm^{-2} to 500 C cm^{-2}. In this case the measured cross sections are much reduced at 10^{-21} cm^2 and 5 x 10^{-22} cm^2.

4. Discussion

The most probable energies measured for the H^+ ion all lie in the range 4.9 eV to 5.5 eV, which is, as noted earlier, the regime generally observed following the rupture of the O-H bond of adsorbed hydroxyl groups. Furthermore, the similarity of the derived cross sections is in keeping with a common source for the OH^+ and H^+. The existence of surface OH groups following room temperature water adsorption at the Si(100) surface is well established [6]. In contrast, adsorption at the Si(111) surface is less clear [7], with some investigations finding evidence of molecular adsorption, which would be consistent with a reduced OH^+ ion signal. However, the evidence from our experiments, notably the promotion of surface oxide under electron irradiation, clearly assigns a role to OH radicals at the non-oxide surface. The desorption efficiency for OH ions in this case is necessarily low, but as oxidation proceeds the emergence of OH ions points to changes in the electronic structure of the surface. It is of interest to note that with H^+ and OH^+ desorbed from water at aluminium [8] the desorption cross sections were again similar (1.1 x 10^{-18} cm^2 and 6.8 x 10^{-19} cm^2), but smaller than the values for silicon.

The reaction of an oxidised silicon (111) surface to a large electron flux shows a relatively complex response. Initially the O^+ signal is enhanced while both the H^+ and OH^+ signals are depleted. This may be attributed to the conversion of surface OH to O via the electron stimulated desorption of hydrogen. Here, this generation process competes with the desorption of surface oxygen. Once the generation rate falls below the removal rate through ESD, the O^+ signal begins to decrease. Following the depopulation of OH groups (0.14 C cm^{-2}), the response is purely that of O^+ desorption characterised by a cross section of 1.9 x 10^{-18} cm^{-2}. At higher electron exposures the depletion curve is fitted by an exponential of cross section 6.4 x 10^{-19} cm^{-2}. Throughout the O^+ depletion the most probable energy remains constant suggesting that the desorption mechanism itself is unchanged although, perhaps, the initial excitation cross is reduced, or the probability of reneutralisation increased.

Analysis of the silicon fine structure around the intersection of the two decay curves of fig. 5, shows that the transition from the higher cross section to the lower occurs where the surface layers are composed of the sub-maximal valence oxide, SiO_x ($x \sim 1$). This is in line with the predictions of the Knotek-Feibelman mechanism [9] and would suggest that a process of this type is, at least in part, responsible for oxygen desorption from the oxidised surface. Ion energy distributions for the O^+ ions desorbed from the (100) and (111) surfaces have peak energies of 12.1 eV and 7 eV respectively. These relatively high kinetic energies are also a feature of the Knotek-Feibelman mechanism where a large degree of charge transfer is involved.

Most strikingly, the total desorption cross sections deduced from measurement of the ESD ion current are significantly larger than those measured by AES. Although AES samples several atomic layers below the surface it would seem unlikely that this can account for a difference of two to three orders of magnitude in the measured values. It must thus be concluded that surface sites giving rise to O^+ desorption form only a fraction of the surface oxygen present, or alternatively, that there is conversion to ESD inactive sites under irradiation.

References

1. For example, T. E. Madey, D. E. Ramaker, and R. Stockbauer, Ann. Rev. Phys. Chem. **35**, 215 (1984); E. M. Williams and J. L. de Segovia, Vacuum, **39**, 633 (1989)
2. M. L. Knotek, Rep. Prog. Phys. **47**, 1499 (1984).
3. S. L. Bennett and E. M. Williams, Vacuum, **38**, 923 (1988).
4. M. Q. Ding, E. M. Williams, J. P. Adrados, and J. L. de Segovia, Surf. Sci. Lett. **140**, L264 (1984).
5. R. H. Stulen and P. A. Thiel, Surf. Sci. **157**, 99 (1985).
6. S. Ciraci and H. Wagner, Phys Rev **B27**, 5180 (1983); H. Ibach, H. Wagner and D. Bruchmann, Solid State Commun. **42**, 457 (1982); Y. J. Chabal and S. B. Christman, Phys. Rev. **B29**, 6974 (1984)
7. J. A. Schaefer, J. Anderson, and G. J. Lapeyre, J. Vac. Sci. Technol. **A3** 1443 (1985); J. A. Schaefer, F. Stucki, D. J. Frankel, W. Goepel, and G. J. Lapeyre, J. Vac. Sci. Technol. **B2**, 359 (1984).
8. M. Q. Ding and E. M. Williams, Surf. Sci. **160**, 189 (1985).
9. M. L. Knotek and P. J. Feibelman, Phys. Rev. Lett. **40**, 964 (1978); Surf. Sci. **90**, 78 (1979)

Part V

Condensed Gases

Ejection of Argon Dimers from Solid Argon Films Electronically Excited by MeV He Ions

C.T. Reimann[1], *W.L. Brown*[2], *M.J. Nowakowski*[2,*], *S.-T. Cui*[3], *and R.E. Johnson*[3]

[1]Naval Research Laboratory, Washington, DC 20375, USA
[2]AT&T Bell Laboratories, Murray Hill, NJ 07901, USA
[3]Department of Nuclear Engineering and Engineering Physics, University of Virginia, Charlottesville, VA 22901, USA
* Present address: Department of Materials Science and Engineering, University of Pennsylvania, Philadelphia, PA 19104, USA

Abstract. We have examined the spatial and temporal characteristics of the 11.3-eV luminescence band emitted from electronically excited Ar films. This luminescence arises primarily from vibrationally excited Ar triplet dimers which are ejected into a spherical plume with a kinetic energy of about 0.13-eV. This energy is larger than that calculated for ejection during cavity formation around excited surface species. We therefore conclude that dimer ejection occurs after repulsive recombination of a self-trapped hole followed by dimerization of the moving excited atom.

1. Introduction

When solid argon is electronically excited by ions, electrons, or photons, little free exciton luminescence [1] and no electron-hole recombination luminescence is seen. Instead, the lattice relaxes around excited species, and a luminescence spectrum arises which consists of bands that are red-shifted relative to the exciton levels. Two broad molecular-type bands are seen at 9.8-eV (M-band) and 11.3-eV (W-band) due to radiative decay of vibrationally relaxed and excited $Ar_2{}^*$ $({}^{1,3}\Sigma_u{}^+)$ respectively [2].

Lattice relaxation and subsequent decay of electronic energy causes ejection of material from the Ar film. Under 1.5-MeV He^+ excitation, a net desorption yield of about 40 atoms per incident ion has been measured [3]. The ejection yield is mostly ground state atoms [3,4,5], but excited state atoms [6,7], ions [7], and dimers [3,8] are also ejected under electronic excitation. In the present work, we focus on the

Springer Series in Surface Sciences, Vol. 19
Desorption Induced by Electronic Transitions, DIET IV
Editors: G. Betz · P. Varga

molecular-type luminescence bands and on the desorption of Ar_2^* ($^{1,3}\Sigma_u^+$).

The film thickness dependences of the luminescence features indicate that 9.8-eV luminescence comes dominantly from the bulk of an Ar film [3,9] while 11.3-eV luminescence comes from *either* a surface-trapped species [1,9,10] or a desorbed species [3,8]. We recently showed that at least *some* of the triplet 11.3-eV luminescence arises from a plume of desorbed dimers [3]. However, we were not able to determine how much of the luminescence comes from the desorbed species; nor could we determine a value for the desorbed dimer kinetic energy (K.E.). An improved experimental examination of the spatial and temporal characteristics of the molecular-type luminescence bands is the subject of this paper.

2. Experiment

The experimental arrangement consists of a polished Cu target block cooled to 12K; a pulsable 1.5 MeV He^+ beam; and a vacuum-ultraviolet monochromator with f/4.5 collection optics whose axis is at about 135° with respect to the incident ion beam (see inset of Fig. 2). The surface normal of the Cu target is rotatable in the plane defined by the ion beam and the monochromator input axis. Ar films are grown by orienting the cold Cu target in front of a micro-channelplate nozzle and admitting a known quantity of Ar gas. A 2000-Å thick film is grown in about one minute. The base pressure of the ultra-high vacuum system is 1×10^{-9} torr so codeposition of impurities is negligible [3].

The spatial extent of the plume is measured by monitoring the luminescence signal as the Cu target is rotated, blocking first bulk and surface luminescence and then successively greater amounts of plume luminescence. The temporal characteristics of the plume are measured by exciting the Ar films with 50-nsec ion beam pulses and using a time-to-amplitude converter to record the luminescence time decays.

3. Results

The dependence of the luminescence spectrum on target angle is shown in Fig. 1. Two resolution-limited atomic lines are seen around 1048-Å and 1067-Å. (In the "head-on" spectrum there is broadening on the red side of the 1067-Å

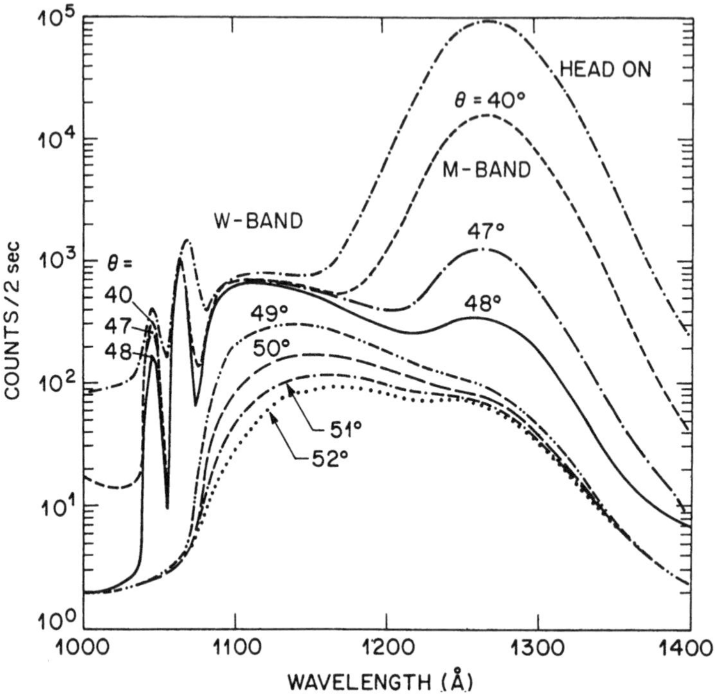

Fig. 1 Vacuum-ultraviolet luminescence spectra observed as target is rotated to hide portions of the luminescent plume.

feature, due to an unresolved contribution from bulk atomic-type self-trapped excitons [1].) These atomic lines are attributed to desorbed excited Ar atoms undergoing allowed radiative transitions to the ground state [8]. The lifetimes of these states are less than 10-nsec [11]. Thus, even for a K.E. as high as 1-eV, the light from these emitting atoms would be blocked by turning the edge of the Cu target only about 0.1° past the input axis of the monochromator. The sudden disappearance of the atomic lines near Θ=48.5° indicates that the light from bulk- or surface-trapped species is completely hidden from the view of the monochromator for Θ>48.5° (this angle corresponding to the direction of the axis of the monochromator with respect to the Cu target center).

The smooth drop in W-band luminescence intensity as the target angle is increased past 48°, Fig. 1, leads to the clear conclusion that the W-band is dominantly due to desorbed species [3] rather than to surface-trapped species [1,10]. A change in the shape of the W-band spectrum occurs as the plume is blocked. The extent of the plume is much smaller at a wavelength of 1100-Å than it is at 1200-Å. In contrast to the W-band, the M-band is dominantly from the bulk of the film.

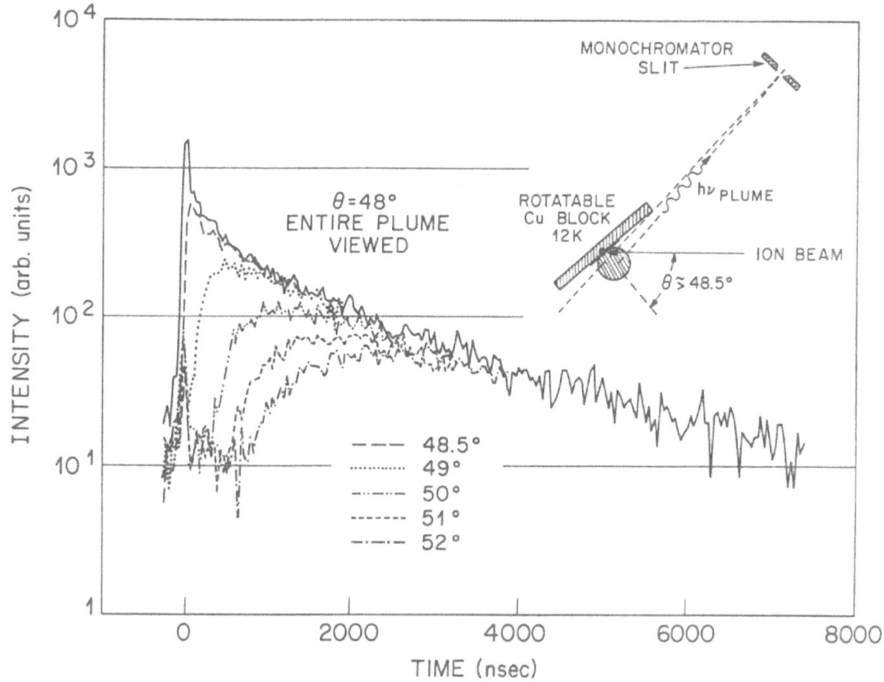

Fig. 2 Time-resolved luminescence decays as a function of target angle. The wavelength is in the W-band. The inset is a view of the experimental geometry.

Less than 0.1% of the peak M-band luminescence is seen at angles greater than 48.5° for the film of Fig. 1.

Luminescence decay curves for a wavelength of 1185-Å (10.5-eV) in the W-band are shown in Fig. 2. The triplet decay, identified by its μsec decay time, is nonexponential, probably a consequence of the fact that the dimers are populated in various vibrational levels which have different lifetimes. It is likely that more than one vibrational level can contribute to luminesence at a given wavelength, as is the case for Ne [1]. Singlet luminescence can also be identified in Fig. 2 by the presence of a time resolution-limited prompt feature visible at $\Theta<48.5°$.

The result of monitoring luminescence time-decays as a function of target angle is shown in Fig. 2. As the base of the plume is increasingly hidden, the triplet luminescence decay curves display a delayed onset of up to 2000-nsec. The delayed onset is the time-of-flight for ejected dimers to leave the shadow of the edge of the Cu target block and become visible to the monochromator. The rounded nature of the onset primarily reflects the various pathlengths ejected dimers can take to come out from the shadow. Normalization of the data in Fig. 2 is by cumulative incident ion fluence. The close correspondence of the curves at long times confirms that the triplet W-band luminescence is emitted exclusively from desorbed species. The singlet luminescence decay feature disappears by $\Theta=48.5°$. Its lifetime, less than 10-nsec [10], is so short that decay occurs too near to the surface for us to distinguish between surface-trapped and desorbed singlet species.

The data in Fig. 2 can also be expressed as a ratio of each time dependent decay curve to the $\Theta=48°$ decay curve, removing the overall nonexponential time dependence. The result of this operation is shown in Fig. 3 along with the predictions of two simple analytical models of the plume. In one model, a spherical (independent of polar angle) plume is assumed along with a single characteristic velocity of the ejected dimers. A reasonably good fit to the delayed,

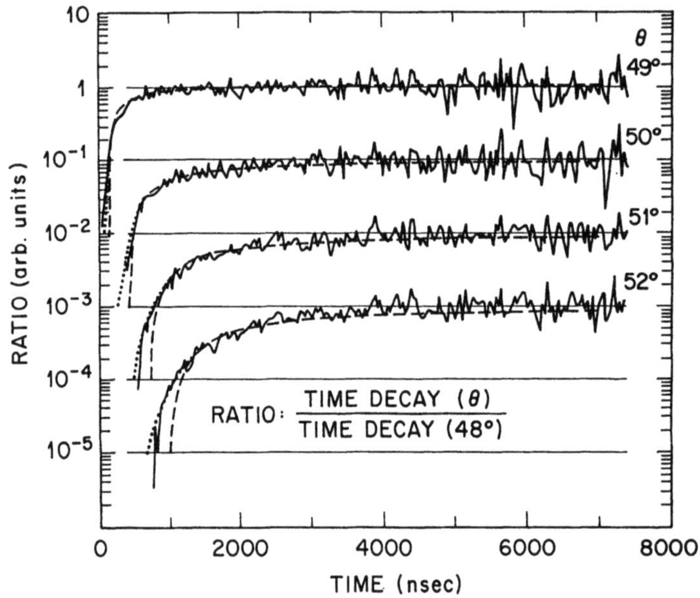

Fig. 3 Time-resolved luminescence decay with the nonexponential time dependence removed by normalization. Dashed curve is a model assuming a spherical plume and a single ejection velocity, $v=5.5\times10^4$-cm/sec. Dotted curve is the same model assuming in addition a gaussian velocity 1/e-width of 4.1×10^4-cm/sec.

"rounded" onsets seen in Fig. 2 is obtained with a velocity of 5.5×10^4-cm/sec, or a dimer K.E.=0.13-eV. The fit of the model to the data at the onset of the curves is improved by including a gaussian spread in velocity with a full-width at 1/e of 4.1×10^4-cm/sec.

Using a more directed angular distribution rather than a spherical one, or a polar angle dependent velocity, results in fits which are poorer than the ones shown in Fig. 3. Though the time dependence of the luminescence is a strong function of wavelength, the fitted velocity does not vary significantly with wavelength in the W-band. From the time-dependent data on the 9.8-eV plume (not shown), a K.E. of the same order of magnitude as for the 11.3-eV plume results. More complete details of the model and fit will be published elsewhere [12].

4. Discussion

One model of dimer ejection is the cavity ejection model [3,8,13,14]. Solid argon is characterized by a negative electron affinity [15]; therefore, the Rydberg electron associated with an excited center such as an excited atom or an excimer causes the formation of a cavity around the center. If the excited species is on the surface of the Ar film, the formation of the cavity may cause the excited species to be ejected.

Molecular dynamics calculations have been carried out to see if cavity ejection is feasible. Potential energy interactions relevant to the solid state are constructed using linear combinations of gas-phase ab-initio potential interactions between pairs of atoms [13,14]. Shifts in absorption energies are predicted which agree favorably with experiment, establishing the general validity of the potentials.

The results of these calculations for Ar dimers [13] and excited atoms [14] are shown in Table 1. Unrelaxed triplet dimers on the surface promptly eject before vibrational relaxation takes place, but the predicted K.E. is very low compared to the experimental result, 0.13-eV. Although the use of an average pair potential is reasonable for describing cavity formation in the bulk, corrections should be made for lack of symmetry and the dynamic behavior associated with the surface ejection process. Also, many-body interactions may need to be incorporated into the model potentials. However, it is not clear whether these corrections could

Table 1 Predicted triplet dimer kinetic energies from molecular dynamics calculations of the cavity ejection mechanism. Data from Refs. 13 and 14.

Crystal Face	Atoms	Vibrationally Excited Dimers
(100)	0.024-eV	0.02-eV
(111)	0.055-eV	0.02-eV

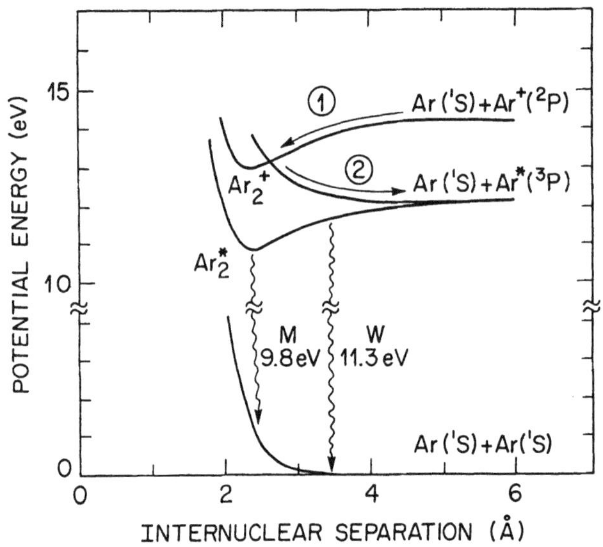

Fig. 4 Solid state potential curves adapted from Ref. 3. Process #1 represents self-trapping of a hole; process #2 represents repulsive recombination of an electron with the self-trapped hole.

account for the differences between the calculated and measured dimer K.E., especially as metastable ejected atoms have been detected with K.E.'s of the order of 0.04-eV, consistent with the calculated K.E.'s in column two of Table 1. Therefore, although cavity ejection clearly occurs it does not seem to account for the ejection of dimers studied here.

Energetic ejection of dimers may arise as a consequence of repulsive recombination. The idea is summarized in Fig. 4. Electron recombination with a self-trapped hole [3] results in a repulsive separation of a ground state and excited state atom. The excited state atom may be ejected directly, accounting for the high K.E. atomic ejection feature that has recently been observed [7]. Alternatively, without equilibration with the solid, the moving excited Ar atom may dimerize near the surface with the resulting dimer being ejected with considerable K.E.

In conclusion, we have shown that the W-band luminescence comes dominantly from desorbed species [3], in contrast to some reports in the literature [1,10]. We have also determined the K.E. of desorbed triplet Ar dimers. We find the average K.E. to be around 0.13-eV with a spread between 0.07-eV and 0.29-eV. This K.E. is too high to be accounted for by cavity ejection [8,13]. Ejection due to repulsive electron-hole recombination [3] can give much larger K.E. and simulations are in progress to determine the feasiblilty of this mechanism.

Acknowledgments. The work at the University of Virginia was supported by NSF grants DMR-86-00469 and CBT-86-13614. One of us (CTR) is a National Research Council Research Associate at the Naval Research Laboratory.

6. References

1. I.Ya. Fugol', Adv. Phys. 37, 1 (1988).
2. N. Schwentner, E.-E. Koch, and J. Jortner, *Electronic Excitations in Condensed Rare Gases* (Springer-Verlag, Berlin, 1985) p. 126
3. C.T. Reimann, W.L. Brown, and R.E. Johnson, Phys. Rev. B37, 1455 (1988).
4. D.J. O'Shaughnessy, J.W. Boring, S.-T. Cui, and R.E. Johnson, Phys. Rev. Lett. 61, 1635 (1988).
5. P. Feulner, T. Muller, A. Puschmann, and D. Menzel, Phys. Rev. Lett. 59, 791 (1987).
6. T. Kloiber and G. Zimmerer, Physica Scripta (in press).
7. I. Arakawa, M. Takahashi, and K. Takeuchi, J. Vac. Sci. Technol. A7, 2090 (1989).
8. F. Coletti, J.M. Debever, and G. Zimmerer, J. Physique Lett. 45, L467 (1984).
9. A. Hourmatallah, F. Coletti, and J.M. Debever, J. Phys. C: Sol. State Phys. 21, 1307 (1988).
10. E. Roick, R. Gaethke, P. Gurtler, T.O. Woodruff, and G. Zimmerer, J. Phys. C: Sol. State Phys. 17, 945 (1984).

11. W.L. Wiese, M.W. Smith, and B.M. Miles, *Atomic Transition Probabilities*, Natl. Stand. Ref. Data Ser., Nat. Bur. Stand. (U.S.) 22, 192 (1969).
12. W.L. Brown, S.-T. Cui, R.E. Johnson, M.J. Nowakowski, and C.T. Reimann, to be published
13. S.-T. Cui, R.E. Johnson, C.T. Reimann, and J.W. Boring, Phys. Rev. B39, 12345 (1989).
14. S.-T. Cui, R.E. Johnson, and P.T. Cummings, Phys. Rev. B39, 9580 (1989).
15. J. Jortner, L. Mayer, S.A. Rice, and E.G. Wilson, J. Chem. Phys. 42, 750 (1965).

Erosion of Pure and Alloyed Ices of Rare Gases and Small Molecules by Core Excitation

R. Scheuerer, P. Feulner, G. Rocker, Zhu Lin, and D. Menzel

Physik-Department E20, Technische Universität München, D-8046 Garching, Fed. Rep. of Germany

Because of their fundamental interest, electronic sputtering processes of Van-der-Waals bound solids have been investigated by many researchers /1/. Most of these studies dealt with pure solids of only one component, though for a variety of applications (e.g. space science or cryogenic devices) the investigation of electronically stimulated reactions and erosion of *alloys* might be even more important. Moreover, studies of such layers could shed light on the mechanisms by which the electronic excitation is transferred to nuclear motion, especially if resonant excitation by monochromatic photons from synchrotron sources is investigated. Tuning the wavelength to a suitable absorption line makes the localisation of the primary excitation on selected constituents of the "mixed ice" possible. We have therefore investigated total electron yield (TEY), and neutral and ionic photon stimulated desorption yields (n-PSD, i-PSD) of pure and mixed ice layers of Ar, O_2, CO, and H_2. Erosion of Ar has been studied before by desorption and luminescence spectroscopy and the prevailing sputtering mechanisms for valence excitation are well understood (see, e.g., ref.2 and references therein). O_2 and CO are interesting candidates because the sputtering mechanisms seem to be different in the two cases. For O_2, a high relaxation energy of up to 7 eV due to dissociative recombination of the molecular ion is expected to be the main "power source" for desorption /3/. For CO, a quadratic dependence of the erosion yield on the electronic stopping power has been found and interpreted in terms of interaction of adjacent excitations /4/. Finally, mixtures with H_2 have been investigated because of their possible importance for space molecules /5/.

The experiments were performed under UHV conditions at the SX-700 beam line of BESSY, Berlin. Thick layers ($\approx 10^{17}$ particles/cm^2) were dosed by a needle doser onto the substrate cooled to about 12K. Alloyed layers were prepared by adjusting partial pressures according to the desired ratios of the particular constituents in the gas handling system under constant flow conditions, and dosing these gas mixtures onto the sample. TEY spectra were recorded by a current-voltage converter directly connected to the sample. n-PSD and i-PSD data were recorded with quadrupole mass spectromters (QMS; see ref.6 for details). All spectra shown here were taken at grazing incidence (surface normal inclined vs. the $\vec{E}$-vector by 7°). The axis of the QMS for n-PSD was oriented parallel to $\vec{E}$, the axis of that for i-PSD parallel to the surface normal. The photon flux was obtained from a gold coated photo diode, and independently from a bolometer of high sensitivity /7/.

NEXAFS data were used to obtain information on electronic and geometric properties, and on radiation induced reactions for the various ices. TEY-excitation spectra from pure Ar, Ar + CO and Ar + O_2 layers are shown in Fig.1,1-5. Four maxima are obtained for the Ar 2p threshold region, which for Ar gas are due to resonant excitations of the Ar $2p_{3/2}$ and $2p_{1/2}$ electrons to the 4s (A, A') and 3d levels (B, B')/6,8/. The excitation energies in the solid are shifted with respect to the gas phase values by $\approx$ 1 eV to higher energies /6,8/. With admixtures of CO, the energy of peak B increases (peak A' becomes observable by this) while that of peak A stays constant. This is

Desorption Induced by Electronic Transitions, DIET IV
Editors: G. Betz · P. Varga

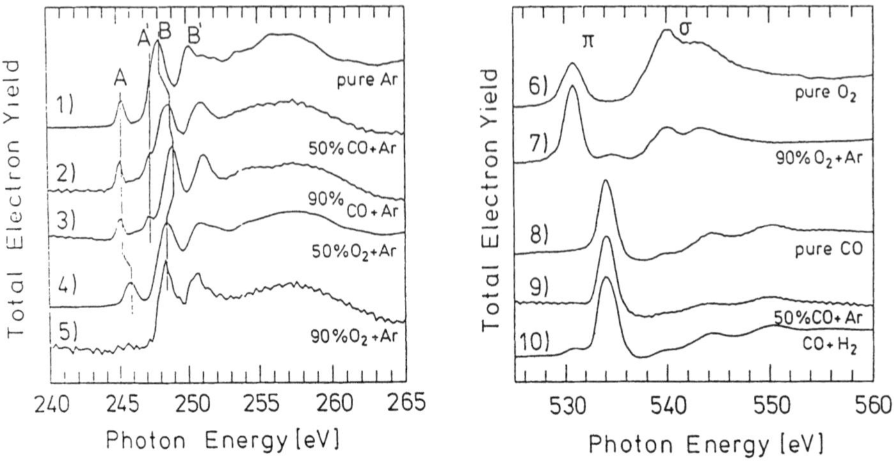

Fig.1: TEY for Ar2p excitation - left - and O1s excitation - right

probably caused by interaction of the Ar 3d final states with empty states on CO which are known to lie only slightly higher /9/. For Ar 4s such an interaction is not possible since there are no states on CO close in energy. Such an effect has been found before for N_2+Ar /8/. For O_2 which has its conduction band close to the 4s energy /9/, peak A is indeed broadened and shifted to higher energy with 50% O_2 admixture, and is essentially quenched in 90%. The fact that all these shifts appear to be gradual with concentration supports the assumption of homogeneous layers without segregation. O1s excitation spectra are shown in Fig.1,6-10. For O_2 and CO, the positions of the π resonances is unchanged compared to the gas phase /10,11/. Only for O_2, the σ peaks are shifted to higher energies in the solid /11,12/. For CO ice, an additional peak at 544 eV which is not present in the gas phase has been interpreted in terms of an EXAFS feature. This maximum is shifted to slightly lower energy for CO diluted with Ar. No other peak shifts are observed for the alloys. For grazing incidence, the intensity ratio of the π and σ maxima is reversed for pure solid oxygen (compare Fig.1,6 and ref.10). For normal incidence, the π peak dominates and the σ maxima are decreased /13/. This clearly shows that at least for the top-most layers of our oxygen films the O_2 molecules are predominantly oriented perpendicular to the surface. This orientation is destroyed if 10% of Ar is added, and π/σ peak ratios as for gaseous oxygen are obtained (see Fig.1,7). No such ordering is seen for solid CO (see Fig.1,8).

An additional maximum 2.8 eV below the CO π resonance is obtained for CO + H_2 ices (Fig.1,9). It is close to the energy of the π resonance of H_2CO which lies 3.4 eV below the CO resonance /14/. Indeed, H_2CO is found in subsequent TPD. Detailed studies showed that it is formed by irradiation and not by co-condensation. Formation of H_2CO from mixtures of CO and $^3H_2$ has been observed before /15/. We believe that we see a primary photo absorption in OC-H_2 neighbours to a final state resembling OCH_2 (but at too large C-H distance, which explains the energy shift). This does not mean that all H_2CO formed in this transient state will survive as a molecule, because dissociation following the Auger decay of the core hole should be strong, and in fact H_2CO^+ ions are observed from these layers (see Tab.1).

Desorption of neutrals, fragment ions, molecular ions and cluster ions was found. Our i-PSD results are in rough agreement with those of Rosenberg et al. /12/. For solid CO however, we observed cluster ions up to $(CO)^+_{11}$, which have not been obtained in ref.12 (see Tab.1). The desorption of neutrals (Ar*,

Table.1: Ion signals and their relative intensities (in []) in % of the largest signal obtained from different layers (hν = 600eV).

pure **CO**:	C^{2+}[2], O^{2+}[2], C^+[100], O^+[29], C_2^+[8], CO^+[15], C_3^+[4], C_2O^+[30], C_4^+[4], C_3O^+[15], $(CO)_2^+$[36], C_4O^+[4], $C_3O_2^+$[42], C_5O^+[2], $C_4O_2^+$[5], $(CO)_3^+$[9], $C_5O_2^+$[1.5], $C_4O_3^+$[2], $(CO)_4^+$[7], $(CO)_5^+$[5], $(CO)_6^+$[3], $(CO)_7^+$[2], $(CO)_8^+$[1], $(CO)_{9\ 11}^+$[<0.5].
50%CO+Ar:	C^{2+}[2], O^{2+}[2], C^+[100], O^+[30], C_2^+[11], CO^+[25], C_3^+[3], C_2O^+[21], C_4^+[2], C_3O^+[4.5], $(CO)_2^+$[17], $C_3O_2^+$[10], $(CO)_3^+$[1], Ar^+[1.5].
$CO+H_2$:	C^{2+}[2], O^{2+}[1], C^+[100], O^+[44], C_2^+[3], CO^+[6], C_2O^+[5], $(CO)_2^+$[6], CH_2^+[18]
50%O_2+Ar:	Ions with masses up to 200amu have been observed. Because of the interference of Ar_n-O_m with O_n clusters, the assignment of most of the maxima was impossible. The desorption of O_2^+[100], Ar^+[23], ArO^+[7], and ArO_2^+[7], however, was unambiguous.

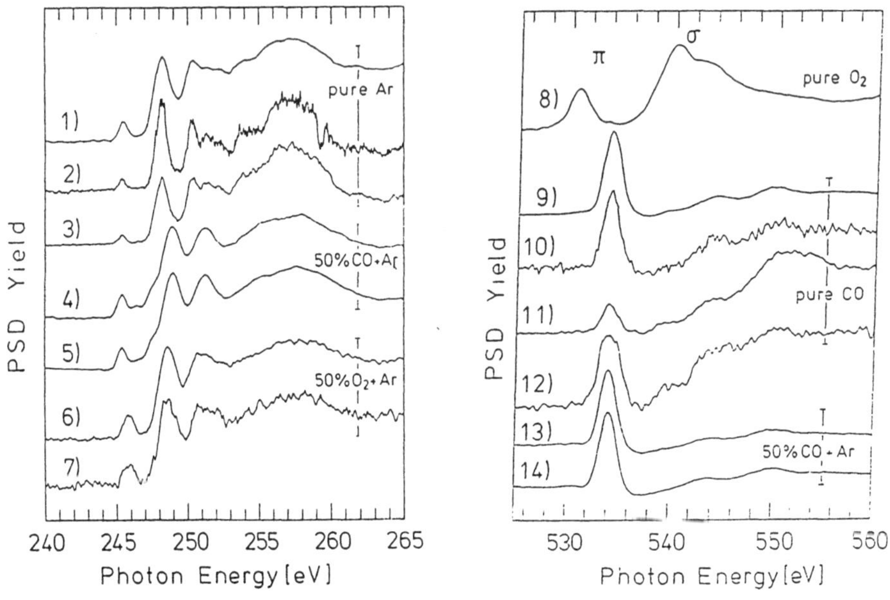

Fig.2: n-PSD and i-PSD for Ar2p excitation - left side: 1) Ar^0, 2Ar^+, 3) Ar^{2+}, 4) Ar^0, 5) CO^0, 6) Ar^0, 7) O_2^0 - and O1s excitation - right side: 8) $O_2^{\bullet}$, 9) $CO^{\bullet}$, 10) CO^+, 11) O^+, 12) C^+, 13) $CO^{\bullet}$ and 14) $Ar^{\bullet}$.

$O_2^{\bullet}$ and $CO^{\bullet}$) is insensitive to the location of the primary excitation (see Fig.2,1,4-9,13,14). From the alloys, desorption of both constituents is observed at each threshold. The ratios of the relative desorption yields (e.g. Y_{CO}/Y_{Ar}) and the relative amounts of the constituents (N_{CO}/N_{Ar}) depend only weakly on the localisation of the excitation, see Fig.3, and the overall desorption yields normalised to photon absorption/monolayer are comparable (1 : 1.6 : 5.5 for Ar2p, C1s, and O1s excitation of a 50%CO+Ar layer). This is surprising, because C1s excitation of gaseous CO even in the π-resonance

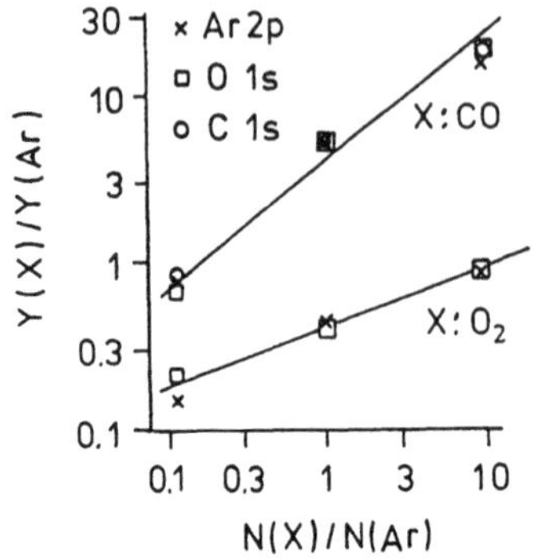

Fig.3: Dependence of the n-PSD yield ratios on the relative density of the constituents for CO+Ar and O_2+Ar mixtures.

leads to dissociation with a probability of ≈90%, and the average kinetic energy of the ions formed is ≥ 14 eV /11/, i.e. much more kinetic energy is deposited into the solid than by the decay of vibrationally "cold" Ar ionomers or excimers, which is thought to be the main mechanism for erosion of rare gas solids /1/. Quenching of these dissociation processes by rapid delocalisation of the excitation to the solid (as found for adsorbates on metals /16/) can be excluded, since the desorbing ions show abundances (Tab.1) and NEXAFS spectra (Fig.2,10-12) similar to gas phase data /11/. Because the n-PSD and i-PSD NEXAFS pattern for molecular and cluster ions closely resemble the TEY spectra (this is also true for the pure layers), we conclude that most of the neutral desorption yield is due to non-selective secondary processes like ionisation by electrons, or chemical ionisation, and subsequent nuclear motion. The larger desorption yield at the O1s threshold then must be due the more energetic Auger and autoionisation electrons which can create more valence excitations, and partly due to the higher total ion yield per core excitation /11/ (see below).

There are however some small but interesting differences between the n- and i-PSD (Fig.2,1-3), and the TEY (Fig.1,1). For ions and neutrals, the desorption yield for the Ar $2p^{-1}4s$ excitation is slightly smaller compared to TEY than for the excitation to the continuum. It is well known that for this excitation formation of doubly charged ions is reduced due to autoionisation /6/. Desorption due to dissociative recombination of vibrationally relaxed multiply ionised complexes like Ar_n^{2+} is unlikely. At least for the gas phase, such clusters seem to be unstable or only weakly bound /17/. A mechanism which might be considered is Coulomb repulsion of ions formed close together. The valence ionisation cross section is so large for electrons of 200 eV (which are created by the Auger decay of an Ar2p hole) that the probability for the ionisation of a nearest neighbour Ar is ≈ 18% /18/. Such a process could deliver strong repulsive forces if the lifetime of the separation of positive and negative charge is long enough (≈10^{-13} s), i.e. until both particles will have accumulated sufficient kinetic energy. This can be taken as granted for primary transitions to continuum states well above the threshold. Desorption need not be ionic, because neutralisation could occur in a later stage of the desorption process. From the data for pure Ar, we estimate this mechanism to contribute 10 to 30 % to the total desorption yield of neutrals. This mechanism can easily explain the high number of molecular and cluster ions desorbed by core excitations, which for Ar ice irradiated with Al Kα radiation was found to be ≈1% of the neutral yield /19/, and the finding that the NEXAFS pattern of all ions except of those from the fragments (which stem from the decay of the primarily excited molecule) closely resemble the Auger yield.

It is unclear from our data why n-PSD of CO for Ar+CO, and of Ar for Ar+O_2 is dominating (see Fig.3). XPS spectra taken under different angles do not indicate different segregation behaviour for these layers (in both cases, however, the concentration of the molecules seemed to be enhanced near the surface). For Ar+CO, only CO-derived ions and ionic clusters are obtained (see Tab.1), and segregation can not be excluded from i-PSD. For Ar+O_2 however, a variety of Ar-O clusters is formed. The evaporation enthalpies for Ar, CO, and O_2 in the pure solids are similar, and similar sputtering rates would be expected for non-selective processes like mini-cascades following dissociative recombination, Coulomb repulsion or excimer decay. One explanation for the lower yield of oxygen might be that dissociation of O_2 and desorption as neutral atoms occurs, which are not detectable with our set-up. We speculate that the high CO yield from Ar+CO could stem from ordering processes in the solid initiated by the relaxation processes mentioned above. It is well known that Ar+CO in equilibrium does not form a homogeneous crystal like Ar+O_2 /20/ - anealing of our amorphous layers would have led to segregation into domains of Ar and CO alone.

In conclusion, we have shown that n-PSD and i-PSD of molecules and clusters after core excitation are dominated by secondary processes and strongly related to the total photoabsorption. Photoabsorption spectra are different for alloys and pure ices, and indicate chemical interaction for the final state even for these Van-der-Waals bound solids. The n-PSD yield can be modified considerably by matrix effects. Desorption of ionic clusters was always strongest for the pure layers, and decreased by impurities.

We gratefully acknowledge discussions with W.Wurth and technical assistance by the staff of BESSY. This work was supported by the German BMFT (05366 CAB).

References

/1/ J.Schou, Nucl.Instr.Methods B27 (1987) 188.
/2/ C.T.Reimann, W.L.Brown, and R.E.Johnson, Phys.Rev. B37 (1988) 1455.
/3/ F.L.Rook and R.E.Johnson, Surf.Sci. 164 (1986) 625; O.Ellegard, J.Schou, H.Sorensen, and P.Borgesen, Surf.Sci. 167 (1986) 474.
/4/ W.L.Brown, L.J.Lanzerotti, K.J.Marcantonio, R.E.Johnson, and C.T.Reimann, Nucl.Instr.Methods B14 (1986) 392.
/5/ V.Pirronello, W.L.Brown, L.J.Lanzerotti, K.J.Marcantonio, and E.H.Simmons, Astrophys.J. 262 (1982) 636.
/6/ G.Rocker, P.Feulner, R.Scheuerer, ZHU Lin, and D.Menzel, Physica Scripta (in press).
/7/ C.Andelfinger, F.Mast, G.Schramm, P.Vallet, and G.Weber, to be published.
/8/ R.Haensel, N.Kosuch, U.Nielsen, U.Rössler, and B.Sonntag, Phys.Rev. B7 (1973) 1577.
/9/ G.Bader, G.Perluzzo, L.G.Caron, and L.Sanche, Phys.Rev. B30 (1984) 78.
/10/ A.P.Hitchcock and C.E.Brion, J.Electron Spectr. 18 (1980) 1.
/11/ A.P.Hitchcock, P.Lablanquie, P.Morin, E.Lizon, A.Lugrin, M.Simon, P.Thiry, and I.Nenner, Phys.Rev. A37 (1988) 2448.
/12/ R.A.Rosenberg, P.J.Love, P.R.LaRoe, V.Rehn, and C.C.Parks, Phys.Rev. B31 (1985) 2634.
/13/ F.Sette, J.Stöhr, and A.P.Hitchcock, J.Chem.Phys. 81 (1984) 4906, and references therein.
/14/ W.Wurth, G.Rocker, J.Stöhr, and D.Menzel, to be published.
/15/ W.H.Beattie, Int.J.Chem.Kinetics 5 (1973) 333.
/16/ R.Treichler, W.Riedl, W.Wurth, P.Feulner, and D.Menzel, Phys.Rev.Lett(1985) 462.
/17/ L.J.Kieffer and G.H.Dunn, Rev.Mod.Phys. 38 (1966) 1.
/18/ Kenneth B.McAfee, Jr., and R.S.Hozack, Phys.Rev. A32 (1985) 810.
/19/ Y.Baba, E.Hudel, P.Feulner, and D.Menzel, to be published.
/20/ C.S.Barrett and L.Meyer, J.Chem.Phys. 43 (1965) 3502; C.S.Barrett, L.Meyer, and J.Wassermann, J.Chem.Phys. 44 (1966) 998.

Role of Valence Double Photoionization Processes in Ion Desorption from Condensed Gases

L. Hellner[1], *M.J. Besnard-Ramage*[1], *G. Dujardin*[1], *and R. Azria*[2]

[1]Laboratoire pour l'Utilisation du Rayonnement Electromagnétique, Bât. 209d, Université Paris-Sud, F-91405 Orsay, France and Laboratoire de Photophysique Moléculaire, Bât. 213, Université Paris-Sud, F-91405 Orsay, France

[2]Laboratoire des Collisions Atomiques et Moléculaires, Bât. 351, Université Paris-Sud, F-91405 Orsay, France

Abstract. Relations between valence double photoionization and ion desorption from condensed gases are investigated. Double photoionization either by direct processes or via the formation of singly charged ion satellite states is shown to produce ion desorption from solid nitrogen and rare gas solids respectively. In contrast, single ionization of two separate atoms observed in double photoemission studies of rare gas solids does not seem to induce detectable ion desorption.

I. Introduction

The role of multiple electronic excitation and multiple ionization processes in ion desorption is a long standing problem for both core excitation [1,2] and valence excitation [3-5], studies. We focus here on valence processes induced by photon impact on condensed gases. These latter systems have only weak van der Waals interactions, which allows a more direct comparison with gas phase processes than chemisorbed atoms and molecules. The aim of this paper is to provide a new insight on ion desorption mechanisms based on the comparison between recent ion desorption results and double photoionization studies both in the gas and condensed phases.

2. Experimental

Briefly, synchrotron radiation from Super-ACO in Orsay is used as a photon source of variable energy in the 25-150 eV range. The experiments are performed in an ultra high vacuum chamber (residual gas pressure below 10^{-10} mbar). Gases are condensed at low temperature (15 K) on a copper substrate cooled with a liquid-helium flow cryostat. The sample can be rotated either in front of a quadrupole mass analyser, which enables the number of desorbed ions to be counted as a function of the photon energy, or in front of a double time of flight spectrometer [6], which allows detection of pairs of emitted electrons in coincidence.

Springer Series in Surface Sciences, Vol. 19
Desorption Induced by Electronic Transitions, DIET IV
Editors: G. Betz · P. Varga

3. Double photoionization observed in rare gas solids by photoemission of pairs of electrons.

Photoemission of pairs of electrons from rare gas solids (Ar, Kr and Xe) was observed by using a photoelectron-photoelectron coincidence (PEPECO) method [6]. Previous results [6] shown in figure 1 indicate two onset energies (I_1 and I_2) for double photoionization processes. Taking argon as an example, the first onset at I_1 = 28.6 eV clearly corresponds to the single photon ionization of two separate argon atoms. We recall that the lowest ionization energy of a single atom in solid argon is 13.8 eV [6].

The second onset at I_2 = 40 eV is more difficult to assign since it may correspond either to the single photon ionization of three separate atoms or to the direct double ionization of a single argon atom. Recent experiments performed in solid xenon [7], in which the kinetic energy of both emitted electrons has been analysed, have demonstrated that both three atom single ionization and single atom double ionization processes exist.

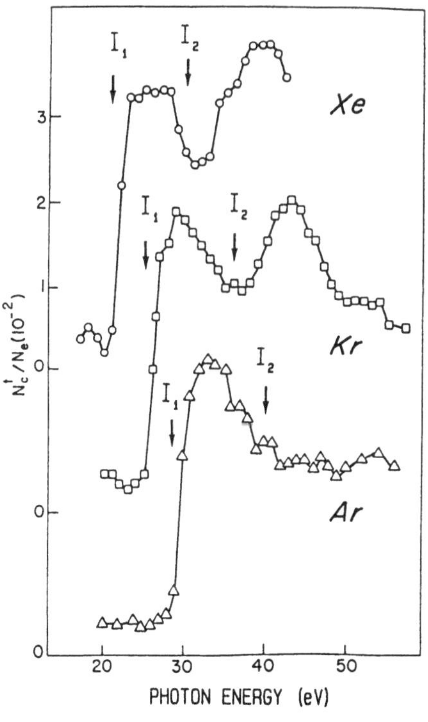

Figure 1 : Ratio of the number of true electron-electron coincidences N_c^t to the number of electrons detected by each detector N_e as a function of the photon energy. I_1 and I_2 indicate the onset energies of the double photoemission processes. The sample thickness is 3000 Å

4. Ion desorption from solid argon

Ion desorption was previously observed by electron impact [8] and by core photoexcitation [9] of rare gas solids. We have recently studied ion desorption produced by valence photoexcitation of rare gas solids [10]. The results on solid argon are shown in figure 2.

The threshold energies observed in figure 2 for ion desorption lie in the 33.4 - 38.8 eV photon energy range. They are well above the first onset (28.6 eV) and below the second onset (≃ 40 eV) observed in the double photoionization experiment (figure 1). By comparing these data with known ionization data in the gas phase, the various features of figure 2 could be assigned [10] to the formation of singly charged ion satellite states ($(Ar^+)^{**}$) in the solid. The relaxation of these satellite states by charge exchange with a neighbour atom would then lead to the formation of a pair of singly charged atoms Ar^+ - Ar^+. The Coulomb repulsion between these two neighbour ions could explain the ion desorption observation.

The ion desorption based on the primary formation of satellite states would also lead to the ejection of pairs of electrons. However there is no evidence of these processes in the curves of figure 1, probably because they are much less intense than other double ionization processes. On the other hand, we could expect the two atom ionization process (onset energy at 28.6 eV in solid argon) to give rise to ion desorption because of the Coulomb repulsion between singly charged ions. However, no signal was observed in this energy range in the ion desorption experiment (figure 2). This may be due to a very rapid delocalization of the excitation by charge exchange. Finally, the direct single atom single photon double ionization process observed in

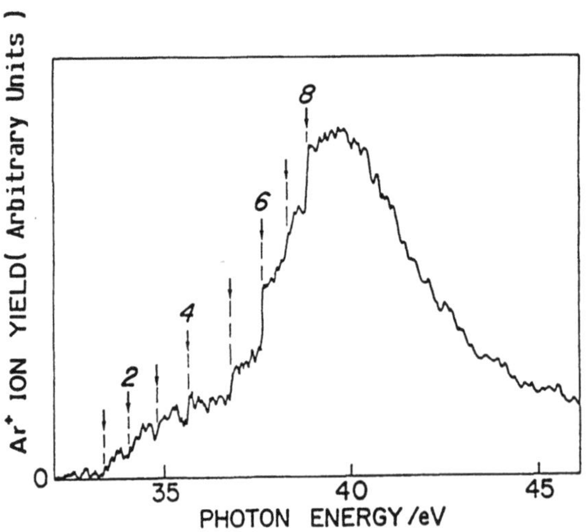

Figure 2 : Ar^+ ion yield from solid argon. The arrows indicate the threshold energies of the various structures observed in this curve.

double photoionization experiments [7] was also observed to produce ion desorption in solid argon and xenon [10] but at higher photon energy than what is seen in figure 2.

5. Ion desorption from solid nitrogen

Ion photodesorption from solid nitrogen was previously studied by Rosenberg et al. [4] in the 20-33 eV photon energy range. This energy range is too low to allow a comparison with gas phase double photoionization processes which are known to occur in the 43-60 eV energy range [11]. We then started some new investigations of ion desorption from solid nitrogen in the higher 25-75 eV photon energy range [12]. Preliminary results are shown in figure 3.

Data in the low photon energy range are quite similar to what was observed by Rosenberg et al. [4]. At higher energies several new features are visible in figure 3. We note in particular some inflection followed by a steep increase in the N^+, N_2^+, N_3^+ and N_4^+ curves of figure 3 at a photon energy of about 45.4 eV. Comparing this energy with the energies of dissociative states of $N_2^{++} \rightarrow N^+ + N^+$ at 45.2 eV, 46.1 eV, 46.9 eV, 47.9 eV, 49.6 eV and 53.0 eV [11] leads to the conclusion that direct double photoionization of a N_2 molecule in the solid may produce an ion (N^+) desorption via a <u>unimolecular dissociation</u> very similar to that occurring in the isolated

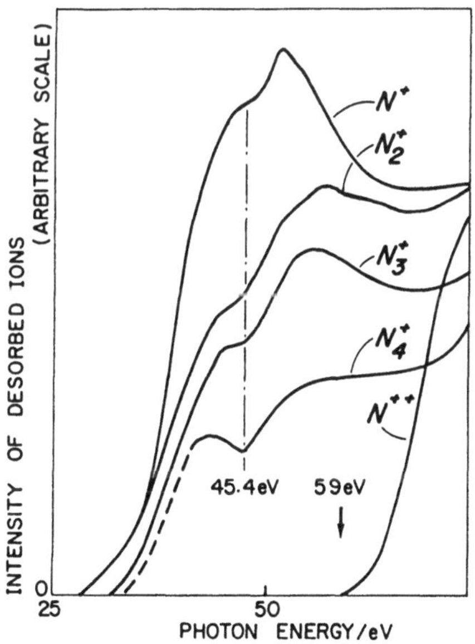

Figure 3 : Ion yield for desorption of N^+, N_2^+, N_3^+, N_4^+ and N^{++} from solid nitrogen (thickness = 1000 Å) as a function of the photon energy.

molecule. Such a desorption mechanism would also be valid for the N^{++} desorption since its threshold energy at 59 eV exactly coincides with that of the electronic states of N_2^{++} in the gas phase which dissociate into $N^{++} + N$ [13].

6. Conclusion

Three types of double photoionization processes in condensed gases have been encountered in this paper : (i) ionization of two separate atoms or molecules, (ii) formation of a singly charged ion satellite state followed by an inter-atomic or inter-molecular charge exchange, (iii) direct double photoionization of a single atom or molecule. Only processes (i) and (iii) have been observed in double photoemission experiments. Process (i) seems to be absent in ion desorption, maybe due to a strong delocalization of this type of excitation. Processes (ii) and (iii) which correspond to two electronic holes on the same species are expected to be more localized and they are indeed observed to give rise to ion desorption in rare gas solids and solid nitrogen. Much further work is needed to extend double photoemission studies to molecular solids, to assign the primary electronic processes producing ion desorption, to elucidate the various unimolecular and inter-molecular desorption processes and finally to achieve state-to-state studies of ion desorption in which both the primary excited electronic states and the final states of the products are well defined.

References :

[1] R. Franchy and D. Menzel, Phys. Rev. Lett. 43, 865 (1979).
[2] D.E. Ramaker, C.T. White and J.S. Murday Phys. Lett. 89A, 211 (1982).
[3] D.E. Ramaker, J. Chem. Phys. 78, 2998 (1983).
[4] R.A. Rosenberg, V. Rehn, A.K. Green, P.R. LaRoe and C.C. Parks, Phys. Rev. Lett. 51, 915 (1983).
[5] S. Auer, P. Feulner and D. Menzel, Phys. Rev.B, 35, 7752 (1987).
[6] H.W. Biester, M.J. Besnard-Ramage, G. Dujardin, L. Hellner and E.E. Koch, Phys. Rev. Lett. 59, 1277 (1987).
[7] G. Dujardin, H.W. Biester and L. Hellner, to be published.
[8] I. Arakawa and Y Tuzy, J. Nucl. Mat. 128-129, 779 (1984).
[9] G. Rocker, P. Feulner, R. Scheuerer, L. Zhu and D. Menzel, Physica Scripta, in press.
[10] G. Dujardin, L. Hellner, M.J. Besnard-Ramage and R Azria, submitted to Phys. Rev. Lett.

[11] M.J. Besnard, L. Hellner, G. Dujardin and D. Winkoun J. Chem. Phys. 88, 1732 (1988).
[12] G. Dujardin, L. Hellner, M.J. Besnard-Ramage and R. Azria, to be published.
[13] W. Eberhardt, E.W. Plummer, I.W. Lyo, R. Carr and W.K. Ford, Phys. Rev. Lett. 58, 207 (1987).

Photon Stimulated Desorption of Excited Neutrals from the Surface of Solid Ar

I. Arakawa[1] *and M. Sakurai*[2]

[1]Department of Physics, Gakushuin University, Toshima, Tokyo 171, Japan
[2]National Institute for Fusion Science, Chikusa, Nagoya 464-01, Japan

1. Introduction

The electron and photon stimulated desorption (ESD and PSD) of ions and molecules from the surface of van der Waals solids or physisorbed layers has been investigated by several groups for two decades in relation to the fundamental studies of physisorbed system by low-energy electron diffraction, Auger electron spectroscopy, etc. and the practical problem of outgassing in a cryopump. Recently, ESD and PSD phenomena at the surface of solid rare gases attracted the interest of many people. It became obvious that excitons play an important and main role in DIET(desorption induced by electronic transitions) of rare gas solid.

Feulner et al.[1] showed that the excitation of bulk and surface excitons with the order of n=1 and 2 led to desorption of neutral Ar. Correlation between the exciton excitation and PSD of Ne was also reported by Kloiber et al.[2]. The kinetic energy measurements of desorbed particles have been published by a few groups[3-5]. The time-of-flight(TOF) measurement of the overall desorbed species, which is thought to mainly consist of ground state neutrals, shows peaks in the kinetic energy distribution at about 0.5 eV and 0.05 eV for Ar[3]. The direct detection of desorbed metastables and its TOF measurement of ESD of Ar by Arakawa et al.[5] showed clear features of the kinetic energy distribution with two peaks at 0.6 eV and 0.04 eV. For rare gas solids, a number of desorption channels involving excitation and decay of excitons or excimers have been proposed[6]: conversion of electronic energy to kinetic energy by the decay and dissociation of excited dimers[6,7]: ejection of an excited atom from the cavity around an exciton at a surface[8]. Combination of the selective excitation technique using monochromatic light and the measurement of the kinetic energy distribution of desorbed particles is one of the most promising ways to reveal the DIET mechanism.

Desorption of Ar metastables (Ar') from the surface of solid Ar induced by the irradiation of VUV light has been studied using the beam line BL5B of UVSOR facility at Institute for Molecular Science (IMS), Okazaki. In our experimental study, only excited neutrals, which are detectable by a secondary electron multiplier, are detected and their kinetic energy distributions are obtained by TOF measurements. The purpose of the present study is to reveal the origin of two groups of desorbed Ar' with the kinetic energy of 0.6 eV and 0.04 eV, which were found in the previous ESD study on Ar[5], by selective excitation using monochromatic synchrotron radiation(SR) light as an excitation source.

Springer Series in Surface Sciences, Vol. 19
Desorption Induced by Electronic Transitions, DIET IV
Editors: G. Betz · P. Varga

2. Experimental

We only give an outline of the experiment here as the details will be published elsewhere[9]. Metastable atoms are detected by a secondary electron multiplier(Ceratron, Murata Corp.) with usual pulse counting electronics. Exact quantum efficiency of the Ceratron for the detection of Ar metastables has not been calibrated, but direct comparison with Channeltron showed no large difference between their detection efficiencies, which are thought to be in the range 0.1 - 1[10]. The kinetic energy of desorbed species is determined by TOF measurements. SR light from the storage ring is mechanically chopped by a rotating disc with narrow slits and is shaped to a pulsed beam with a width of 20μs and a period of 20 ms, typically. Solid argon is formed on a copper substrate which is cooled to 25$\pm$2 K by a mechanical refrigerator. The distance between the substrate and the Ceratron detector, namely the flight length, is 315$\pm$3 mm. A retarding potential of 20 V is applied to the grids in front of the detector to reject desorbed ions.

The diagram of the TOF measurement system is shown in Fig. 1(a). TOF spectrum, as shown in the upper part of Fig. 1(b), is obtained in a multichannel scaler by signal accumulation for 10^4 - 2×10^5 scans. The gate of the counter is controlled as shown in the lower part of Fig. 1(b) so that photon signals are excluded from the metastable signals.

The base pressure of the chamber is 3×10^{-7} Pa after a bake-out at 400 K for 24 hours. This condition, unfortunately, is not satisfactory for the purpose of the present study. Signal intensity of metastables is significantly reduced in 10 minutes or so after the deposition of an Ar layer. This is probably because of impurities, H_2O, CO, etc. adsorbed on the surface of solid Ar. During the TOF measurement, therefore, Ar gas is continuously introduced to the chamber at pressures about 10^{-5} Pa to maintain an uncontaminated surface.

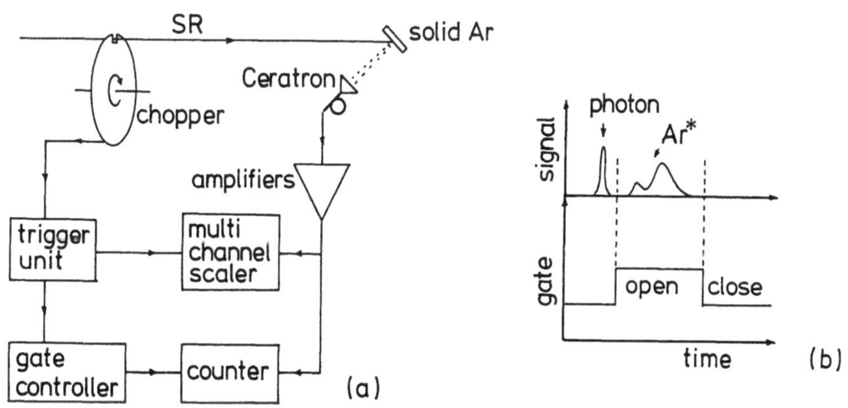

Fig. 1. Schematic diagram of the TOF measurement system.

3. Results and Discussion

A time-of-flight spectrum of PSD Ar^* from the solid Ar excited by 13.8 eV (λ = 90 nm) photon is shown in Fig.2. The number of accumulations is 2×10^5 (~1 hour) and the thickness of Ar layer is 1100 atomic layers at the beginning of the run and 1500 atomic layers at the end. The

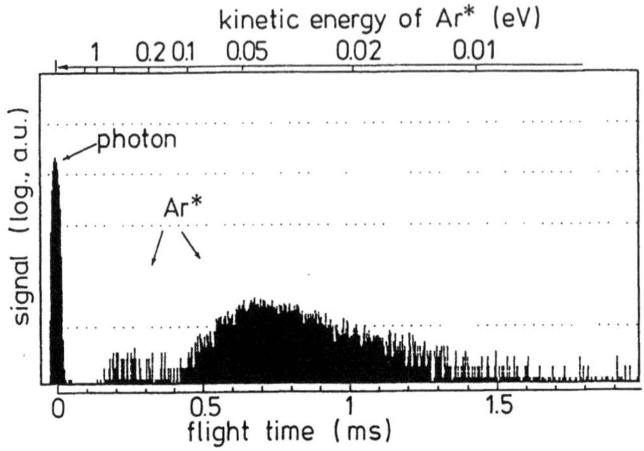

Fig. 2. Ar' PSD TOF spectrum at a photon energy of 13.8 eV. The center of the photon peak is chosen as the origin for the flight time measurement. Signal intensities (counts/channel) are plotted in a logarithmic scale.

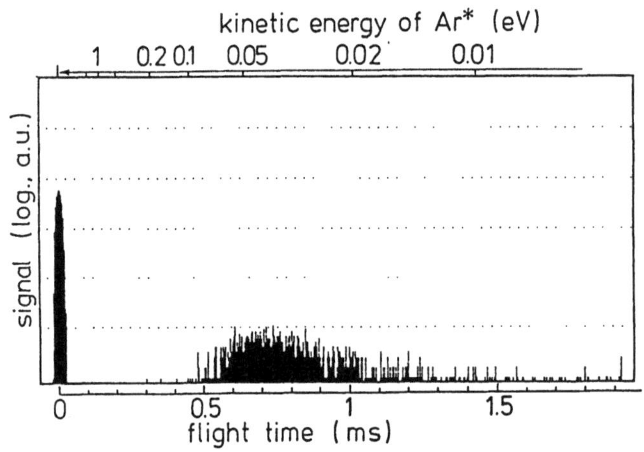

Fig. 3. Ar' PSD TOF spectrum at a photon energy of 12.2 eV. The number of scans is 2×10^4. The layer thickness is about 500 atomic layers.

kinetic energies of Ar' at the two peaks, namely 0.6 eV and 0.04 eV, and their shapes of distribution seem to be almost identical with those of ESD Ar'[5], except for the relative intensity of the two peaks. The Ar' peak with the higher energy (0.6 eV) disappears when the excitation photon energy is 12.2 eV (λ = 102 nm) as shown in Fig. 3. When the photon energy is lower than 11.7 eV (λ > 106 nm), the peak with the lower energy (0.04 eV) also disappears and no Ar' signals are observed.

The dependence of Ar' PSD yield on the photon energy is shown in Fig. 4 for the wavelength range from 80 nm to 110 nm. This spectra exhibits a clear correlation between the photon energies corresponding to peaks in the Ar' signal and the excitation energies of the first and second surface (1s and 2s) and bulk (1b and 2b) excitons presented in the absorption spectra by Saile et al.[11]. The photon energy depen-

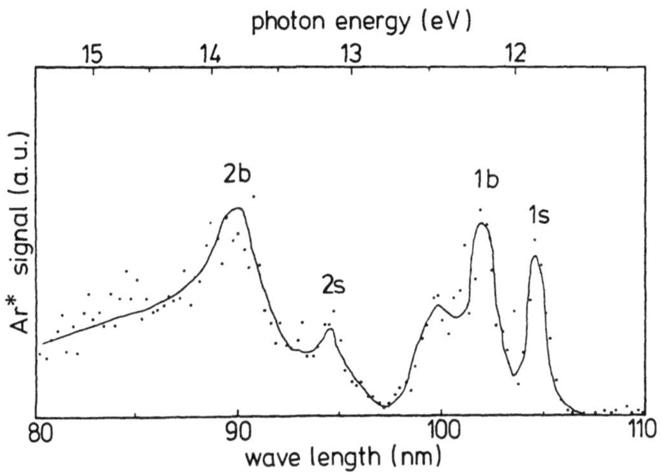

Fig. 4. Dependence of the Ar' signal intensity on the wavelength of light from 80 nm to 110 nm. The line is drawn as a visual aid.

dence of the Ar' yield obtained in the present study is almost identical with that of Feulner et al.[1], though there is a difference in the detection scheme of desorbed particles: the direct detection of excited neutrals in the present study and the total desorption yield in the latter case.

The TOF spectra of Ar' shown in Figs. 2 and 3 were obtained at energies corresponding to the excitation of the second bulk (2b) and the first bulk (1b) exciton, respectively. The comparison of the TOF spectra at these photon energies shows that the excitation of the second order exciton leads to desorption of Ar' with the kinetic energy around 0.6 eV and that the Ar' with the lower kinetic energy around 0.04 eV originates from the first order exciton. The exact threshold energy for the higher energy Ar', however, is hard to determine because of its weak signal intensity around 13 eV corresponding to excitation of the 2s exciton. This result suggests a different desorption channel for each Ar' group (0.6 eV and 0.04 eV) and supports the following two models of desorption mechanism. The lower kinetic energy Ar' (0.04 eV) originates from an exciton trapped at the surface where an excited atom is ejected by a repulsive potential around an exciton. Cui et al.[8] have estimated its kinetic energy at 0.1 eV or less. The desorption of the higher energy Ar' (0.6 eV) is induced by the relaxation of excited dimer: $Ar_2' \rightarrow Ar + Ar$. In the condition of this experiment, where the excited neutrals were detected, the excitation of the second (or more) order exciton is necessary to desorb the excited neutrals: $Ar_2'' \rightarrow Ar' + Ar$.

4. Summary

Photon stimulated desorption of Ar metastables from the surface of solid Ar was studied using monochromatic synchrotron radiation as an excitation source. The kinetic energy distribution of PSD Ar' were determined by time-of-flight measurement. The fundamental feature of the TOF spectrum is almost identical with that of ESD of Ar'[5]: two peaks in energy distribution at 0.6 eV and 0.04 eV. The measurement of the correlation between Ar* yield or TOF spectrum and the excitation photon energy revealed that the desorption of Ar* was initiated by the excitation of excitons. Two peaks in the Ar* TOF spectrum are attributed to two different desorption mechanisms respectively: the ejection of excited atoms from the cavity around an exciton which open at the surface leads to the 0.04 eV Ar* and the electronic decay of a highly excited dimer leads to the 0.6 eV Ar*.

References

1. P. Feulner, T. Müller, A. Puschmann, and D. Menzel, Phys. Rev. Lett. **59**, 791 (1987).
2. T. Kloiber, W. Laasch, G. Zimmerer, F. Coletti, and J.M. Debever, Europhys. Lett. **7**, 77 (1988).
3. D.J. O'Shaughnessy, J.W. Boring, S. Cui, and R.E. Johnson, Phys. Rev. Lett. **61**, 1635 (1988).
4. R. Pedrys, D.J. Oostra, A. Haring, A.E. de Vries, and J. Schou, Nucl. Instrum. Methods B 33, 840 (1988).
5. I. Arakawa, M. Takahashi, and K. Takeuchi, J. Vac. Sci. Technol. A 7, 2090 (1989).
6. C.T. Reimann, W.L. Brown, and R.E. Johnson, Phys. Rev. B **37**, 1455 (1988).
7. R.E. Johnson and M. Inokuti, Nucl. Instrum. Methods **206**, 289 (1983).
8. S.T. Cui, R.E. Johnson, and P.T. Cummings, Phys. Rev. B **39**, 9580 (1989).
9. M. Sakurai, T. Hirayama, and I. Arakawa, Vacuum, to be published.
10. I. Arakawa, M. Kim, and Y. Tuzi, J. Vac. Sci. Technol. A **2**, 168 (1984).
11. V. Saile, M. Skibowski, W. Steinmann, P. Gürtler, E.E. Koch, and A. Kozevnikov, Phys. Rev. Lett. **37**, 305 (1976).

Electron-Stimulated O^- Desorption from O_2 Condensed on a Rare-Gas Film: Evidence for an Indirect "Bounce" Trajectory

H. Sambe and D.E. Ramaker

Department of Chemistry, George Washington University,
Washington, DC 20052, USA

1. INTRODUCTION

The electron-stimulated O^- dissociation of O_2 gas is well understood [1]. In the energy range from 4 to 10 eV, O^- ions are produced by dissociative attachment (DA) via the transient $O_2^-(^2\Pi_u)$ state. This same DA process has been identified in electron-stimulated desorption (ESD) of O^- ions from O_2 condensed on thick rare-gas (Ar, Kr, or Xe) films [2]. However, electron-energy-dependence curve of the O^- yield is quite different from that for gaseous O_2 as shown in Fig. 1. The gas-phase curve rises slowly at the threshold and has a symmetric bell shape, while the condensed-phase curve rises sharply at the threshold and has an asymmetric triangular shape. Furthermore, the threshold energy of the condensed-phase curve depends on the rare-gas substrate as shown in Table 1. In this note, we offer an explanation for the threshold changes and suggest a trajectory for the O^- ion desorption via the $O_2^-(^2\Pi_u)$ intermediate state. We shall explain the different energy dependence elsewhere [3].

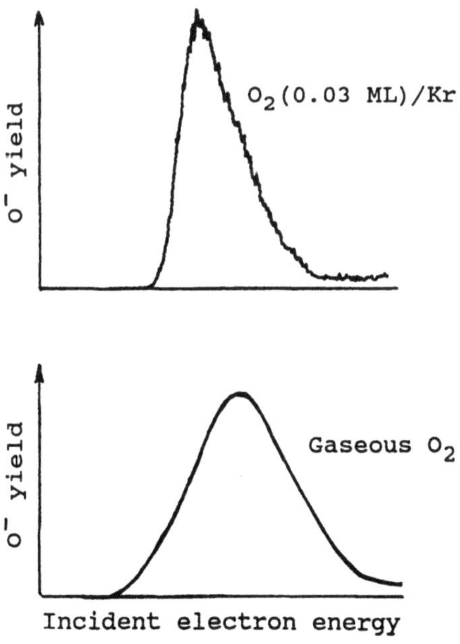

Fig. 1 Comparison of the ESD O^- yields from (a) gaseous O_2 and (b) O_2(0.03 ML)/Kr(20 ML)/Pt.

Table 1. Threshold energies (eV) for the O^- yield from O_2/**RG**.

RG =	Ar	Kr	Xe
Experiment	4.3	4.3	4.5
IB trajectory	4.3	4.4	4.5
DP trajectory	3.7	3.7	3.7

2. DISCUSSION

2.1 Two possible trajectories

Figure 2 illustrates two possible O^- desorption trajectories for an inclined O_2 on a rare-gas substrate. In one trajectory, DA yields an O^- ion moving away from the surface (we shall call this the direct-path (DP) trajectory), while in the other it

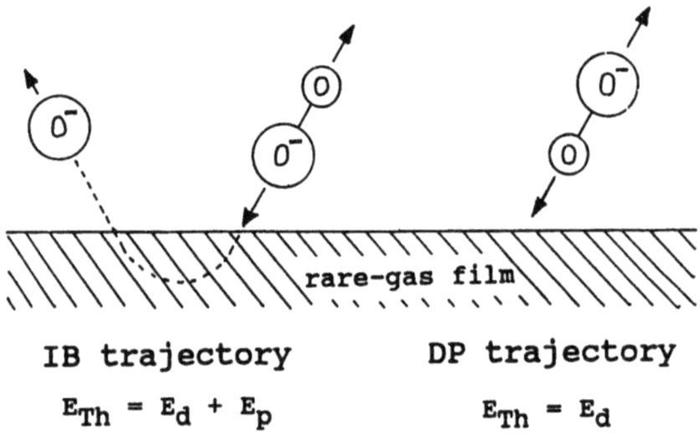

Fig. 2 Schematic illustrations of the direct-path (DP) and indirect-bounce (IB) O^- desorption trajectories for an inclined O_2 on a rare -gas substrate.

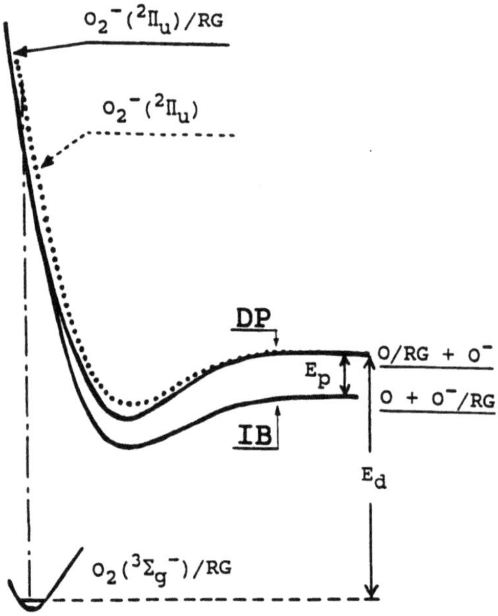

Fig. 3 Schematic $O_2^-(^2\Pi_u)$/RG potential curves corresponding to the DP and IB trajectories. Only the IB curve is valid.

yields an O^- ion moving initially towards the surface requiring it to "bounce" off the surface (we shall call this the indirect-bounce (IB) trajectory). The schematic potential curves which lead to these two trajectories are illustrated in Fig. 3. In the IB case, the $O_2^-(^2\Pi_u)$ intermediate state dissociates into the $O+O^-$/RG limit, which means that the O^- ion initially ends on the rare-gas (RG) surface, while in the DP case, it dissociates into the O^-+O/RG limit, which indicates that the O^- ion proceeds directly from the surface. The $O+O^-$/RG limit is lower than the O^-+O/RG limit by the surface-polarization energy (E_p) of the RG substrate. In the Franck-Condon region, the potential curves for the DP and IB trajectories become the same and are lowered by E_p with respect to the gas-phase potential curve as shown in Fig. 3.

2.2 Threshold energy

The threshold energies for O^- desorption via the DP and IB trajectories are given by E_d and E_d+E_p respectively, where E_d (=3.65 eV) is the dissociation energy (i.e. difference between O_2 and $O+O^-$). The threshold energy (E_{Th}) for the IB trajectory is derived from the equation

$$E_p = [E_{Th}-(E_d-E_p)]/2, \qquad (1)$$

which means that the O^- ion created on the surface must have its kinetic energy larger than the surface-polarization energy E_p to desorb from the surface. Threshold energies calculated with the E_p's obtained by Sanche et al. [5] (i.e., 0.62, 0.72, and 0.89 eV for Ar, Kr, and Xe, respectively) are listed in Table 1. The good agreement between the IB threshold energies (as opposed to the DP threshold energies) and the observed threshold energies suggests that the O^- desorption via the $O_2^-(^2\Pi_u)$ intermediate state proceeds exclusively along the IB trajectory. We note that the DA mechanism involving the IB trajectory is distinctly different from the Antoniewicz "bounce" mechanism [4].

2.3 Potential curve of $O_2^-(^2\Pi_u)$/RG

How does the predominance of the IB trajectory follow from the intuitive $O_2^-(^2\Pi_u)$/RG potential curves in Fig. 3? Based on these intuitive potential curves, one might infer that $O_2^-(^2\Pi_u)$/RG should dissociate equally into the $O+O^-$/RG and O^-+O/RG limits and predict that the branching ratios for the DP and IB trajectories should be about equal for high electron energies. Based on theoretical rules, we shall show below that the $O_2^-(^2\Pi_u)$/RG state dissociates only into the $O+O^-$/RG limit, predicting the predominance of the IB trajectory.

The lowest dissociation limit for O_2^- is $O(^3P)+O^-(^2P)$. According to the Wigner-Witmer rules, two $O_2^-(^2\Pi_u)$ states and two $O_2^-(^2\Pi_g)$ states arise from the interaction of the $O(^3P)$ and $O^-(^2P)$ atomic states. The equal number of the $^2\Pi_u$ and $^2\Pi_g$ states is a consequence of the degeneracy of the $O(^3P)+O^-(^2P)$ and $O^-(^2P)+O(^3P)$ limits. When O_2 is placed on a rare-gas substrate, the two O atoms are no longer identical, e.g., O_F and O_N [farther (F) and nearer (N) from surface]. According to

Wigner-Witmer rules, two $O_FO_N^-(^2\Pi)$ states arise from the $O_F(^3P)+O_N^-(^2P)$ atomic states and the other two $O_FO_N^-(^2\Pi)$ states, from the $O_F^-(^2P)+O_N(^3P)$ limit. According to the non-crossing rule, the lowest two $O_FO_N^-(^2\Pi)$ states connect to the $O_F(^3P)+O_N^-(^2P)$ limit, which is the lowest limit since O_N is closer to the surface. The two lowest $^2\Pi$ states are the O_2^- ground state and the $O_2^-(^2\Pi_u)$ state. Therefore, the $O_2^-(^2\Pi_u)$/RG state connects to the $O+O^-$/RG limit. In other words, the $O_2^-(^2\Pi_u)$/RG state dissociates exclusively into the $O+O^-$/RG limit.

2.4 Cl^- desorption from condensed Cl_2

Cl^- desorption from Cl_2 condensed on Pt provides a clear test of the above arguments. For this case, only four Cl_2^- states, namely, the $^2\Sigma_u^+$, $^2\Pi_g$, $^2\Pi_u$, and $^2\Sigma_g^+$ states, arise from the lowest dissociation limit, $Cl^-(^1S)+Cl(^2P)$. Analysis similar to that given above predicts that Cl^- desorption via the $Cl_2^-(^2\Pi_g)$ and $Cl_2^-(^2\Pi_u)$ intermediate states should follow the IB and DP trajectories, respectively.

The Cl^- ions in the IB trajectory must bounce off the Pt metal for Cl_2(1 ML)/Pt or off the Cl_2 film for Cl_2(4 ML)/Pt, where ML=monolayers. Colliding with the Pt metal, the Cl^- ions are most likely neutralized; colliding with the Cl_2 film, the Cl^- ions lose most of its kinetic energy so that it cannot overcome the polarization barrier. [The polarization barrier arises from the image charge potential (≈1.5eV) on the Pt metal substrate or the polarization potential (≈1 eV) for the thicker Cl_2 film.] Therefore, we expect that Cl^- desorption via the $Cl_2^-(^2\Pi_g)$ state, which must follow the IB trajectory, is quenched for both Cl_2(1 ML)/Pt and Cl_2(4 ML)/Pt.

The ratio of the Cl^- yield peaks via the $Cl_2^-(^2\Pi_g)$ and $Cl_2^-(^2\Pi_u)$ states is about 1/2 for gaseous Cl_2 [6]. For Cl_2(1 ML)/Pt and Cl_2(4ML)/Pt, this ratio nearly vanishes [7] showing that Cl^- desorption via the $Cl_2^-(^2\Pi_g)$ state is indeed quenched. This quenching cannot be attributed to angular discrimination, since the Cl^- ions were collected around 70° in the experiment [7] and the $^2\Pi_g$ peak should disappear only if we collect the Cl^- ions exclusively at 90° with respect to the incident electron beam [8]. This quenching also cannot be a discrimination due to a high threshold energy, since the expected peak position (≈2 eV) of the $Cl_2^-(^2\Pi_g)$ state is well above the expected threshold energy (≈0.0 eV).

The Cl^- desorption through the DP trajectory is expected to be similar to that from gaseous Cl_2: that is, the Cl^- ion kinetic energy as a function of the incident electron energy is expected to be a straight line with a slope of 1/2. This has been observed experimentally [7] for the Cl^- ions from the $Cl_2^-(^2\Pi_u)$ state, which proceeds through the DP trajectory.

3. CONCLUSIONS

Our study suggests the following generalizations: The ESD of A^- ions from physisorbed homonuclear diatomic molecules, A_2, are often produced by DA via a transient A_2^- state. The trajectory (DP or IB) of the desorbing A^- ions is uniquely determined by the potential curve of this A_2^- state. Furthermore, the A^- desorption via the IB trajectory is often quenched or severely reduced except on rare-gas substrates. In contrast, the A^- desorption via the DP trajectory proceeds, on most surfaces, similarly to A^- dissociation from gaseous A_2^-.

ACKNOWLEDGEMENTS

This work was supported by the Office of Naval Research.

REFERENCES

[1] G.J. Schulz, Rev. Mod. Phys. **45** (1973) 423.
[2] H. Sambe, D.E. Ramaker, L. Parenteau, and L. Sanche, Phys. Rev. Letters **59** (1987) 236 ; 505.
[3] H. Sambe, D.E. Ramaker, L. Parenteau, and L. Sanche, to be published.
[4] P.R. Antoniewicz, Phys. Rev. **B 21** (1980) 3811. In the Antoniewicz "bounce", an ion A^+ created on a metal surface moves towards the surface due to the image-charge attraction; and the neutralization of the A^+ leads to desorption of the neutral A because of the repulsion between the A and the metal surface.
[5] L. Sanche, L. Parenteau, and P. Cloutier, J. Chem. Phys. **91** (1989) 2664.
[6] M.V. Kurepa and D.S. Belic', J. Phys. **B 11** (1978) 3719.
[7] R. Azria, L. Parenteau, and L. Sanche, J. Chem. Phys. **87** (1987) 2292.
[8] T.F. O'Malley and H.S. Taylor, Phys. Rev. **176** (1968) 207; G.H. Dunn, Phys. Rev. Letters **8** (1962) 62; R. Azria, R. Abouaf, and D. Teillet-Billy, J. Phys. **B 15** (1982) L569.

Lower Limits on the Absolute Dissociative Electron Attachment Cross Section from O_2 Condensed on Rare-Gas Films

H. Sambe [1], *D.E. Ramaker* [1], *M. Deschenes* [2], *and L. Sanche* [2]

[1]Department of Chemistry, George Washington University, Washington, DC 20052, USA

[2]Medical Research Council Group in Radiation Sciences, Faculty of Medicine, University of Sherbrooke, Sherbrooke, Quebec, Canada J1H 5N4

Abstract. We summarize our approach for experimentally obtaining a lower limit on the absolute cross section for production of O^- by dissociative electron attachment from O_2 condensed on a Kr film. Comparison of this result with the dissociative attachment cross section for gas phase O_2 reveals a factor of 17 enhancement on the surface. It is semiquantitatively shown that a decreased autoionization rate, arising from the electronic polarization of the neighboring Kr atoms, can account for most of this large enhancement.

1. Introduction

It could be claimed that the simplest system imaginable for studying surface effects is closed-shell molecules condensed on a rare-gas film. This claim is made because of the weak interactions involved. In this paper we summarize previously published work, which taken together accomplishes to our knowledge the first direct comparison of absolute electron dissociation cross-sections in the gas phase and on the surface of a solid. Section 2 summarizes the experimental procedure for obtaining a lower limit on the absolute dissociative electron attachment cross section from O_2 condensed on rare gas films. Section 3 summarizes a semi-quantitative theory for explaining the large enhancement seen in the cross-section on the surface.

2. Experiment

Using low-energy electron transmission spectroscopy (LEETS), two of us (M.D. and L.S.) along with Marsolais [1] have devised a new method to measure surface charges on a thin dielectric film deposited on a metal substrate. In LEETS [2], the electron current arriving at the metal substrate is measured as a function of the potential applied between the substrate and the electron source. When the dielectric film is not charged, the electron current rises sharply at the zero-energy reference (ZER) of the vacuum level as seen in Fig. 1. When electrons are trapped in the dielectric or on the surface, this ZER shifts to a higher accelerating voltage because the trapped negative charges retard the incoming electrons. This ZER shift, ΔV, can be related to an average surface charge density Q by the relation $\Delta V = \varepsilon^{-1}LQ$, where ε and L are the dielectric constant and thickness of the film, respectively [1].

If the dielectric film is covered by molecules with a small surface density, σ_0, and only these molecules can trap electrons, then the average surface-charge density, Q(t), trapped by the molecules after small time, t, can be given, $Q(t) = J\sigma_0\mu t$. Here J is the average electron current density, and μ is the electron trapping cross section. The corresponding

Springer Series in Surface Sciences, Vol. 19
Desorption Induced by Electronic Transitions, DIET IV
Editors: G. Betz · P. Varga

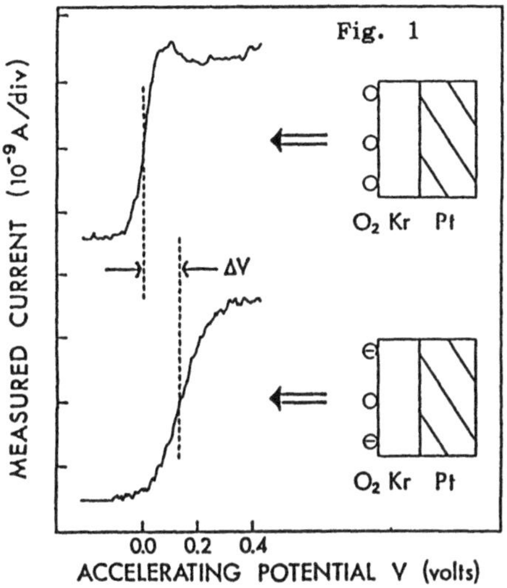

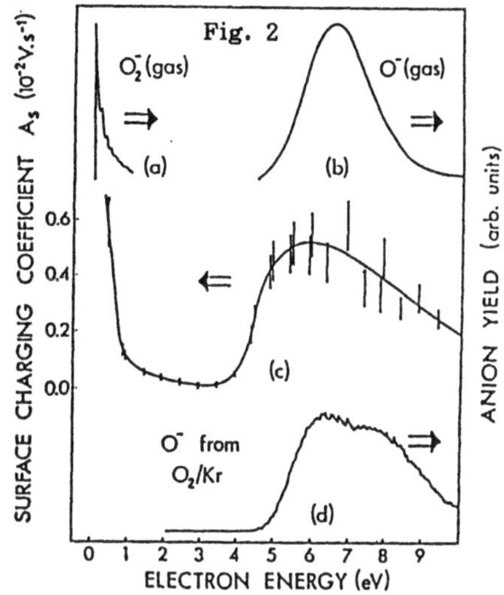

Fig. 1 Current transmitted through a clean Kr layer (top) and that through a Kr layer covered with 0.1-monolayer (ML) of O_2 (bottom) as a function of the the incident electron beam voltage, V. (Data from reference 1; used with permission).

Fig. 2 Anion yields produced by 0–10 eV electron impact on gaseous O_2 (a and b) and on 0.1 ML O_2/Kr (d). The electron energy dependence of the surface charging coefficent A_s for 0.1 ML O_2/Kr is shown in (c). (Data from reference 1; used with permission).

ZER shift $\Delta V(t)$ is then $\Delta V(t) = \varepsilon^{-1} L J \sigma_0 \mu t$. Its time derivative at t=0, denoted by A_s, is

$$A_s = d\Delta V(t)/dt]_{t=0} = \varepsilon^{-1} L J \sigma_0 \mu. \quad (1)$$

The proportionality of A_s with L, J, and σ_0 has been experimentally verified [1]. No charge accumulation is observed on pure Kr films for thicknesses of 1–20 monolayers (ML) in the electron energy range 4–10 eV [1], indicating that indeed only the O_2 species trap charge.

Figure 2 shows the electron energy dependence of A_s between 0–10 eV [1]. The results are comparable to the energy dependence of the anion yields derived from O_2 in the gas phase. No signal has been reported between 1.2 and 4.5 eV in the gas-phase. Curve d represents the electron stimulated desorption (ESD) signal from a 0.1 ML O_2/Kr. The similarity with the gas phase data shows that the surface charging in the 0–2 eV range is due to vibrational stabilization via the $^2\Pi_g$ O_2^- shape resonance,

$$e^- + O_2(X^3\Sigma^-_g)/Kr \rightarrow O_2^-(^2\Pi_g)/Kr. \quad (2)$$

The trapping cross-section in the region 4–10 eV is due to dissociative attachment via the intermediate $^2\Pi_u$ state of O_2^- [3],

$$e^- + O_2(X^3\Sigma^-_g)/Kr \rightarrow O_2^-(^2\Pi_u)/Kr \rightarrow O + O^-/Kr. \quad (3)$$

Since the ESD of O^- from the surface is also known to proceed through this process, the O^- ESD and charging proceed via the same mechanism in this energy region. The small shift in energy between the peaks for the ESD and charge trapping yields is due to the polarization force of the Kr film (see below).

We wish to make a quantitative comparison between the DA cross-sections in the gas phase and on the surface; i.e. comparison of the peak heights in the O^- yield from O_2 gas and the charge trapping ratio [4]. To do this quantitatively, estimates must be made of several variables. The film thickness of 20 ML is estimated within 15% accuracy by monitoring the interference structures in LEETS [5]. The O_2 coverage, σ_0 = 1/7 ML, is determined within an accuracy of 15% by verifying that the sticking coefficient of O_2 on Kr and Kr on Kr are the same at 20K. Thus $L=20\times3^{-1/2}a$ and $\sigma_0=(1/7)\times4\times3^{-1/2}a^{-2}$, and the dielectric constant, ε, of Kr is known to be $(1.78 \mp 0.01)\times8.85\times10^{-12}$ Farad/M [4]. The absolute current density, J = $(1.18 \mp 0.08)\times10^{-2}$ A/M^2, is measured by allowing all electrons, including those reflected, to return to the surface. The maximum in A_s around 6 eV is $(1.1\mp0.1)\times10^{-2}$ volt/sec. This gives a value from Eq. 1 of μ = $(2.2 \mp 0.7)\times10^{-17}$ cm^2. This can be compared with the cross-section in the gas phase at 6.7 eV of $(1.3 \mp 0.2)\times10^{-18}$ [6]. Apparently an enhancement of the DA cross-section by at least a factor of about 17 occurs on the surface.

3. Theory

The origin of this enhancement is explained using Fig. 3 which schematically shows how a Kr film influences the DA process [4]. The upper curves illustrate the accepted interpretation of the DA process for gaseous O_2, where the intermediate $O_2^-(1\pi_u^3 1\pi_g^4)^2\Pi_u$ state dissociates into

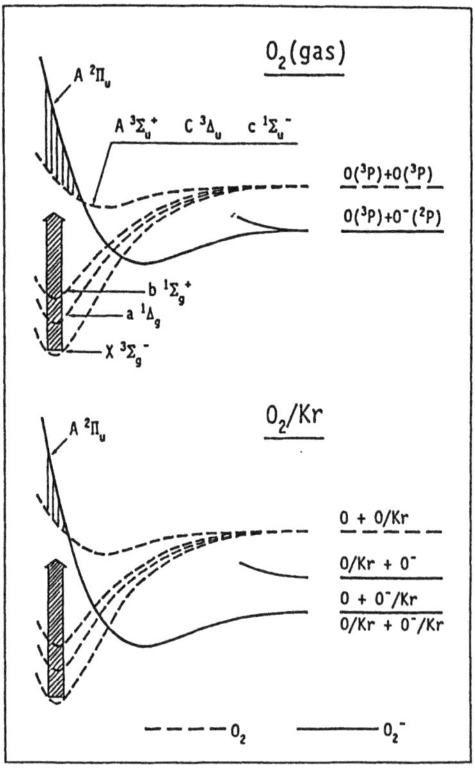

Fig. 3 Schematic potential energy curves (from ref. 4) of O_2^- (solid curves) and O_2 (dashed curves) for gaseous O_2 (top) and O_2 physisorbed on a Kr film (bottom). Autoionization of the $O_2^-(A^2\Pi_u)$ and $O_2^-(A^2\Pi_u)$/Kr states is indicated by the downward arrows.

the lowest O+O$^-$ limit, competing with autoionization into the $A^3\Sigma_u^+$, $C^3\Delta_u$, and $c^1\Sigma_u^-$ states arising from the excited configuration, $1\pi_u^3 1\pi_g^3$ and into the $b^1\Sigma_g^+$, $a_1\Delta_g$, and $X^3\Sigma_g^-$ states arising from the ground configuration, $1\pi_u^4 1\pi_g^2$. The three states arising from the excited configuration are close to each other and hence are represented by a single potential curve in Fig. 3. According to O'Malley [7], less than 1.3% of the $O_2^-(^2\Pi_u)$ states dissociates into the O+O$^-$ limit. About half of the $O_2^-(^2\Pi_u)$ states autoionize into the three states arising from the excited configuration, and the remainder into the three states arising from the ground configuration.

Theoretical expressions for the DA process in O_2 have been derived previously by O'Malley et al.[7]. The potential curve for $O_2^-(^2\Pi_u)$ and its survival probability against autoionization have also been calculated by O'Malley [7]. From this information, the cross-section can be evaluated using the expression [4]

$$\sigma_{DA} \propto \exp\left[3.4(R_C-R_E)^2 - 376(R_E-R_e)^2 - 41(R_C-R_E)^{3/2}\right], \qquad (4)$$

where R_E is the turning point (in A°) of the $O_2^-(^2\Pi_u)$ intermediate state at the incident-electron energy E; R_C (=1.44 A°) is the crossing point between the potential curves of the $O_2^-(^2\Pi_u)$ state and the $O_2^*(1\pi_u^3 1\pi_g^3)$ states; and R_e is the equlibrium internuclear distance (1.21 A°) of the O_2 ground state. The first two terms in brackets in eq. 4 arise from the Franck-Condon overlap integral between the lowest vibrational state of the O_2 electronic ground state and the vibrational continuum of the repulsive $O_2^-(^2\Pi_u)$ state. The delta-function approximation yields the second term and the first term arises from a first order correction. The third term represents a survival probability of the $O_2^-(^2\Pi_u)$ intermediate state against autoionization. The proportionality constant is nearly independent of the electron energy E, so that it is also nearly independent of R_E. The σ_{DA} reaches a maximum at R_E = 1.244 A°, which corresponds to E = 6.8 eV or the energy for the maximum cross-section.

In the presence of the Kr film, the potential curve of the charged $O_2^-(^2\Pi_u)$ state is shifted downward in parallel, because of an electronic polarization of the neighboring Kr atoms [3]. The dissociation limit O+O$^-$ is split into the O/Kr+O$^-$ and O+O$^-$/Kr limits, separated by the polarization energy. R_e and the three exponential coefficients in eq. 4 are not expected to be severely altered by the downward shift, however, R_E and R_C are altered. To calculate σ_{DA} for O_2/Kr, we need just to estimate the polarization energy.

The surface-polarization energy of the Kr film appears to be about 0.9 eV. The observed shift in the threshold energies (4.4 eV vs 3.5 eV) of the O$^-$ yield from O_2 gas and electron trapping seen in Fig. 2 give 0.9 eV. A theoretical estimate of this energy gave 0.86 eV [8]. The value of R_C has been obtained from an analytical fit of Fig. 3 giving R_C equal to 1.295 eV [4]. Insertion of these values into eq. 4 gives an enhancement factor of 21 on the surface. Nearly all of the enhancement comes from the survival factor, the third term in Eq. 4.

It is known that some of the O$^-$ produced via DA on the surface escapes or desorbs from the surface (i.e. the source of curve d in Fig. 2). If this fraction, f, is added to the charge trapping yield, we obtain the total DA cross-section on the surface. Therefore, we can equate 17/(1-f) to 21, and arrive at an estimate of f = 0.19. We believe this to be a reasonable estimate for the probability of desorption from the surface.

We can exclude two other possible explanations for the observed enhancement. The orientation of the O_2 on Kr is fixed, unlike in O_2 gas. The maximum possible enhancement due to orientation dependencies is estimated to be 1.3 [4]. Finally, we have previously discussed an enhancement of O$^-$ desorption from Kr due to a coherent scatterng phenomena or surface resonance [3]. However, in the current experiment, the incident electron energy is too low for normal-incident electrons to play a role in surface state resonances on Kr. Therefore, the evidence indicates that the polarization effect of the Kr lattice dramatically reduces the autoionization rate, and hence dramatically increases the DA branching ratio on the surface.

Acknowledgement HS and DER acknowledge support from the U.S. Office of Naval Research and MD, ADB and LS from the Medical Research Council of Canada.

References

1. R.M. Marsolais, M. Deschenes and L. Sanche, Rev. Sci. Instr. (in press).
2. L. Sanche, J. Chem. Phys. **71**, 4860 (1979).
3. H. Sambe, D.E. Ramaker, L. Parenteau, and L. Sanche, Phys. Rev. Lett. **59**, 236 (1987); 505 (1987).
4. H. Sambe, D.E. Ramaker, M. Deschenes, A.D. Bass, and L. Sanche, to be published.
5. R.M. Marsolais, M. Michaud, and L. Sanche, Phys. Rev. **A35**, 607 (1987).
6. G.J. Schulz, Phys. Rev. **128**, 178 (1962).
7. T.F. O'Malley, Phys. Rev. **150**, 14 (1966); **155**, 59 (1967).
8. F.J. Himpsel, N. Schwentner, and E.E. Koch, Phys. Stat. Sol. **B71**, 615 (1975).

A Comparative Study of Core-Induced Photon Stimulated Desorption and Electron-NEXAFS of Condensed H_2O, NH_3, and CH_4

G. Rocker, D. Coulman, P. Feulner, R. Scheuerer, Zhu Lin, and D. Menzel

Physik Department E20, Technische Universität München, D-8036 Garching, Fed. Rep. of Germany

Photon stimulated desorption of neutrals and ions from condensed films of H_2O, NH_3 and CH_4 has been investigated in the C 1s, N 1s and O 1s core excitation regimes, respectively, and compared to absorption data obtained via electron-NEXAFS. Desorption of neutral molecules and of H^+ as well as fragment and cluster ions has been observed. The PSD yields of neutrals, molecular ions and ionic clusters, as well as the corresponding total-yield and/or Auger-NEXAFS spectra show identical structures. This suggests that X-ray induced ESD, probably via double and multiple valence ionization, is the dominant mechanism for desorption of these species. The PSD yield of H^+ ions, however, exhibits significant discrepancies from these XESD derived NEXAFS spectra. On H_2O and NH_3, where it is by far the most abundant ion, it is dominated by a narrow peak right at the onset of desorption. The intensity of this threshold peak shows a remarkable polarization dependence. On CH_4, the H^+ signal is small, and no enhancement is observed at any photon energy; there is an overall shift of 0.2 eV of the H^+-NEXAFS structure to lower energy as compared to absorption. The mechanisms of these effects are discussed.

1. Introduction

A combined Auger-NEXAFS and H^+-desorption study performed on condensed and adsorbed H_2O/D_2O on Ru(001) /1,2/ has shown that excitation of an O 1s core electron into the antibonding $4a_1$ molecular orbital of water leads to an unusually high efficiency of H^+/D^+ formation. This effect and its unexpected angular behaviour have been tentatively explained by comparable time scales of electronic decay and nuclear motion, and by symmetry breaking through the influence of hydrogen bonding and/or the coupled excitation-deexcitation-dissociation sequence. We have then investigated the corresponding behaviour in multilayers of the isoelectronic molecules NH_3 (at N 1s) and CH_4 (at C 1s), including mixed NH_3/CH_4 layers /3/. While pure NH_3 shows essentially the same behaviour as H_2O, CH_4 does not. In mixed NH_3/CH_4 layers, the H^+ "threshold peak" is quenched, presumably due to rapid diffusion of CH_4 molecules to the surface of the mixed condensate and preferential photodestruction of NH_3 in the surface region. In the present paper, we briefly review these results, supplement them with additional data, in particular concerning the photodesorption of molecular ions and cluster ions, and discuss these results in context.

The UHV apparatus used has been described before /1-3/. It contained facilities to measure total and partial yields of emitted electrons, electron spectra, and neutral and ionic desorbates. Synchrotron light from BESSY (Berlin) was used employing the SX-700 monochromator in the energy range 200-800 eV.

Springer Series in Surface Sciences, Vol. 19
Desorption Induced by Electronic Transitions, DIET IV
Editors: G. Betz · P. Varga

2. H^+ Desorption From Condensed H_2O, NH_3, and CH_4 Films

Fig. 1 shows the comparison of Auger-NEXAFS to cluster ion and H^+ emission from a H_2O multilayer (more than 50 monolayers thick) in the O 1s core region. The Auger NEXAFS (AEY) which corresponds to absorption was found to be independent of both angles of photon incidence and of electron take-off which must be due to the amorphous structure of these layers /1,2/. The molecular absorption spectrum from Ref. 4 is also indicated in the upper part of Fig. 1.

Under all conditions, the H^+ ion was by far the most abundant ion detected (by factors between 50 and 1000, depending on conditions). Its emission was at a maximum around the surface normal. Molecular ions and cluster ions, such as $(H_2O)_kH^+(k=1,\ldots6)$, were detected at much lower intensities, as will be discussed below. The H^+ yield spectrum for glancing incidence ($\alpha=10^\circ$) and normal detection shows the threshold peak at 533.6 eV fully developed; only a weak shoulder is seen in the Auger-NEXAFS (=absorption) spectrum at its position. For polarization corresponding to increasingly normal incidence, the H^+ threshold peak decreases relative the the "main peak" between 535 and 542 eV (i.e. that paralleling absorption), corresponds to a mere shoulder at about 60° incidence /5/ and disappears totally for normal incidence ($\alpha=90^\circ$) /1,2/.

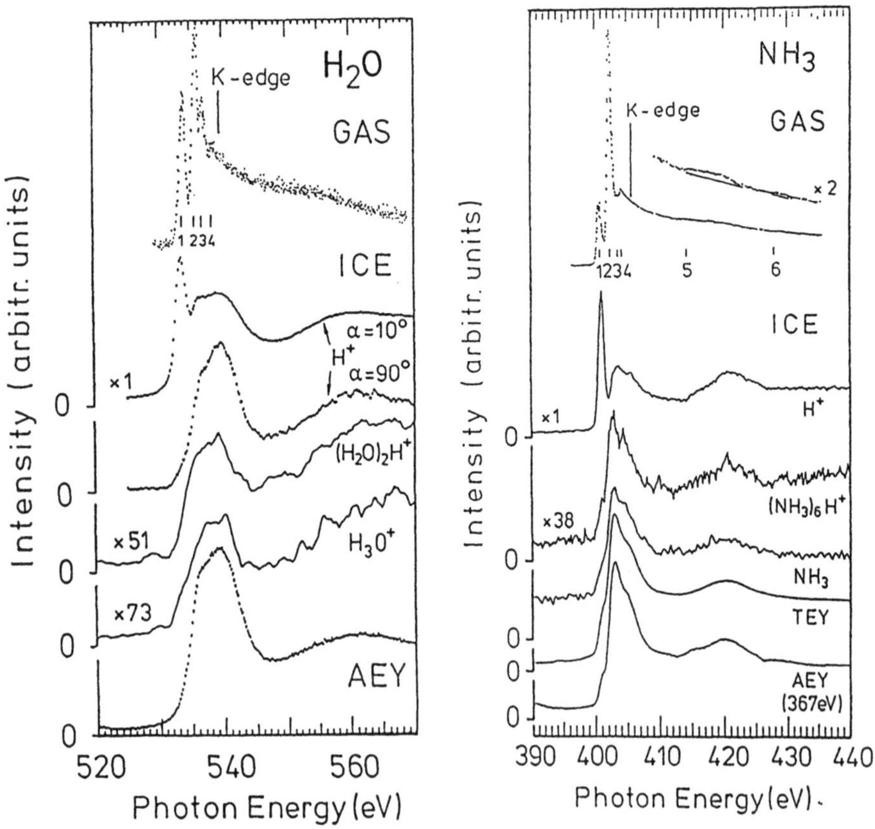

Fig. 1 (left) and Fig. 2 (right): Comparison of the molecular absorption spectrum (from Ref. 4) to the yields indicated from condensed layers for H_2O (1) and NH_3 (2), in the O 1s and N 1s regions, respectively

The corresponding situation for NH_3 multilayers, now at N 1s, is shown in Fig. 2. The absorption spectrum is given by both the total-electron yield (TEY) and the Auger electron yield (AEY, CFS energy: 367 eV). As in the case of H_2O, H^+ is by far the most abundant ion detected, but neutral NH_3 molecules as well as cluster ions of form $(NH_3)_nH^+$ (n=1...7) are also found and will be discussed below. The H^+ threshold peak is well developed at 401.0 eV. At this energy, the absorption spectra (TEY, AEY) show a relatively weak, but clearly visible shoulder. As for H_2O, the "main peak" corresponding to TEY/AEY is roughly explainable as a broadened version of the molecular absorption spectrum shifted to higher energies /3,6/. Also, the relative strength of the threshold peak decreases with increasing angle of incidence (not shown in Fig. 2).

The situation is quite different for CH_4 multilayers at C 1s. The H^+ yield is much lower in this case and is not the largest ion signal (see below). Besides ions, also neutral yields and total-electron-yield spectra were recorded. These exhibit features very similar to those found in the H^+ spectrum as illustrated in Fig. 3. A typical gas-phase absorption spectrum /4/ is indicated in the upper part of this figure. The overall small H^+ yield and the absence of a threshold peak demonstrate that CH_4 condensates behave very differently from H_2O and NH_3 ice in terms of core-induced photodissociation. Also, the shift and the

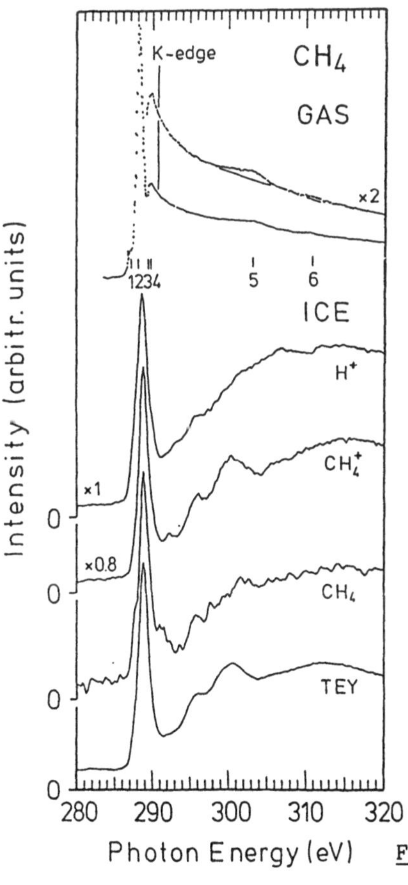

Fig. 3 As Figs. 1 and 2 but for CH_4 at C 1s

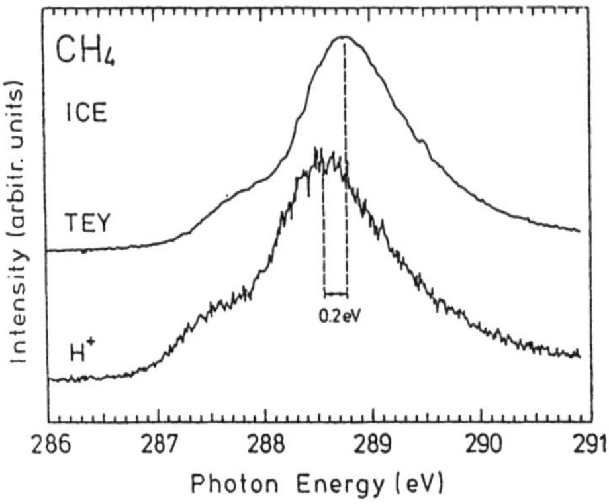

Fig. 4 Comparison of total electron yield (TEY) and H^+-yield from condensed CH_4 ice in the C 1s threshold region measured simultaneously (glancing incidence)

broadening of the absorption spectrum relative to the gas phase are small. However, a number of experiments confirmed that the entire H^+-yield spectrum monitored in the C 1s region is shifted by ~0.2 eV to lower excitation energies, if compared to all the other yield spectra recorded. This is illustrated in Fig. 4 which displays well resolved H^+-yield and total-electron-yield spectra at the C 1s threshold, taken simultaneously (so no shift of energy scale can be responsible). The most probable explanation for the origin of this shift is that the TEY spectrum measures the bulk absorption while in the H^+ spectrum the surface absorption is more strongly weighted. Clearly, there is no enhanced dissociation anywhere in the NEXAFS range.

Another interesting effect was observed for mixed NH_3/CH_4 layers in 1:1 composition prepared by co-condensation. The H^+ threshold peak monitored at N 1s was found to rapidly decrease as a function of measuring time. This may be indicative for preferential photoinduced removal of NH_3 from the surface and/or CH_4 diffusion to the surface of the co-condensate, even at temperatures as low as 10 K.

We conclude that the lowest core excitation in H_2O and NH_3 ices is particularly dissociative, while this is not the case for CH_4. Also, the energies of these threshold peaks lie below the main absorption bands which indicates that the absorption energies of the surface molecules are lower than those of the bulk molecules (see also below). The bulk absorption can in all cases be represented by broadened and shifted replicas of the gas-phase absorption spectra. The shifts are different for the three substances (H_2O:~2-3 eV, NH_3:~1.5 eV, CH_4:~0.7 eV). The broadening of the overall absorption spectrum is strongest for H_2O (about a factor of 2), intermediate for NH_3 (about x 1.5), and for CH_4 can be explained by our poorer resolution compared to Ref. 4. These observations suggest that the broadening as well as the energy shifts depend on the strength of the hydrogen bonding in the condensates.

The very intense H^+ threshold peaks for H_2O and NH_3 (compared to weak shoulders in the other yield spectra), their energetic difference from the bulk excitations, and their polarization dependences indicate that the small asymmetry introduced into the OH (or NH) bonds of H_2O (NH_3) surface molecules by hydrogen bonding of those H atoms not sticking out of the surface leads to splitting of the excitations. The localization of the excitation on the OH (NH) bond which is not hydrogen-bonded into the bulk, is then further increased by motion of that H atom during core lifetime, totally concentrating the excitation energy on this bond, as has been discussed before /1,2/. Autoionization can then leave the hydrogen charged /3/.

3. Desorption of Neutrals, and of Fragment and Cluster Ions

Besides electrons and H^+ ions, neutral NH_3 and CH_4 molecules as well as fragment and cluster ions, such as $(H_2O)_kH^+/(NH_3)_kH^+$ (k=1,...7) and $C_mH_n^+$ (m=1,2,3,;n=2,4,..12), respectively, were detected. Some representative yield spectra of these desorbing species are also shown in Figs. 1-3. The spectra exhibit essentially the same features as those found in the corresponding electron-NEXAFS (i.e. absorption) curves. In particular, a threshold peak was not detected in any of the neutral- or ion-yield spectra. This behaviour suggests that X-ray-induced electron-stimulated desorption (XESD), probably via double and multiple valence ionization, is the dominant mechanism for desorption of these species.

The yield intensities of the desorbing ionic species were determined at photon energies of 350, 500 and 600 eV for CH_4, NH_3 and H_2O, respectively (i.e. above the threshold regions). They were referenced to corresponding H^+ signals (≙100) and are given in Table 1. From this table it can be seen that the H^+ ion is by far the most abundant ion in the cases of NH_3 and H_2O, even far away from the threshold region with its additional enhancement of H^+. Only a few percent of the total intensity is derived from cluster ions whose signals maximize at k=2 for H_2O and k=5 for NH_3.

In the case of CH_4, the H^+ yield is not the largest ion signal observed, at any energy. Species such as CH_4^+, $C_2H_2^+$ and $C_2H_4^+$ dominate the ion mass spectrum taken at hν=350 eV. Other ionic clusters and fragment clusters, such as $C_2H_8^+$ and $C_3H_n^+$, contribute significantly to the total ion yield. Furthermore, it seems as if only $C_mH_n^+$ ions desorb with m=1,2,3

Table 1. Relative PSD intensities of fragment and cluster ions measured on solid ammonia, water and methane at hν=500, 600 and 350 eV, resp.

AMMONIA / WATER			METHANE	
k	$(NH_3)_kH^+$	$(H_2O)_kH^+$	m, n	$C_mH_n^+$
0	100	100	0, 1	100
			1, 2	46.2
1	0.9	2.1	1, 4	114
			"1, 6"	6.8
2	2.0	3.2	2, 2	112
			2, 4	102
3	2.5	2.8	2, 6-8	24.7
			3, 2	18.0
4	5.1	2.0	3, 4	15.1
			3, 6	6.1
5	7.1	1.0	3, 8	4.4
			3,10	3.0
6	4.4	0.7	3,12	4.9
7	4.1	-		

and even numbers for index n. No detectable amounts of ionic species with masses above 50 amu were observed. These findings again demonstrate the completely different behaviour of CH_4 condensates, if compared to H_2O and NH_3. As pointed out above (see also ref. 3), weaker hydrogen bonding and/or even distribution of the excitation energy over the four H atoms may account for the missing threshold peak. In the latter case, not enough energy would be concentrated on one CH bond to cause dissociation and thus desorption of hydrogen species. We note that the well-known weakness of the C 1s to $3a_1$ excitation in CH_4 (which is due to its being dipole-forbidden to first order and becoming allowed by vibronic coupling only /7/) cannot explain the missing H^+ threshold peak since an enhancement in desorption would still have become obvious (see fig. 4). The ionic clusters appear to derive from ion-molecule reactions in the solid phase for CH_4, while hydrogen-bonded clusters dominate for H_2O and NH_3.

4. Conclusion

We conclude that the strong dissociativeness of the lowest core excitation and its particular properties for condensed H_2O and NH_3 can be explained by excitations localized to the surface and by comparable time scales of core lifetime and atom motion, so that symmetry breaking by the surface or/and the coupled excitation/deexcitation sequence occurs. The relative importance of these two aspects should become clearer from gas phase experiments. This channel is totally absent in CH_4 which shows a much smaller H^+ contribution for all excitations. Strong differences become obvious also for cluster ion formation between H_2O and NH_3 on the one hand (hydrogenated molecular cluster ions of increasing size) and CH_4 (ion-molecule reaction products) on the other. The latter appear to be due to secondary processes in the layers which need further clarification.

5. Acknowledgements

The cooperation of A. Puschmann with the H_2O measurements and valuable discussions with W. Wurth are gratefully acknowledged. The support given by H. Peterson (BESSY) was very helpful. This work has been supported by the German Ministry of Research and Technology (BMFT) through grant 05 366 and 466 CAB.

References

1. D. Coulman, A. Puschmann, W. Wurth, H.-P- Steinrück, and D. Menzel Chem. Phys. Lett. 148, 371 (1988).
2. D. Coulman, A. Puschmann, U. Höfer, H.-P. Steinrück, W. Wurth, P. Feulner, and D. Menzel, submitted to J. Chem. Phys.
3. D. Menzel, G. Rocker, D. Coulman, P. Feulner, and W. Wurth, Physica Scripta, in press.
4. G.R. Wright and C.E. Brion, J. Electr. Spectr. Rel. Phen. 4, 25 (1974)

5. The first measurement of H^+ desorption from ice used about this condition and showed only a shoulder at this energy: R.A. Rosenberg, P.R. LaRoe, V. Rehm, J. Stöhr, R. Jaeger, and C.C. Parks, Phys. Rev. B 28, 3026 (1983).
6. G. Rocker, L. Zhu, R. Scheuerer, P. Feulner, and D. Menzel, to be published.
7. P.S. Bagus, M. Krauss, and R.E. LaVilla, Chem. Phys. Lett. 23, 13 (1973).

Part VI

Insulators

Stimulated Desorption from Alkali Halides

M. Szymoński

Institute of Physics, Jagellonian University, Reymonta 4,
30-059 Krakow, Poland

Abstract. Charged particle and high energy photon irradiation of alkali halides stimulates various desorption processes, studied already for several decades. In general, those processes could be classified as electronic- and collisional- type mechanisms. In this review I will concentrate on the electronic processes and discuss experimental and theoretical data available on desorption yields, angle-resolved and mass selected energy spectra of ejected particles and luminescence intensities. A problem of excited alkali atoms and ions from ESD and PSD of alkali halides will also be mentioned in order to set limits for alternative explanations existing in the literature. A brief description of the valence exciton mechanism for desorption in irradiated alkali halides is given based on the current theoretical knowledge in this field. A comparison of the theoretical predictions with the existing experimental data allowed us to sketch out some future developments in the area of stimulated desorption from alkali halides.

1. Introduction

Ion(ISD)-, electron(ESD)- and photon(PSD)- stimulated desorption from alkali halide surfaces has regained much interest in recent years. The reasons for this are threefold. The older experimental work, summarized thoroughly by Townsend /1/, was hampered by a lack of sufficiently refined experimental techniques for studying such complex processes in rather difficult, insulating materials. Recent development of sophisticated experimental methods, however, such as a mass selected time-of-flight spectroscopy (TFS) and a Doppler shift laser-induced fluorescence spectroscopy (DSLIF) combined with angle-resolved detection provided the appropriate tools for studying various desorption processes.

Furthermore, theoretical knowledge on elementary excitations in ionic crystals was limited and only recent development of advanced computational methods with the use of fast computers began to deliver the required information.

Finally, there is an increasing interest in desorption of insulators because of many technological applications. Alkali halides are considered as model materials because they have simple crystallographic and electronic structures, and because desorption processes in these materials are very efficient /1,2/.

Although stimulated desorption processes in alkali halides have been studied for more than two decades, there exist several unanswered questions and controversies. The purpose of the present review is to discuss experimental and theoretical data available, and to address the following problems:

- conversion of electronic energy deposited by the stimulating beam into nuclear motion of the desorbing particles,
- relation between desorption and the luminescence of alkali halides,
- directional emission of desorbing halogen,

Springer Series in Surface Sciences, Vol. 19
Desorption Induced by Electronic Transitions, DIET IV
Editors: G. Betz · P. Varga

- surface stoichiometry changes due to stimulated desorption,
- the role of hot holes and surface excitons in desorption,
- gas-phase versus core-hole excitation mechanisms in desorption of excited neutrals and ions,
- ion-stimulated desorption of alkali halides.

2. Desorption of Neutrals

Ground state neutral particles are predominantly emitted due to ISD, ESD or PSD. Both halogen (X) and alkali atoms (M) are observed as well as up to 20% of halogen dimers, and in some cases a very small yield of MX molecules /1,2/. It has been conclusively shown /3,4/, however, that only the halogen component is desorbed due to the interaction with a stimulating beam, whereas alkali atoms evaporate thermally from the decomposed surface.

2.1 Energy Spectra

The first experimental results on energy distributions of atoms desorbed from alkali halides by electron bombardment were obtained for compressed powder samples /4/. Recently we have measured such distributions with both time-of-flight and DSLIF methods /5,6/. A set of time-of-flight distributions of Br atoms leaving a (100) KBr target along the normal to its surface at various surface temperatures is shown in fig.1. A double structure of the low temperature spectra is clearly visible. The broad peak with a maximum at 1 ms corresponds to thermal particles, while the second

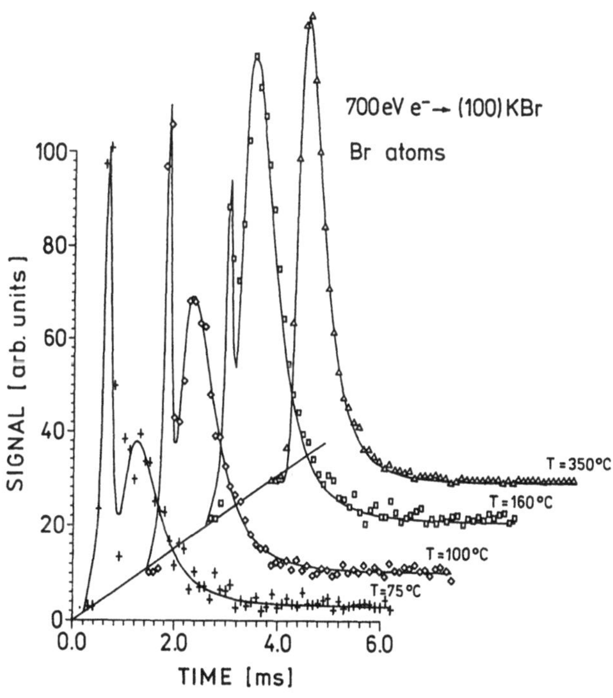

Fig.1. Time-of-flight distributions of Br atoms leaving a (100) KBr target along the normal to its surface at various surface temperatures. The spectra are normalized to their maximum values.

narrow one with a maximum at 0.4 ms corresponds to faster, non-thermal particles. It is seen that with increasing sample temperature the thermal part of the spectrum grows and at 350 °C the number of non-thermal particles accounts only for a small fraction of all emitted Br atoms.

We have found that the Maxwell-Boltzmann distribution, dN/dE, for 3-dimensional evaporation described accurately the shape of the thermal part of the spectrum:

$$dN/dE \sim E \exp(-E/k_BT), \quad (1)$$

where k_B is the Boltzmann constant and T is the temperature. The spectra for alkali atoms were Maxwellian at all temperatures as reported in our previous work /2/. The measurements for single crystals, generally agree with the previous study for compressed powders by Overeijnder et al. /4/, except that Overeijnder had measured the thermal distribution as proportional to $E^{1/2}$ but not to E (eq.(1)). It has been shown, however, that observation of the $E^{1/2}$ factor was caused by a systematic problem with the slit width used in the time-of-flight spectrometer /7/.

The 3-dimensional Maxwell-Boltzmann distributions were accordingly found with DSLIF technique for ESD of Na atoms from a (100) NaCl.

The emission of non-thermal particles takes place only for certain alkali halides. They were clearly visible in the energy distributions from RbCl, KBr, RbBr, KI and RbI below 150°C /4,8/. On the other hand, NaBr, CsBr, NaI and CsI give rise to a purely thermal halogen energy distribution even below 100°C. Overeijnder et al. /4/ have correlated the fraction of desorbed halogen atoms with a geometric parameter S/D determining the space available in the lattice to receive the halogen atom. A new approach to

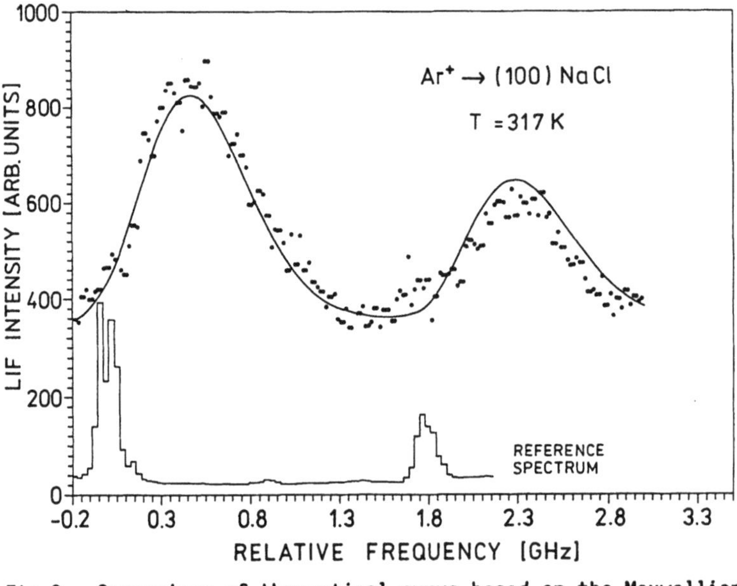

Fig.2. Comparison of theoretical curve based on the Maxwellian distribution (solid line) with experiment (dots) for the DSLIF scan of the Na atoms desorbed from a (100) NaCl with a 5 keV Ar ion beam. The reference signal from the polarization spectroscopy setup is given in the lower part of the plot.

this so called "Rabin-Klick parameter" has been proposed by Williams and coworkers /9,10/ and it is discussed in the next subsection.

For the case of ISD all measurements performed on compressed powder samples showed at room temperature a typical double structure: a high energy component well described by the collision cascade distribution /7,11,12/ and a thermal part well described by the Maxwellian spectrum, eq.(1). No significant differences were seen for spectra of alkali or halogen components.

For single crystal alkali halides only the energy distributions of Na atoms from ISD of NaCl were measured with the DSLF method /5,13,14/. Such a spectrum obtained in our laboratory for 5 keV Ar^+ bombardment along the normal to a (100) surface is shown in fig. 2. All 4 hyperfine components of the D_1 line corresponding to a Doppler shifted and broadened transitions from the sodium ground state are analysed in the figure. The spectrum can be very well reproduced by assuming that the desorbing flux is described by the Maxwellian distribution (1) only. This indicates that ISD of single crystalline alkali halides is dominated by the same electronic processes as in the case of ESD. One should note, however, that with the DSLIF method only the neutral ground state atoms can be detected. Further experimental data taken at low sample temperatures and in particular for halogen component would be of great importance for a better understanding of the ion stimulated desorption of alkali halides.

2.2 Excitonic Mechanisms for Desorption of Neutrals

It is generally accepted that valence electronic excitation (excitons or holes) is responsible for PSD, ESD and to a large extent ISD of ground state neutrals from single crystalline alkali halides /1-5,14/. Both free excitons and holes evolve in a time range of picoseconds /10/ into localized defects called "self-trapped excitons (STE)", i.e. an electron coulombically bound to the site of a covalently bounded diatomic halogen molecular ion. The last configuration has been usually called a V_k center lying along the <110> crystal axis and centered on the midpoint between two nearest-neighbor halide sites (see fig.3). One way of recombination of electron with a V_k center via states of the STE results in emission of an σ and/or π polarized luminescence with respect to the V_k symmetry axis. This process dominates at low temperatures of the sample.

It has been proposed by Pooley /15/ and Hersh /16/ that alternatively to the π -luminescence emission, the lowest state of the STE can decay nonradiatively forming a halogen vacancy (an F-centre) and an interstitial halogen atom (H-center) removed through a chain of replacement collisions along the <110> axis. The first suggestions of electron sputtering from the Pooley/Hersh model came from Townsend /1,3/.

It has been observed, however, that the delay in F-center formation for KCl and KBr crystal is much smaller than the lifetime of the π -luminescent state. This contradicted the lowest STE state as the precursor of the Frenkel defect formation and consequently of the halogen desorption /17/. Recently Williams et al. /18/, based on the self-consistent theoretical calculations of the STE relaxation, have distinguished two types of F-H pair production: - a diffusive, thermally activated process and a dynamic one.

The diffusive channel starts from the lowest STE electronic state which is unstable against a shift along <110> axis to "off-center" equilibrium position as seen in fig. 3. The off-center position of the relaxed STE is in fact equivalent to a nearest-neighbor F-H pair. It has been proposed that thermally activated F and H center formation proceeds by diffusion of a halogen atom out of the relaxed STE in its lowest, π-luminescent state

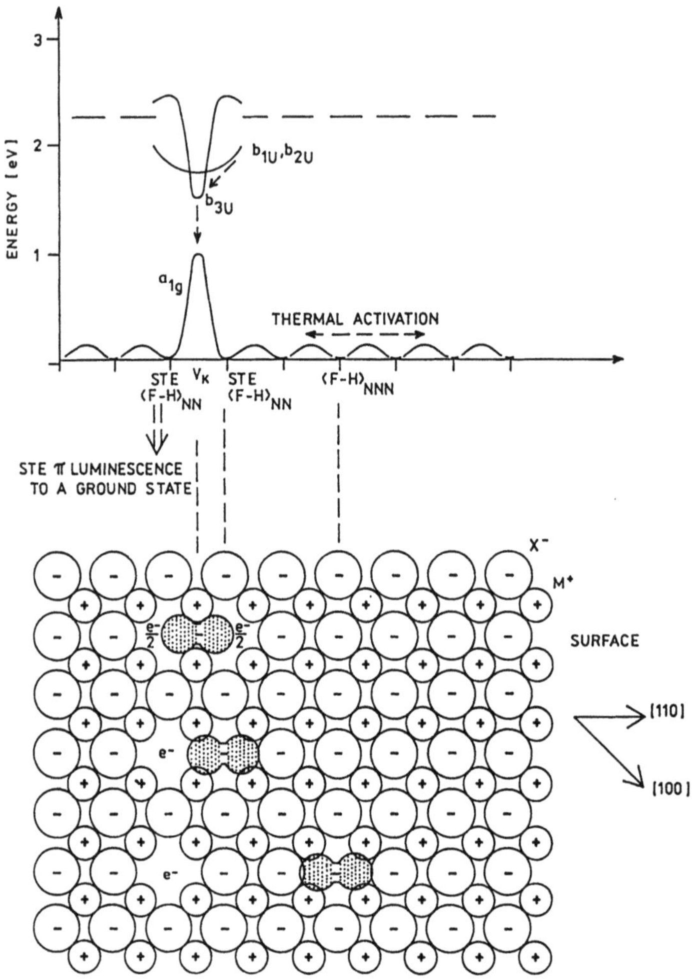

Fig.3. The change of the lattice energy relative to the perfect lattice of KCl versus the traslational coordinate along the <110> axis. The STE states are designated by the electron´orbitals. In the lower part of the figure, pictorial representations of lattice configurations for the "on-center" and "off-center" STE and F-H pair in a next nearest-neighbour position are shown. Due to R.T.Williams et al./18/.

/18/. At temperatures equal to or above room temperature, emission of thermal halogen atoms can be expected from the diffusive mechanism occurring near the surface. The thermally activated motion of the halogen can also cause a recombination of the F-H pair into the STE with a subsequent emission of the π-luminescence photon. Thus, the π-luminescence should be in anticorrelation with the desorption yield, if the diffusive channel of defect production dominates. Such an anticorrelation effect has been observed for NaCl by Townsend et al. /3/ and it was registered for a somewhat different temperature range in our laboratory as seen in fig. 4.

In the dynamic process the energy for F-H pair formation and separation is derived from stored energy of higher electronic states of the STE. As

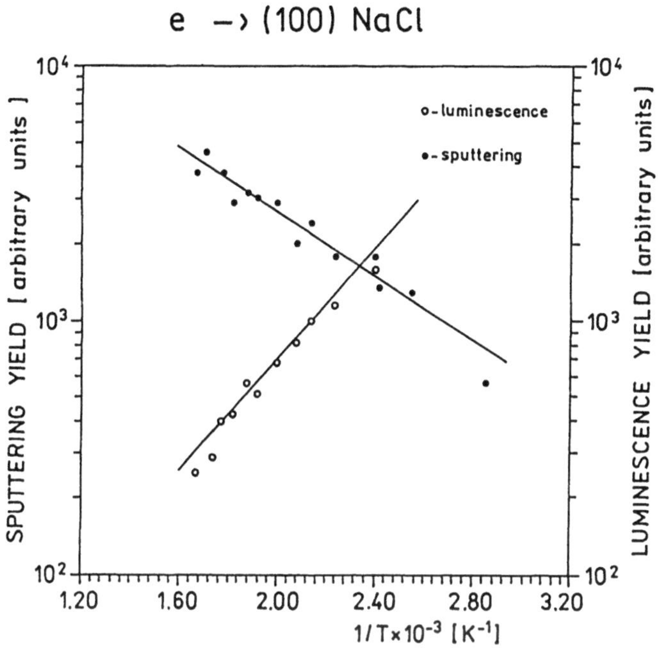

Fig.4. The temperature dependence of luminescence and desorption of a (100) NaCl crystal bombarded with 500 eV.

indicated in fig.3 the STE relaxing through those higher levels, should cross into the lowest a_{1g} potential surface near the configuration with "on-center" (V_k-like) symmetry. Since the a_{1g} surface at this point is unstable, the H center may be ejected at sufficient velocity to continue on for several lattice spacings along <110> direction of the crystal before thermalizing /1,10,18,19/. The proposed mechanism is constrained to those specific crystals for which the energy drop associated with translation of the H center along <110> from the "on-center" position is relatively high allowing for stable F-H separation /10/. If this event occurs in close proximity of the surface, ejection of the non-thermal halogen atoms with energies up to about 1 eV for KBr can be observed as seen in fig. 1. Apparently this efficient process of non-thermal halogen ejection causes a considerable reduction of thermally migrating H-centers at low temperatures and subsequently it reduces the recombination luminescence intensity as seen for (100) KBr in fig. 5.

2.3 Directional Ejection of Non-thermal Halogen Atoms

The recent experimental results on angle-resolved energy distributions of halogen atoms desorbed from a (100) surface of a KBr single crystal subjected to electron bombardment /6/ are reproduced in fig. 6. The angular scan was made in a (010) plane at several observation angles. It is seen that the ejection of non-thermal Br atoms is strongly forward peaked along the normal to the (100) surface. If we assume that the evaporating thermal Br atoms have a standard cosine distribution, the ratio of non-thermal to thermal yields for the target temperature of 120°C is roughly $\cos^{2.5}\theta$.

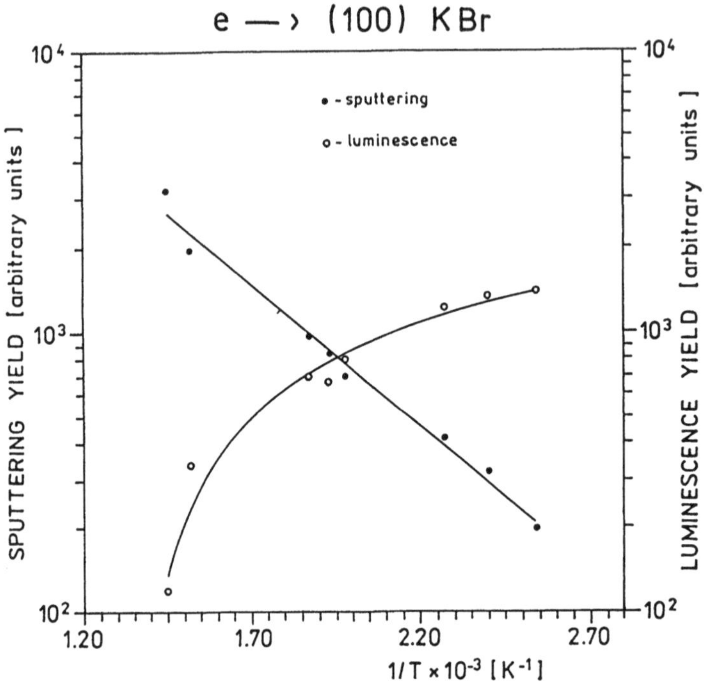

Fig.5. The temperature dependence of luminescence and desorption of a (100) KBr crystal bombarded with 700 eV.

but for surface temperature lower than 75°C it approaches $\cos^4\theta$. This demonstrates a strong directional effect in the ejection process of desorbed particles which is contrary to the expectations based on the excitonic model. The symmetry of the dynamic process of F-H pair formation indicates that a <110> direction of the crystal (i.e. $\theta = 45^0$) should be prefered in the halogen ejection. One possibility would involve the concept of a surface exciton and it will be discussed in the next subsection.

An alternative explanation takes into account the fact that the excitonic mechanism of desorption activates only one target component, namely the halogen sublattice. This process results in formation of halogen deficient surface layers and consequently in accumulation of alkali metal at the surface depending on their evaporation rate at the given temperature. This strong nonstoichiometry of electronic sputtering was already discussed by Al Jammal and Townsend /20/ and experimentally confirmed by Auger electron spectroscopy /2/ and electron energy loss spectroscopy /21/. The build up of an excess metal layer at the surface of the LiF single crystal was also recently observed by Wurz and Becker /22/. Therefore, it is quite plausible, that the energetic halogen atoms have to penetrate in the ejection process through the damaged, irregular surface overlayer. It is known that atoms originating from deeper layers are preferentially ejected into a narrow cone around the surface normal /23/. Interestingly, in the quoted experiment a pronounced difference in angular distributions was observed between the two groups of the same desorbed atoms. The first group of fast, nonthermal halogen atoms (relatively large escape depth) showed the strong forward peaking. The second group of thermal Br atoms evaporates from the very surface and thus should obey the cosine angular distribution. The thickness of the altered layer increases with lowering of the surface

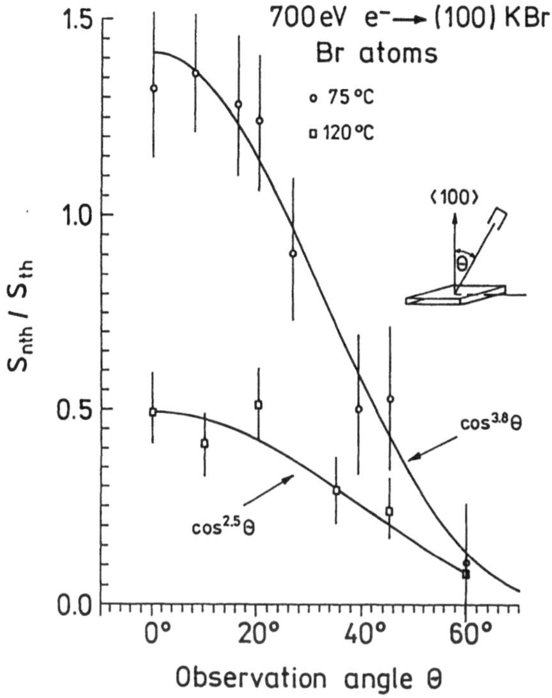

Fig.6. Dependence on the observation angle 0 of nonthermal to thermal halogen desorption yield ratio, S_{nth}/S_{th}, for Br atoms emitted from a (100) KBr surface at 120°C and 75°C.

temperature, thus causing the angular distribution of nonthermal halogen atoms to be even more forward peaked as seen in fig. 6.

Thermal evaporation of alkali atoms due to ESD involves the problem of charge neutralization of the originally positive alkali ions in the altered surface layers. Green et al. /24/ suggested that the F centers must diffuse from their place of creation to the surface, neutralizing metal atoms which then may thermally desorb. The authors argued that this model could explain a considerable delay (of the order of seconds) in desorption of ground state alkali atoms as observed experimentally for LiF crystals by Loubriel et al. /25/. In general, the F centers, however, are rather immobile compared to H centers and such slow processes would hardly account for very high desorption yields of alkali atoms observed experimentally /2/.

In addition, one has to note that the primary energy deposition profile has its maximum in a close proximity of the fastly eroding surface /8/. Therefore, it means that the concentration of the created F centers in the surface layers is highest if the centers are quite stable. Furthermore, Overeijnder et al. /26/ have shown that for compressed powder RbI, RbBr and NaCl the delay times for thermal atoms were at most of the order of ms independent on the sample temperature. Our measurements for single crystal NaCl, KBr and CsI showed the same result. More likely, one should associate the observation by Loubriel et al. with the low vapour pressure of Li compared to other alkalis at given temperature. This may lead to formation of large Li aggregates which disintegrate thermally after the stimulating beam is switched off on a rather slow time scale. Accordingly, measurements of the electron energy loss spectrum of LiF in the band gap region by Singh and Gallon /27/ reported peaks consistent with surface and plasmon excitation in free alkali metal.

2.4 Hot Holes and Surface Excitons

It is well established that holes and excitons self-trap rapidly (in less than ~ 1 ps) in alkali halides /10/. Elango et al. /28,29/, however, have argued that holes generated in, e.g., NaCl by X-ray or electron irradiation are distributed according to the valence-band density of states and have a typical initial hole kinetic energy of ~ 1.5 eV. Considering a random walk type diffusion of such hot holes with an energy loss via longitudinal-optical phonons, they estimated the diffusion length for self-trapping as equal to about 70 nm (at 80 K). This would provide an efficient transport mechanism for the bulk electron-hole excitations to the surface, resulting in relatively high numbers of surface self-trapped excitons. As already indicated in the previous subsection, the surface excitons could have a symmetry axis along the <100> direction of the crystal and account for the directional emission of non-thermal halogen atoms observed in our experiment. Although almost nothing is known theoretically about the surface STE, some experimental evidence has been found in EELS spectroscopy measurements for LiF, NaF, NaCl, KCl, NaBr and KBr by Roy et al. /21/. Each of the alkali halide loss spectra showed a peak towards the upper end of the band gap region which was independent of irradiation time and specimen temperature. This peak has been identified as arising from a surface exciton. Further angle-resolved ESD yield and energy distribution measurements for various surfaces of alkali halides could clarify the role of surface excitons in stimulated desorption from alkali halides.

3. Desorption of Ions and Excited Alkalis

According to the mechanisms discussed above only neutral, ground state particles could be desorbed. It is improbable that the valence excitonic process could explain the experimental observations on desorption of excited alkali atoms and ions. Instead, core-exciton and/or core-hole creation are frequently invoked in ESD and PSD of such species /10,30,31/. Such core-excitations could decay via an interatomic Auger transitions from a neighboring anions resulting in ion expulsion due to repulsive Coulomb interactions. The kinetic energies of particles ejected from the surface by such a mechanism are expected to be higher than thermal. This has been observed experimentally for Li^+ and F^+ desorbed from LiF /32/, but so far not for excited alkalis. In the last case only thermal energies were observed by Walkup et al. /33/. Recent trajectory calculations, however, indicate that ionic desorption can not take place from a perfect alkali halide surface because of fast lattice relaxation and trapping effect /34/. Thus, the ion desorption observed by Yasue et al. /32/ could only take place from desorption sites at the altered surface which can easily be created in the electron bombarded LiF due to a low vapour pressure of lithium. The core-exciton mechanism is extensively discussed by Wurz et al. in these Proceedings.

Many experimental observations on ESD of positive ions /33,37/ and excited neutrals /2,33,35,36/ suggest that a direct excitation and ionization of alkali atoms can take place above the surface in collisions with the flux of primary and/or secondary electrons. The gas-phase excitation of desorbed Na was firstly proposed by Rutkowski et al. /38/ and subsequently confirmed by more elaborate experiments reviewed by Walkup at the previous DIET meeting /33/. Efficient gas-phase excitation requires a large flux of desorbed alkalis together with a relatively high excitation cross section. Both these requirements are met in the ESD experiments of NaCl ar even LiF, kept at sufficiently elevated temperatures. In contrast, one should expect that a core-excitation mechanism might play an important role in PSD of excited atoms and ions at low surface temperatures. In fact experiments by Taglauer et al. /39/ showed the striking difference between ESD and PSD of excited sodium. The temperature dependence of photon stimulated yield of excited Na was in a clear anticorrelation with the yield of ground state atoms. It would be desirable to measure the kinetic energy distributions of excited alkalis under circumstances in which the core-excitation processes are expected to show up.

4. Concluding Remarks

It has been shown that the recent progress in theoretical understanding of defect creation processes in alkali halides combined with the refined angle resolved and mass selected energy distribution measurements as well as yield versus luminescence observations, provide a consistent description of PSD and ESD of ground state neutrals in terms of the valence-exciton mechanism.

ISD mechanisms still require further experimental effort to understand the balance between electronic- and cascade- type processes in alkali halides. In the case of PSD and ESD of excited alkali atoms and ions one can likely determine the experimental circumstances under which the gas-phase processes are dominant but core-hole and/or core-exciton mechanisms still await further experimental and theoretical understanding.

5. Acknowledgements

This work was supported by the Polish Ministry of National Education under Programs CPBP 01.06 and CPBP 01.09.

6. References

1. P.D.Townsend: In Sputtering by Particle Bombardment II, ed.R.Behrish (Springer-Verlag, Berlin, Heidelberg 1983) p.147.
2. M.Szymonski, J.Rutkowski, A.Poradzisz, Z.Postawa and B.Jorgensen: In DIET II, Springer Series in Surface Science 4 (Springer-Verlag, Berlin, Heidelberg 1984) p.160.
3. P.D.Townsend, R.Browning, D.J.Garland, J.C.Kelly, A.Mahjoobi, A.J.Michael and M.Saidoh, Rad.Effects 30,55(1976).
4. H.Overeijnder, M.Szymonski, A.Haring and A.E.de Vries, Rad.Effects 36, 63(1978).
5. M.Szymonski, P.Czuba, T.Dohnalik, L.Jozefowski, A.Karawajczyk, J.Kolodziej, R.Lesniak and Z.Postawa, Nucl.Instrum.Methods, in press.
6. Z.Postawa and M.Szymonski, Phys.Rev.B 39,12950(1989).
7. M.Szymonski and A.E.de Vries, Rad.Effects 54,135(1981).
8. Z.Postawa, P.Czuba, A.Poradzisz and M. Szymonski, Rad.Effects and Defects in Solids 109,189(1989).
9. K.S.Song, C.H.Leung and R.T.Williams, presented at Int.Conf.Defects in Crystals (Parma 1988), to be published.
10. R.T.Williams, Rad.Effects and Defects in Solids 109,175(1989).
11. W.Husinsky, R.Bruckmuller, P.Blum, F.Viehbock, D.Hammer and E.Benes, J.Appl.Phys. 48,4734(1977).
12. M.Szymonski, H.Overeijnder and A.E.de Vries, Rad.Effects 36,189(1978).
13. M.L.Yu, D.Grischhowsky and A.C.Balant, Appl.Phys.Lett. 39,703(1981).
14. W. Husinsky, P.Wurz, K.Mader, E.Wolfrum, B.Strehl and G.Betz, Nucl. Instrum.Meth.B 33,(1988).

15.D.Pooley, Proc.Phys.Soc. 87,245(1966).
16.H.N.Hersh, Phys.Rev. 148,928(1966).
17.R.T.Williams, J.N.Bradford and W.Z.Faust, Phys.Rev.B 18,7038(1987).
18.R.T.Williams, K.S.Song, W.L.Faust and C.H.Leung, Phys.Rev.B 33,7232 (1986).
19.M.Szymonski, Rad.Effects 52,9(1980).
20.Y.Al Jammal and P.D.Townsend, J.Phys.C: Sol.State Phys. 6,955(1973).
21.G.Roy, G.Singh and T.E.Gallon, Surf.Sci. 152,104(1985).
22.P.Wurz and C.H.Becker, Surf.Science, in press.
23.P.Sigmund, A.Oliva and G.Falcone, Nucl.Instrum.Methods 194,541(1982).
24.T.A.Green, G.M.Loubriel, P.M.Richards, N.H.Tolk and R.F.Haglund,Jr., Phys.Rev.B 35,781(1987).
25.G.M.Loubriel, T.A.Green, P.M.Richards, N.H.Tolk, R.H.Albridge, R.F.Haglund, K.J.Snowdon, L.T.Hudson, D.W.Cherry, R.K.Cole, M.H.Mendenhall, D.M.Newns and P.M.Savundararaj, Phys.Rev.Lett. 57, 1781(1986).
26.H.Overeijnder, R.R.Tol and A.E.de Vries, Surf.Sci. 90,265(1979).
27.G.Singh and T.E.Gallon, Solid State Commun. 51,281(1984).
28.V.N.Kadchenko and M.Elango, Phys.Status Solidi (a) 46,315(1978).
29.M.A.Elango, V.N.Kadchenko, A.M.E.Saar and A.P.Zharakovski, J.Lumin. 14, 375(1976).
30.M.L.Knotek and P.J.Feibelman, Phys.Rev.Lett. 40,964(1979).
31.P.Wurz, E.Wolfrum, W.Husinsky, G.Betz, L.Hudson and N.H.Tolk, Rad. Effects and Defects in Solids 109,203(1989).
32.T.Yasue, A.Ichimiya and S.Ohtani, Appl.Surf.Sci. 33/34,167(1988).
33.R.E.Walkup, Ph.Avouris and A.P.Ghosh: In DIET III, Springer Series in Surf.Sci. (Springer-Verlag, Berlin, Heidelberg 1987) p.248.
34.R.E.Walkup and Ph.Avouris, Phys.Rev.Lett. 56,524(1986).
35.R.E.Walkup, Ph.Avouris and A.P.Ghosh, Phys.Rev.Lett. 57,2227(1986).
36.J.Rutkowski, A.Poradzisz, Z.Postawa and M.Szymonski, Nucl.Instrum. Methods B 34,27(1988).
37.R.E.Walkup, Ph.Avouris and A.P.Ghosh, Phys.Rev.B 15,(1987).
38.J.Rutkowski, A.Poradzisz and M.Szymonski: In Proc.Symp.on Surface Science, Obertraum,1983, edited by P.Braun et al., p.157.
39.E.Taglauer et al., Surf.Sci.164,(1986).

Hot Hole Diffusion in the Electron-Stimulated Desorption of Metal Atoms from Alkali Halides

T.A. Green[1], *G.M. Loubriel*[1], *P.M. Richards*[1], *L.T. Hudson*[2], *P.M. Savundararaj*[2], *R.G. Albridge*[2], *A.V. Barnes*[2], *and N.H. Tolk*[2]

[1]Sandia National Laboratories, Albuquerque, NM 87185, USA
[2]Center for Atomic and Molecular Physics at Surfaces, Department of Physics and Astronomy, Vanderbilt University, Nashville, TN 37235, USA

Abstract. Within the framework of a realistic theoretical model, diffusion of hot holes in alkali-halide crystals prior to self trapping is shown to have a substantial influence on the time dependence of the flux of desorbed ground state neutral metal atoms resulting from pulsed electron beam bombardment of these crystals. Comparison of results from the model with new time dependent measurements of Li ESD from LiF suggest that substantial hot hole diffusion occurs in LiF, and that, depending on its preparation, the LiF surface can be an absorber or reflector of hot holes.

1. Introduction

It is well known that the desorption of adsorbed species from metal surfaces by electron and photons occurs via the absorption of energy right at the surface.[1] The same is true of ion desorption from maximally valent ionic solids via the inner-shell Knotek-Feibelman mechanism.[2] By contrast, in the rare-gas solids and alkali-halides, desorption occurs following the absorption of energy in the bulk as well as on the surface. The primary mechanism for energy deposition is the production of electron-hole pairs leading to the formation of valence holes. The specific role of hole creation and migration has been discussed as a factor in characterizing absorption and luminescence, but not desorption, in several treatments of alkali halides.[3-6] Works dealing with desorption have ignored hot holes by assuming that the holes self trap immediately. These works then dealt with the subsequent defect formation and migration.[7-10] Recent work on the desorption of atoms from rare gas solids has demonstrated that hot hole diffusion plays a significant role in these electronically similar systems.[1] In this paper, we report on first theoretical and experimental studies which point out the importance of hot hole diffusion in desorption from alkali halide surfaces.

2. Discussion of Theory

In alkali halides, holes created in the valence band self trap in sub-picosecond times to form V_k centers. When the electrons return to the V_k centers, self-trapped excitons (STE) are formed.[12] Recent experimental and theoretical studies have considerably clarified the electronic and lattice relaxation processes by which the

Desorption Induced by Electronic Transitions, DIET IV
Editors: G. Betz · P. Varga

STE decays.[13,14] In a fraction of the decays, pairs of separate F and H centers are formed, and these lead to the desorption (sputtering) of ground state neutral halogen and metal atoms at sufficiently high temperatures.[9,15-18] In order to discuss the desorption process in detail it is necessary to know the spatial distribution of F-H pairs produced by the energy depositing beam. As has been stated above, it has been customary in the alkali halides to assume that the diffusion of a newly created hole in the valence band before self-trapping can be neglected.[7-10]

The luminescence experiments mentioned above show that hot holes in alkali halides diffuse a considerable distance before self trapping. In reference 3, the hot hole diffusion length before self trapping in NaCl is found to be 700Å, and a (presumably contaminated) surface at 10^{-5} Torr is found to be an absorber of hot holes. The experiments in references 5 and 6 do not yield a value for the hot hole diffusion length. Measurements of electronically induced sputtering and luminescence from solid argon films has led to an estimate of the hot hole diffusion length of 230Å.[11] In this case, a reportedly clean surface was found to be nearly a perfect reflector of hot holes. These results show that hot hole diffusion should be considered in investigations of desorption from alkali halides .

Hot hole diffusion, prior to self trapping and the production of F center-H center pairs, can have a drastic effect on the time dependence of the alkali metal atom desorption signal produced by a pulsed beam of electrons incident on an alkali halide target. At low electron energies the hot hole diffusion length can be comparable to or even larger than the electron range. It is necessary, consequently, to extend the F center diffusion theory presented in Ref. 10 to include hot hole diffusion. In addition, new time dependent ESD experiments on LiF by the authors are reported, which strongly suggest the need to take hot hole diffusion into account.

The inclusion of hot hole diffusion into the F center diffusion equations of Ref.10 is straightforward, since the hot hole diffusion is effectively instantaneous on the time scales for desorption. We utilize the hot hole diffusion theory of Refs. 3, 4 and 11. Hot hole diffusion transforms the energy deposition distribution dE/dx, which describes the distribution of electron-hole pairs, into a new spatial distribution for self-trapped holes, STE, and F center-H center pairs. This provides the initial F center spatial distribution for the time-dependent diffusion calculation described in Ref. 10. Hot hole diffusion in the theory, as discussed in Refs. 3 and 11, is characterized by two parameters, a hot hole diffusion length L , and a hot hole surface reflectivity parameter K, which specifies the ratio of the hot hole density to the hot hole density gradient at the surface. K goes to zero for an absorbing surface and goes to infinity for a reflecting surface. K and L interact in a somewhat complicated way in the hot hole diffusion solution to determine a surface plane occupation probability P(0) for self trapped holes.

The physical picture can be summarized as follows. Electrons stop in the crystal, resulting in a distribution of electron-hole pairs described by the appropriate dE/dx energy loss curve. The hot hole diffusion leads from this to a new distribution of self trapped holes, STE, and separated F center-H center pairs, on picosecond time scales. In the hot crystal, the interstitial H centers quickly diffuse to the crystal surface and desorb as neutral halogen atoms, leaving the less mobile F centers behind. The F centers diffuse, and eventually give their electron to a surface Li^+ ion.

The resulting ground state neutral Li atom can then desorb. The F center diffusion is described by a hopping frequency W; the surface reactions are described by first order kinetic rate equations. In Refs. 10 and 18, F center diffusion turned out to be the rate limiting process, and the surface thermal desorption rate constant γ (~105 sec) did not play a very important role in the discussion.

3. Comparison with Experiment

Figure 1 shows how hot hole diffusion modifies the initial dE/dx distribution for 400 eV electrons when the hot hole diffusion length L = 275 Å. The resulting curves describe the new distribution of V_k centers, STE and separated F center-H center pairs, assuming that the bulk and surface formation probabilities are the same. The curve marked by squares (□), for L = 0, is of course just dE/dx. In this model distribution, the electron range corresponds to the largest depth for which dE/dx is non-vanishing. The curve marked by triangles (Δ) corresponds to a reflecting surface

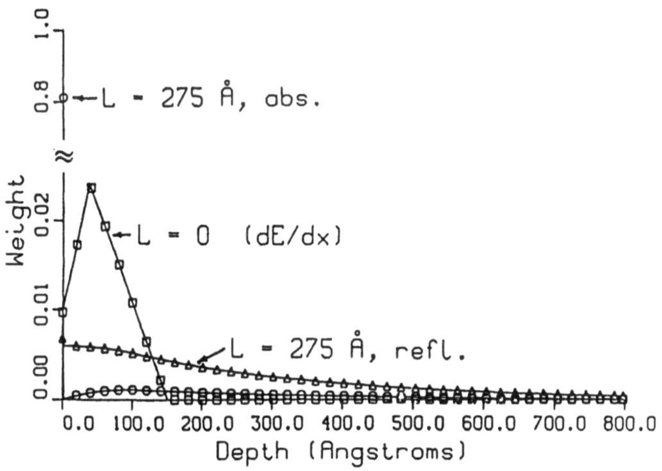

Fig. 1. Theoretical treatment of the influence of hot hole diffusion on the initial distribution of separate F center-H center pairs resulting from the creation of electron-hole pairs by 400 eV electrons stopping in a LiF crystal after impinging on a (100) surface. (□): Model dE/dx electron energy loss curve. The dE/dx curve is the source distribution for hot hole diffusion, and it is also the distribution of F-H pairs if the holes self trap without any diffusion, L= 0.0Å, and produce these defects in place. (Δ): Modification of the dE/dx distribution by picosecond time scale hot hole diffusion before self trapping with a diffusion length L of 250Å and a reflective surface for hot holes. (o): Subsurface modification of the dE/dx distribution with L =250Å and a totally absorptive surface for hot holes. The surface plane occupation weight (not shown) is 0.812. All the distributions are normalized to unity. Thus this latter distribution corresponds to the self trapping of the large majority of the holes on the crystal surface, whereas the surface plane plays a minor role in the other two distributions.

boundary condition for the hot holes; the curve marked by circles corresponds to an totally absorptive boundary condition for the hot holes. In the absorptive case (o) the actual surface occupation probability P(0) (for the surface plane located at x=0.0) is 0.812; this value is purposely set to zero in the figure so that the three curves can be better compared. This value of P(0) is the largest value obtainable for any value of K, given the dE/dx curve and the value of L. All three distributions are normalized to unity. The surface plane plays only a minor role in the first two distributions of Fig. 1; it is very important when the hot holes are totally absorbed and self trap on the surface. As mentioned above, the surface plane occupation depends on both the surface plane reflectivity K and the hot hole diffusion length. For an absorptive surface, the greater L is, compared to the electron range, the larger is the surface F center occupation probability, P(0).

Figure 2 shows calculated ground state neutral Li yields for a LiF crystal at 700 K whose (100) surface is bombarded by a 400 eV electron beam. In the calculation the

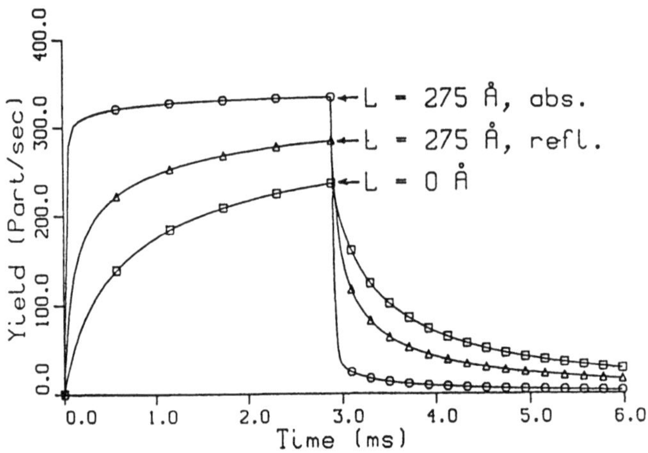

Fig. 2. Calculated time dependence of the desorbed ground state Li atom yield produced by a 400 eV square wave electron beam on a 700 K LiF crystal, when the electron beam is turned on for about 3 ms and then off. Each curve corresponds to the conditions described by the curves in Fig. 1 with the same symbols. In each of the curves, the time delay in the desorption of the Li atoms, which produces the millisecond tails after the electron beam is turned off, reflects the time required for the diffusion of F centers from the place where they are created from self-trapped excitons, to the surface. At the surface, the F center electron neutralizes a Li^+ ion and the resulting ground-state neutral leaves the hot crystal in the order of 10-5 sec. This is the inverse process of additive coloration. In the (Δ) and (▭)plots, the surface plane is not more highly populated with F centers initially than the planes in the bulk. In the (o) plot, over 80 percent of the F centers form on the surface within picoseconds, thanks to hot hole diffusion and consequent self trapping on the surface. The result is a precipitous exponential rise in the flux when the electron beam is turned on, and a comparable fall when the beam is turned off. For this case, the 20 percent of the F centers created in the bulk produce a characteristic tail. The values of the hopping frequency and surface rate constant are given in the body of the paper.

beam is turned on for just under 3 ms and then turned off. The initial F-center distributions and the symbols on the curves correspond to those in Fig. 1. The F center hopping frequency W is equal to W_a in the (□) curve (L = 0), and 75W_a in the other two curves, which correspond to L = 275 A. Here $W_a = 2x10^5$ sec^{-1}, the value obtained from fitting the 699 K experimental data in Refs 10 and 18 where hot hole diffusion was not taken into account. The values W and L are derived from the new measurements as will be discussed. The value of γ in all three cases was $5x10^4$ sec^{-1} also obtained by fitting the new data. Notice that in Fig.2 the (Δ) curve for the reflective hot hole surface boundary condition is qualitatively similar to the (□) curve. The greater depth of the initial F center distribution caused by hot hole diffusion in the (Δ) case is compensated for by the increased hopping frequency. To a good approximation, at large times, all three curves in Fig. 2 are proportional to the factor x_{ave}[1-P(0)]/[W]0.5 , where x_{ave} is the average value of the depth x, taken over the initial F center distribution, excluding the surface plane. The effect of total hot hole absorption and trapping on the surface plane in curve (o) is dramatic. Since about 80 percent of the created holes self trap on the surface, Li atoms are formed on the surface plane and desorb at the surface rate, γ, with an exponential time dependence that produces the relatively fast rise and fall of the Li current. The remaining bulk F centers, resulting from the 20 percent of the hot holes which self trapped in the bulk, produce the much slower tail. *It is thus seen that hot hole diffusion and an absorptive surface for hot holes can be recognized by the time dependence of the desorbed Li atom yield.*

Figure 3 shows the Li atom yield from two of the new experimental ESD runs carried out on a LiF crystal at 700 K, using 400 eV [curve (o)] and 100 eV [curve

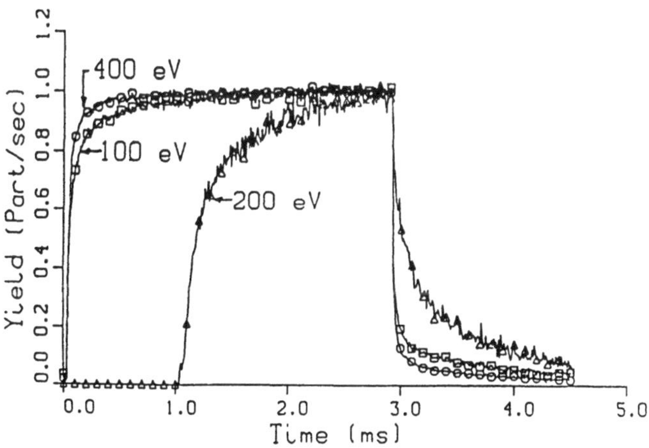

Fig 3. Measured ground-state neutral Li flux (normalized to unit maximum) for LiF crystals resulting from electron stimulated desorption (ESD). The (Δ) curve, corresponding to 200 eV electron bombardment with the LiF crystal at 699 K, is taken from Refs. 10 and 18. The other two curves are recent preliminary measurements corresponding to a LiF crystal at 700 K bombarded by a 400 eV (o) and 100 eV (□) square-wave, pulsed electron beam. The crystals were cleaved in air, inserted into an ultra-high vacuum chamber, heat cleaned and maintained at an elevated temperature during bombardment. There are ten data points between adjacent markers.

(□)] electrons. Also shown in Fig. 3 is a plot, curve (Δ), of the experimental data for 699 K, 200 eV electron energy, taken from references 10 and 18. The reader will note that the beam-off and on times were different in the two experiments. A theoretically determined background was subtracted from these data, and the curves were normalized to agree at their maxima so that their shapes could be compared. The 400 eV curve (o) of Fig. 3 looks like curve (o) in Fig. 2 because L = 275Å and an absorbing surface were chosen to provide a good fit to the data when the theory was fitted to the experimental data at two points, near 3 and 6 ms. The use of a totally absorptive boundary condition in the theoretical fit for curve (o) leads to the smallest possible value of L, which is L = 275Å. Since the electron ranges are about 38Å for 100 eV electrons and 152Å for 400 eV electrons, the magnitude of L is impressive.

In considering the experimental data in Fig. 3, there are two very interesting observations to be made. First, the 100, 200, and 400 eV curves are not in the ascending order as regard the heights of their decay tails as would be expected on the basis of electron range energy considerations. In particular, although both the 100 eV and 400 eV new curves in Fig. 3 rise and fall very quickly, the relatively slower rise and fall exhibited by the 100 eV Li yield as compared to the yield for 400 eV electrons is puzzling. One might have expected that the shorter range of the 100 eV electrons should allow the F centers to arrive at the surface sooner than the 400 eV electrons, in contrast to what is observed. Second, the shapes of the 100 eV and 400 eV curves, with their steep rise and fall, are unlike the earlier experimental 200 eV curve for a 699 K LiF crystal from Refs. 10 and 18, but resemble more the (o) curve of Fig. 2. This is surprising since the experiments were conducted under the same protocols: cleaving the crystals in air, inserting them into the vacuum chamber, heating, and then bombarding with electrons prior to the time dependent runs.

4. Conclusion

Can hot hole diffusion provide an answer to these puzzles? We answer in the affirmative. The solution requires the realization that there are three important characteristics of hot hole diffusion, namely (1) hot holes diffuse very fast (in a sub-picosecond time frame) prior to F center- H center creation, (2) hot holes may have diffusion lengths that are comparable to or greater than electron ranges; and (3) hot holes may either be absorbed (self trapped) at the surface or reflected at the surface, depending on the state of the surface. The 400 eV measurements were carried out early in the experiment and thus represent a surface at an early stage of cleaning and conditioning which, it is reasonable to assume, results in an highly absorptive surface for hot holes. The 100 eV measurements on the other hand were done later in the experiment and represent a cleaner and consequently, more reflective crystal surface. The 100 eV data in Fig. 4 has been fit with the theory using L = 275Å, $W=75W_a$, and a hot hole surface parameter, K, intermediate between complete absorption and total reflection, resulting in a surface plane occupation probabilty P(0)=0.60. Under these assumptions, application of the theory gives excellent agreement with experiment for both electron energies; in fact the theoretical curves are indistinguishable from the measured curves in Fig. 3. Moreover, applying these same parameters to the data in Refs. 10 and 18, under the assumption that the surface is largely reflective also gives excellent agreement. *An inescapable conclusion from this work is that the state of the surface is much more important in determining the time dependence of ground state alkali atom desorption than the energy of the bombarding electrons, for low energy electrons.*

To sum up, the two new measurements presented here are shown to require the assumption of fast hot hole diffusion; with their differences accounted for by the degree of reflectivity of the surface. In addition, the experimental data published in Refs. 10 and 18 have been reconciled to the new theory, modified to include the effect of hot holes, with parameters L close to 275Å and W close to $75W_a$, under the plausible assumption that the crystal surface was highly reflective. Indeed, a consistent treatment of the data discussed here is impossible without utilizing the hot hole concept.

The realization of the importance of the role of fast hot hole diffusion to desorption, coupled with the ideas that (1) hot hole reflectivity can be altered by changing the state of the surface; and that (2) diffusion rates may be influenced by bulk impurities and defects, suggest a wide and fascinating array of experimental and theoretical investigations. These include time, temperature and energy dependent ESD and PSD studies on alkali halides following the introduction of contaminant gasses, deposited alkali metals, and other dopants on the surface and in the near surface bulk.

Acknowledgements

Two of us (TAG and NHT) would like to acknowledge discussions with Professor E. Matthias and the support of the Institut für Atom- und Festkörperphysik, Freie Universität, Berlin. One of us (NHT) gratefully acknowledges support of the Alexander von Humboldt Foundation., We would like particularly to thank Dr. P. West of the Freie Universität for his invaluable help in the computer aspects of production and network transmission of the manuscript. Work performed at Sandia National Laboratories was supported the U.S. Department of Energy under contract Number DC-AC0476DP00789. The work carried out at Vanderbilt University was partially supported by the AFOSR/URI Contract Number F499620-86-C-0125DEF, by the AFOSR under Contract Number AFOSR-86-0150 and by Sandia National Laboratories under contract 65-2377.

References

1. See for example: T. E. Madey and J. T. Yates, Jr., Surface Sci. **63**, 203 (1977); D. Menzel, in DESORPTION INDUCED BY ELECTRONIC TRANSITIONS, eds., N. H. Tolk, M. M. Traum, J. C. Tully and T. E. Madey (Springer-Verlag, Heidelberg, 1983).
2. M. K. Knotek and P. J. Feibelman, Phys. Rev. Letters **40**, 964 (1978).
3. M. A. Elango, V. N. Kadchenko, A. M. E. Saar, and A. P. Zhurakovski, J. Lumi. **14**, 375 (1976).
4. V. N. Kadchenko and M. Elango, Phys. Status Solidi **A46**, 315 (1978).

5. R. G. Kaufman, W. B. Hadley, and H. N. Hersh, IEEE Trans. Nucl. Sci. **17**, 82 (1970).
6. H. B. Dietrich, A. E. Purdy, R. B. Murray, and R. T. Williams, Phys. Rev. **B8**, 5894 (1973).
7. N. Itoh, Nucl. Inst. and Meth. **132**, 201 (1975).
8. H. Overeijnder, M. Szymonski, A. Haring and A. E. De Vries, Rad. Effects, **38**, 21, (1978).
9. M. Szymonski, Rad. Effects, **52**, 9 (1980).
10. T. A. Green, G. M. Loubriel, P. M. Richards, N. H. Tolk, and R. H. Haglund, Jr., Phys. Rev. **B35**, 781 (1987).
11. C. T. Reimann, W., L. Brown, and R. E. Johnson, Phys. Rev. **B37** , 1455 (1988).
12. N. Itoh, Advances in Physics, **31**, 491 (1982).
13. R. T. Williams, K. S. Song, W. L. Faust, and C. H. Leung, Phys. Rev. **B33**, 7232 (1986).
14. R. T. Williams and K. S. Song, J. Phys. and Chem. of Solids (to be published).
15. H. Overeijnder, M. Szymonski, A. Haring, A. E. De Vries, Rad. Effects, **36**, 63 (1978).
16. H. Overeijnder, R. R. Tol, A. E. De Vries, Surf. Sci. **90**, 265 (1979).
17. F. Agullo-Lopez and P. D. Townsend, Phys. Stat. Solidi **B97**, 75 (1980).
18. G. M. Loubriel, T. A. Green, P. M. Richards, R. G. Albridge, D. W. Cherry, R. K. Cole, R. F. Haglund Jr., L. T. Hudson, M. H. Mendenhall, D. M. Newns, P. M. Savundararaj, K. J. Snowdon, and N. H. Tolk, Phys. Rev. Lett. **57**, 1781 (1986).

Different Processes for Desorption of Ground- and Excited-State Atoms Under Electron Bombardment of Alkali-Halides

P. Wurz, J. Sarnthein, W. Husinsky, and G. Betz

Institut für Allgemeine Physik, Technische Universität Wien,
Wiedner Hauptstr. 8–10, A-1040 Wien, Austria

Abstract. Single crystal surfaces of LiF were bombarded with 5 to 500eV electrons at different crystal temperatures (300-800K). The temperature dependence as well as the energy dependence of ground- and excited-state alkali atoms were studied. These measurements strongly suggest different mechanisms for the desorption of ground- and excited-state alkali atoms. An excitation of a valence electron into the conduction band and subsequent decay of the formed excited state into an F-H pair will finally lead to the desorption of a ground-state atom, which is well established in the literature. The creation of a core-hole on an alkali lattice-site close to the surface followed by an inter-atomic Auger decay of that core-hole leads to emission of excited alkali atoms and ions. The measurement of the threshold energies for ground- and excited-state alkali atom desorption is a major piece of evidence. The threshold energies are found at 13 and 60eV for the desorption of ground- and excited-state atoms, respectively. This agrees very well with the band gap of 14.2eV and the core-hole excitation of $Li^+(1s \rightarrow 2s)$ at 60.8eV and for $Li^+(1s \rightarrow 2p)$ at 61.9eVand is clear evidence for a surface intrinsic process for the generation of excited alkali atoms and will be discussed together with the underlying experimental results.

1. Introduction

The interaction of electrons and photons with alkali-halides remains a field of strong current interest. The concept for desorption of neutral *ground-state* atoms is generally agreed on [1-4], discussion being confined to selected topics [e.g. 5]. The situation is worse in the case of the desorption of *excited* atoms and ions from alkali-halides. Conflicting models and experimental results are currently debated in literature [6-9]. In the following we will refer to ground-state atoms as Li^o and to excited atoms as Li^*. As the Li^o atoms are the main fraction of the desorbed particles, we will review their generation mechanism briefly.

The mechanism for ground-state desorption is based on the generation of a self-trapped exciton on a halogen lattice site [1-4]. This excited state can decay by inducing a replacement collision sequence along the chain of halogen atoms in a <110> direction, which separates the H-center from the F-center by some lattice constants. The process can lead either to prompt emission of a halogen atom if the range of the replacement collision sequence extends beyond the distance to the surface, or to delayed emission due to diffusion of H-centers to the surface. As these defect processes

Desorption Induced by Electronic Transitions, DIET IV
Editors: G. Betz · P. Varga

only take place in the halogen sublattice, an alkali metal enrichment results in the near surface layer when the remaining F-centers diffuse to the surface [10]. According to this model the alkali atoms evaporate thermally from the crystal, provided the vapor pressure of the alkali metal is sufficiently high. Otherwise the formation of a metal layer inhibits further desorption.

Basically, there have been two explanations for desorption of excited or ionized atoms from alkali-halides: Firstly, gas-phase excitation of desorbed neutral ground-state atoms by the primary electron beam or by secondary electrons. This accounts for some experimental results [7,8]. Secondly, emission of excited atoms due to a surface intrinsic process via inelastic energy transfer, assumed to start with the creation of an alkali core-hole/exciton as the initial step. Experimental data which can be interpreted by this mechanism have been found for excited atoms [6,9] and ions [11]. It is the aim of this paper to solve this controversy by presenting new experimental data for the desorption of ground- and excited-state atoms.

2. Experimental

LiF single crystal surfaces have been bombarded with electrons of 5eV to 500eV energy and with currents up to 30μA. The electron beam axis was incident on the crystal at 45°. The bombarded area had dimensions of approximately 2 by 4mm. Uncertainties in the electron energy are due to the work-function of the cathode and the work-function of the crystal [12]. Radiation of excited states was collected by a MgF_2 lens into a high resolution one meter grating monochromator viewing the crystal along the surface normal. Ground-state atoms have been measured with a quadrupol mass analyzer also viewing the crystal along the surface normal.

The UHV-system maintained a base pressure of less than $3x10^{-10}$ torr during the measurements. All investigated crystals have been cleaved in air and cleaned in the UHV-system by baking at 400°C overnight. The sample temperature was varied between 20°C and 500°C. Sample charging, which is a crucial problem with insulators especially at low temperatures, was largely reduced by placing a Mo-mesh with a transmission of 80% over the crystal surface and by using low current densities.

3. Experimental Results

As already discussed in the introduction, the evaporation rate of the alkali metal plays an important role in the process of alkali atom desorption. We distinguish three ranges in the temperature dependence of the desorption yield of Li ground-state atoms from LiF (Fig. 1): i) Due to a negligible evaporation rate at low crystal temperatures no desorption of alkali atoms occurs, and alkali metal accumulates on the crystal surface [10]. ii) The desorption rate increases, limited by the exponentially rising evaporation rate. An increase in the electron flux will not lead to an increase of the desorption rate. iii) A plateau in the temperature dependence of the desorption

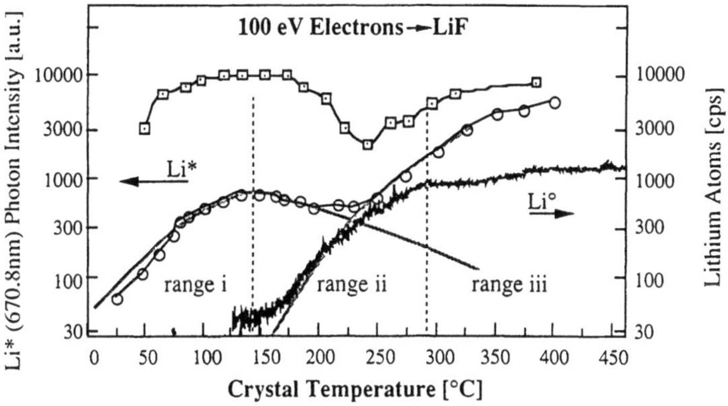

Figure 1: The dependence of the Li^{0} signal upon the crystal temperature during 100eV electron bombardment of LiF. Three temperature ranges are indicated (see text). The open symbols show the Li^{*} yield for 100eV and 400eV, circular and rectangular symbols, respectively. Shaded lines define the domains where surface intrinsic and gas-phase excitation processes are effective.

yield is reached, if defect production limits desorption rather than evaporation. An increase of the electron flux or energy will lead to an increase in yield, but at the same time will shift range iii towards higher temperatures [5].

Comparing the temperature dependence of ground- and excited-state atoms, we note a large difference in the observed behavior. At room temperature (range i), where we have no Li^{0} desorption, there is considerable Li^{*} desorption. This excludes a process involving gas-phase excitation of desorbed atoms in this temperature range. Furthermore, the yield of ground-state atoms increases more than two orders of magnitude, which compares to one order of magnitude for the excited-states in the same temperature range. Thus no proportionality between these two species can be extracted and we have to assume different processes of generation of either state. The yield of Li^{*} atoms at 150°C is a factor of 10^{-3} less than the yield of Li^{0} atoms at 300°C.

For both Li^{0} and Li^{*} desorption we observe well defined and reproducible thresholds in the energy dependence of the yield. For the excited atoms we find the onset for desorption at approximately 60eV. We performed this measurement for two different crystal temperatures: at room temperature (range i), where there is only desorption of excited atoms, and at 360°C, where we are not limited by the evaporation rate (range iii). For both temperatures we find the same energy threshold for the excited-states, as can be seen in Fig. 2a. The threshold correlates with the Li^{+} core-exciton levels of $Li^{+}(1s\rightarrow 2s)$ at 60.8eV and $Li^{+}(1s\rightarrow 2p)$ at 61.9eV in LiF [13]. The valence-hole level $F^{-}(2p)$ at 14.2eV [14] (corresponding to the bandgap) or the core-hole level $F^{-}(2s)$ at 38.2eV [13] lies far below the observed threshold.It is worthwhile to note that we do not observe any pronounced changes in the secondary electron yield for electron energies near the Li^{*} threshold, thus the threshold cannot result from gas-phase excitation.

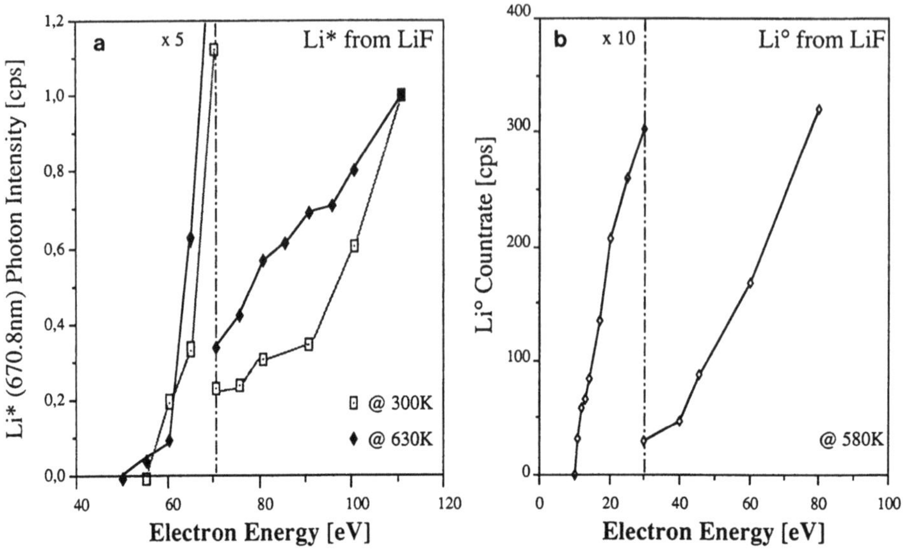

Figure 2: a) The energy threshold for Li* desorption is found at 60eV, independent of crystal temperature. b)The energy threshold for Li^{0} desorption lies around 13eV. The measurement was performed at 580K (range iii) to obtain highest signal/noise ratio, but is valid for all temperatures.

For ground-state Li atoms we find an onset of the Li^{0} desorption around 13eV, which is shown in Fig. 2b. This value corresponds to the energy of the valence exciton of 12.6eV [15] near the band gap energy of 14.2eV [14], thus confirming that the first step in the desorption process of ground-state alkali atoms is the excitation of a valence electron to a bound state (electron-hole pair) or into the conduction band. The total amount of Li* atoms is far less than that of Li^{0} atoms, because the Li atoms originally excited and de-exciting on their way to the mass spectrometer do not add noticeably to the yield of ground-state atoms above the threshold of Li* at 60eV. However, in recent PSD measurements a small peak at 60eV in the energy dependence of Li ground-state atoms has been observed [16].

The dependence of the Li* yield on the current density helps to distinguish a surface intrinsic from a secondary excitation process [7,8]. A linear dependence of the Li* yield on the current density is expected if excitation and ejection of the Li atom is caused by the same incoming electron. A quadratic dependence of the Li* yield on the current density is expected in case of excitation in the gas-phase, because the excitation by primary or secondary electrons is not correlated with the desorption of alkali atoms as a consequence of defect generation. Fig. 3 shows our results of the dependence of the Li* yield on the current density again for two different temperatures: At room temperature (range i) we find a linear relationship, which is not surprising since in the absence of desorbing ground-state atoms only a direct process can be responsible. At 380°C (range iii), we start out with a linear dependence of the Li* yield upon the current density, which changes to quadratic behavior for high current densities.

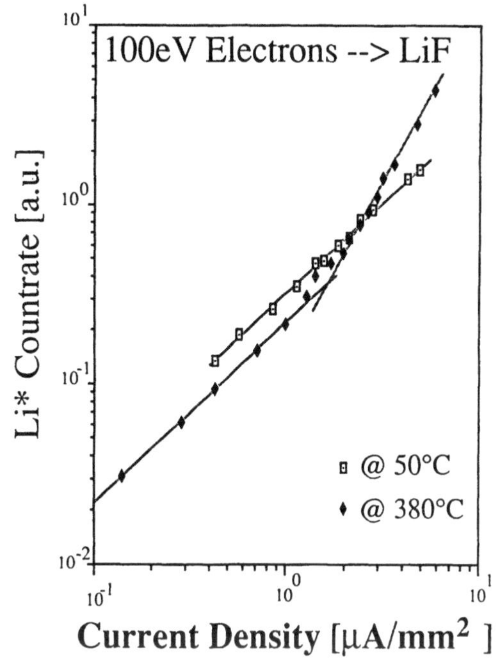

Figure 3: The dependence of the Li* yield on the current density of 100eV electron irradiation is plotted for 50°C and 380°C. Both curves are normalized, the absolute size to each other can be deduced from figure 1. The straight lines have slopes of 1 and 2 according to a linear and a quadratic dependence upon the current density, respectively.

4. Discussion

We propose a new, surface intrinsic mechanism for the generation of excited alkali atoms using concepts introduced by Tolk et al. [17]. The necessity for the new model is based on three major arguments: i) The thresholds for Li^0 and Li^* desorption differ strongly and the one for excited-states coincides well with a core-excitation of the Li ion in the crystal. ii) At room temperature no Li^0 desorption is observed, but Li^* is found in copious quantities. iii) We find a linear and a quadratic regime in the Li^* yield as a function of current density. This new mechanism consists of the following steps:

a) Provided that the primary electron has sufficient energy, the incident electron can create a core-exciton on the alkali-ion lattice site.

b) This core-exciton decays via an interatomic Auger transition from a neighboring anion. Consequently, the anion is left in a singly or multiply positively charged state [12,18], as the excess energy is used for the release of valence electrons originally located on the anion. The electron, which was bound to the core-hole, can be i) released into the conduction band ii) trapped in a surface exciton [19,20] iii) trapped in a surface state.

c) Coulomb repulsion leads to emission of alkali ions, which is observed experimentally [9,11] and predicted by theoretical calculations [21,22].

d) While emerging from the crystal the alkali ion may capture an electron, which is localized on a nearby surface defect [19,20] or surface state, into an excited state or the ground-state. Even in the absence of Coulomb repulsion the neutral atom gains substantial kinetic energy from the reneutralization event. The sudden increase of the radius of the reneutralized ion causes a simultaneous rise of the repulsive energy. This process is referred to as Pauli shock [21].

e) Decay of the excited atomic state in front of the crystal.

Besides the surface intrinsic process also gas-phase excitation contributes to the yield of Li* atoms. In the following we will define the domains where each process is effective.

The linear current dependence of the desorption yields we found for excited Li atoms desorbed from LiF is in good agreement with some previous measurements [6,8,23] but is contradicted by others [7,8], who found a quadratic current density dependence of the yield of excited alkali atoms. The explanation for this discrepancy is that the latter authors used higher current densities and higher temperatures in their experiments. As can be seen in Fig. 3, gas-phase excitation occurs only under these experimental conditions. This is due to a high excitation probability caused by high densities of electrons and atoms in the interaction volume. Higher electron energies, which were used by these authors, result in higher desorption rates, so that the range of gas-phase excitation is reached even for current densities lower than ours. For other alkali-halides with higher alkali vapor pressure, gas-phase excitation will already occur for lower temperatures or lower current densities.

The two excitation processes can be distinguished in the Li* temperature dependence, too (Fig. 1). The surface intrinsic process is dominant in a temperature range from room temperature to 250°C, but is present also at higher temperatures as can been seen in the energy threshold of Li* at 360°C (Fig. 2a) and in the current dependence of the Li* yield (fig. 3). The increase of the Li* yield with temperature in the domain of the surface intrinsic process can be explained by decrease of the alkali metal overlayer [10] and by a higher mobility of the Li atoms on the crystal surface as the melting point of Li metal is reached [19]. This is supported by recent PSD measurements [16]. As the metal overlayer grows with electron bombardment, we observe a decrease of Li* yield with time, until an equilibrium is reached after several minutes. Beginning Li^0 desorption will decrease the amount of adatoms and eventually inhibit the surface intrinsic process. It also enables gas-phase excitation, which we believe to be responsible for the Li* yield increase after the minimum at 250°C.

The similarity between the optical emission spectrum under electron irradiation of alkali-halide crystals at elevated temperatures with an electron impact induced spectrum in an alkali vapour was used to emphasize excitations in the gas-phase in front of the crystal by primary [24] or secondary electrons [8] in this temperature range. The similarities with an electron impact spectrum are not surprising and we agree with these authors that they observe gas-phase excitation in their experiments [7,8,24].

As far as we know, there are three theoretical calculations of the possibility of ion and neutral atom ejection induced by core-electron excitation: A molecular dynamics calculation by Walkup et al. [25], molecular dynamics calculation including all interaction potentials by Elango et al. [21] and a quantum chemical calculation by Itoh et al. [22]. The two latter calculations agree with our model, whereas the first favors gas-phase excitation.The positively charged halogen resulting from the Auger deexcitation of the core-hole can only lead to the ejection of a particle if it is located in the first or second atomic layer [21,22]. In the bulk, the lifetime of the positively charged halogen is too small to gather enough kinetic energy to leave its original lattice site or impart kinetic energy to an adjacent cation. The adatom position of an alkali atom on top of the halogen is a favorable site for desorption [21,25].

In photon-stimulated desorption experiments (PSD), a pronounced peak in the excited Li atom emission at a photon energy of 61.5eV was measured [16], suggesting excitation of a Li core-electron to a discrete level rather than excitation to the continuum of the conduction band.

5. Conclusions

We present new and conclusive experimental results to distinguish the desorption processes for ground- and excited-state atoms. The energy threshold for ground-state atoms supports the current understanding of ground-state alkali desorption from alkali-halides [1-4]. The energy threshold for excited-state atoms, found at the core-hole energy of the alkali atom, is evidence for our proposed mechanism for the desorption of excited states, which is in good agreement with theoretical calculations [21,22]. This work should end the long lasting discussion about the origin of the excited atom desorption of alkali-halides under electron and photon bombardment [6-9].

6. Acknowledgments

The authors would like to gratefully acknowledge financial support by the Österreichische Fonds zur Förderung der wissenschaftlichen Forschung (project Nrs. 5833, 5577).

7. References

[1] P.D. Townsend, R. Browning, G.J. Garlant, J.C. Kelly, A. Mahjoobi, A.L. Michael, M. Saidoh, Rad. Effects 30, 55 (1976)

[2] N. Itoh, Nucl. Instr. Meth. B27, 155 (1987)

[3] M. Szymonski, Rad. Effects, 52, 9 (1980)

[4] R.T Williams, K.S. Song, W.L. Faust, C.H. Leung, Phys. Rev. B33, 7232 (1986)

[5] J. Sarnthein, P. Wurz, W. Husinsky, G. Betz, these proceedings

[6] P. Wurz, E. Wolfrum, W. Husinsky, G. Betz, L. Hudson, N. Tolk, Rad. Eff. Def. Sol. 109, 203 (1989)

[7] Z. Postawa, J. Rutkowski, A. Poradzisz, P. Czuba, M. Szymonski, Nucl. Instr. & Meth. B18, 574 (1987)

[8] R.E. Walkup, Ph. Avouris, A.P. Ghosh, Phys. Rev. Lett. 57, 2227 (1986)

[9] T.R. Pian, N.H. Tolk, J. Kraus, M.M. Traum, J. Tully, W. Collins, J. Vac. Sci. Technol., 20, 555 (1982)

[10] P. Wurz, C.H. Becker, in press in Surf. Sci., (1989)

[11] T. Yasue, A. Ichimiya, Y. Yamada, T. Gotoh, Y. Kawaguchi, M. Kotani, S. Ohtani, Y. Shigeta, S. Takagi, Y. Tazawa, G. Tominaga, Nucl. Instr. & Meth. B34, 357 (1988)

[12] P.J. Feibelman, M.L. Knotek, Phys. Rev. B18, 6531 (1978)

[13] A. Zunger, A.J. Freeman, Physics Letters 60A, 456 (1977)
A. Zunger, A.J. Freeman, Phy. Rev. B 16, 2901 (1977)

[14] A.B. Kunz, Phys. Rev. B12, 5890 (1975)

[15] M. Piancentini and J. Anderegg, Solid State Comm. 38, 191 (1981)

[16] W. Husinsky, P. Wurz, K. Mader, E. Wolfrum, B. Strehl, G. Betz, R.F. Haglund Jr., A.V. Barnes, N.H. Tolk, Nucl. Inst. & Meth. B33, 824 (1988)

[17] N.H. Tolk, M.M. Traum, J.S. Kraus, T.R. Pian, W.E. Collins, N.G. Stoffel, G. Margaritondo, Phys. Rev. Lett., 49, 812 (1982)

[18] M.L. Knotek, P.J. Feibelman, Phys. Rev. Lett. 40, 964 (1978)

[19] G. Roy, G. Singh, T.E. Gallon, Surf. Sci 152/153, 1042 (1985)

[20] P.A. Cox, A.A. Williams, Surf. Sci. 175, L782 (1986)

[21] M. Elango, A.E. Kiv, A.A. Malkin, Surf. Sci. 194, 548 (1988)

[22] N. Itoh, A.M. Stoneham, A.H. Harker, Surf. Sci. 217, 573 (1989)

[23] N.H.Tolk, L.C.Feldman, J.S.Kraus, R.J.Morris, M.M.Traum, J.C.Tully, Phys. Rev. Lett. 46, 134 (1981)

[24] M. Szymonski, J. Ruthowski, A. Poradizs, Z. Postawa, *Desorption Induced by Electronic Transitions,* DIET II, edt. by W. Brenig and D. Menzel, Springer-Verlag, Berlin, 160 (1985)

[25] R.E. Walkup, Ph. Avouris, Phys. Rev. Lett. 56, 524 (1986)

DIET of Neutral Excited State Hydrogen from Alkali Halide Surfaces

L.T. Hudson[1], *A.V. Barnes*[1], *J.L. Rose*[1], *N.H. Tolk*[1], *G. Betz*[2], *W. Husinsky*[2], *E. Wolfrum*[2], *and P. Wurz*[2]

[1]Department of Physics and Astronomy, Vanderbilt University, Nashville, TN 37235, USA
[2]Technische Universität Wien, Institut für Allgemeine Physik, Wiedner Hauptstr. 8–10, A-1040 Wien, Austria

Abstract. The desorption yields of excited hydrogen atoms from the surfaces of KCl, KBr, NaCl, NaF and LiF have been measured as a function of incident photon and electron energy, time of irradiation, and sample temperature. The measurements indicate that the adsorption of hydrogen which leads to the stimulated emission of neutral, excited-state hydrogen is contingent upon surface damage induced by the incident radiation. Specifically, H_2 adsorbs dissociatively at surface sites containing a non-stoichiometric excess of alkali metal. This hydrogen may come from the gas phase or from a source which has accumulated in the bulk of these compounds during the course of the experiment. Incident electrons or photons can then induce a valence excitation to a neutral, antibonding state of the surface alkali hydride molecule leading to the desorption of atomic hydrogen.

1. Introduction

One of the prime objectives in surface physics is to understand the nature of the surface bond and the processes leading to its formation and rupture. As the hydrogen atom serves as a paradigm for atomic physics, so adsorption and desorption of hydrogen is a natural and profitable choice for investigation of the most basic adsorbate-surface interactions. The hydrogen-alkali halide surface system is one such system that exhibits large electronically-stimulated desorption yields of excited hydrogen atoms upon exposure at room temperature to gaseous H_2. The goal of this work is the determination and elucidation of the fundamental interactions of hydrogen with alkali halide surfaces while under electron and photon bombardment which lead, ultimately, to the emission of H*. Specifically, this goal encompasses the identification of the site at which bonding occurs, whether the adsorption is physical or chemical, associative or dissociative, and the bond-breaking processes involved in desorption of excited hydrogen atoms from these surfaces.

2. Experimental Methods

The PSD experiments utilized the Vanderbilt/SRC (Synchrotron Radiation Center) beamline at the University of Wisconsin. This is a 6-meter toroidal grating monochromator which provided well resolved UV radiation from 10 to 200 eV with a

Editors: G. Betz · P. Varga

first-order flux of about 10^{11} photons/sec/mm^2. During these measurements, the entrance and exit slits were opened wide, giving an energy resolution of 100 to 200 meV. Samples were prepared by cleaving in air and heating in vacuum at 400°C for several hours so as to remove adsorbed atmospheric water [1]. The experimental chamber supported a base pressure of 5 x 10^{-11} Torr; during measurements the chamber was backfilled with research grade H_2 gas at pressures not exceeding $1x10^{-6}$ Torr. Visible fluorescence emitted as a result of UV-photon impact was collimated by a lens system parallel to and just in front of the surface and focused onto the slits of a 0.3m McPherson spectrometer (1200/mm grating) and photomultiplier combination with subsequent signal processing with photon-counting electronics. The Balmer alpha (H_α, n=3 to 2) fluorescence yields from desorbing H atoms presented below represent the counting rate above background, normalized to incident photon flux.

The electron irradiation measurements were performed using a low energy electron gun (10 to 350 eV) based on the design of Stoffel and Johnson [2]. Sample charging and irreproducible energy threshold measurements at low energies were eliminated by overlaying the crystal faces with an electrically grounded 80% transmission molybdenum mesh.

3. The Surface Morphology and Stoichiometry

The alkali halide surfaces studied here are far from being atomically smooth checkerboards of alkali and halogen ions. Both cleaving the sample and subjecting it to radiation cause dramatic deviations in the surface stoichiometry and irregularities in the surface structure. In particular, the halogen species is preferentially sputtered, leaving the surface metal-rich. This radiation-induced accumulation of excess metal on the surfaces of alkali halides has been extensively documented using a wide variety of surface-sensitive techniques including AES [3,4], soft x-ray emission spectroscopy [5], EELS [6,7], EPR [8], secondary electron and scanning Auger electron microscopy (SEM and SAM) [9].

4. Time, Energy, and Temperature Dependent Measurements

During the initial irradiation of a freshly cleaved and baked sample and in the presence of H_2, fluorescence spectra exhibit little or no emissions from desorbing hydrogen. The ESD H* yields, however, grow dramatically in time and then saturate. Each of the five alkali halide compounds examined has displayed this effect. The time evolution of H* tracks qualitatively the results of studies of the electron-induced metallization of alkali halide surfaces [4,7]. While the ESD signal of H* saturates on the time scale of one to a few hours, in the photon irradiation experiments it took one to a few days. The flux of incident photons was about two orders of magnitude less than that of the ESD experiments.

The ESD energy thresholds for the desorption of H* are shown in Figure 1. The energy axis has *not* been corrected for the contact charging due to transfer of

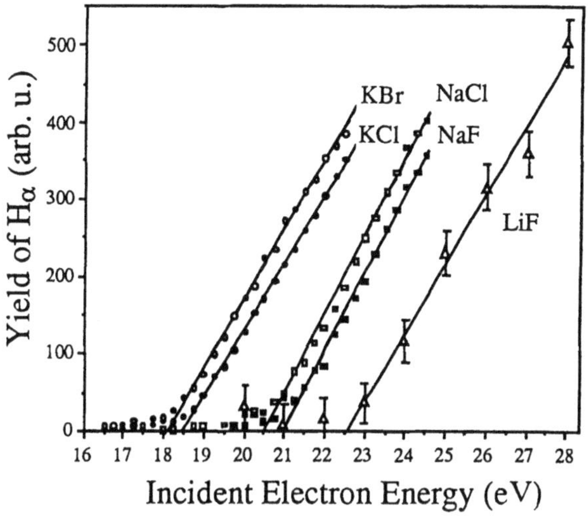

Figure 1. ESD energy thresholds for the desorption of H*(n=3) from five alkali halide surfaces

electrons between the cathode of the electron gun and sub-bandgap surface states of the sample [10]. This is not a factor in PSD and in light of those results, the absolute electron energy calibration is considered accurate to within an electron volt. The thresholds are grouped according to the alkali of the compound. That the thresholds from surfaces containing the same alkali are not identical is attributed to contact potential differences since this shift is not observed in PSD.

In Figure 2(a), the yield of H_α due to photon irradiation of KCl exhibits a prominent resonance centered around 20 eV. To determine the contribution to this signal by incident photons of second or higher order, a tin-foil filter (1500 Å thick) dependent yields of H_α from KCl, KBr and LiF are also shown in Figure 2(b) normalized to the relative flux transmitted by this filter. These results have been scaled for comparison. Scaled error bars are also shown. This shows that the resonance from KCl (black dots) is in fact the primary energy threshold for the emission of H^*(n=3). The resonance from KBr (open squares) is identical in position and width to that from KCl. A resonance from LiF begins around 23 eV and is truncated in this measurement by the transmission of the filter. Without the filter this resonance is found to be centered at about 24 eV. Due to signal strength requirements, the resonant yields from NaCl and NaF were taken without the Sn filter. The counting rates were small, yet reproducible. These two resonances occur at the same energy and with similar widths.

Note that the results from compounds containing the same alkali constituent are essentially identical. The thresholds from the sodium halides occur at incident photon energies between the potassium and lithium halides in agreement with the ESD thresholds. Further, the low energy onset of each resonance agrees to within an electron volt with the corresponding ESD threshold.

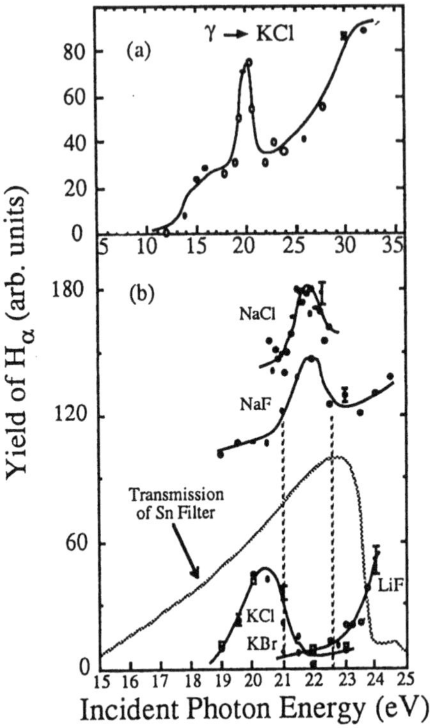

Figure 2.
(a) H*(n=3) desorption yield from KCl without filtering higher order incident photons. (b) H*(n=3) yields from KCl (black dots), KBr (open squares) and LiF with a Sn filter (transmission shown) inserted into the beam. Also shown are yields from NaCl and NaF without use of the Sn filter.

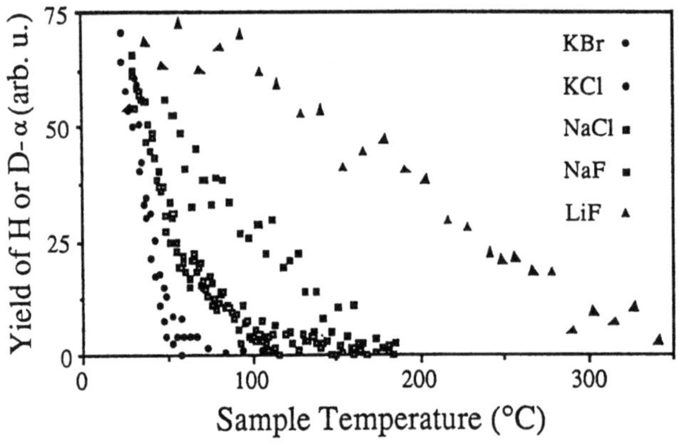

Figure 3. Temperature dependence of the desorption yields of H* from five alkali halide surfaces while irradiating with zero-order photons.

The temperature dependence of desorbing H* is displayed in Figure 3 from KCl, KBr, NaCl, NaF, and LiF surfaces under "white-light" photon irradiation. All of the yields were taken in the presence of about 10^{-7} Torr of H_2 and while increasing the temperature of the samples except that from KCl. The KCl data were acquired as the sample was cooled. The yields have been scaled for purpose of display since the unnormalized desorption rates of H_α from the potassium halides were about three to six times greater than those from the other compounds; this scaling did preserve the relative ordering of these yields as displayed on the temperature axis. Electron irradiation experiments yielded similar results.

The following observations follow by inspection of Figure 3:

(1) in all cases the yield of H* is diminished by heating, but this diminution occurs at different temperatures for different alkali halide compounds, the sequence being KBr < KCl < NaCl < NaF < LiF,

(2) the temperatures at which these desorption yields approach the noise-level are lowest for the potassium halides, higher for the sodium halides and highest for lithium fluoride, and finally,

(3) the ordering described in (2) is the same as the ordering of the heats of vaporization of these three metals, *viz.* K (0.80 eV) < Na (0.92 eV) < Li (1.40) [11].

5. The Desorption Precursor and the Mechanism of Desorption

The temperature regime over which H* yields are depleted is especially telling since many of the possible precursor states of the desorbing hydrogen would be completely unaffected by this thermal change. Chemisorbed atomic hydrogen could reside indefinitely bonded to K or Cl (≈2eV) at 125°C. In contrast, the exceedingly brief residence time (≈1ps) of physisorbed H or H_2 on KCl (≈30meV) [12] implies that these precursors are inconsequential under the conditions of these experiments.

The temperature intervals over which the yields of H* are diminished agree well with the observed temperature intervals found by Roy *et al.* [13] over which electron-induced excess potassium, sodium and lithium are thermally depleted by evaporation from their respective alkali halide surfaces. They measured the temperature dependent peaks in EELS spectra associated with surface and volume metal plasmons from the surfaces of electron-irradiated alkali halide compounds. These peaks from KBr were completely removed at 100°C, those from KCl at 110°C; NaCl, 200°C; NaF, 235°C; and LiF, 315°C (*cf.* Figure 3). This is in good agreement with an AES study by Szymonski *et al.* [14] of NaCl. They measured the ratio of Na to Cl Auger peak heights as a function of sample temperature and found the surface to be metal rich below 200°C compared to the stoichiometric ratio measured at higher temperatures.

These observations argue strongly that the diminishing yield of H* with temperature is not due to the thermal removal of physisorbed or chemisorbed H or H_2. Instead it is related (in first order) to the removal of excess metal from the surface. The role of the alkali constituent of the compound is also apparent in the time and energy dependent measurements. The time dependent measurements of H* suggest that these yields are correlated with the time evolution of surface metallization. The ESD energy thresholds from compounds containing the same alkali constituent agree within uncertainties in contact charging. Furthermore, the PSD energy thresholds are identical for compounds bearing the same alkali.

It is reasonable, therefore, to conclude from the temperature, time, and energy dependent results that H_2 dissociation occurs at and is contingent upon the existence of radiation-induced metal-rich surface sites and that the ultimate desorption precursor is a surface alkali hydride molecule.

When assigning DIET mechanisms to neutral-particle desorption, one must in general consider the possibility of direct resonance or Auger neutralization of desorbing ions as they recede from the surface. In this study, however, the neutral atoms are emitted in an excited state, typically n = 3 is measured. This state in hydrogen is only 1.5 eV below the vacuum. There are no electrons in the surface states of the alkali metals or alkali halides energetically available for neutralization of an ion into such a highly excited state. Therefore mechanisms of *ion* emission are not considered here. Gas-phase excitation of ground state hydrogen atoms is also improbable. Alkali halide surfaces under photon bombardment at the energies which gave threshold H_α resonances emit no secondary electrons energetic enough to excite ground-state hydrogen to n=3 (12.1 eV) [15]. Several mechanisms for ESD and PSD of neutral atoms and molecules correlate the energy thresholds and structure of desorption yields of excited atoms with core-level ionization [16], bulk and surface exciton creation [17], and vibrational excitation [18] of adsorbed molecules. The threshold energies for H* in the present study, however, show no systematic correlation to the core-excitation/ionization or excitonic energies of hydrogen, the alkali metals, the alkali halides or vibrational excitation of the alkali hydrides.

A one-dimensional Franck-Condon model of valence excitation is often applied to desorption of gases from metal surfaces. This so called Menzel-Gomer-Redhead (MGR) model [19] has been applied extensively to the results of photon [20] and electron [21] stimulated desorption of neutral molecules from metal surfaces.

This valence excitation model predicts many features which are consistent with the results of desorption of H* from a surface alkali hydride molecule. In such a model, the first step in the desorption of H* is a vertical excitation within the Franck-Condon region from the alkali hydride ground state to an antibonding state. Previous applications of this model often specified excitation to an ionic repulsive state. Here desorption is expected to follow direct excitation to a neutral antibonding state. This follows since the desorption of protons which are subsequently neutralized as they recede from the surface is unlikely as discussed above. If this excitation is sufficiently localized, the molecule will dissociate emitting H* with a kinetic energy determined by the point on the upper curve to which the molecule was excited and the level of the asymptote of the upper curve at infinite nuclear separation. If all the excited hydrogen atoms produced in this manner escape from the surface, then the H* yield as a function of incident photon energy is a reflection of the square of the ground state alkali hydride wavefunction through the potential energy curve of the antibonding state. This would result in PSD resonance structure at threshold with typical Franck-Condon mediated widths [22] as shown here for desorption from KCl. Such a mechanism is also consistent with non-resonant ESD thresholds due to the inelastic nature of electron-particle interactions.

The time, temperature and energy dependent results strongly indicate that the H* desorption precursor is a surface alkali hydride molecule. The observations presented above argue against desorption of H* which is thermal, initially ionic or ground-state, or desorption which is induced by momentum transfer, core-excitations or excitons. The location in energy of the observed thresholds and the shapes and widths of the features associated with these thresholds are consistent, however, with an MGR-valence-excitation model of desorption.

The desorption mechanism proposed here is primarily a surface interaction and independent of the substrate or the origin of the alkali hydride surface species. To confirm this, preliminary desorption experiments have been performed from a surface on which alkali metal and hydrogen were coadsorbed. A glass slide was dosed with potassium from a SAES getter and then exposed to gaseous hydrogen. The PSD excitation function exhibited a pronounced resonance at the incident photon energies that produced this feature during PSD of the potassium halide samples. The yield was also suppressed as the metal was evaporated from the substrate. These results support the desorption precursor and valence excitation mechanism proposed for the desorption of excited hydrogen atoms from alkali metal surfaces with hydrogen overlayers.

This research has been sponsored by The Air Force Office of Scientific Research under contracts F49620-86-C-0125 and F49620-88-C-0080 and the Office of Naval Research under contract N00014-87-C-0146.

References

1. J. Estel, H. Hoinkes, H. Kaarmann, H. Nahr, and H. Wilsch, Surf. Sci. **54**, 393 (1976), G. Boato, P. Cantini, and L. Mattera, Surf. Sci. **55**, 141 (1976).
2. N. G. Stoffel and P. D. Johnson, Nuc. Instr. Methods A **234**, 230 (1985).
3. P. W. Palmberg and T. N. Rhodin, J. Phys. Chem. Solids **29**, 1917 (1968).
4. H. Tokutaka, M. Prutton, I. G. Higginbotham, and T. E. Gallon, Surf. Sci. **21**, 233 (1970).
5. K. L. Tsang, C. H. Zhang, T. A. Callcott, and D. L. Ederer, Phys. Rev. B **35**, 8374 (1987).
6. G. Singh and T. E. Gallon, Solid State Commun. **51**, 281 (1984).
7. L. S. Cota Araiza and B. D. Powell, Surf. Sci. **51**, 504 (1975).
8. G. C. Fryburg and R. A. Lad, Surf. Sci. **48**, 353 (1975).
9. Q. Dou, D. W. Lynch, and A. J. Bevolo, submitted to Surf. Sci.
10. L. Ernst, Solid State Commun. **19**, 311 (1976), L. Ernst, Nuovo Cimento **39**, 797 (1977).
11. *CRC Handbook of Chemistry and Physics*, R. C. Weast, ed. (CRC Press, Inc., Boca Raton, 1988), p. F-65.
12. H. Hoinkes, Rev. Mod. Phys. **52**, 933 (1980).
13. G. Roy, G. Singh, and T. E. Gallon, Surf. Sci. **152/153**, 1042 (1985).
14. M. Szymonski, J. Ruthowski, A. Poradzisz, Z. Postawa, and B. Jørgensen, *Desorption Induced by Electronic Transitions - DIET II*, edited by W. Brenig and D. Menzel (Springer, New York, 1985), p. 160.
15. W. Pong and J. A. Smith, Phys. Rev. B **9**, 2674 (1974).
16. N. H. Tolk, M. M. Taum, J. S. Kraus, T. R. Pian, W. E. Collins, N. G. Stoffel, and G. Margaritondo, Phys. Rev. Lett. **49**, 812 (1982).
17. R. F. Haglund, Jr., A. V. Barnes, M. H. Mendenhall, N. H. Tolk, and D. E. Ramaker, to be submitted to Phys. Rev. B, P. Fuelner, T. Müller, A. Puschmann, and D. Menzel, Phys. Rev. Lett. **59**, 791 (1987), F. Coletti, J. M. Debever, and G. Zimmerer, J. Physique Lett. **45**, L-467 (1984).

18. J. Heidberg, H. Stein, E. Riehl, and A. Nestman, Z. Phys. Chem. **121**, 145 (1980), T. J. Chuang, J. Chem. Phys. **76**, 3828 (1982).
19. P. A. Redhead, Can. J. Phys. **42**, 886 (1964), D. Menzel and R. Gomer, J. Chem. Phys. **41**, 3311 (1964).
20. R. O. Adams and E. E. Donaldson, J. Chem. Phys. **42**, 770 (1965).
21. D. Menzel, Ber. Bunsenges. Physik. Chem. **72**, 591 (1968); P. Feulner, R. Treichler, and D. Menzel, Phys. Rev. B **24**, 7427 (1981); P. Feulner, D. Menzel, H. J. Kreuzer, and Z. W. Gortel, Phys. Rev. Lett. **53**, 671 (1984).
22. M. Higo and T. Ogawa, Chem. Phys. **44**, 279 (1979), J. Kurawaki, K. Ueki, M. Higo, and T. Ogawa, J. Chem. Phys. **77**, 3071 (1983), T. Ogawa, J. Kurawaki, and M. Higo, Chem. Phys. **61**, 181 (1981).

Observation of Excitonic Effects in Excited Neutral Alkali Desorption from Alkali Halides

P.H. Bunton, R.F. Haglund, Jr., and J.L. Rose [1]

Department of Physics and Astronomy, Box 1807-B, Vanderbilt University, Nashville, TN 37235, USA

Abstract. We report simultaneous measurements of excited state alkali desorption and photoelectron emission from NaCl irradiated by ultraviolet light from a synchrotron source. The excitation function shows structure near known exciton energies. Incident photons of energy sufficient for core-exciton production are seen to channel energy away from broad-band bulk luminescence and toward electron and excited alkali atom emission. This suggests that excitons can play a direct role in excited-atom, neutral desorption.

1. Introduction

The origin of excited alkali atoms desorbed from alkali halide surfaces during low energy photon or electron irradiation has remained an area of controversy since its discovery [2]. Some of the excited-atom yield is certainly accounted for by secondary-electron excitation of desorbed ground-state atoms, as proposed for electron-stimulated desorption [3]. Until recently, the possibility of direct photon-stimulated desorption (PSD) in excited states - in contrast to indirect mechanisms such as secondary electron excitation - has remained an open question. Conclusive evidence for electron-stimulated desorption of excited alkali atoms from alkali halides due to core-hole excitation has recently been published [4]. Our results indicate that excitons, or more generally core-hole excitations, also play a significant role in the PSD of excited atoms, and that exciton formation competes with broad-band luminescence for flux channeled into excited state desorption and electron emission.

2. Experiment

These data were taken at the SRC-Vanderbilt beamline at the Synchrotron Radiation Center of the University of Wisconsin in Stoughton, Wisconsin. The zero-order light from the Aladdin beam was dispersed by a 6-meter toroidal-grating monochromator (TGM). Single crystal alkali halide samples produced by Harshaw Chemical Company were cleaved in air and cleaned by heating to temperatures of approximately 350°C under ultrahigh vacuum conditions.

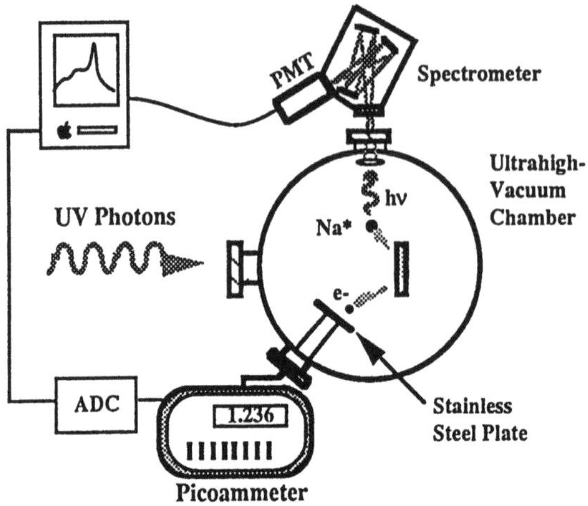

Figure 1: Sketch of the experimental arrangement used in these experiments.

Figure 1 shows the experimental arrangement used in these measurements. Desorbed excited state atoms were detected by a McPherson 0.3 meter Czerny-Turner monochromator with a photomultiplier tube monitoring the first resonance lines of the desorbing alkali atoms. Data were collected and stored in a Macintosh microcomputer for subsequent analysis. The secondary electron yield striking a biased, square piece of stainless steel was measured with a Keithley electrometer. No precautions were taken to shield the collector plate from ultraviolet fluorescence of the irradiated crystals. The collector voltage was adjusted for maximum current without the picoammeter going off scale. Thus some scans were weighted toward lower energy electrons; however, the location of peaks in the excitation function did not appear to depend upon the choice of voltage. Scans were taken with and without the sample grounded and for NaCl with the sample biased to both + 12 Volts and - 12 Volts; again the results seem to be unaffected. Data are not corrected for higher-order photons transmitted by the monochromator.

3. Results

Figure 2 shows the luminescence of NaCl from 200 to 800 nm both on and off a known exciton peak. At the photon energy corresponding to a bulk exciton, the bulk fluorescence is reduced while the excited-state yield is increased. The core exciton - which is associated with both excited atom and secondary electron emission - apparently competes for incident photon flux with the precursor channels for the bulk fluorescence.

Figure 3 shows the excitation function for excited atom and total electron yields from NaCl over the photon energy range 30 - 50 eV. These results were obtained simultaneously. Peaks in the excited-atom yield generally correspond to the peak locations in the secondary electron excitation function. The slope of the electron background is not completely reproducible. The lineshape for the 2p core exciton

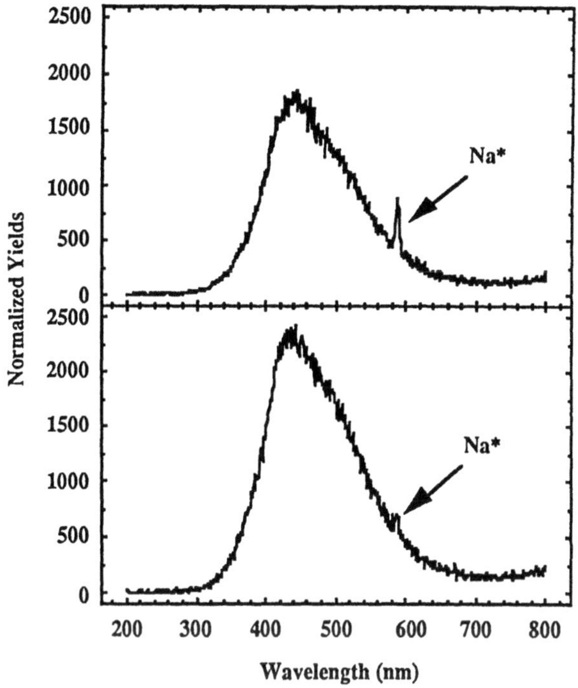

Figure 2: Fluorescence of NaCl with the incident photon energy tuned on and off the core exciton. The top graph is with the incident photon energy tuned to the 33.5 eV core-exciton peak and the bottom graph with the photon energy tuned to 34 eV or just above the core exciton. Units for yield in both groups are counts per half second normalized to synchrotron beam current.

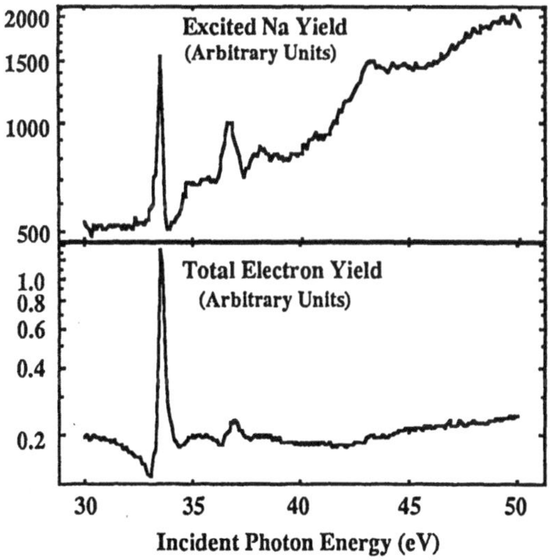

Figure 3: Excited sodium and total electron yields from NaCl under ultraviolet photon irradiation.

in the total-electron yield exhibits an antiresonance indicating two interfering processes which is not the case for the excited atom yield.

4. Discussion

The enhancement of the excited-state, alkali yield at the energies of core exciton formation is evidence that core excitons can play a role in excited-state, neutral desorption. The differing lineshapes for the total electron yields and the excited alkali yields can provide further insight into the actual mechanism for excited state desorption. The electron yield has a lineshape indicative of interfering processes. Explanation of this requires comments on Auger decay of excitons. Three processes contribute to the electron yield at the Na 2p core exciton energy: two types of Auger decay and the normal photoemission process [5]. The first Auger decay is the $L_{2,3}$(exciton)-VV whereby the core hole is filled by an electron from the valence band and an Auger electron is emitted from the valence band, thus leaving two valence-band holes. The second is the $L_{2,3}$(exciton)-V process whereby an electron in the excitonic state fills the core hole (essentially recombination of the electron and hole of the exciton) and simultaneously an electron is removed from the valence band. In this process the final state is the same as that for direct photoemission from the valence band thus leading to interference between the two processes. The total electron yield represents a superposition of the three processes, but the interference is still evident. Since the excited sodium yield from NaCl does not show the antiresonance of the electron yield we conclude that the dominant decay process leading to excited Na emission is the $L_{2,3}$(exciton)-VV process. This is consistent with a Knotek-Feibelman type mechanism for $L_{2,3}$(exciton)-VV decay leading to neutral desorption. Localization of both valence-band holes on the same ion converts a negative halogen ion into a positive ion. In this case Cl^- becomes Cl^+. This leaves two nearest-neighbor, positive ions either of which may have enough energy to desorb. As the Na ion leaves it is then neutralized into an excited state, consistent with previous observations that excited Na yields resemble both Na^+ yields and total electron yields. All three functions, in fact, are related to the core-hole formation probability.

5. Acknowledgements

We thank Dr. Richard Rosenberg of the Synchrotron Radiation Center (SRC) of the University of Wisconsin-Madison for lending equipment during the experiments and for stimulating discussions, and Dr. Wolfgang Husinsky of the Technische Universität, Vienna and Professor Norman Tolk of Vanderbilt for helpful suggestions. This work was partially supported by the Air Force Office of Scientific Research under the University Research Initiative (Contract Number F49620-86-C-0125). One of us (PHB) was supported by the National Aeronautics and Space Administration under Training Grant number NGT-50126. The SRC is supported by the National Science Foundation.

References

[1] Current address: Rayovac Corporation, 601-T Rayovac Dr., Madison, WI. 53711 U.S.A.

[2] N. H. Tolk, M. M. Traum, J. L. Kraus, T. R. Pian, W. E. Collins, N. G. Stoffel and G. Margaritondo, Phys. Rev. Lett. 49, 812 (1982).

[3] R. E. Walkup, Ph. Avouris, and A. P. Ghosh, Phys. Rev. Lett. 57, 2227 (1986).

[4] P. Wurz, E. Wolfrum, W. Husinsky, G. Betz, L. Hudson, and N. Tolk, Rad. Eff. Def. Solids 109, 203 (1989).

[5] K. Ichikawa, O. Aita, and K. Tsutsumi, Phys. Rev. B 39, 1307 (1989); M. Kamada, O. Aita, K. Ichikawa, and K. Tsutsumi, Pyhs. Rev. B 36, 4962 (1987).

Electron-Stimulated Desorption of Neutral Ground-State Lithium Atoms from LiF and the Influence of Stored Defects

J. Sarnthein, P. Wurz, W. Husinsky, and G. Betz

Institut für Allgemeine Physik, Technische Universität Wien,
Wiedner Hauptstr. 8–10, A-1040 Wien, Austria

Abstract. We have investigated the ESD of neutral ground-state Li atoms from LiF single crystal <100> surfaces caused by 100eV electron irradiation at varied crystal temperatures between 300K and 800K. During irradiation desorption increases exponentially with temperature, limited only by the evaporation of the alkali metal layer on the surface. The obtained activation energy varies between 0.8eV and 1.6eV with current density and reflects the size of the metal clusters on the surface. Above 300°C a plateau is reached where the electron flux becomes the rate-limiting quantity. Above 450°C the evaporation of the crystal adds significantly to the desorption yield.

We have also studied the time-dependence of the desorption which continues after the end of irradiation. This delayed emission can be divided into two components by their different time constants: A fast decay of signal in less than 20ms, which is followed by a slow decay with a time constant of several seconds. The fast decay has been attributed to F-center diffusion. We propose to explain the slow decay in terms of aggregation of F-centers and subsequent disintegration of the aggregates.

1. Introduction

Electron Stimulated Desorption (ESD) from alkali-halides has been studied for more than 20 years. Although the mechanism leading to desorption of halogen atoms and consequent evaporation of the alkali metal atoms is generally agreed on [1-4], some questions on the details of the process are still open. As has recently been summarized by Williams [1], the interaction of electrons with an alkali-halide lattice is restricted to the halogen sublattice. If the energy deposited by the incoming electron suffices to overcome the bandgap, an electron of the valence band may be excited into the conduction band, thus forming an electron-hole pair, which may lead to production of an F-H-center pair via a self-trapped exciton. A F-center is an electron on a halogen (X^-) vacancy and a H-center is comparable to a X_2^--molecule on an anion lattice site. While H-centers are mobile and will diffuse already at temperatures as low as 30K [5], F-center diffusion is limited to temperatures above a few hundred K [6]. If a H-center reaches the surface it will decay and emit a neutral halogen atom. The surface is left stoichiometric as the F-center remains at the original site of the defect pair formation. Whereas the H-centers are responsible for halogen emission, F-centers cause alkali desorption. F-centers formed at the surface or having diffused to the surface will decay in neutralizing a surface alkali atom and thus form an alkali

Springer Series in Surface Sciences, Vol. 19
Desorption Induced by Electronic Transitions, DIET IV
Editors: G. Betz · P. Varga

overlayer [7] at the surface. If the vapor pressure of the alkali metal is sufficiently high, alkali atoms evaporate from the crystal. At low temperatures the metal overlayer inhibits desorption. Our temperature dependent measurements focus on this effect.

Time resolved measurements on the delayed emission of alkali atoms from LiF after the end of the irradiating electron pulse have been explained by Green et al. [8-10] in terms of F-center diffusion to the surface, followed by Li evaporation. We have performed similar measurements, but on an expanded time scale (delayed emission up to seconds). We will explain the results by taking into account not only F-center diffusion but also F-center aggregation [6,11-13] and subsequent slow disintegration of these aggregates with time.

2. Experimental

Single crystal <110> surfaces of LiF at different target temperatures (300-800K) were irradiated with 5eV to 500eV electrons in pulses of 1s to 100s with current densities up to $25\mu A/mm^2$. The angle of incidence of the electron beam was 45° with respect to the crystal normal. The irradiated area had dimensions of approximately 2 by 4mm. In the time-resolved measurements the electron beam was deflected from the crystal by applying a negative extraction voltage within less than 1msec. We have measured neutral ground-state alkali atoms with a quadrupole mass analyzer oriented normal to the surface. The pulses from the secondary electron multiplier were processed digitally in pulse counting mode with a time resolution of up to 20ms but usually 1s. The UHV system maintained a base pressure of less than $3\cdot10^{-10}$mbar during the measurements. The crystals were heated at a rate of 8.6K/min and the temperature was measured with a thermocouple close to the surface. The crystals have been cleaved in air and cleaned in the vacuum system by prolonged heating at 700K [14]. The countrate of Lithium atoms was calibrated to monolayers per second [ML/s] [15] by comparison to three independent quantities: i) the vapor pressure of solid Li metal with an activation energy of 1.6eV [16]; ii) the evaporation of a LiF crystal with an activation energy of 2.7eV [17]. In our setup no evaporating molecules were observed, they disintegrate in the ionizer. iii) Alkali enrichment on the surface at room temperature is limited to one monolayer [18]. Desorption of this monolayer at a higher temperature gives a number of SEM-counts which is corrected for the solid angle. Each approach gives the same detection efficiency of 10^{-5} counts per atom for the quadrupol in our setup. We estimate a desorption efficiency of 0.1 atoms per incoming electron of 100eV by comparing the calibrated desorption rate to the irradiation dose. This value is in reasonable agreement with previous estimates [19].

3. Results

Figure 1 shows the temperature dependence of the Li desorption from a LiF crystal surface under electron irradiation. Four temperature ranges can be distinguished: i) At low temperatures the vapor pressure of the alkali metal is negligible and no alkali

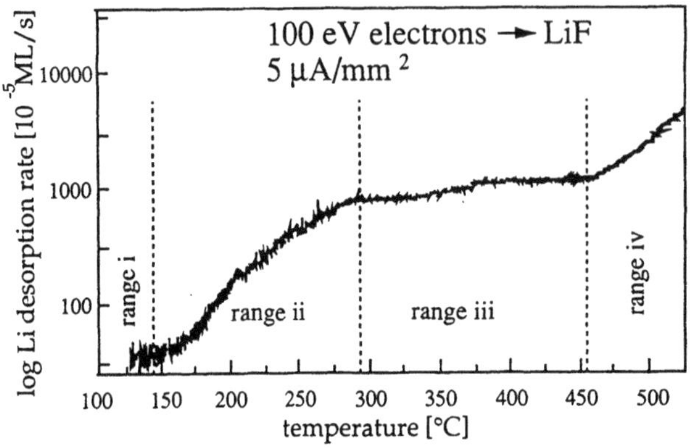

Figure 1: The Li^0 desorption yield as a function of crystal temperature during 100eV electron bombardment of LiF. Four temperature ranges can be distinguished: i) no desorption below 430K due to negligible vapor pressure of the Li layer on the surface; ii) exponential increase limited by the evaporation rate of Li; iii) desorption rate independent of temperature, limited only by irradiation intensity; iv) evaporation of the LiF crystal adds significantly to the desorption yield.

evaporation takes place. A surface metal film will develop and finally prevent further halogen emission. No ESD of ground-state Li atoms was observed below 450K at all irradiating energies; ii) As the temperature is increased alkali atoms can evaporate from the surface and an exponential increase in the yield is observed. The desorption flux is limited only by the evaporation rate in this range; iii) As soon as the evaporation rate becomes larger than the Li metal formation rate at the surface, a constant yield is observed even with increasing temperature (plateau). The height of the plateau is proportional to current density. The onset of the plateau is shifted to higher temperature with higher current density; iv) At temperatures above 700K the evaporation of the crystal adds significantly to desorption. Separate measurements for an unirradiated surface verified this point.

Figure 2 shows three Arrhenius-plots of the ranges ii) and iii) at different current densities yielding different activation energies. The results were the same for both directions of temperature change. The onset of range iii) is shifted to higher temperature with higher current density, because a higher temperature is needed to make the evaporation rate exceed the formation rate of Li metal on the surface.

In an additional set of experiments we have studied delayed emission of Li from LiF after the electron beam was turned off (fig. 3). Contrary to experiments discussed by Green et al. [8-10], we focused on delayed emission in the time range of 0.1 to 10 sec. As the main feature, two types of delayed Li emission could be distinguished for all energies and temperatures (fig. 3): i) A fast decay, where the desorption yield drops faster than our time resolution (20ms); ii) a slow decay resembling a combination of exponentials with decay rates in the order of seconds. The shape depends strongly on irradiation time and temperature.

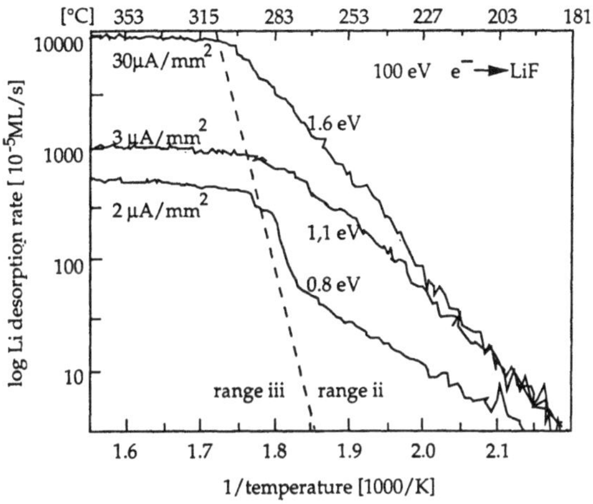

Figure 2: Ranges ii) and iii) for three different current densities. The activation energies smaller than 1.6eV (bulk Li metal) are attributed to lower Li metal coverages of the LiF surface (clusters).

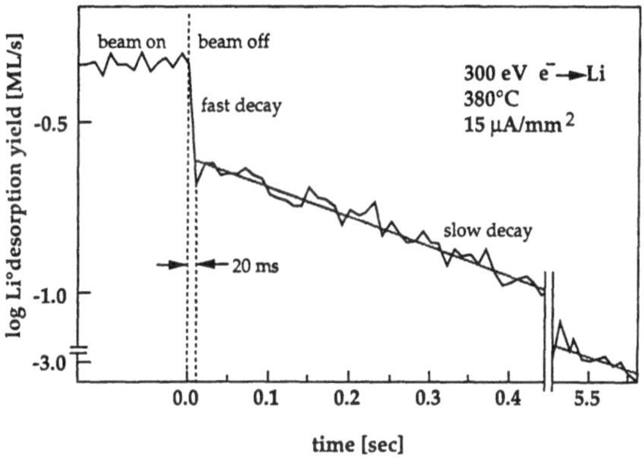

Figure 3: After each interruption of the electron beam the signal shows a fast decay within 20ms followed by slow decay in the order of seconds.

At 670K the signal ratio fast/slow decay was about one, decreasing for higher as well as for lower temperatures. The ratio of the two types of delayed emission was essentially constant with energy. A peak in the delayed emission signal is observed (fig. 4) at temperatures around 700K for short irradiation times at current densities in the order of 10^{14}electrons/(cm^{2}sec). This peak could be observed at all irradiating energies but was dependent on the current density. With increasing electron dose the peak becomes less pronounced and disappears for electron irradiation longer than 150s.

4. Discussion

We propose to explain the different activation energies found in range ii) with formation of Li metal clusters on the surface. Experiments with Na clusters on a LiF surface have shown a proportionality of cluster size to the activation energy for evaporation [20]. The activation energies attributed to evaporation of bulk lithium is 1.6eV [16], smaller activation energies should result from smaller sizes of Li metal

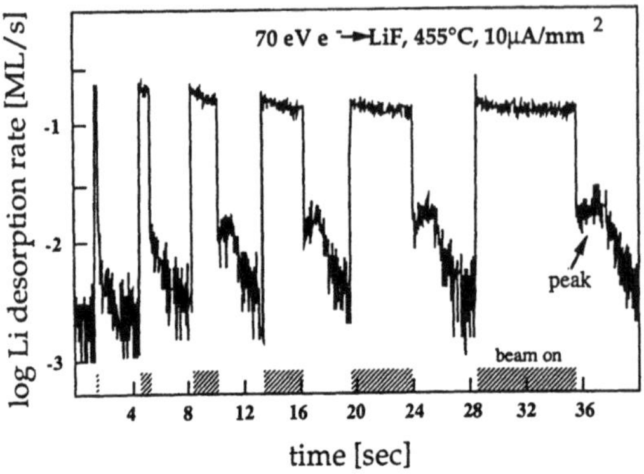

Figure 4: Delayed emission of Li^0 from LiF after the electron beam is repeatedly interrupted. The electron current density was 10^{14} $e^-/(mm^2 sec)$. Under special experimental conditions a peak may appear in the delayed emission.

clusters in analogy to [20]. The step around 570K observed in curves with an activation energy of 0.8eV may be attributed to a sudden formation or evaporation of small droplets. As the current density determines the rate of Li enrichment on the surface, it is proportional to the metal cluster size and thus the activation energy.

The delayed emission is explained in terms of aggregation of F-centers. Aggregate formation in the bulk is a general phenomenon in ionic crystals [6,11-13]. In our temperature range F-centers are believed to be mobile [6], whereas F^n-aggregates (M-centers n=2, R-centers n=3) are immobile. They may serve to store radiation damage. During electron irradiation F-centers are formed in a near surface layer of about 10nm. We have two loss channels for F-centers during irradiation: i) Formation of F^n-aggregates ii) Diffusion to the surface and trapping of F-centers at the surface followed by alkali metal evaporation. As soon as irradiation is stopped, no more F-centers are produced. The remaining F-centers will diffuse to the surface and are responsible for the fast decay [8-10]. The slow decay is due to the radiation damage stored by aggregates. Slow disintegration of aggregates to F-centers and subsequent diffusion to the surface is followed by desorption. The decay in our experiments resembles an exponential shape (fig. 3). This conforms with the fact, that the ratio of F-centers to aggregates is given by thermal equilibrium leading to an exponential rate of disintegration. For macroscopic aggregates (colloids) one would expect delayed emission at a constant rate. The details of the time dependence in delayed emission reflect the interplay of the initial defect distribution profile, aggregate formation and disintegration and diffusion of F-centers to the surface, which may in some cases lead to the peak seen in fig. 4.

5. Conclusions

We have performed temperature dependent and time-resolved measurements of the neutral Li atom desorption yield. The temperature dependence confirmed that the evaporation of a Li metal layer on the surface is the rate limiting quantity for desorption at low temperatures. We showed that the current model has to be extended by metal cluster formation on the surface. The observed delayed emission after irradiation strongly suggests that one has to take into account the formation and disintegration of F-center aggregates.

6. Acknowledgment

The authors gratefully acknowledge financial support by the Österr. Fonds zur Förderung d. wiss. Forschung (project Nrs: 5833,5577).

7. References

[1] R.T. Williams, Rad. Eff. a. Def. in Solids **109** (1989) 175
[2] M. Szymonski, these proceedings
[3] N. Itoh, Nucl. Instr. Meth. **132** (1976) 201
[4] P.D. Townsend, Nucl. Instr. Meth.**198** (1982) 9
[5] J.H. Crawford, L.M. Slifkin, Point defects in solids, Plenum Press, New York (1972)
[6] P.Durand, Y.Farge, M.Lambert, J.Phys.Chem.Sol. **30** (1969) 1353-1374
[7] G. Singh, T.E. Gallon, Sol. State Comm. **51** 5 (1984) 281-284
[8] T.A. Green, G.M. Loubriel, P.M. Richards, N.H. Tolk and R.F. Haglund Jr, Phys. Rev. **B35** (1987) 781
[9] T.A. Green, M.E. Riley, P.M. Richards, G.M. Loubriel, D.R. Jennison and R.T. Williams, Phys. Rev. **B39** (1989) 5407
[10] T.A. Green, these proceedings
[11] U. Jain and A.B. Lidiard, Philosoph. Mag. **35** (1977) 245
[12] A.E. Hughes and S.C. Jain, Adv. Phys. **28** (1979) 717
[13] A.E. Hughes and A.B. Lidiard, Harwell Report, AERE-R13319 (1989)
[14] T.E. Gallon, I.G. Higginbotham, M. Prutton, H. Tokutaka, Surf.Sci. **21** (1970) 224-232
[15] R. Kelly, Surf.Sci. **90** (1979) 280-318
[16] R.E. Honig, RCA Review **23** (1962) 567-586
[17] C.T. Ewing, K.H. Stern, J. Phys. Chem. **79** 19 (1975) 2007
[18] P. Wurz, C. Becker, Surf. Sci. in print
[19] N. Itoh, Nucl. Instr. Meth. **132** (1976) 201-211
[20] M. Vollmer, F. Träger, Surf. Sci. **187** (1987) 445-462

Molecular Cluster Calculations of the Electronic Structure of the (111) Surface of CaF_2

E. Westin [1], *A. Rosén* [1], *and E. Matthias* [2]

Department of Physics, Chalmers University of Technology and University of Göteborg, S-41296 Göteborg, Sweden
[2]Fachbereich Physik, Freie Universität Berlin, D-1000 Berlin 33, Germany

Molecular cluster calculations within the Local Density Approximation have been performed to analyze the electronic structure of stoichiometric and different non-stoichiometric (111) surfaces of CaF_2. The effect of the surrounding crystal ions, i.e. the long range electrostatic potential, have been included by a Fourier summation. Calculations for clusters representing the bulk and stoichiometric surfaces give similar results while calculations for non-stoichiometric surfaces show the existence of occupied surface states in the upper half of the bandgap. Existence of these types of occupied surface states are supported by experimental EELS studies on CaF_2 as well as by observation of laser induced emission of ions and electrons from surfaces of BaF_2 .

1. Introduction

Understanding the nature of surfaces of ionic crystals and semiconductor/ ionic crystal interfaces is today an area of great interest, both with respect to fundamental concepts and practical applications. An overview of the development of the field can be found in earlier DIET workshops [1]. The knowledge of ionic surfaces is however limited in comparison with the very extensive studies of metal surfaces [2]. This is mainly due to the charging up effect when using LEED and/or AES although some XPS and UPS studies have been performed with synchrotron radiation [3]. The access to new laser spectroscopic methods such as Second Harmonic Generation (SHG) and Sum Frequency Generation (SFG) have however opened new possibilities not only to characterize the geometry but also to get information about the electronic structure of these surfaces and ionic/ semiconductor interfaces [4-6]. For the analysis of the mechanism behind different observed effects it is of great help to have access to theoretical predictions of the electronic structure of these surfaces and interfaces.

In our earlier study we calculated the electronic structure of the (111) surface of the divalent ionic compound BaF_2 to analyze the laser induced emission of electrons and ions [7]. Motivated by many recent experimental studies of the CaF_2 crystal, especially surface and interface properties [3,8], we have here focused our attention on this lighter but similar system. Surfaces of CaF_2 are more appropriate as model systems for these ionic crystals since in the case of BaF_2 the electronic structure is influenced by relativistic effects. The aim has been to investigate the electronic structure of stoichiometric and non-

Springer Series in Surface Sciences, Vol. 19
Desorption Induced by Electronic Transitions, DIET IV
Editors: G. Betz · P. Varga

stoichiometric surfaces by means of molecular clusters embedded in an external crystal potential.

2. Computational approach

The (111) surface of CaF_2 consists of layers of Ca^{++} and F^- ions with stacking sequence F^--Ca^{++}-F^----F^--Ca^{++}-F^- etc., where the F^- ions are located either 1.49 a.u. above or below the Ca plane. The distance to the next-neighbour Ca plane inside the crystal is 5.96 a.u. To model a local area of the (111) surface, small molecular clusters (≤ 13 atoms) are used. In figure 1, two types of planar surface clusters are shown, one Ca centered i.e. $Ca^{++}{}_1Ca^{++}{}_6F^-{}_3F^-{}_3$, and one F centered i.e. $F^-{}_1Ca^{++}{}_3F^-{}_3F^-{}_6$. The subscripts indicate symmetrically equivalent atoms in the C_{3v} point group (shown in figure with different shadings). These clusters will in general be charged and it is therefore necessary to add the effect of exterior crystal charges. In calculations for the bulk the molecular cluster is normally embedded in an external potential [9]. In this case, as a comparison, we have also used a F centered $F^-{}_1Ca^{++}{}_4F^-{}_6$ cluster treated in T_d symmetry, since this is the point group of the crystal itself.

Extension of the embedding approach to surfaces is not so straightforward. In our earlier calculations of the electronic structure for BaF_2 surfaces we limited ourselves to certain fixed embeddings to get a neutral system. These restrictions were however inconvenient and limited straightforward comparisons between, for instance, anion and cation centered clusters. A more general method was therefore required as outlined in figure 1. A local molecular cluster of a certain size was constructed as discussed above and placed in a crystal field, i.e. an embedding potential derived from frozen ionic charge distributions and a sufficient number of point like ions surrounding the cluster. To evaluate this long range electrostatic potential shifts a method by Parry [10] has been extended to general layered ionic systems [11]. Briefly, we split the external

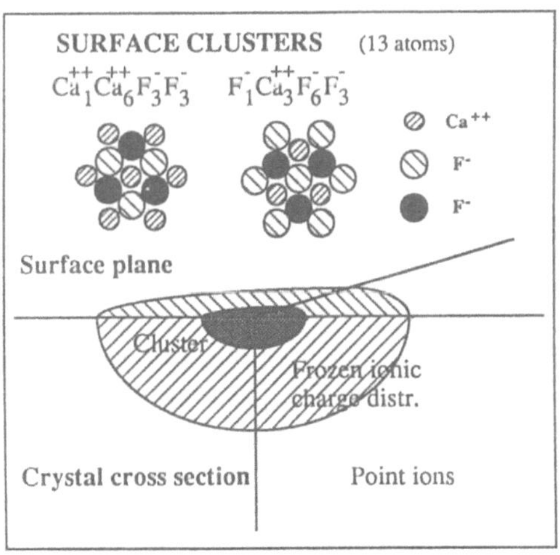

Figure 1 Geometry of the planar clusters used in the calculation to model a patch of thc (111) surface of CaF_2. Below is shown a cross section of the surface, where the local molecular cluster, frozen ionic charges and point ion regions are indicated.

electrostatic potential into a long range part and a short range part by means of the error and complementary error function [12], i.e

$$V^{ext}(\bar{r})=V^{sr}(\bar{r})+V^{lr}(\bar{r}),$$

where

$$V^{sr}(\bar{r})=\sum_i q_i \sum_l \frac{1}{R}\mathrm{erfc}(wR),$$

with erfc replaced by erf for $V^{lr}(\underline{r})$. The summation is over sublattices i, with charge q_i, and direct lattice l. $R=|\underline{l}+\underline{r}_i-\underline{r}|$, where $\underline{r}_i$ is the relative position of the ion within the unit cell. The crystal is treated as consisting of infinite two dimensional planes. The contribution from each layer can be identified and a crystal surface environment can be constructed by adding displaced parallel planes until convergence is reached. For each of these planes, parametrized by $z_i=r_{iz}-r_z$ the relative distance to sub-lattice plane i perpendicular to the surface, the Fourier coefficients for V^{lr} can be expressed as

$$V^{lr}_{\underline{k}\neq 0}=\frac{\pi}{A}\frac{1}{k}\left\{e^{kz_i}\mathrm{erfc}\left[\frac{k}{2w}+z_i w\right]+e^{-kz_i}\mathrm{erfc}\left[\frac{k}{2w}-z_i w\right]\right\},$$

$$V^{lr}_{\underline{k}=0}=\frac{2\pi}{A}\left\{\sqrt{z_i^2+\frac{A}{\pi}}+\mathrm{erf}[z_i w]-\frac{1}{\sqrt{\pi}w}e^{-(z_i w)^2}\right\},$$

where A is the area of the planar unit cell, k the corresponding two dimensional wave vector. w, the same for both expressions above, is a parameter which can be chosen within a broad range to get fast convergence. Close to the cluster, the point like ions are replaced by frozen charge distributions.

Knowing the fixed external potential, the electronic structure of the surface is evaluated by solving the Schrödinger equation, within the local density approximation [13], for the chosen molecular cluster. From a set of atomic wavefunctions, generated and given numerically on a grid, a symmetry adapted basis set for the cluster is constructed using a LCAO approach. The matrix elements in the secular equation are calculated by use of the discrete variational method (DVM) as developed by Ellis and co-workers [14]. To evaluate the potential in explicit form, the electron charge density is fitted to a multi-center multipolar model charge density by a least square fit [14].

We have used an ionic basis set i.e. Ca^{++}(1s2s2p3s3p4s4p3d) and F^-(1s2s2p) The unoccupied Ca^{++} atomic basis functions are generated in a well to produce contracted cluster valence functions needed to describe the ionic crystal in a realistic way. This basis has also been extended by contracted Ca^+(4s4p3d) and F^-(2s2p3s3p) functions. The extra F functions do lower the F(2p) band somewhat but have negligible effect on the lower conduction band and were therefore omitted. The choice of the degree of contraction is discussed below.

Occupied atomic basis functions as well as the frozen charge distributions in the embedding are generated in the bulk Madelung potential, i.e. -19.94 eV for Ca^{++} and 10.72 eV for F^-. To block the transfer of electrons by extended valence wavefunctions into the deep wells outside the cluster a pseudo potential is added which cuts the potential close to exterior atoms to an arbitrary value [7,15]. This pseudo potential, within ionic radii of Ca^{++} and F^-, is here chosen to coincide with the value of the Fermi level obtained from the bulk calculation. This is a reasonable choice but has strictly to be inferred from calculations of larger clusters.

3. Results and discussion

The embedding scheme described above enables us to compare the local electronic structure of arbitrary clusters of widely different charges without having to compensate for large absolute shifts of energy levels. Thus we have evaluated the electronic structure for bulk CaF_2 as well as for different stoichiometric and non-stoichiometric surface configurations. Here, we have assumed that inter-ionic bond lengths do not differ significantly between bulk and surface clusters.

Figure 2 shows the molecular ground state eigenvalues of the cluster configurations treated in this study. The extended basis set has not been used for the Ca centered cluster, since these clusters already contain a large number of Ca basis functions. For the F centered surface clusters with few Ca ions, we have used a double basis for all of them in order to get a sufficient variational freedom for the conduction band structure. The position of the F(2p) band is rather insensitive to the choice of basis set, and other parameters, such as the degree of contraction of ionic basis functions and pseudopotential cut off exterior to the

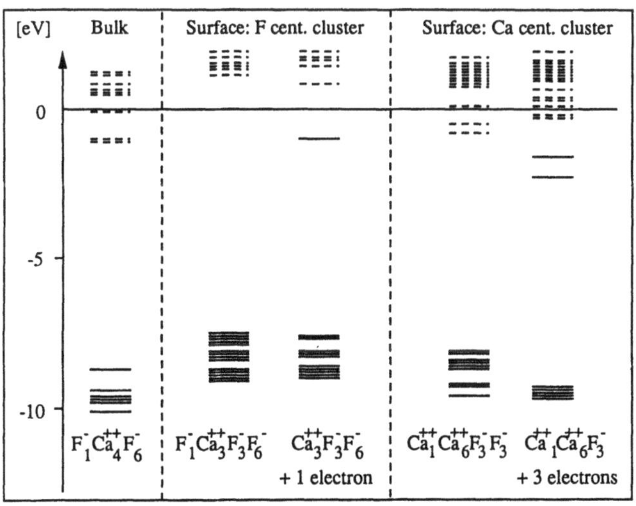

Fig. 2 Comparison of ground state eigenvalues for bulk and surface clusters. The cluster symbols are explained in the text.

cluster. This is also true for the internal structure of the conduction band within different cluster configurations. However, the position of the conduction band and thus the value of the bandgap can be quite sensitive to these parameters. We have therefore adapted the parameters to give a realistic description of the bulk electronic structure and used these for all subsequent calculations. This gives a bandgap around 8 eV which is too small compared to the experimental value of about 12 eV, but we cannot expect to get a very good description bandgap due to neglect of certain many-body effects in LDA theory. Keeping this in mind we can however compare the different cluster models and investigate effects of small changes i.e. the removal of a single neutral F atom etc.

Desorption of a single neutral F atom from the uppermost layer leaves an excess electron on the surface. Our calculations indicate that this electron occupies a surface state of a_1 symmetry with a binding energy accessible for laser light in the visible and UV. The picture is not essentially altered when three symmetrically equivalent F atoms are removed from the surface. States of this type are also found in a recent EELS study by Saiki et al. [8]. They see a gradually increasing intensity from intrinsic surface states as they increase the electron dose. The experimentally found energy loss is 1.8 eV, which they interpret as arising from an electronic transition from an F type vacancy state to the conduction band edge. As can be seen in figure 2 our ground state results are close to this value, 2.0 eV for a single F deficient surface and 1.4 - 1.9 eV if three neutral F atoms are removed. Increasing the F-deficient spot further will probably lead to a gradually increasing metallic behaviour at the surface. The occupied levels for the $Ca^{++}{}_1Ca^{++}{}_6F^-{}_3$ cluster, due to the excess electrons, would then for larger clusters develop into the conduction band of a Ca-metal overlayer. Saiki et al. saw indications of this, at higher electron doses, when they detected surface plasmons.

In conclusion, the convergence of the F (2p) band among different bulk and surface clusters (figure 2) and the existence of surface states in the band gap for non-stoichiometric surfaces, as supported by EELS and photoionization data, show that the computational procedure presented is realistic.

Acknowledgements

This work was supported by the Swedish Natural Science Research Council and the Deutsche Forschungs-gemeinshaft, Sfb 6.

References

[1] R.H. Stulen, M.L.Knotek, *Desorption Induced by Electronic Transitions, DIET III* , Springer Series in Surface Science 13, Springer-Verlag, Berlin 1988 and references therein to earlier DIET workshops

[2] G.A. Somorjai,*Chemistry in two dimensions*, Cornell U.P., 1981

[3] F.J. Himpsel, U.O.Karlsson, F.R.McFeely, J.F. Morar, D. Rieger, A Taleb-Ibrahimi and J.A. Yarmoff, Mat. Sci. Eng., **B1** (1988)9 and references therein

[4] Y.R. Shen, *Principles of Non-linear Optics*, Wiley, New York 1984
[5] J. Reif, P. Tepper, E. Matthias, E. Westin and A. Rosén, Appl. Phys. **B46** (1988)131
[6] T.F. Heinz, F.J. Himpsel, E. Palange and E.Burstein, Phys. Rev.lett. **63** (1989) 644
[7] A. Rosén, E. Westin, E. Matthias, H.B. Nielsen , J. Reif, Physica Scripta **T23**, (1988)184
[8] K. Saiki, Y. Sato, Y. Ando and A. Koma, Surf. Sci. **192** (1987)1
[9] Xia Shangda, Gou Changxin, Lin Libin, D.E. Ellis, Phys. Rev. **B35** (1987) 7671
[10] D.E. Parry, Surf.Sci.**49** (1975)433; **54** (1976)195
[11] E. Westin, to be published.
[12] M. Abramowitz, I.A. Stegun, *Handbook of Mathematical Functions*, Dover, New York 1965
[13] J.P. Dahl, J.Avery, *Local density Approximations in Quantum Chemistry and Solid State Physics*, Plenum Press, New York 1984
[14] B. Delley, D.E. Ellis, J.Chem. Phys. **76** (1982)1949
[15] D.E. Ellis, G.A. Benesh, E. Byrom, Phys. Rev. **B20** (1979)1198

ESD of Multiply Charged Ions from SiO_2: A Search for Mechanisms

R.A. Baragiola[1], *T.E. Madey*[1], *and A.-M. Lanzillotto*[2]

[1]Department of Physics and Astronomy and Laboratory for Surface Modification, Rutgers University, Piscataway, NJ 08855, USA

[2]David Sarnoff Research Center, Princeton, NJ 08543, USA

Abstract. We report ESD of singly- and multiply-charged substrate species (Si^+, Si^{2+}, Si^{3+}, O^+, O^{2+}, and SiO^+) and adsorbed species (H^+ and F^+) from thermally grown SiO_2 and from quartz. We have measured ion yields *Y(E)* and energy distribution curves (EDCs) for electron energies E between 100 and 5000 eV. *Y(E)* curves have onsets related to the Si*2p* and O*1s* thresholds, but do not follow core ionization cross sections. EDCs of Si ions are a few eV wide, while those of O^+ and H^+ have tails extending to >25 eV. The results suggest ESD is initiated by Auger decay from a core hole in the presence of additional electronic excitations.

1. Introduction

Energetic electrons decompose most oxides by mechanisms similar to those of ESD of adsorbed layers [1,2]. Unlike the case of metals, ESD of substrate cations from insulators is observed [3], because of relatively low neutralization rates between the tightly bound valence electrons and the ion escaping the surface. Recently, Lanzillotto and Magee [4] observed ESD of multiply charged substrate ions (MCI) from BF_2-implanted SiO_2, which is surprising since many channels for neutralizing highly excited ions are expected to occur inside the solid. In fact, doubly charged ions were previously observed only from adsorbates [5,6], and following deep core ionization. In this work we verify the ESD of MCI from SiO_2 (Si^{2+}, Si^{3+}, O^{2+}), measure energy thresholds and kinetic energy distributions, and propose emission mechanisms.

2. Experimental

Electrons are incident at 70° with respect to the surface normal of SiO_2 films thermally grown on Si. Ions emitted normal to the surface are analyzed in energy and mass with a SIMS system equipped with a high sensitivity quadrupole mass spectrometer (QMS) [7], operating in the 10^{-11} Torr range. The sample is kept at +20 V so ions receive an extra energy of q.20 eV before analysis, where q is the charge of the ion. We measure energy distributions curves (EDCs) for oxides thinner than 65 Å, where charging effects are very small and cause little uncertainty in the origin of the EDC. The instrumental broadening of the EDCs is ~$0.04(E_i + 20\ \text{eV}.q)$, where E_i is the ion energy. The EDCs

Springer Series in Surface Sciences, Vol. 19
Desorption Induced by Electronic Transitions, DIET IV
Editors: G. Betz · P. Varga

are not corrected for the energy-dependent transmission function of the energy analyzer, or the QMS.

The electron beam current is measured on a conducting surface biased to suppress secondary electrons. We measure ion count rates at the peak of the EDCs, and normalize them by the incident electron current to obtain ion yields *Y*. The data are not corrected for the different efficiency of the QMS for different ions.

3. Results

The typical mass spectrum in Fig. 1 shows Si^+, Si^{2+}, Si^{3+}, SiO^+, O^+, O^{2+}, O_2^+ and the ubiquitous H^+ and F^+ impurities. The identification of Si^{q+} is aided by the presence of its isotopes in the correct ratios. We made several checks to exclude experimental artifacts in the measured MCI signals. MCI are observed from air-grown oxides and also from crystalline quartz but not from the stainless steel sample holder or a phosphor screen. When the electron beam is deflected by a small magnet, ion signals drop several orders of magnitude disappearing into the background, ruling out desorption induced by impact of negative ions which could conceivably be emitted by the filament in the electron gun. Fig. 2 shows that ion signals are proportional to electron current proving that they do not arise from gas-phase ionization of previously desorbed atoms. The probability of this process should be proportional to the square of the electron current. The observed linearity shows that ions are emitted by the impact of a single electron.

To observe the very low yields of substrate ions we need μA electron currents over the ion collection area of a few mm^2. At such current densities, detectable damage to the oxide occurs in a few seconds. This damage, consisting essentially of O vacancies and free Si [8,9], causes the ESD signals to drop monotonically with bombarding time. All signals drop at about the same rate, except for Si^{2+}, which decreases more slowly with time. Representative *Y(E)* curves for Si^+ and Si^{2+} are shown in Fig. 3; the yields for Si^{3+} and O^+ show the same structure but grow faster with electron energy. The positions

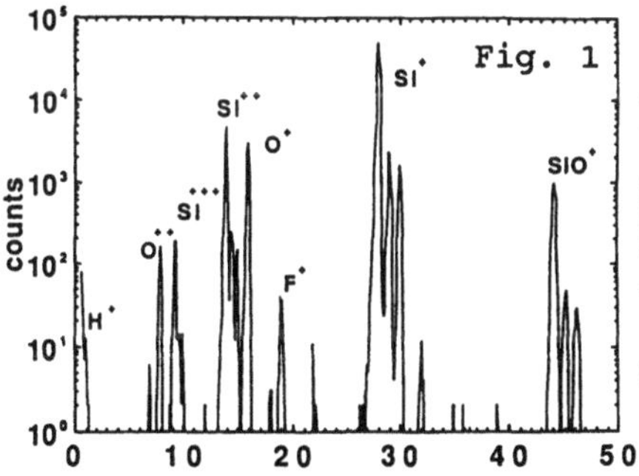

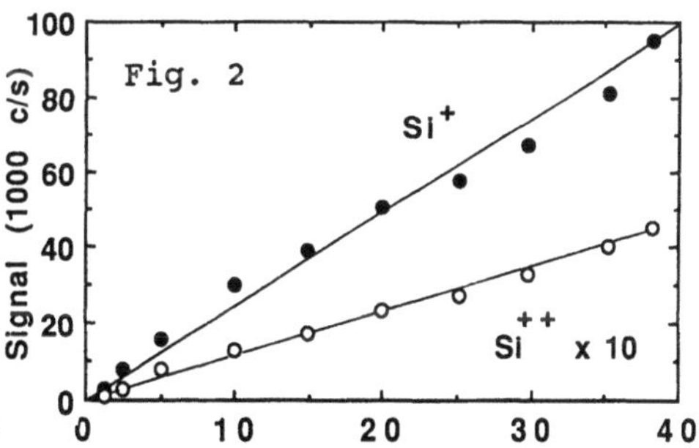

Fig. 1. ESD ion mass spectra from a 5000 Å SiO_2 film bombarded with 2 keV electrons, at a pass energy corresponding to the peak of the Si^+ EDC.

Fig. 2. ESD ion signals versus electron beam current.

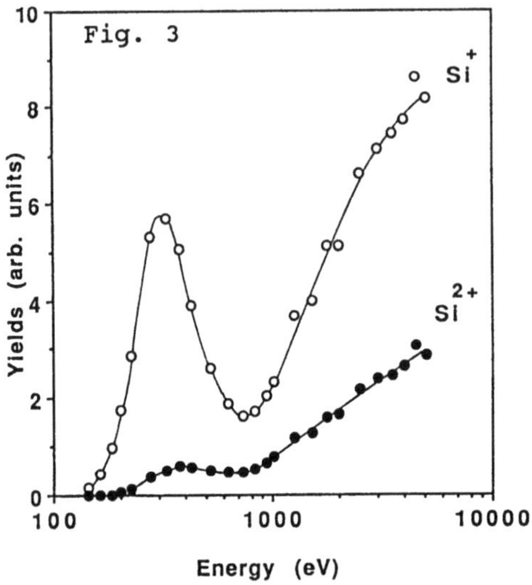

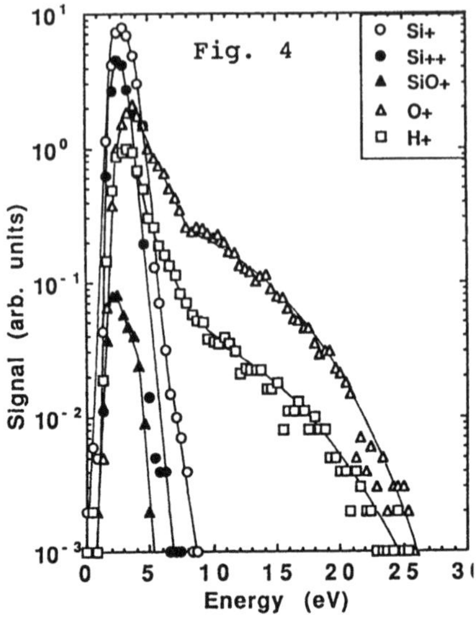

Fig. 3. ESD ion yields from a 65 Å SiO_2 film versus electron beam energy.

Fig. 4. Ion energy distribution curves for a 65 Å SiO_2 film bombarded with 2 keV electrons.

of the thresholds and the main features in *Y(E)* and the EDCs do not depend on electron dose or film thickness ranging from 17 Å to 5000 Å. No yield measurements could be made on the thicker films below 300 eV, due to sample charging.

Fig. 4 shows that EDCs are narrow (i.e. a few eV FWHM), as typical of ESD. The log plot exposes a high energy tail for H^+ and O^+, which extends to tens of eV. These tails become longer when changing the electron energy from 0.5 keV to 1 keV and remain approximately constant at higher energies.

4. Discussion

The structures in the *Y(E)* curves near the Si*2p* and O*1s* edges suggest that Auger decay following core excitation is the origin of ESD. In current models of Auger stimulated desorption [1,10], the Auger decay produces a 2-hole (2h) repulsive final state in a bonding orbital which can lead to desorption. For the specific case of SiO_2 bombarded with 1 keV electrons, Ramaker et al. [10] calculate that the probability of desorption should be more than two orders of magnitude larger from initial Si*2p* than O*1s* excitations. The reason is that, compared to Si*2p*, O*1s* has a much smaller core excitation cross section and a smaller probability that its Auger decay results in 2h in a bonding orbital. The shapes of *Y(E)* are conspicuously different from the energy dependence of core excitation cross sections, as would be predicted by theory. Since theoretical predictions agree with recent results of the energy dependence of electron induced damage [9], the difference with our results must be due to the ion desorption mechanism not included in the model.

The lack of proportionality between ESD yields and core excitation cross sections clearly shows that an initial core hole is not sufficient to produce significant ESD in our studies. We propose that to desorb ions, the incident electron needs to create additional holes in shallow levels. The multiple holes created by the decay of these double excitations should favor desorption by two effects: a) by increasing the lifetime of the repulsive final state [11] to make it comparable with ejection times, and b) by limiting the inhibiting effect on desorption caused by long range lattice relaxation [2].

Initial 2-electron excitations have been considered previously to be important only for covalently bonded adsorbates; they cause a "delayed" onset of yields near core thresholds in DIET from adsorbates [12]. Our *Y(E)* curves show that this effect also occurs in the partially ionic bonds in bulk silica; the first onset occurs at ~140 eV, delayed from the Si*2p* threshold at 108 eV, and *Y(E)* increases more slowly than reported Auger yields [13]. The second onset occurs several tens of eV delayed with respect to the O*1s* threshold at 535 eV. Excitation of bonding O*2s* electrons is possibly responsible for the delayed onsets. The requirement of additional excitations can explain the finding of the lower efficiency of Si*2p* excitations for ESD, compared to O*1s*. Most short-lived (10^{-15}-10^{-14} s) excitations in shallow levels will have extinguished before the decay of the Si*2p* hole due to its long lifetime ($\tau = 10^{-13}$ s [14]), and not be effective in desorption. This is not the case for the fast decaying O*1s* core holes ($\tau = 4.5 \times 10^{-15}$ s).

The high kinetic energies of H^+ and O^+ and the increase in the energies of Si^{q+} with ion charge also support Coulombic repulsion in the final state. To obtain repulsion energies of 25 eV, these ions need to be repelled from a Si^{3+} ion, with little energy lost to the lattice. The high energy O^+ are probably from adsorbed species where the energy loss to the lattice accompanying the motions of substrate ions [2] should be small.

Regarding the origin of the observed ions, we note that three or four final holes following the Auger decay of Si*2p*-O*2s* or O*1s*-O*2s* excitations can readily produce the degree of ionization and the bond breaking required to explain our observations. Ions can capture electrons during ejection; this process is unlikely for Si^+ since its hole state lies opposite the unoccupied band gap of the solid.

In summary, these first observations on the energy dependence of yields and EDCs for ESD from SiO_2 show that yields are related to core excitations but do not follow the corresponding cross sections. The phenomenon is initiated by a core vacancy in a Si or O atom, accompanied by additional shallow electronic excitations. These excitations produce a higher ion yield if the Auger decay occurs in times of the order of or shorter than lifetimes of shallow holes. O^+ and H^+ have energies extending to a maximum which increases with primary electron energy.

Work partially supported by the New Jersey Commission for Science and Technology, through the Advanced Technology Center for Surface Engineered Materials.

References

[1]M. L. Knotek, Rep. Prog. Phys. **47**, 1499 (1984); T. E. Madey, D. E. Ramaker, and R. L. Stockbauer, Ann. Rev. Phys. Chem. **35**, 215 (1984).

[2]P. Avouris and R. E. Walkup, Ann. Rev. Phys. Chem. **40**, 173 (1989).

[3]Y. X. Wang, F. Ohuchi and P. H. Holloway, J. Vac. Sci. Technol. A2, 732 (1984); D. S. McPhail, M. G. Dowsett, and E. H. C. Parker, J. Appl. Phys. **60**, 2573 (1986)

[4]A-M. Lanzillotto and C. W. Magee, J. Vac. Sci. Technol. (in pres).

[5]P. Feulner, R. Treichler, and D. Menzel, Phys. Rev. **B24**, 7427 (1981).

[6]P. Williams and G. Guillen, Surf. Sci. **180**, L109 (1987).

[7]C. W. Magee, W. L. Harrington and R. E. Honig, Rev. Sci. Instr. 49, 477 (78).

[8]H. Poppa, and A. G. Elliot, Surf.Sci.24, 149 (1971); S. Thomas, J. Appl. Phys. 45, 161 (1974); J. S. Johannessen, W. E. Spicer, and Y. E. Strausser, J. Appl. Phys. 47, 3028 (1976).

[9]L. Calliari, M. Dapor, L. Gonzo, and F. Marchetti, these proceedings.

[10]D. E. Ramaker, C. T. White, and J. S. Murday, Phys. Lett. **89A**, 211 (1982).

[11]P. J. Feibelman, Surf. Sci. **102**, L51 (1981)

[12]R. Jaeger, J. Stöhr, R. Treichler and K. Baberschke, Phys. Rev. Let. **47**, 1300 (1981).

[13]J. H. Neave, C. T. Foxon, and B. A. Joyce, Surf. Sci. **29**, 411 (1979); J. E. Rowe and S. B. Christman, Solid. State Commun. **13**, 315 (1973); E. J. McGuire, Phys. Rev. **16**, 73 (1977).

[14]D. L. Walters and C. P. Bhalla, Phys. Rev. **A4**, 2164 (1971).

Positive Ion Desorption from MgO by Low Energy Electron Irradiation

T. Gotoh, S. Takagi, and G. Tominaga

Department of Physics, Toho University, Funabashi-shi 274, Japan

Abstract. Observations of low energy (up to 1.5 keV) electron stimulated desorption and light emission have been performed on the MgO (001) surface. Emitted species have been detected by means of a quadrupole mass spectrometer (QMS) and by time-of-flight (TOF) methods for the desorbed ions, and a photomultiplier tube through a monochromator for light emission (wavelength about 400 nm). Onset energies of the light emission and the desorption of O^+ ions with respect to primary electron beam energy have been observed at 80 eV and these onset energies correspond to the binding energy of 2p electrons in Mg^{2+}. The energy distribution of the desorbed H^+ ions has also been observed as a function of the primary electron energy.

1. Introduction

Photon and electron stimulated ion desorptions from LiF have been explained by the Knotek-Feibelman mechanism [1,2]. In the LiF system, F^+ is created by an interatomic Auger transition, i.e. one electron undergoes a transition from F^- to the 1s hole of Li^+ in LiF and another electron is emitted from the flourine.

It is of great interest whether or not a similar process with two-electron emission occurs in an alkaline oxide for O^+ desorption. It is more probable that one electron is emitted by Auger decay than that two electrons are emitted. A neutral oxygen atom is created by one-electron emission decay of the Auger transition and the oxygen atom can move toward O^{2-}. Then the F^{2+}-center may remain in the void of the oxygen atom. It is known that deexcitation of the F^{+*}-center created by electron trapping of the F^{2+}-center emits 400 nm light [3]. We have observed the luminescence of 400 nm light from the MgO by low energy electron irradiation. The creation of the anion vacancy results in excess oxygen in the lattice. However, two-electron emission by the Auger decay should have a small probability. If this process occurs in the vicinity of the surface the desorption of the positive oxygen ions should be observed.

2. Experimental Apparatus

Irradiation of MgO (001) with low energy electrons has been performed in two different UHV chambers (base pressures $p \leqq 4 \times 10^{-7}$ Pa). One of them has a continuously emitting electron gun with a light emission detector and a quadrupole mass spectrometer (QMS). The other one has an ion flight tube and a pulsed electron gun.

Desorption Induced by Electronic Transitions, DIET IV
Editors: G. Betz · P. Varga

2.1 Light Emission Detector

The first of the UHV chambers is shown in Fig.1. It has an electron gun (EG 1), a sample holder (SH 1), and a photomultiplier tube (PMT) connected through a Kovar glass window (KW), monochromator (PSM) and rotary shutter (RS). Signals of the PMT were measured by lock-in detection. The MgO (001) surface is prepared by cleavage in air and mounted on the SH 1. The SH 1 is rotatable so that the angle of electron incidence and the detection angle of the emitted species (light and ions) were 45° (see inset of Fig.1).

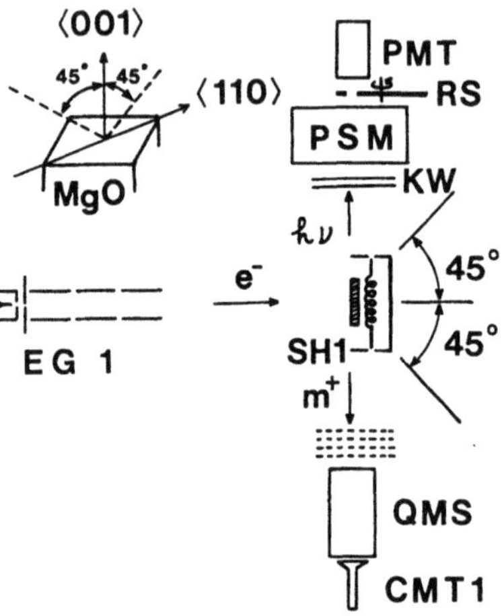

Fig.1 Schematic diagram of the light emission detector and the quadrupole mass spectrometer (QMS). Dotted lines in the inset indicate the directions of the electron incidence, and light and ion detection. Electron gun (EG 1), Rotary shutter (RS), Specimen holder (SH 1), Kovar glass window (KW), Monochromator (PSM), Photomultiplier (PMT), Quadrupole mass spectrometer (QMS), Channeltron-type detector (CMT 2)

2.2 Quadrupole Mass Spectrometer

Observation of positive ion desorption from MgO (001) with the QMS has been performed in the same chamber as used for the observation of the light emission. The channeltron-type detector (CMT 1) is mounted on the end of the quadrupole to count the ions. The ionization cage is detached from the QMS and a system of the four grids is mounted at the entrance of the QMS to analyze the kinetic energy of the desorbed ions. Scanning of the QMS is controlled by a programmable power source (PPS). The period taken to scan from m=1 to m=45 is 100 s. Signals of the CMT 1 are accumulated for 50 mass scans in the multichannel analyzer (MCA).

2.3 Time of Flight

Measurements of the time of flight (TOF) of the desorbed ions have been performed in the second chamber to determine the kinetic energies of the ions precisely. The electron gun (EG 2) in Fig.2 emits an electron beam pulse (pulse width ≦100 ns) every 20 μs. The directions of the electron incidence and ion detection are the same as described in the previous section. The starting time of the ion flight is

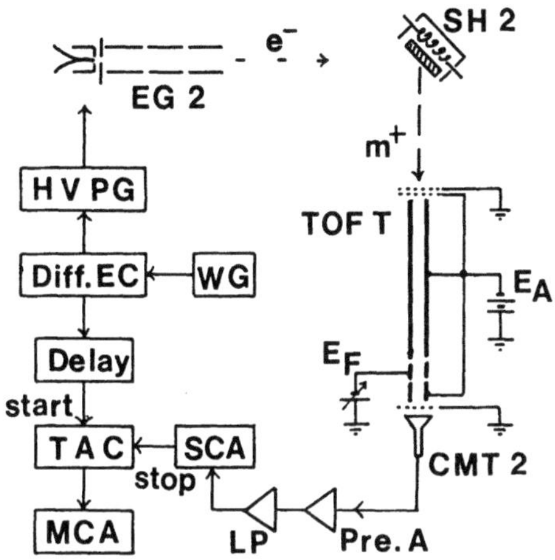

Fig.2 Apparatus for time-of-flight (TOF) measurements. Pulsed electron gun (EG 2), Specimen holder (SH 2), TOF drift tube (TOFT), Ion acceleration voltage (E_A), Ion focusing voltage (E_F), Canneltron-type detector (CMT 2), Square wave generator (WG), Electronic differential circuit (Diff.EC), High voltage pulse generator (HVPG), Electronic delay circuit (Delay), time-to-amplitude converter (TAC), Pre-amplifier (Pre.A), Linear amplifier (LA), Single channel analyzer (SCA), Multichannel analyzer (MCA)

determined by the arrival time of soft x-rays at the detector (CMT 2). The peak position of the x-rays is shifted to about the 1000th channel of the accumulated spectra in the multichannel analyzer by an electronic delay circuit (Delay) for convenience of the measurements.

3. Result and Discussion

The intensity of the emitted light with wavelength about 400 ± 3 nm is shown in Fig.3 as a function of the incident electron energy, normalized to the incident electron intensity. The curve contains two components of different onset energies, at 80 and 210 eV (indicated by arrows in the figure). The onset energy of the desorption of the O^+ is also observed at 80 eV and explained by Auger decay of the 2p-hole of Mg^{2+} with two-electron ejection [4]. The emission mechanism of the 400 nm light is explained as deexcitation of an excited state of the F^+-center generated by electron trapping of the F^{2+}-center. Hole creation of the 2p state in Mg^{2+} followed by the interatomic Auger decay of Mg^{3+} creates the F^{2+}-center and neutralizes O^{2-}. The neutral oxygen atom makes an H-center with O^{2-}. The binding energy of a 2p electron in Mg^{2+} is 80.1 eV and in Mg^{3+} is 109.3 eV [5]. Since Mg^{3+} is the recognized initial state of the Auger decay, the lifetime of the Mg^{3+} should be very short in the MgO matrix. The light with the onset energy 210 eV may be emitted by the same process but the initial state is Mg^{4+}, which may create two F^{2+}-centers.
Typical mass spectra of the desorbed ions are shown in Fig.4. Temperatures of the MgO are indicated in the figure and the incident electron energies are fixed at 400 eV. Peaks of O^+, H^+ and H^{2+} are prominent peaks and other small peaks existing at mass numbers 12, 18 and 28 originated from residual gases.

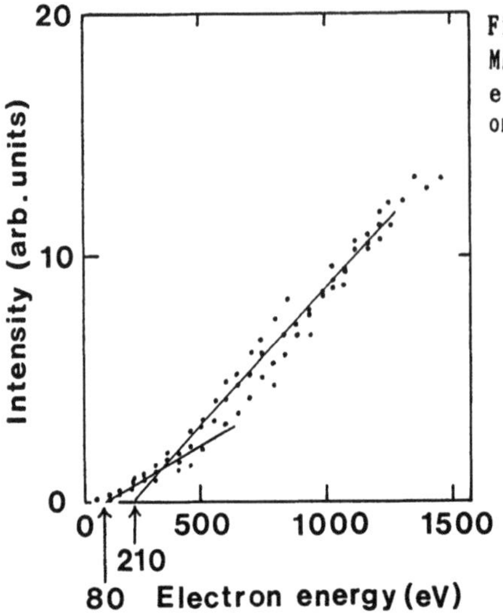

Fig.3 Intensity of 400 nm light emitted from MgO(001) as a function of the incident electron energy. 80 and 210 indicate the onset energies

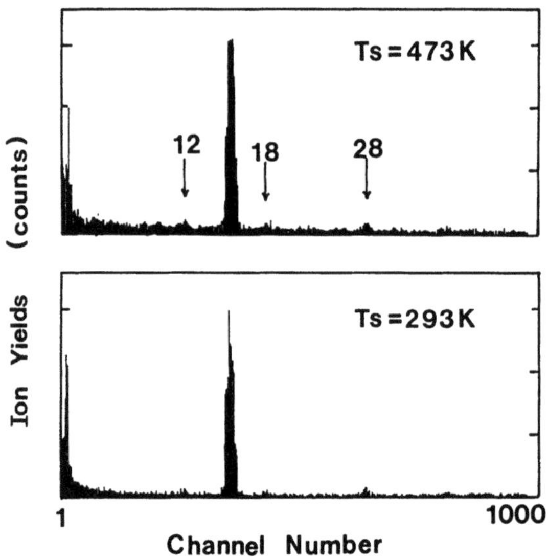

Fig.4 Desorbed ion mass spectra from MgO(001) at temperatures of 293 K and 473 K induced by 400 eV electron irradiation. 12, 18 and 28 indicate the masses of ions

The TOF results as a function of the residual gas pressure are shown in Fig.5. The upper one was obtained at 1×10^{-6} Pa and the lower one at 5×10^{-6} Pa. The two peaks of the oxygen ions with different emitting energies from the MgO surface are present in both spectra. The faster one (a in the figure) originates from the MgO matrix and the slower one (b) from the adsorbed species [4]. The excess H^{2+} peak exists for the higher residual gas pressure.

TOF spectra are shown in Fig.6 as a function of incident electron energy. The intensity of the faster O^+ (a) does not change with the electron energy, but the

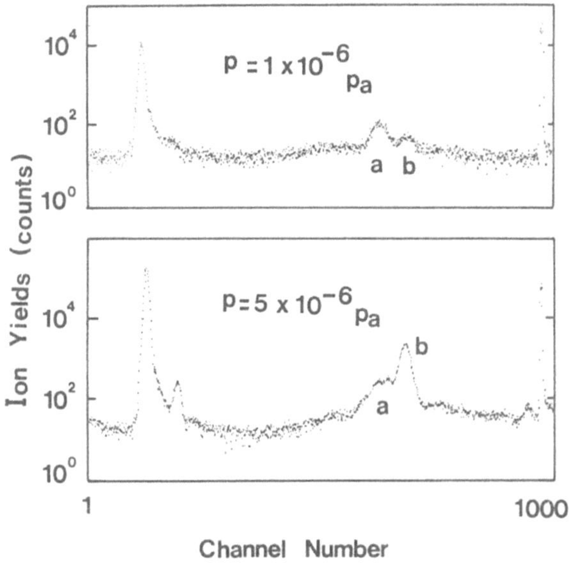

Fig.5 Time-of-flight spectra of the desorbed ions at different residual gas pressures. The irradiation electron energy is 200 eV. The faster and slower oxygen ions are indicated by a and b

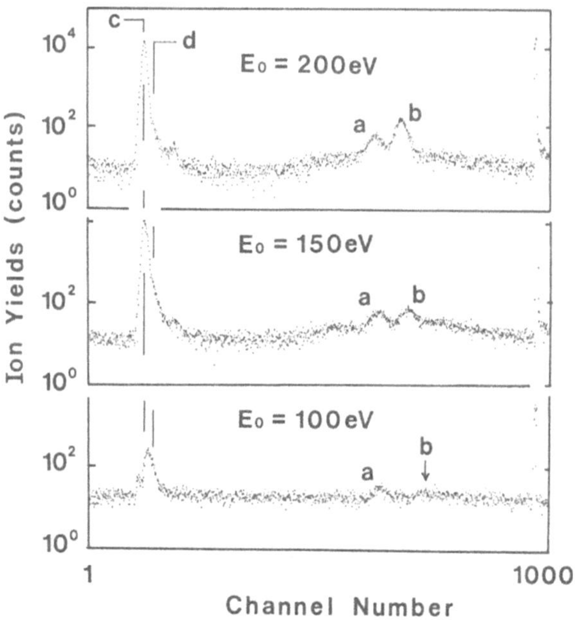

Fig.6 TOF spectra of desorbed ions at various irradiation electron energies. Two oxygen peaks are indicated by a and b; c and d indicate peaks of hydrogen ions at 4 and 6 eV respectively

slower O^+ (b) increases and also the emitting energy decreases with the electron energy. There are two peaks of H^+ ions at different emitting energies, 4 and 6 eV. The slower one (d) does not change in intensity with electron energy. The faster one (c) increases with the electron energy and has no onset or threshold energy with respect to the incident electron energy in this region. The origin of the hydrogen ions is probably solved hydrogen molecules in the MgO matrix, magnesium hydroxide or adsorbed gas molecules. It is difficult to discuss the ionization mechanism of the hydrogen from the experimental results.

References

[1] T. Yasue, T. Gotoh, A. Ichimiya, Y. Kawaguchi, M. Kotani, S. Ohtani, Y. Shigeta S. Takagi, Y. Tazawa and G. Tominaga, Jpn. J. Appl. Phys. 25 (1986) L363.

[2] T. Yasue, A. Ichimiya, Y. Yamada, T. Gotoh,Y. Kawaguchi, M. Kotani, S. Ohtani, Y.Shigeta, S.Takagi, Y. Tazawa and G. Tominaga, Nucl. Instr. and Meth. B34 (1988) 357.

[3] Y. Chen, J.L. Kolopus and W.A. Sibley, Phys. Rev. 186 (1969) 865.

[4] T. Gotoh, S. Takagi and G. Tominaga, Proc. Intern. Conf. Solid Surface, Cologne (1989)

[5] W. Lotz, J. Opt. Soc. Am., 58 (1968) 915.

Part VII

Gas Phase Processes

Single Photon Multiple Photoionization of Gaseous CO, C_6F_6, and trans-1,2-C_2H_4BrI

J. Delwiche[1,*], *M.-J. Hubin-Franskin*[1,*], *K. Ibrahim*[1,**], *B. Kempgens*[1], *P. Lablanquie*[2], *I. Nenner*[2,***], *J.-M. Robbe*[3], *and G. Gandara*[3]

[1]Université de Liège, Institut de Chimie, B6, Sart Tilman, B-4000 Liège 1, Belgium

[2]LURE, Bât. 209d, F-91405 Orsay, France

[3]Université de Lille-Flandres-Artois, Institut de Physique P5, F-59650 Villeneuve d'Ascq, France

* "Maître de Recherches du FNRS" of Belgium

** On leave from: Chemistry Department, Technical College of Kashgar, People's Republic of China

*** Also at: CEA, CEN de Saclay, F-91191 Gif-sur-Yvette, France

We report new results obtained by photon impact mass and photoion-photoion coincidence (PIPICO) spectrometry on CO, C_6F_6, and *trans*-1,2-C_2H_4BrI. For CO we have investigated the mechanisms of production of metastable CO^{2+} ions and of the $C^+ + O^+$ ion pair and identified the lowest energy electronic states of CO^{2+} with the help of *ab-initio* calculations. The mass and PIPICO spectra of *trans*-1,2-C_2H_4BrI show some site selectivity in the bond breaking pattern at wavelengths corresponding to iodine 4d $\rightarrow \sigma^*$ atomic resonances. The results on C_6F_6 show that bond breaking is strongly dependent on the photon energy and reveals the existence of long lived $C_6F_6^{2+}$ ions.

1. Introduction

As already pointed out in the preceding DIET Workshop [1], some new insight on the sequential and competing processes that lead to desorption induced by electronic transitions may be gained through comparison with results obtained on gas phase molecules. The fundamental problem to be solved is the knowledge of the mechanisms by which a primary electronic excitation can be transfered to nuclear motion, inducing bond breakings either between the adsorbed molecule and the surface or within the adsorbed molecule itself.

Thanks to the increasing availability of synchrotron radiation facilities and the development of sophisticated coincidence techniques, the decay pathways of electronically excited species in the gas phase is now understood to some extent, at least for primary excitations near the core edges.

Springer Series in Surface Sciences, Vol. 19
Desorption Induced by Electronic Transitions, DIET IV
Editors: G. Betz · P. Varga

The situation is much more complex for valence shell excitation due to a great congestion of electronic states.

We report here some new results on the single photon double photoionization of CO and C_6F_6 in the energy region from threshold up to 120 eV. In addition, the selectivity in bond breakage after excitation of core shell electrons has been studied in the case of *trans*-C_2H_4BrI.

2. Experimental

The experimental setup has been described in detail previously [2,3]. Very briefly, the synchrotron radiation from ACO and SuperACO Orsay's storage rings is wavelength-selected by means of a toroidal holographic monochromator, focused with a toroidal mirror at the center of an interaction chamber where it crosses an effusive jet of gas. The ions created are detected either with a linear time-of-flight or a reflectron type mass spectrometer [4]. Both spectrometers are used to get conventional mass spectra or coincidences between pairs of ionic fragments arising from the fragmentation of doubly charged parent ions.

3. Results and Discussion

3.1 Carbon monoxide

CO is a model molecule for case studies of molecular absorption. In the gas phase, we have measured the cross sections for the production of metastable CO^{2+} ions as well as for the ion pair $C^+ + O^+$, from their respective thresholds up to about 150 eV. We also performed *ab initio* calculations to obtain the potential-energy curves of CO^{2+}. Partial results are shown on Fig. 1. The ionization energy for the formation of metastable CO^{2+} ions was determined to be 40.75±0.5 eV and is assigned to the first $^1\Pi$ and $^1\Sigma^+$ states. As shown by the calculated potential-energy curves, these latter are lying too close together in the Franck-Condon zone to be experimentally distinguishable. Most interesting is the fact that the $C^+ + O^+$ ion pair appears at a lower energy than the lowest $^3\Pi$ state of CO^{2+}. We must therefore consider a two-step mechanism involving the autoionization of highly excited $(CO^*)^+$ ions, probably some of the numerous satellite states observed by photoelectron spectroscopy [5-8].

We wish to stress specially the comparison between the cross sections for the production of CO^{2+} and $C^+ + O^+$ with the photon stimulated ion desorption spectra (PSID) of surfaces which, in the valence region, reveal strong energy dependence of the ionic (as well as neutral) species [1]. In Fig. 2, we

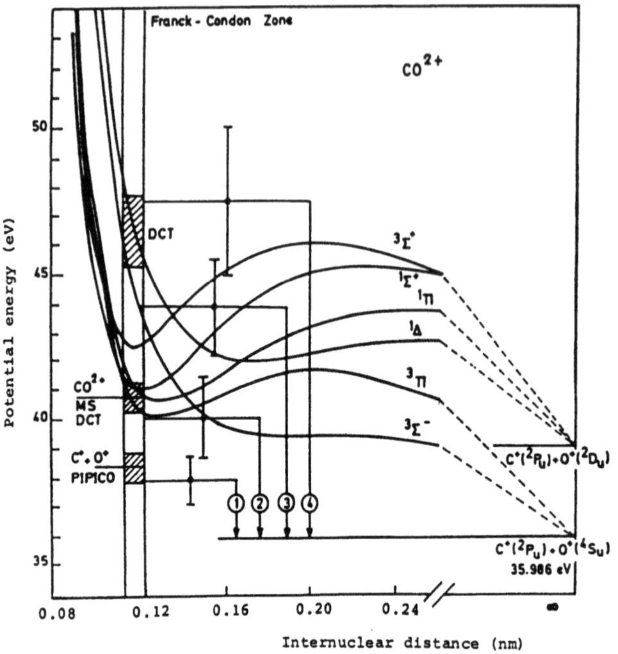

Figure 1: Calculated potential-energy curves for the six lowest energy electronic states. Vertical arrows 1 to 4 position the four lowest-energy KERD's with respect to the first ionization limit. Error bars figure the FWHM of the respective KERD's.

report our gas phase results together with the photodesorption curves of CO molecules absorbed on Ru(001) as measured by Auer *et al* [9], spectrum (7).

3.2 trans-1,2-C_2H_4BrI

The excitation of a core electron into a low lying valence unoccupied orbital has been shown to produce selective ionic fragmentation patterns in CH_3Br [10]. We have tried to investigate how far from the absorbing atom the breaking selectivity goes by recording mass spectra and PIPICO spectra of *trans*-1,2-C_2H_4BrI at selected photon energies corresponding to the 4d $\rightarrow \sigma^*$ or 3d $\rightarrow \sigma^*$ transitions of the iodine or the bromine atom located at 51.45/53.2 eV and 72.1 eV, respectively. At the energies of the iodine resonances, we observe in the mass spectra a decrease of the relative intensity of the I^+ ions and an increase of the relative intensity of the $C_2H_nI^+$ ions, while in the PIPICO spectra, all the ions pairs show a resonant behavior and the relative intensity of the CH_n^+ + CH_mBr^+ is enhanced. On the contrary, there seems to be no special behavior at 72 eV.

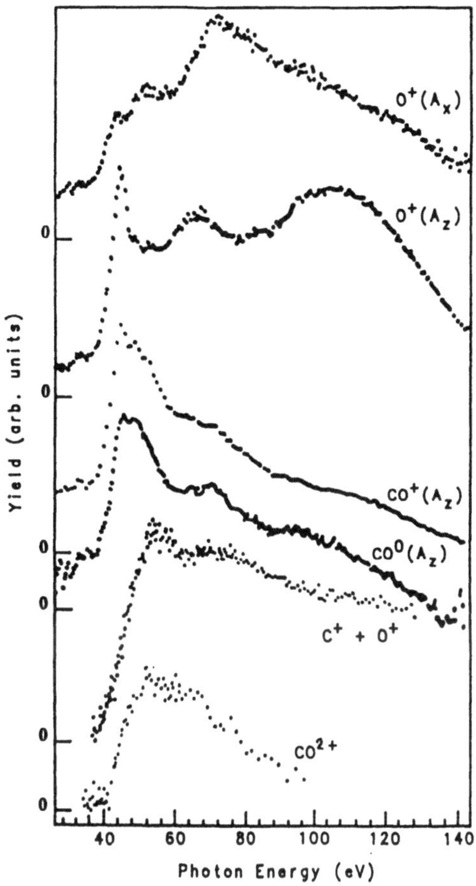

Figure 2: CO^0 from $C^{18}O$, and CO^+, O^+ from $C^{16}O$ on Ru(001) (θ=0.33) for A_z and mixed (A_x:A_z=1:0.6) polarization from Ref. 9, and CO^{2+}, C^+ + O^+ from gaseous CO (this work) vs photon energy. The yield scales are arbitrary and shifted.

3.3 Hexafluorobenzene

We wish to illustrate in the C_6F_6 molecule, taken as a model polyatomic system, the complexity of the dissociative double photoionization as a function of photon energy. In Fig. 3, we show two mass spectra and two PIPICO spectra obtained at 35 and 120 eV, respectively. At 35 eV, i.e. near the lowest-energy double ionization threshold, we observe in the mass spectrum that the molecule is efficiently dissociated into numerous singly charged fragments (Fig. 3a). From the corresponding PIPICO spectrum (Fig. 3a'), we see that only the following dissociative channel is populated through double photoionization:

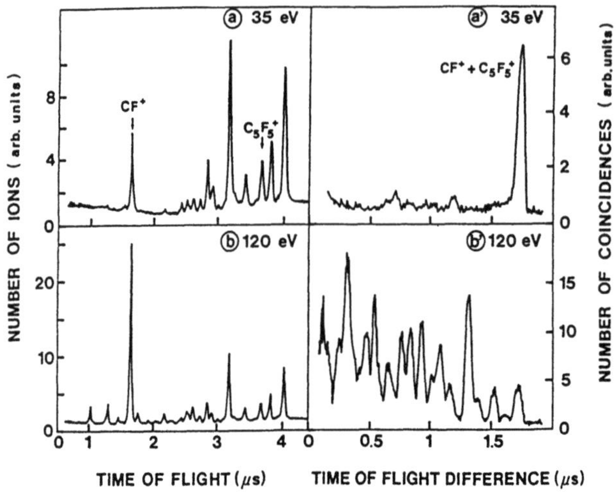

Figure 3: Mass spectra of C_6F_6 at (a) 35 eV and (b) 120 eV photon energy. PIPICO spectra of C_6F_6 at (a') 35 eV and (b') 120 eV photon energy.

$$C_6F_6 + h\nu \longrightarrow C_5F_5^+ + CF^+ + 2e^-$$

At 120 eV, we observe (Fig. 3b) a large increase of the number and intensity of the fragments, especially CF^+ . The PIPICO spectrum (Fig. 3b') reveals many coincidence peaks from dissociative double and probably triple ionization channels. So, multibond breaking is very efficient at large excess energies.

Additional dynamical information is revealed by the PIPICO spectra recorded with photons in the 32 to 35 eV energy range: an asymmetry (11) toward short differences of time-of-flight of the $C_5F_5^+$ + CF^+ ion pair peak indicates the formation of long lived $C_6F_6^{2+}$ ions that dissociate in the acceleration region of the spectrometer.

Acknowledgments

We gratefully thank the staff of the LURE and of the LINAC for their technical support. Two of us (J.D. and M.-J. H.-F.) express their gratitude to the FNRS, the FRFC and the Ministère des Finances (Loterie Nationale) for their financial support. We are also very pleased to acknowledge the support from NATO (grant N°484/87).

REFERENCES

1. D. Menzel in *Desorption Induced by Electronic Transitions*. DIET III, edited by R. H. Stulen and M. L. Knotek (Springer-Verlag, Berlin, 1988), p. 2.
2. P. Lablanquie, I. Nenner, P. Millie, P. Morin, J.H.D. Eland, M.-J. Hubin-Franskin, and J. Delwiche. J. Chem. Phys., **82**, 2951 (1985).
3. P. Lablanquie, J. Delwiche, M.-J. Hubin-Franskin, I. Nenner, P. Morin, K. Ito, J.H.D. Eland, J.-M. Robbe, G. Gandara, J. Fournier, and P.G. Fournier. Phys. Rev. A, (1989), in press.
4. M.-J. Besnard-Ramage, P. Morin, I. Nenner, M.-J. Hubin-Franskin, J. Delwiche, P. Lablanquie, and J.H.D. Eland. Rev. Sci. Instr., **60**, 2182 (1989).
5. A. W. Potts and T. A. William. J. Electr. Spectrosc. Relat. Phenom., **3**, 3 (1974).
6. J.L. Gardner and J.A.R. Samson. J. Chem. Phys., **62**, 1447 (1975).
7. S. Krummacher, V. Schmidt, F. Wuilleumier, J.-M. Bizeau, and D. Ederer. J. Phys. B, **16**, 1733 (1983).
8. D. Winkoun, G. Dujardin, L. Hellner, and M.-J. Besnard. J. Phys. B, **21**, 1385 (1988).
9. S. Auer, P. Feulner, and D. Menzel. Phys. Rev. B, **35**, 7752 (1987).
10. I. Nenner, P. Morin, P. Lablanquie, and G.G.B. de Souza in *Desorption Induced by Electronic Transitions*. DIET III, edited by R.H. Stulen and M.L. Knotek (Springer-Verlag, Berlin, 1988), p. 10.
11. K. Ibrahim, M.-J. Hubin-Franskin, J. Delwiche, P.Lablanquie, and I. Nenner. to be published.

Electron Induced Vibrational Excitation of Gaseous Polyatomic Molecules: The NH_3 Case

M.-J. Hubin-Franskin, M. Furlan, J. Delwiche*, and J.E. Collin*

Laboratoire de Spectroscopie d'Electrons diffusés, Université de Liège, Institut de Chimie, B6, Sart Tilman par, B-4000 Liège 1, Belgium
*Maître de Recherches FNRS

Excitation of the vibrations of gaseous ammonia has been studied by High Resolution Electron Energy Loss Spectroscopy in the 12-50 eV impact energy region where it proceeds mainly through a direct mechanism. At 25 and 50eV, the vibrational differential cross sections have angular behaviour strongly dependent on the specific vibration excited despite all modes being both infrared and Raman active.

1. INTRODUCTION

Electron induced vibrational excitation of gaseous molecules is not a newly developed research field. However, in the last decade much attention has been given to the mechanisms, especially the indirect ones. Only a small amount of work has been done on the direct process, centered mainly on diatomics and triatomics [1-9]. For polyatomics, the data are even more scarce and largely incomplete [10,11]. At low impact energy the angular differential cross sections have been shown to be strongly forward peaked for infrared active modes, long range interaction prevailing, whereas they are nearly isotropic for Raman active ones [5,6]. The data presently available are too scarce to allow the formulation of general rules for the cross sections, their angular and energy behaviour. Such generalisation would be helpful, especially as vibrational Electron Energy Loss Spectroscopy is widely used in surface physical chemistry. Ammonia has quite interesting physical properties: it has a permanent electric dipole moment and vibrations which are all both infrared and Raman active but two of them ν_3 and ν_4 have not been observed in Raman spectra due to the expected weaknesss of the Raman bands for such non-totally symmetric vibrations [12,13]. In addition the vibrational spectroscopy of this molecule adsorbed on various substrates has been reported [14].

The purpose of this work is to report on the high resolution vibrational spectroscopy of gaseous ammonia and the angular differential cross sections for 12 to 50 eV impact electrons, outside the resonance region. The angular behaviour is discussed according to the specific mode excited and to the nature of the

Springer Series in Surface Sciences, Vol. 19
Desorption Induced by Electronic Transitions, DIET IV
Editors: G. Betz · P. Varga

electron-molecule potentials prevailing in the vibrational excitation.

2. THE EXPERIMENT

This study has been performed using a Vacuum Generator electron energy loss spectrometer SEELS 400 modified for gases and described previously [15]. Briefly, it is made of an electron gun, an electron monochromator fitted with a three-element lens, a collision region and an electron analyser rotatable around the forward direction. Both analysers are electrostatic spherical sectors operating in the constant pass energy mode. The experimental resolution has been between 22 and 35 meV according to the type of measurements. It did not allow the rotational transitions to be resolved. The electron energy scale has been calibrated using the elastic scattering signal. The gas sample purchased from "L'Air Liquide" was of commercial grade N45.

3. THE 50 eV ELECTRON ENERGY LOSS SPECTRUM

As shown in the spectra displayed in Fig.1, the lowest energy inelastic peak located at 0.120 ± 0.003 eV is to be assigned to excitation of the ν_2 symmetric out-of-plane vibration. The feature at 0.205 ± 0.003 eV is due to one quantum of the asymmetric out-of-plane ν_4. The highest energy peak at 0.420 ± 0.003 eV includes two components not resolvable within the resolution: the ν_1 and ν_3 modes, i.e the symmetric and asymmetric stretching vibrations respectively. As in infrared spectroscopy [12,13], the low angle spectrum (Fig. 1, top spectrum) is dominated by excitation of ν_2, which has the largest variation of the electric dipole moment with internuclear distance ($d\mu/dr$) [16]. The relative intensities are slightly different from those in the infrared spectrum, very likely due to the non-zero momentum transfer. They are quite different from those for the adsorbed molecule (0.05 L on Fe(110)) [14] for which the ν_4 mode is not excited at least in the 2.4 eV spectrum recorded in specular reflected direction with a 70° incident angle.

At higher scattering angles, the relative intensities are quite different, the ν_4 mode having the largest cross section at 60° (Fig. 1, bottom spectrum).

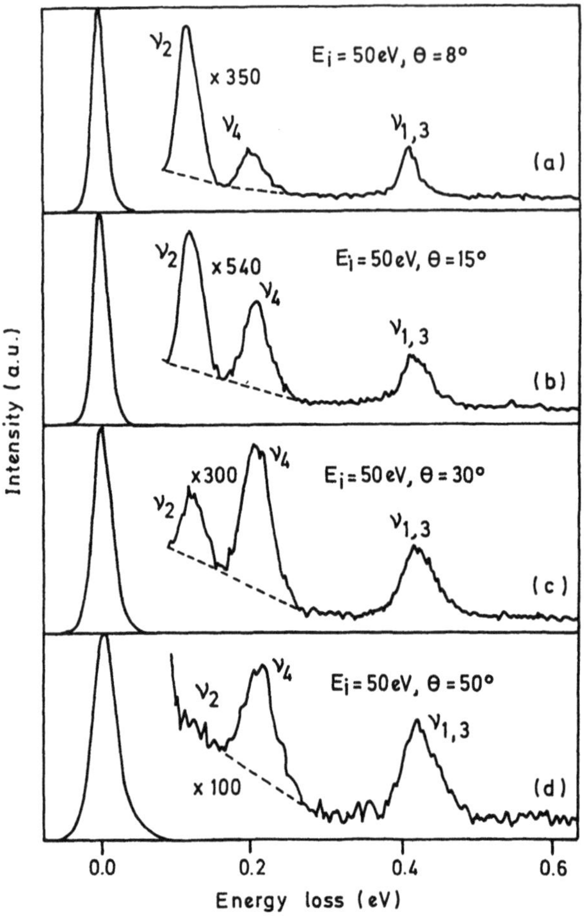

Fig. 1 : Vibrational Electron Energy Loss Spectrum of ammonia at 50 eV and 8°, 15°, 30°, 50° scattering angles.

4. THE EXCITATION MECHANISMS ACCORDING TO THE IMPACT ENERGY

At 30° the vibrational differential cross sections have been plotted with respect to the elastic scattering at various impact energies (Fig. 2). Above 12 eV the ratios vary quite smoothly, indicating that excitation proceeds mainly through the direct process in this energy region. Below this energy, the cross section ratio for $\nu_{1,3}$ and ν_4 is strongly enhanced around 7.5 eV through the shape resonance reported previously [17]; this effect is largest for the $\nu_{1,3}$ modes which are characterized by the lowest $d\mu/dr$ [16]. Excitation of the ν_2 symmetric bending mode is not perturbed by the shape resonance (Fig.2). Its cross section with respect to the elastic one is strongly peaked at low impact energies, suggesting very likely a virtual state mechanism.

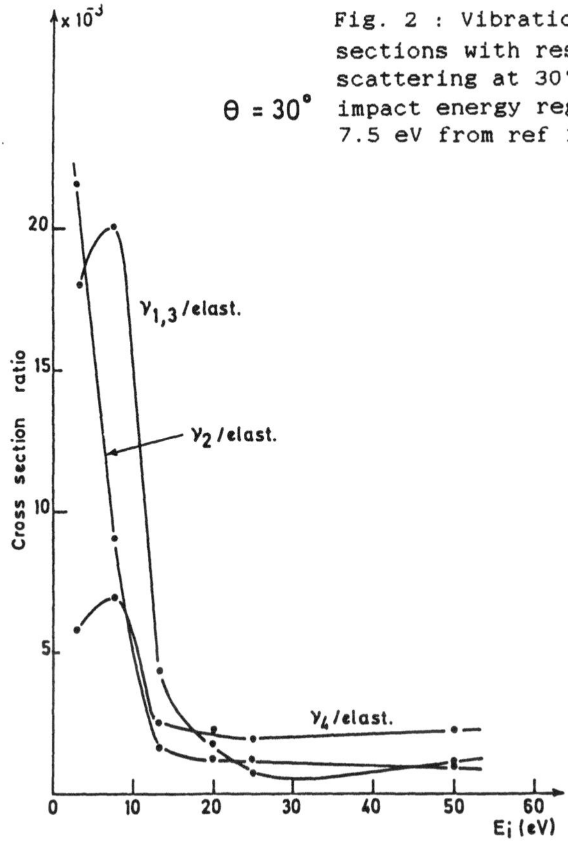

Fig. 2 : Vibrational differential cross sections with respect to the elastic scattering at 30° in the 3 to 50 eV impact energy region. Data at 3 and 7.5 eV from ref 17.

5. RELATIVE ANGULAR DIFFERENTIAL CROSS SECTIONS

The vibrational differential cross sections measured as a function of the scattering angle at 50 eV and 25 eV are displayed in Figs. 3 and 4 respectively.

5. a. The ν_2 symmetric out-of-plane vibration

At 50 eV the differential cross section is strongly peaked at low scattering angles (Fig. 3). Above 30° no data can be given as the cross section vanishes at larger angles (Fig.1). The angular behaviour is quite similar to that reported for the ν_3 asymmetric stretching mode in CO_2 and the $\nu_{1,3}$ symmetric and asymmetric stretching vibrations in H_2O [3-5, 7], all three being strongly infrared active and non-observed, except ν_1, in Raman spectroscopy. For these triatomics, the interaction has been shown to be mainly long range dipolar below 30° [6,8].

At 25 eV, the differential cross section also peaks at low angles. In addition it exhibits a minimum at 30°. This behaviour may be interpreted as due below 30° to a long range dipole interaction and at larger angles to the contribution of shorter

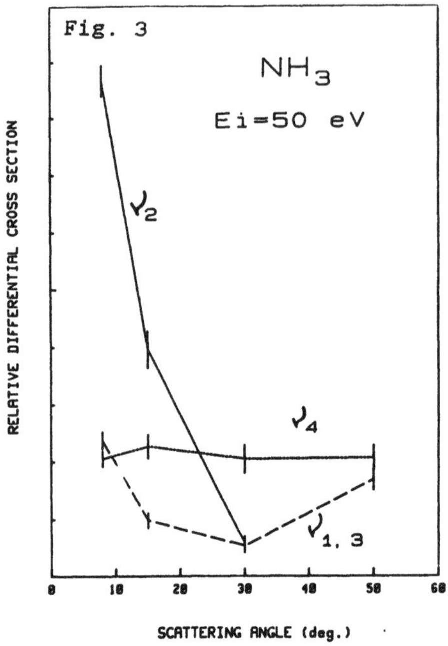

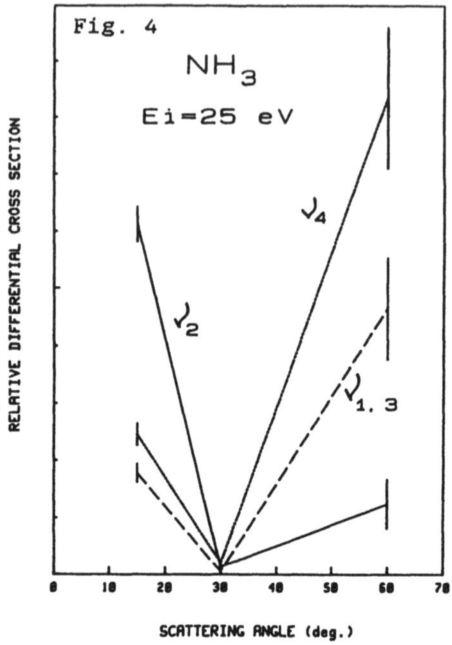

Fig. 3 : Relative vibrational differential cross sections as a function of the scattering angle at 50 eV.

Fig. 4 : Relative vibrational differential cross sections as a function of the scattering angle at 25 eV.

range potentials, probably a quadrupole one, which may be responsible for the minimum.

5. b. The ν_4 asymmetric out-of-plane vibration

The ν_4 asymmetric bending mode has a differential cross section nearly isotropic at 50 eV as shown in Fig. 3. This vibration, which is strongly infrared active with the second largest $d\mu/dr$ [16], has not been observed in Raman spectra, due to the expected weakness of Raman bands for such non-totally symmetric vibration. According to this angular behaviour of the cross section, excitation must proceed through short range interactions. Such a behaviour has been reported previously for the ν_1 symmetric stretching mode of CO_2, strongly Raman active and infrared inactive, at lower impact energy and has been interpreted in terms of polarisation interaction [5,18]; however, above 20 eV, it is no longer isotropic, but definitely forward peaking [4]. For ammonia, in addition to the long range dipole potential, quadrupole interaction might play a non-negligible role at 50 eV. Indeed, in the H_2O case the polarisation potential vanishes in this energy range contrary to the quadrupole cross section [8, 18].

At 25 eV, the differential cross section exhibits a minimum at 30° and gains intensity at large scattering angles (Fig. 4). This suggests different contributions to the electron-molecule potentials according to the impact energy.

5. c. The $\nu_{1,3}$ symmetric and asymmetric stretching vibrations

At 50 and 25 eV, the differential cross section of the non-resolvable symmetric and asymmetric stretching vibrations is only very weakly peaked at low angles and has a minimum at 30° (Fig. 3,4). These vibrations are both infrared and Raman active, the ν_3 one being expected to lead to very weak Raman bands due to its symmetry. Long range dipolar interaction is expected to play a minor role as these modes have the lowest value for the transition dipole matrix element and for $d\mu/dr$. The angular behaviour is similar to that for nitrogen and carbon monoxide for which quadrupole and polarisation effects have been invoked [1,2].

ACKNOWLEDGMENTS

The authors are indebted to J.Heinesch for his highly valuable technical assistance. This work was sponsored by the "Action de Recherche Concertée" at the Service du Premier Ministre, by the Patrimoine de l'Université de Liège and by the Fonds National de la Recherche Scientifique of Belgium.

REFERENCES

1. D.G. Truhlar, M.A. Brandt, S.K. Srivastava, S. Trajmar and A. Chutjian, J.Chem.Phys. 66, 655 (1977).
2. A. Chutjian and H. Tanaka, J.Phys. B 13, 1901 (1980).
3. Z. Pavlovic, J.W. Boness, A. Herzenberg and G.J. Schulz, Phys.Rev. A 6, 676 (1972).
4. D.F. Register, H. Nishimura and S. Trajmar, J.Phys. B 13, 1651 (1980).
5. A. Andrick, D. Danner and H. Ehrhardt, Phys.Lett. 29, 346 (1969).
6. Th. Antoni, K. Jung, H. Ehrhardt and E.S. Chang, J.Phys. B 19, 1377 (1986).
7. S. Trajmar, W. Williams and A. Kuppermann, J.Chem.Phys. 58, 2521 (1972).
8. Y. Itikawa, J.Phys.Soc.Jap. 36, 1127 (1974).
9. C.E. Kuyatt, S.R. Mielczarek and M. J. Weiss, J.Chem.Phys. 65, 3481 (1976).

10.H. Tanaka, M. Kubo, N. Onodera and A. Suzuki, J.Phys. B 16, 2861 (1983).
11.P.J. Curry, W.R. Newell and A.C.H. Smith, J.Phys. B 18, 2303 (1985).
12.D.C. McKean and P. Schatz, J.Chem.Phys. 24, 316 (1985).
13.G.H.Herzberg, "Infrared and Raman Spectra of Polyatomic Molecules and Ions", D.Van Nostrand Company Inc., New York, (1945).
14.W.Erley and H.Ibach, Surface Science 119, L357 (1982).
15.M.Furlan, M.-J.Hubin-Franskin, J.Delwiche, D.Roy and J.E.Collin, J.Chem.Phys., 82, 1797 (1985).
16.A.A. Radzig and B.M. Smirnov," Reference Data on Atoms, Molecules and Ions", Springer-Verlag, Berlin (1985).
17.M. Ben Arfa and M. Tronc, J.Chim.Phys. 85, 889 (1988).
18.Y. Itikawa, Phys.Rev. A 3, 831 (1971).
19.M. Furlan, M.-J. Hubin-Franskin, J. Delwiche and J.E. Collin, J.Chem.Phys. on press.

Part VIII

Applications and Related Processes

Useful Macroscopic Phenomena due to Laser Ablation

R. Dreyfus

IBM Research Division, T.J. Watson Research Center, P.O. Box 218, Yorktown Heights, NY 10598, USA

Abstract. The laser ablation of surfaces encompasses material removal in amounts spanning more than 10 decades. At high power levels, $> 4 \times 10^8 W/cm^2$ and primarily with inorganic materials ablation is a plasma process. The laser light is deposited in the plume of ablated material by free electrons via inverse bremsstrahlung absorption. The energy is transported to the surface by thermal conduction within the plasma. A contrasting pair of processes occurs at somewhat lower power densities, 10^6 to $\sim 5 \times 10^7 W/cm^2$; photochemical or thermal processes within the surface monolayers then control the ablation which depend on the specific chemical and morphological makeup of the surface.

A number of useful processes involve the use of lasers to ablate or deposit thin films. The deposition is facilitated by the fact that the plasma ablation process yields a reasonably well collimated beam (due to a quasi supersonic expansion) at high (> > 1eV) kinetic energy. Two features which, however, have limited the applicability of these films is that splashing produces large clusters or particulates in the beam, while the high density of the plasma inhibits sorting or deflecting the beam by electric or magnetic fields. There also exist interesting photochemical-type processes in the world of organic materials, including even the specialized areas of medicine and ophthalmology. The fact that lasers can remove material by DIET (electronic desorption) type processes means minimal thermal damage to surrounding tissue. Two examples that epitomize this include reshaping of the lens of the eye by "laser planing", and destruction of cancerous cells in a specific location by first making them sensitive to photochemical processes by addition of reactive dye molecules followed by excitation of the molecules locally with laser light. In this latter case, it is the surface (i.e. cell wall) that is attacked by activated oxygen radicals and ions.

1. Introduction

Even before the laser was completely developed, the resultant photon beam was recognized for its processing capabilities [1]. The concepts were derived directly from the basic nature of a laser, e.g. a monochromatic light source, high brightness and short pulse-width capability. The possible useful combinations of the above characteristics are sufficiently broad that surface treatments and modifications span many decades in material removal or power density, see Fig. 1. The modification of surfaces at adjustable depths is a natural consequence of the energy deposition within time, τ, ranging from cw to subpicoseconds; i.e. depths from < 100Å to ∞. Besides depth profiling, in-plane patterning is also possible by image projection. This implies the capability for "parallel processing"; not a minor consideration when 10^6 or 10^7 points must be separately addressed for many microelectronic purposes. Along with the above advantages of laser processing is the fact that vacuum is not required, particularly for the laser ablation of surfaces. Surfaces, as contrasted to bulk material, are emphasized as laser photons are relatively valuable as will be seen later.

The laser energy may be absorbed directly in the surface , or alternately by absorption in a free-electron plasma [2] above the surface, see Fig. 2. In the latter case, it is

Springer Series in Surface Sciences, Vol. 19
Desorption Induced by Electronic Transitions, DIET IV
Editors: G. Betz · P. Varga

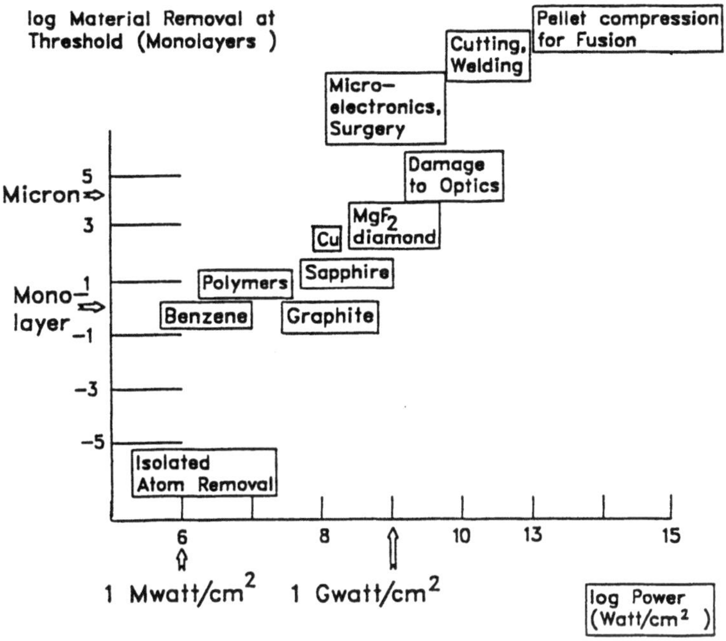

Fig. 1. Schematic representation of the breadth of laser-surface interactions. Note the many decades of parameter space in both power density and amount of material being removed from a surface.

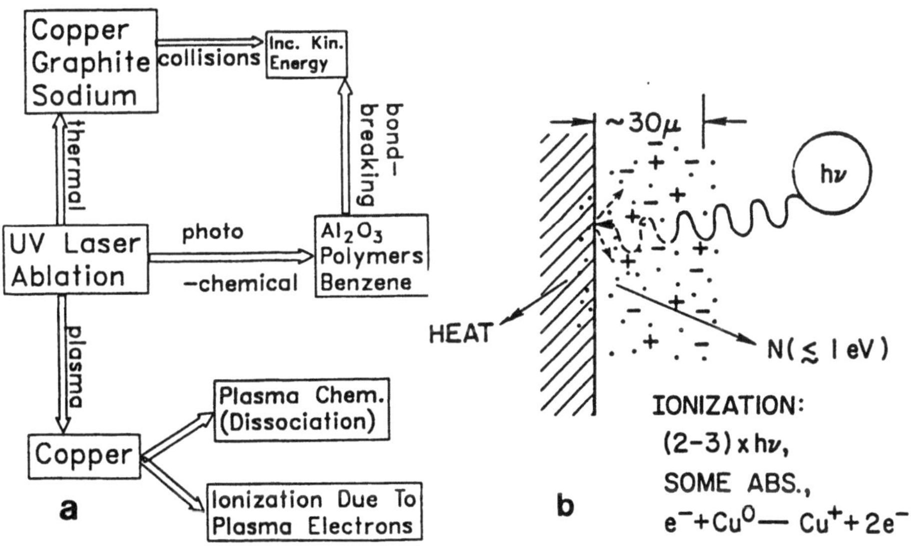

Fig. 2. (a) Schematic representation of the various pathways between laser energy leading to removal of surface atoms. (b) Representation of the primary interaction zones when dealing with power densities in the 50 to 1000 MW/cm^2 range. Note that even though inverse bremsstrahlung produces energetic free electrons, their density is sufficiently low (typically $\lesssim 10^{18}$ cm^{-3}) that the laser beam still strikes the surface, as this plasma density is insufficient to move the plasma cut-off frequency (from the infra red) into the UV range. There is, however, the possibility of some of this energy reaching the surface by conduction or shock waves within the plasma, see Ref. [2].

the thermal conductivity of the plasma which transfers the energy into the surface. In the direct absorption case, there are two general pathways leading to surface modifications [3]. First there are the obvious thermal effects, in or near the surface. The depth, ℓ, is given by the well known expression

$$\ell^2 = D_{th} \cdot \tau, \tag{1}$$

where the thermal diffusivity, D_{th}, commonly ranges from 10^{-4} to 0.3 cm^2/s. For a ns pulse from an excimer laser, ℓ then ranges from $\sim 10^{-2}$ to 0.5μ. Only with picosecond pulses is ℓ much smaller than the optical penetration depths of $\sim$ 100 to 200Å typical for metals.

Besides thermal energy, the surface etching may appear as a result of photochemical modification, sometimes referred to as photoablation. In this case, the electronic rearrangements, brought on by visible or UV photons, result in completely altered bonding or even repulsive states. When the fractional density of rebonded atoms becomes dominant, ablation follows. Examples exist where the size of the area in question extends from single atom defects (in the case of V_k color centers [4]) to nearly explosive vaporization [5] of macroscopic volumes in the case of polymers. Between these two extremes, there are numerous cases [3] where the surface layers appear to decompose sequentially even during a 10 ns pulse. Returning to the specific area of polymer ablation, the surface removal evidences a combination of photochemistry and heat with photochemistry clearly dominating at wavelengths of 193 nm and shorter, i.e. a DIET type process.

While there exist examples of soft X-rays generated in the plume directly giving rise to DIET type processes, the usual effect of plasmas is to transfer energy to the surface by either thermal conduction or a shock wave.

Besides the above microscopic processes leading to surface removal, there also exists several processes which involve separating macroscopic fragments from the surface. Some examples of these are splashing droplets off of molten layers, and exfoliation (fracturing off of sheets). Quartz and polymethylmethacrylate with 10 ns, 308 nm pulses [6] appear to undergo the latter process, possibly due to subsurface arcing.

While we shall primarily emphasize processes in the nanosecond range, various works have described laser ablation phenomena ranging from cw to subpicoseconds [6].

One specific limitation on laser ablation that is difficult to overcome is the fact that laser photons are expensive photons. Only in the case of the somewhat unique CO_2 laser is there any semblance of efficiency. In the more common cases such as YAG, noble gas ion or excimer lasers, the overall efficiency drops into the 0.01 to 0.1% range. Such a low efficiency is indicative not only of the cost of electricity, but more importantly of a high cost of rare gases, frequent servicing and equipment replacement.

2. Useful Phenomena Involving Laser Ablation

The preceding comments on the high cost of laser photons implies some compelling reasons for their utilization. Indeed, often there exist such reasons. First, consider the cases of materials for which the only practical means is utilization of lasers; i.e. conventional thermal, plasma or chemical etching are ineffectual. Two such cases are the etching of sapphire and polyimides.

Etching of Inorganic Material. Sapphire is practically impossible to etch by either chemical or thermal processes. Chemical etching involves hours of boiling in phosphoric plus chromic acid in order to remove layers of much less than 1μ. Thermal etching is obviously impossible as atmospheric vapor pressures are only achieved near 4000K. In contrast to these impractical approaches, even mm's of sapphire are removed in seconds with either 248 or 193 nm UV [7]. The reason is that there are photo sensitive surface sites which efficiently channel the laser photon energy into a photochemical ablation [8]. This process brings out two points about the uniqueness of laser ablation. First, the etching of sapphire requires the high ($P > 100$ MW/cm^2) pulsed power of a laser acting through a nonlinear process, otherwise the isolated electronic excitations would simply relax back to their original state. Second, the wavelength selectivity is crucial since short wavelength UV readily ablates Al_2O_3, while in the *visible range* this same material *protects* surfaces against laser damage. The difference is that the shorter wavelengths of UV access surface states within the band gap [8], and even this is unique as the bulk crystals are transparent well into the vacuum UV. The tendency at present is to associate these surface states with dislocation networks produced in the polishing of sapphire wafers. The mechanical strain and dislocations, initially present in the surface and contributing to the energy pick-up, presumably propagate during ablation, and hence, self-catalyze the etching. Along with this and the possible loss of anions due to the short UV radiation, non-stoichiometry may also contribute to the surface absorption. Atmospheric gases do not appear to alter this surface absorption although they do reduce the yield of ions as compared to neutral species. Since laser ablation of sapphire proceeds by a photochemical process, the dimensional resolution can be of nearly atomic size. Thus, projection masking gives sharp features in the plane of the films. (Since the plasma etching threshold for mask materials, e.g. metals, is only $\sim$500MW/cm^2, contact masking is difficult as hot spots in the laser beam can ablate the mask also.) A disadvantage associated with the threshold-like behavior for most laser etching is that depth profiling is not easily controlled. Even a 20% to 30% increase in fluence may double the etch rate and hence produce undesirable surface roughness due to laser-beam hot-spots.

Ablation of Organic Materials. The laser ablation of polyimide films has many similarities with ablation of sapphire. This polymer is not readily removed by chemical or thermal treatment. True, oxygen plasma etching is possible but this may not be convenient. This material is known to form a passivating layer which can only be broken down when ions at hundreds of eV's of energy [9]. are a significant fraction of the plasma species.

The laser ablation of polyimide brings up one of the disadvantages of laser ablation, however, and this is the area of *debris*. With polyimide the debris commonly consists of micron sized particles of carbonaceous lumps scattered over the surface, particularly in the area immediately surrounding the etched area [10]. The carbonaceous particles appear to coalesce around large carbon polyatomic groups (perhaps the unusually stable C_{60} molecules) within the dense expanding plume. The existence of such large molecules and particles demonstrates the rapidity of clustering in a vapor of density equivalent to tens or hundreds of atmospheres.

Material Transfer. The use of laser vaporization for the deposition of thin films has often been proposed as having technologically interesting possibilities [11-13] . Some of the attractive features are that the automatic collimation [14] of an intense ablative source provides for efficient material deposition. Also, the collimation indicates a highly directed, high kinetic energy (in the 10 to even 100 eV range) plume, and this energy tends to the production of dense films. Part of the reason for this may lie in the fact that interstitial atoms tend to diffuse much more rapidly than vacancies, hence relaxation by expansion is favored over densification when aiming for an equilibrium crystallographic structure. The natural collimation of the laser plume is also efficient for avoiding waste of a rare or valuable source material.

Some of the counter balancing features of laser plume deposition as compared to conventional ion sputtering or thermal vaporization are as follows: First and foremost, when metals are laser sputtered, there is usually some splashing of micron sized droplets onto the deposited area. This appears to result from hydrodynamic splashing [15] on the source surface and not necessarily coalescence as in the case of metals. If electromagnetic deflection of the plume were possible, then collecting the droplets onto the deposition area could be avoided. Unfortunately, the requirement that the (plasma) plume be electrically neutral (since the Debye shielding length is much smaller than the plume dimensions) means that deflection is impractical. A counterbalancing area is the fact that laser microprobe ablation and mass analysis (LAMMS) does indeed work [16]. However, in this case, the small ablated volume ($\sim 10^{-11}$cm^3) means that even with complete ionization, the plume readily expands to the point of having the Debye distance *greater* than the plume dimensions. Thus, while a micro (LAMMS) plume can be accelerated and deflected by electrostatic fields and even introduced into a mass spectrometer, as soon as the ablation plume becomes usefully large ($> 10^{14}$ atoms) the plume can no longer be deflected. The LAMMS technique does form the basis of useful machines which can rapidly analyze surface areas of ($\sim 1\mu)^2$. One limitation to the technique is that depth control is difficult, hence analyzing monolayers is usually not practical [17]. The reason is again that the etch depth is threshold-like in its dependence on fluence, and hence overly dependent on the precise fluence.

An area where laser ablation/deposition (to fabricate thin films) has had unexpectedly great success is the deposition of high temperature superconductors (HITC) [18]. It is not surprising that these oxides are readily ablated; possibly by a process similar to the previously discussed sapphire, namely, anion removal as a first step. It should also be noted that the laser deposition is commonly carried out in an oxygen atmosphere, which presumably encourages reforming the stoichiometric solid. Just as in the case of metal films, the high kinetic energies involved in laser sputtering may be helpful in promoting good crystallographic properties. Beyond these two features, however, it is not understood how so disruptive a process as laser ablation, including plasma decomposition of radicals [19] can produce an optimum superconductor.

The above discussion brings us to the point of asking where the dividing line lies between the understood and yet-to-be deciphered areas. While not all areas of laser ablation are quantitatively understood, most areas of atom and ion vaporization have been reasonably well investigated and analyzed. Even though one accepts that electronic bonding changes, as well as thermal processes, can remove atoms from surfaces, the exact electronic states involved are not clear. For instance, how can one describe the detailed electronic state of a sapphire crystal when the first few atomic layers have an exciton-like state at nearly every site [8]? In the case of CaF_2, it appears that the surface becomes non-stoichiometric. How can the new electronic bonding be described when it again pertains to a majority of the surface sites [20]? Similarly, can the state of polymers be described when nearly every monomer group has sustained an electronic excitation? Throughout all this, it should be recalled that laser etching of a monolayer commonly occurs in the 10^{17} to 10^{18} photons/cm^2 range, hence, any electronic transition (including those originating in an excited state) with even a small oscillator strength will be driven to saturation by the incident laser beam. Presumably, along with detailed knowledge of electronic states would come two areas of understanding, e.g. the role of non-stoichiometry in such materials as Al_2O_3 and CaF_2 in accelerating etching at above threshold fluences or multishot irradiations, and second, the ballistic process involved in either single atom or ion ejection [21] or "explosive" decomposition [5].

Medical Applications. The areas of useful excimer ablation are not limited to inert organic and inorganic surfaces, but there are also many areas of medical treatment and surgery that are effective. Again, note that while laser photons are expensive, it is often their unique specificity which makes them the technique of choice.

The eye is very sensitive to damage by heat or mechanical intervention. This gives rise to the laser having several different applications. First, wavelengths such as ruby at 694 nm or YAG at 1.06μ are transmitted clear to the back of the eyeball and this permits a technique such as "spot-welding" to correct a retina which is detached from the rear surface to succeed. Next, by using 193nm UV pulses, the cornea (front surface) of the eye can actually be reshaped to correct the focus of the lens. Thickness changes as small as 30μ are said to be sufficient for most corrections [22]. Such corrections can be made by "laser-planing" or alternately by a sequence of radial incisions to alter the stress pattern. The former can be carried out under "real-time" computer control to minimize uncertainty.

Laser activated photochemotherapy for malignancies has been expanding for several years [23]. This is a significant improvement over conventional chemotherapy for the following reason. In the latter case, the entire body is flooded with toxin that is only marginally more destructive to malignant cells than normal ones. This difference in toxicity is apparently derived from the fact that malignant cells metabolize more rapidly, and hence tend to incorporate the toxin more rapidly than normal cells. While this is a desirable result, one must recall that the entire body is being subjected to this treatment with often serious side effects. In a sense, lasers have opened the possibility of localizing the toxicity to a specific region of the body; this is known as photodynamic or

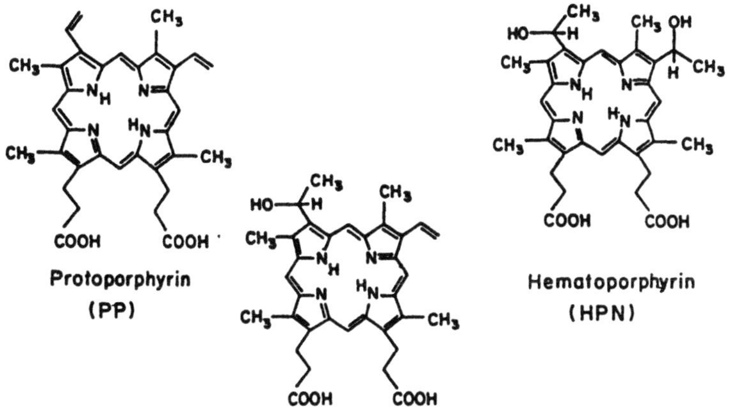

3'-OH ethyl-8-vinyl deuteroporphyrin
(HVD)

Fig. 3. Chemical formulas for compounds of types used for photochemotherapy. Excitation with red (500-600 nm) light liberates reactive oxygen ions or radicals from these compounds.

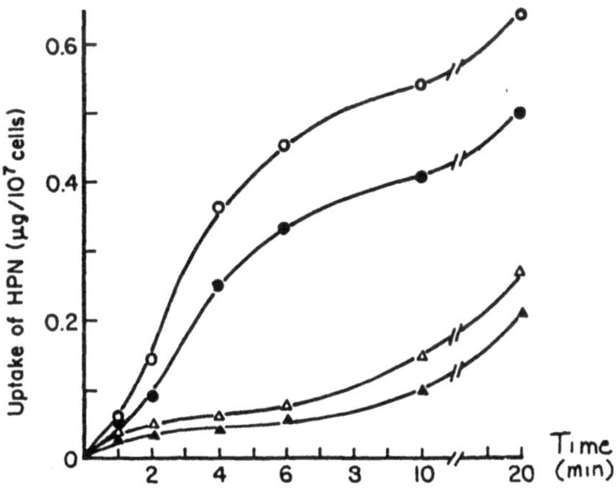

Fig. 4. Relative rate of incorporation of HPN in normal or acute myeloblastic (leukemia related) red blood cells. Note that for dose periods of ~6 min., there is a six-fold difference.

photochemotherapy and operates as follows: A relatively inert chemical such as HPN, see Fig. 3, is administered to the person. While these types of compounds are not normally toxic to either malignant or nonmalignant cells, they are nevertheless absorbed more rapidly by malignant cells, see Fig. 4. The important next step is to irradiate the HPN with light near 600 nm, in which case highly reactive and thus toxic oxygen ions and radicals are liberated within the cells. Since the malignant cells have absorbed up to 6 times more HPN, they are more readily killed. The step involving *photoactivation can be localized to essentially the site of the malignancy.* Thus, only a small fraction of the body is subjected to the detrimental side effects in the case of photochemotherapy and not the entire body. While it is obvious how the photoactivation can be applied to skin cancer, even exotic illumination schemes such as introducing (noble-gas ion) laser

Effect of In Vitro Destruction of Lymphocytes from AML Patient with HPN and Laser Irradiation (625nm)

Incubation System	Cytochrome P-450 Dependent Enzyme (Arylhydrocarbon hydroxylase, pmol/min x 10^7 cell)	Heme (nmol/ 10^7 cell)	Heme Oxygenase (pmol bilirubin/ 10^7 cells x hr)
Control	1.2	1.06	1750±140
HPN	1.3	0.86	1730±50
HPN + Laser	0.55	0.41	1950±80
Laser	1.1	0.89	1940±300

light through an optical fiber in the lungs has shown to be effectual in lung cancer cases [23].

We have carried out quantitative experiments on treating red blood cells from acute myeloblastic leukemia patients using photochemotherapy in order to quantify the process [24]. There appear to be several channels by which these malignancy-related red blood cells are affected. First, apparently a surface, namely the cell wall, is attacked by treatment with the HPN followed by exposure to about 125 J of 625 nm pulsed dye laser light. When the cell wall is penetrated completely, the desired malignant cell destruction is accomplished; it is also interesting how specific this cell treatment can be. Besides attacking the outer cell wall, a few specific sites on the mitochondria are also attacked. The heme oxygenase site remains unchanged; on the other hand, the impurity elimination site known as P-450 is attacked, see Table I. The above site is primarily responsible for eliminating toxic impurities such as benzene from the cell and the body. One can now even conceive of a third step in the photochemotherapy where not only the malignant cell walls are attacked directly, but also the ability of these cells to metabolize impurities is intentionally impaired. A toxic compound which could normally be metabolized both by normal and malignant cells would now be lethal to the malignant cells. As a further extrapolation, it even appears possible to generate these toxic (chemotherapy) compounds locally by a second photochemical reaction at the same or another wavelength. We now see how the localization of photochemotherapy by lasers possesses the ability to destroy malignant cells in both a metabolic and geometric selective process while grossly reducing the side effects of conventional chemotherapy.

3. Summary.

We have seen several examples of the usefulness of laser ablation. While laser photons are relatively expensive, nevertheless, their ability to provide temporal, spatial and wavelength selectivity has given rise to many practical applications in widely different settings. It is not unreasonable to derive from the variety of these applications that there now exist many, and in the future additional applications, for laser radiation interacting with surfaces.

This work was aided by NATO Research Grant 5-2-05/RG No.0020/88. The author would like to express his appreciation of many fruitful conversations with the following persons: N. Abraham, E. Matthias, R. Srinivasan and R. von Gutfeld.

References

1. Gordon Gould, notebook, on or before 13 Nov. 1957.
2. C. R. Phipps, T. P. Turner, R. F. Harrison, G. W. York, W. Z. Ozborne, G. K. Anderson, X. F. Corlis, L. C. Haynes, H. S. Steele, and K. C. Spicochi, J. Appl. Phys. **64**, 1083 (1988).
3. E. Matthias and R. W. Dreyfus, Topics in Current Physics, **47**, editor P. Hess, Springer-Verlag, Heidelberg (1989).
4. N. Itoh, Nucl. Instr. and Meth. **132**, 201 (1976).
5. R. Srinivasan, B. Braren, K. Casey and M. Yeh, to be published, and T. Zyung, H. Kim, J. Postlewaite and D.Dlott, J. Appl. Phys. **65**,4548 (1989).
6. R. Srinivasan, E. Sutcliffe and B. Braren, Appl. Phy. Lett. **51**, 1285 (1987).
7. Roger Kelly, J. J. Cuomo, P. A. Leary, J. Rothenberg, B. E. Braren and C. F. Aliotta, Nucl. Instr. and Meth. **B9**, 329 (1985).
8. R. W. Dreyfus, F. A. McDonald and R. J. von Gutfeld, Appl. Phys. Lett. **50**, 1491 (1987), and J. Vac. Sci. Technol. **B5**, 1521 (1987).
9. G. S. Selwyn, J. Appl. Phys. **60**, 2771 (1986).
10. R. von Gutfeld and R. Srinivasan, Appl. Phys. Lett. **51**, 15 (1987).
11. A. O'Keefe, M. M. Ross and A. P. Baronavski, Chem. Phys. Lett. **130**, 17 (1986).
12. G. R. Keen, J. A. O'Neill, D. Seddon and G. Paul, Conf. on Lasers and Electro-Optics, Baltimore, MD, 24-28 April 1989.
13. M. Hanabusa, S. Moriyama and H. Kikuchi, Thin Solid Films **107**, 227 (1983).
14. Roger Kelly and R. W. Dreyfus, Nucl. Instr. and Meth. **B32**, 341 (1988), and Surface Science **198**, 263 (1988).
15. Roger Kelly and J. Rothenberg, Nucl. Instr. and Meth. **B7/8**, 755 (1985).
16. Fourth International Workshop on Laser Microprobe Mass Spectrometry (LAMMS), Annual Meeting of the Microbeam Analysis Society, Ashville, NC, July 16-21, 1989.
17. "Are Laser-Ablated **Monolayers** Accurately Characterized by Their Ion Emission?", R. W. Dreyfus, edited by Phillip E. Russell, Microbeam Analysis 1989 (publ. by San Francisco Press, CA, 1989) p. 261.
18. D. Dijkkamp, T. Venkatesan, X. Wu, S. Shaheen, N. Jisrawi Y. Min-Lee, W. McLean and M. Crott, Appl. Phys. Lett. **51**, 619 (1988), and C. Guarnieri, R. Roy, K. Saenger, S. Shivashankar, D. Yee and J. Cuomo, *ibid.* **53**, 532 (1988).
19. R. W. Dreyfus, Roger Kelly, R. E. Walkup and R. Srinivasan, Proc. of Society Photo Optical Instrumentation Engineers **710**, 46 (1987).
20. S. Petzoldt, to be published.
21. P. Avouris and R. Walkup, Ann. Rev. Phy. Chem. **40**, 173 (1989).
22. T. F. Deutsch, Physics Today, **41**, 56 (1988) and references therein.
23. T. Dougherty, J. Kaufman, A. Goldfarb, A. Weishaupt, D. Boyle and A. Mittleman, Cancer Res. **38**, 2628 (1978).
24. "Preferential Elimination of Leukemic Cells by Porphyrin-Induced Laser Photosensitization", N. Abraham, Ann Brown, R. D. Levere and R. W. Dreyfus, Laser Studies of Engineering Materials, Medicine and Biology, Physics and Chemistry Conf., Laser Center of the Rhone-Alps Area, Lyon-Villeurbanne, France, July 7-9, 1987.

Electron Stimulated Desorption of Nuclear Spin Polarized Alkali-Metal Atoms

J. Chrost, M. Detje, M. Riehl-Chudoba, and D. Fick

Philipps-Universität, Fachbereich Physik und Wissenschaftliches Zentrum für Materialwissenschaften, D-3350 Marburg, Fed. Rep. of Germany

DIET was used successfully to control the residence time of nuclear spin polarized Na atoms on a clean Si(111) surface of about room temperature. The various nuclear relaxation processes which may cause depolarization during the rather long residence times of the atoms on cold surfaces are discussed in detail.

1. Introduction and Background

In surface science a wealth of interesting data have been accumulated for monolayer to submonolayer deposits of adatoms. However, experiments on dilute systems for which the mutual interaction of the atoms can be neglected are rather sparse. Here, specific experimental tools are required which explore the local environment of adsorbates. An increasing number of experimental studies are being reported which use nuclear probes, i.e. nuclear moments [1]. The necessity to heat the surface beyond 1000K in order to determine the nuclear spin polarization of the thermally desorbed alkali-metal atoms is one of the limitations of some of these experiments. However, in a contribution to DIET II [2] it was speculated that an extension of such experiments to low temperature surfaces might be possible using DIET to control the residence time of the polarized nuclei on the surface. The elementary process underlying DIET occurs on a time scale of about 10^{-13}s, much shorter than the Larmor precession time. Thus, any depolarization caused by the desorption process itself can be neglected. But there may be a large number of 'trials' preceding the final desorption process itself. How much these might depolarize the nuclear spin of the adsorbed atoms is not known and has to be investigated experimentally.

However, the interaction of the nuclear moments with fluctuating electromagnetic fields always leads to nuclear spin relaxation (depolarization) in particular for long residence times of the atoms on the surface. In experiments with hot surfaces such mechanisms were observed [2-4], but, as we will see, may become sufficiently small for sufficiently low surface temperatures.

The principle features of such experiments are illustrated in fig.1. The source can produce atomic beams of alkali-metal atoms (Li, Na) in various sublevels m_I of the nuclear spin I. Such a thermal-velocity atomic beam impinges on the heated sample surface which is placed in UHV (electron bombardment for DIET, as indicated also in fig.1, was not used

Springer Series in Surface Sciences, Vol. 19
Desorption Induced by Electronic Transitions, DIET IV
Editors: G. Betz · P. Varga

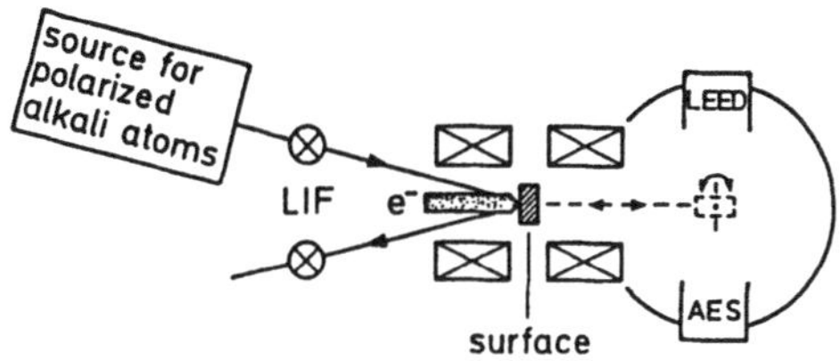

Fig.1
Schematic view of the experimental arrangement.

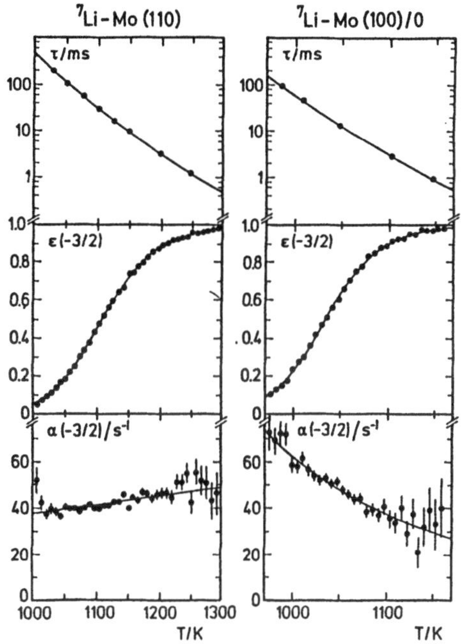

Fig.2
Nuclear spin relaxation of ^{7}Li adsorbed on a clean Mo(110) and partly oxygen covered Mo(100) surface.

in the experiments described in this section). The surface coverage with alkali-metal atoms was always less than 10^{-3} of a monolayer, permitting their mutual interaction to be neglected. The sample is situated in a static magnetic field aligned along the surface normal. It serves as the quantization axis. The adsorbed alkali-metal atoms desorb from the surface partly as positive ions, partly as neutral atoms. From the latter the nuclear m_I-substate population can be determined by laser induced fluorescence (LIF) [5].

Typical results which were obtained for ^{7}Li (I=3/2) adsorbed on a clean Mo(110) surface are plotted in fig.2. Other examples can be found in refs.[1,3]. The middle graph displays as a function of surface temperature T the polarization effect ϵ measured for the sublevel m_I=-3/2.It displays the difference of the sublevel population for a polarized beam to the one for an unpolarized beam normalized to the value at high surface temperatures where no depolarization occurs. The mean residence time τ was always measured separately by chopping the atomic beam mechanically.

The time dependence of the effect ϵ can be described by an exponential $\epsilon=\epsilon_o \cdot \exp(-\alpha t)$, α being a temperature dependent effective relaxation rate (for details see refs.[3,6,7]). Since for an adsorbed atom the probability of staying on the surface for a time t is proportional to $\exp(-t/\tau)$, τ being the mean residence time (fig.2), the temperature dependence of ϵ is related to the relaxation rate α by

$$\epsilon = \tau^{-1} \int \epsilon_0 \cdot \exp(-\alpha t) \cdot \exp(-t/\tau) dt = \epsilon_0/(1+\alpha\tau).$$

Thus measuring ϵ/ϵ_o and τ for each temperature the relaxation rate α can be extracted (lower graph of fig.2). The observed relaxation rate increases linearly with surface temperature T.

The same experiments were performed with other clean surfaces as well. For all of them, a linear increase of α with temperature was found. However, for various surfaces the absolute relaxation rates vary considerably. A linear increase of the NSR rate with temperature is a unique hint at spin lattice relaxation [3,6]). It decreases linearly with decreasing temperature and will in general not prohibit surface physics experiments with nuclear probes in connection with DIET to control the residence time.

Similar experiments were performed with a partly oxygen covered Mo(100) surface (fig.2). The temperature dependence of the mean residence time τ and the effect ϵ look at first glance similar to the result obtained for the clean Mo(110) surface. And indeed the NSR rate α is of the same order of magnitude as for the clean Mo(110) surface. But the extracted NSR rate α increases now with decreasing temperature. This indicates another NSR process which can be identified by plotting the NSR rate in an Arrhenius plot (fig.3). The NSR rate α reflects a thermally activated process ($E_d \approx 0.55$eV) which can be identified with diffusion [3,4,7]. It is important to notice that for diffusion as the mechanism the NSR rate depends strongly on temperature [7]: $\alpha \propto \tau_c/[1+(\omega_L\tau_c)^2]$, with ω_L the Larmor frequency and τ_c the so-called correlation time, which for diffusion as the underlying process is directly related to the diffusion rate Γ by $1/\tau_c = \Gamma = \Gamma_o \exp(-E_d/kT)$. For $\omega_L\tau_c \ll 1$, the regime of the experiments displayed in fig.2, α is proportional

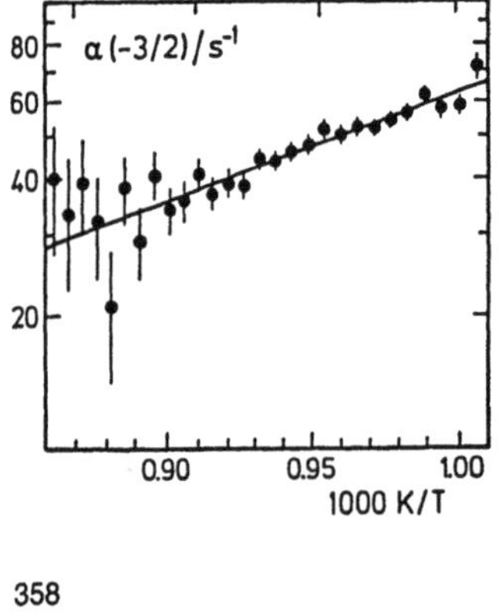

Fig.3
Arrhenius plot of nuclear spin relaxation rate of ^{7}Li adsorbed on partly oxygen covered Mo(100).

to τ_c in agreement with the results in figs. 2 and 3. On lowering the temperature, α reaches its maximum for $1/\tau_c = \omega_L$. Then the NSR rate can be quite large! At still lower temperatures α becomes smaller again and reaches its minimum value, which is proportional to $1/\tau_c$, for $\omega_L \tau_c >> 1$. Using DIET this is the regime which has to be reached for experiments with nuclear probes on cold surfaces.

2. Experimental Results Using DIET

In experiments using DIET, instead of being heated, the "cold" surface is bombarded with electrons (gun with a BaO cathode, $j_e \le 40$ mA/cm^2, $U \le 350$V) (fig.1). In order to deal with large enough cross sections for DIET instead of a metal surface a Si(111) surface was studied on which ^{23}Na (I=3/2) was adsorbed.

Fig.4 displays results with unpolarized ^{23}Na atoms. The surface was exposed to the atomic beam for a few seconds, which leads to a coverage of about 0.1 monolayer. Thereafter, the electron beam was switched on (upper part of fig.4). During the measurement the surface attained a temperature of about 250°C, which increased slightly due to the electron bombardment. The time dependence of the LIF signal (D1 transition, F=2 → F'=1) is plotted in the lower part of fig.4. It displays above a constant background a well developed LIF signal caused by DIET. It decreases with an average residence time of about 20s. Experimental runs were performed varying both energy and density of the electron beam. The cross section Q as a function of electron energy is plotted in fig.5. It turns out to be roughly 1 Mb ($Q = (0.94 \pm 0.24) 10^{-18}$ cm^2).

To what extent a "polarization signal" can be preserved during the rather long residence times is the remaining crucial question using DIET in experiments with nuclear spin polarized adsorbates. In order to minimize the absolute residence time of atoms on the surface the atomic beam and electron beam were now switched on simultaneously after the cleaning flash (fig.6).

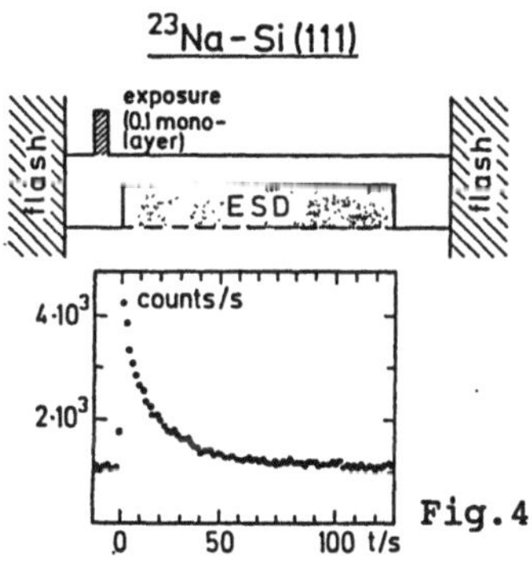

Fig.4

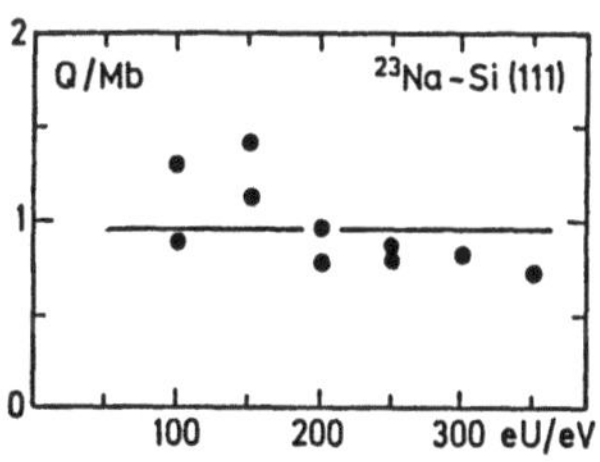

Fig.5

Fig.4
DIET of ^{23}Na adsorbed on a Si(111) surface (electron energy 200eV, $j \approx 13$ mA/cm^2).

Fig.5
Cross section for electron stimulated desorption of ^{23}Na from a Si(111) surface as a function of electron energy.

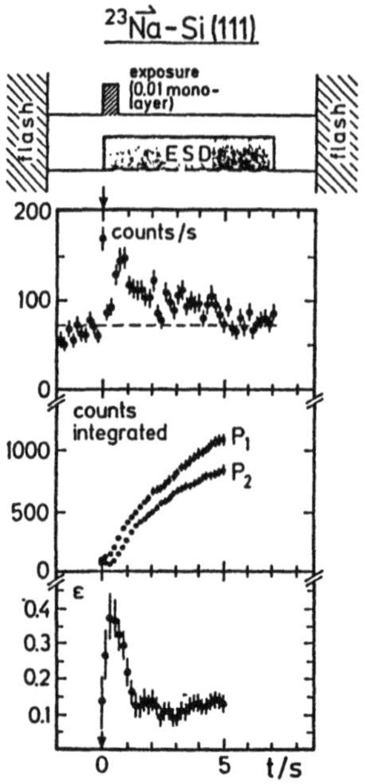

Fig.6
DIET of nuclear spin polarized ^{23}Na adsorbed on a Si(111) surface.

The atomic ^{23}Na beam, however, was interrupted after 0.6s, which resulted in a maximum coverage below 1/100 of a monolayer. To gain information on the nuclear polarization of the desorbing atoms the LIF signal stems now furthermore from one single transition between two hyperfine levels only [5] (observed $m_I=3/2$ by the $F=2$, $m_F=2 \rightarrow F'=1$, $m_F=1$ transition of the D1 line in a magnetic field of about 50mT). This, together with the much lower coverage led to a much weaker LIF signal caused by DIET (2nd part of fig.6). In the 3rd graph of this figure the signal is integrated and plotted as a function of integration time t.

To learn about NSR and depolarization the experiment was performed with two different preparations of the atomic beam. In one, called P_1, half of the atoms impinging on the surface were in the sublevel with $m_I=3/2$, in the other, P_2, the $m_I=3/2$ sublevel was empty. The high frequency transitions of the source allow one to switch between these preparations without changing the atomic beam intensity and the beam profile [9]. The third graph in fig.6 displays for both preparations the integrated LIF signals. They exhibit a clear polarization effect! The normalized difference of the integrated spectra for preparation P_1 and P_2, called the asymmetry ϵ, is plotted at the bottom of fig.6. It displays asymmetries of about 0.1 which increase for short enough integration times t. This is reasonable, since for short enough residence times the depolarization caused by hyperfine interaction with fluctuating fields is certainly the smallest (fig.2).

3. Outlook

The essential information for the feasibility of using DIET in surface physics experiments with nuclear probes is now available. However, the experimental results presented in sect.2 can be improved considerably concerning statistics and definition of the experimental conditions. At present a cooling system is being installed which allows one to control the temperature of the surface but also to lower its temperature below room temperature. This is not only necessary to reduce the NSR rate (see sect.2), but also to study NSR and thus spin lattice relaxation [1,3,6] and surface diffusion [1,3,9] in detail. Furthermore, for NMR experiments small coils have still to be installed in the vicinity of the surface to expose the sample to a radio frequency field [10].

References

[1] D. Fick, Surface physics with nuclear probes, Appl. Phys. **A49** (1989) 343

[2] D. Fick, R.F. Haglund jr. and N.H. Tolk, Proc. 2nd Int. Workshop on Desorption induced by electronic transitions, DIET II, p.252, ed. W. Brenig and D. Menzel (Springer Berlin 1985)

[3] M. Riehl-Chudoba, PhD Thesis Marburg 1989 and to be published

[4] B. Horn, W. Dreves and D. Fick, Z. Physik **B48**, 335 (1982)

[5] H. Jänsch, E. Koch, W. Dreves and D. Fick, J.Phys. **D17**, 231 (1984), U. Memmert and D. Fick, J.Phys.E, **21**, 208 (1988)

[6] K. Waßmuth and D. Fick, Phys.Rev.Lett. **59**, 3007 (1987)

[7] A. Abragam, Principles of Nuclear Magnetism (Oxford University Press 1961, reprinted 1986)

[8] D. Krämer et al., Nucl.Instr.Methods **220**, 123 (1984)

[9] U. Memmert, M. Bickert, M. Riehl, K. Waßmuth, D. Fick, and R.F. Haglund, jr., Hyperfine Int. **39**, 295 (1988)

[10] U. Memmert and D. Fick, Europhys.Lett. **5**, 203 (1988) and Surf.Sci. **203**, 276 (1988)

Mechanisms of Laser Ablation of Biological Tissue

W. Husinsky[1], *G. Grabner*[2], *I. Baumgartner*[2], *F. Skorpik*[2], *S. Mitterer*[1], *and T. Temmel*[1]

[1]Institut für Allgemeine Physik, Technische Universität Wien, Wiedner Hauptstr. 8–10, A-1040 Wien, Austria

[2]Universitätsaugenklinik, Universität Wien, Alserstr. 8, A-1090 Wien, Austria

Abstract. The application of laser ablation for tissue removal - mostly in the UV-regime - has recently stimulated considerable experimental efforts among research groups. Beyond this, clinical applications of laser ablation have already been reported. An understanding of the desorption and ablation mechanisms and potential side effects of the radiation is, however, still in at the infant stage, but indispensable for a useful and responsible application.

1. Introduction

A crucial factor for the evaluation of the usefulness of laser ablation and its potential for future applications is the *"incision quality"*, resulting from the specific ablation-desorption mechanism involved [1-3]. Experimental ablation studies of human corneal tissue have been performed with different laser wavelengths and radiation fluxes with native and "*doped*" tissue. Three different ablation mechanisms - *thermal, explosive and photochemical* - have been identified. *Effective, "good quality"* ablation can be obtained whenever the photochemical or "explosive" mechanism dominates [3].

We have measured the photo-ablation rates as a function of the laser fluence in different systems: a) 193 nm light and native corneal tissue ; b) 193 nm light and doped corneal tissue; c) 308 nm light and native corneal tissue ; d) 308 nm light and corneal tissue doped with oxybuprocain; e) 485 nm light and corneal tissue doped with fluorescein dye.

The UV light was generated by an excimer laser (Lambda Physik EMG 101MSC). 485 nm light was generated by an excimer pumped dye laser (Lambda Physik FL 3002). Whereas the excimer laser beam was focused to a spotsize of approximately 1 mm in diameter, the dye laser beam could be focused to spotsize smaller than 0.5 mm in diameter. For all experimental results shown the repetition rate of the laser was 20Hz. Variation of the repetition rate, furthermore, showed a linear dependence of the incision time on the repetition rate. For the doping, the corneas were incubated in a solution containing different concentrations of the dopant. After incubation-times between 45 and 60 minutes apparent saturation with the dopant was achieved. The ablation rates were determined by measuring the thickness of the corneas and the time needed to generate a through-and-through perforation in the cornea with the laser beam. During the ablation process the corneas were exposed to air. During the exposure of the corneas to the laser beam (typical times for the process ranged from 20 seconds to 2 minutes)

Springer Series in Surface Sciences, Vol. 19
Desorption Induced by Electronic Transitions, DIET IV
Editors: G. Betz · P. Varga

no drying or thermal heating was observed. The thickness was determined prior to the exposure to the laser beam at the proper spot of the cornea by ultrasonic pachymetry.

2. Ablation Mechanisms

Studies on model systems as well as measurements of corneal tissue ablation have established three different mechanisms [3,4] for how the energy deposited by the laser photons into electronic transitions of the tissue molecules can be converted into kinetic energy of the desorbing molecules. The specific characteristics of these mechanisms then critically determine the damage induced in the tissue surrounding the interaction zone of the laserlight with the tissue. These three situations are illustrated in Fig. 1 - Fig. 3.

"Thermal ablation" is observed whenever the photon flux is sufficiently high to heat the substrate by conversion of electronic excitations into vibrational energy. An explicit threshold flux for ablation and strong damage induced in the surrounding tissue is characteristic for this process. For practical use, situations in which thermal ablation occurs should be avoided. "Photochemical ablation", on the other hand, as pictured in Fig. 2, shows extremely low damage, due to the strong localization of the energy. The direct conversion of electronic energy into fragmentation happens on a time scale short enough to avoid thermalisation and does not create vibrating ("heated") molecules. This process in its pure form should not exhibit a flux-threshold for ablation. However, it is not quite understood whether the fragmentation can happen only at the topmost layer or whether bond-breaking in several layers of the tissue can take place accompanied by immediate ejection of the molecules. In all cases rather low ablation rates are expected. This seems to be confirmed by experimental data, if one compares the ablation rates with those achieved by other mechanisms. However, typical ablation rates of a few µm / pulse demand the bond-breaking to evolve in a corresponding large volume of the tissue. Sutcliffe and Srinivasan [5] have developed a model which can account for a threshold in the case of photochemical ablation.

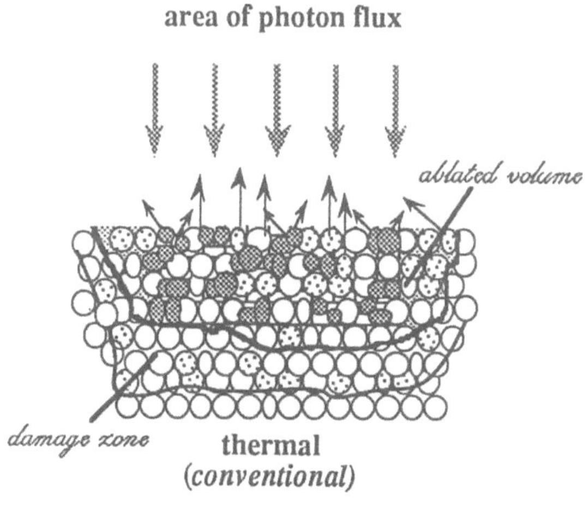

Fig. 1.: Thermal ablation: Electronic excitations of the tissue material are converted into vibrational energy of the molecules. Above a certain threshold the material is heated sufficiently, so that evaporation can take place. In the time scale of this process enough energy can be transferred to the adjacent tissue. Strong damage of the tissue is the consequence.

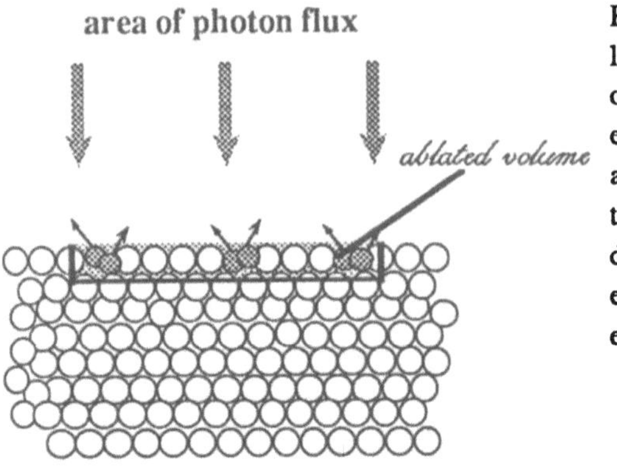

Fig.2: Photochemical Ablation: Provided the energy of the photons is sufficient, electronic excitation into antibonding states of the tissue-molecules results directly in a conversion of electronic into kinetic energy.

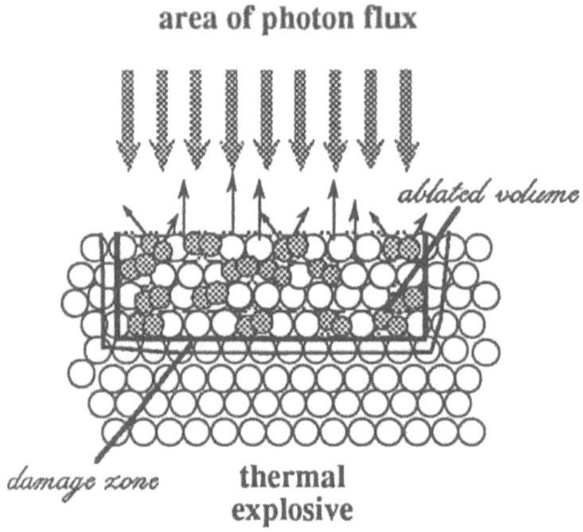

Fig. 3: Explosive Ablation: Similar to thermal ablation energy is converted into vibrational energy, however, the energy density in the irradiated volume is so high that the material is ablated explosively before thermalisation is reached.

A third mechanism has been experimentally identified, which is basically of thermal nature, but differs from the conventional thermal ablation in that it requires extremely high heating (energy deposition), resulting in "explosive ablation" of the irradiated material. The main difference of this mechanism as compared to the thermal situation is the fast time scale in which the energy is transformed to kinetic energy, not allowing the unwanted thermal energy transfer to not-irradiated tissue. This makes this mechanism interesting for practical applications.

3. Experimental Results and Discussion

With regard to experimentally obtained ablation rates, all of the mechanisms described in the previous section can be observed for different experimental parameters. In particular, we will address the influence of the laser-wavelength and possibilities of enhancing the energy absorbance by introducing dopants into the tissue. Without addi-

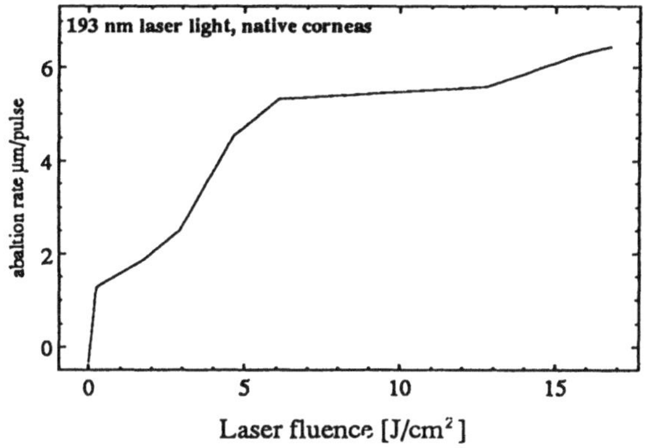

Fig. 4: Fluence dependence of the ablation yield of undoped corneal tissue under 193 nm laser irradiation. Only the averaged data points are shown in the figure.

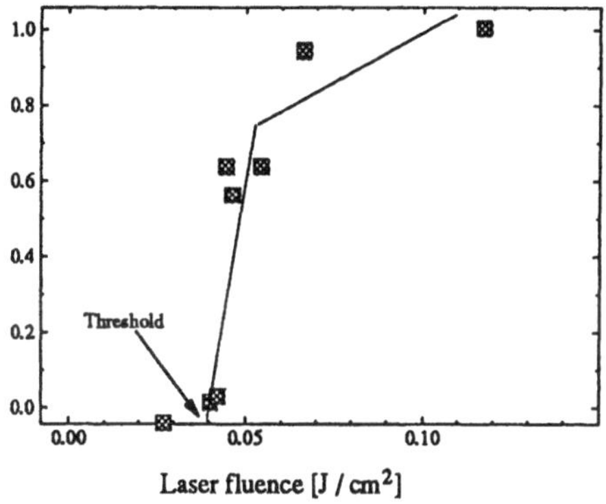

Fig. 5: Ablation threshold for 193 nm irradiation of corneal tissue around 0.05 Jcm^{-2}.

tional dopant, the only case where laser ablation has proven to exhibit "good incision quality" is ablation with 193 nm UV light. In this case it is generally assumed that the photo chemical ablation mechanism plays an important role. The energy of the photons is high enough to break molecular bonds. In Fig. 4 the ablation rate for 193 nm light ablation is shown as a function of laser fluence. The most interesting regime for practical application lies between the fluence threshold (0.04J/cm²) and approximately 2 J/cm². Above the threshold the ablation rate increases until it saturates around 0.5 J/cm² at an ablation rate of 2 µm / pulse. For higher fluences, a further increase of the ablation rate can be achieved, but this can be attributed to uncontrollable heating effects.

Below 0.04 J/cm² no ablation was observed (Fig. 5). The physical meaning of the threshold implies that the ablation via the photochemical mechanism requires some threshold energy deposition in the volume to convert the bond-breaking into particle ejection [5]. Another indication for these additional effects is the yield dependence of the water content of the tissue, observed by several groups [3].

As will be shown later, doping of the cornea can enhance the energy absorption quite drastically and thus can initiate the ablation process. 193 nm radiation, however, is so efficiently absorbed by the tissue, so that no further ablation enhancement could be achieved by adding UV absorbant dopants in low concentrations.

An example of the explosive desorption case is shown in Fig. 6. In this case the tissue was ablated with visible light. The tissue itself is transparent for this radiation.

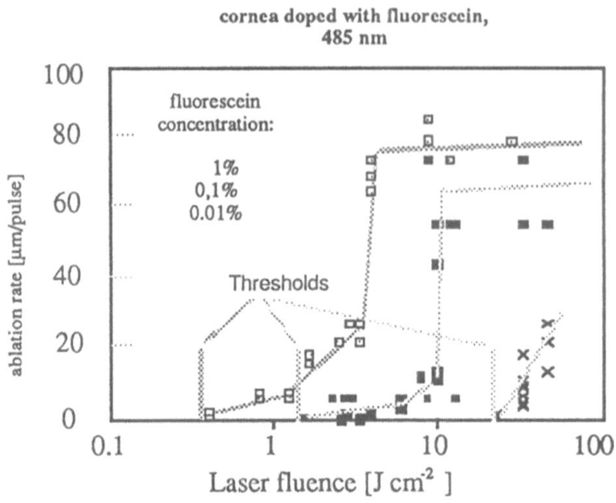

Fig. 6: Fluence dependence of the ablation yield of fluorescein-doped corneal tissue under 485 nm laser irradiation. Different dopant concentrations exhibit different fluence thresholds.

However, the addition of fluorescein (an organic dye) into the tissue as an energy absorbant dopant leads to effective ablation by visible 485 nm light. The features of the fluence dependence shown in Fig. 6 are a strong indication that explosive desorption, as described earlier, is the mechanism responsible in this situation. Distinct fluence thresholds and high ablation rates, which themselves are a function of the dopant concentration, are observed.

4. Conclusions

Finally, the mechanisms and their influence on the cutting quality can be clearly seen in Fig. 7, where an electron micrograph of the zone adjacent to the laser-ablated material

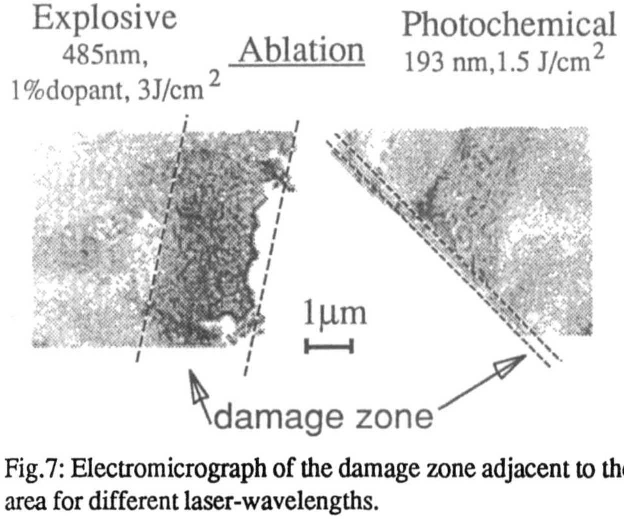

Fig.7: Electromicrograph of the damage zone adjacent to the ablated area for different laser-wavelengths.

is shown for explosive and photochemical ablation. For comparison the damage zone in the case of thermal ablation is several hundred μm wide.

As a consequence, thermal ablation is rather useless for practical applications , in particular for corneal surgery, as demonstrated here. But this finding seems to be true for many other applications too. On the other hand, the other two ablation mechanisms provide promising aspects for surgical applications, whenever thermal heating of the tissue is critical. In both cases the damage zone, identified as a darker area in the electron micograph, seems to be tolerable in most applications.

Acknowledgements

We would like to acknowledge financial support from the "Österreichische Fonds zur Förderung der Wissenschaftlichen Forschung" (Project Nr. P6386PH and P2717MED) and from the "Hochschuljubiläumfonds der Stadt Wien".

References

[1] R. Srinivasan, V. Mayne-Banton, Appl. Phys. Lett. 41, 576 (1982)
[2] S.L. Trokel, R. Srinivasan, R. Braren , Am. J. Ophthalmol. 96, 710 (1983)
[3] W. Husinsky, G. Grabner, I. Baumgartner and S. Mitterer, Appl.Phys.B 49,463 (1989)
[4] T. J. Chuang, H. Hiroka and A. Mödl, Appl. Phys.A45, 277 (1988)
[5] E. Sutcliffe and R. Srinivasan, J. Appl Phys. 60, 3315 (1986)

Electron Damage of a CH_3OH Layer Adsorbed on Nb(110)

S. Rey, I. Colera, and J.L. de Segovia

UEI de Física de Superficies, Instituto de Ciencia de Materiales, Serrano 144, 28006 Madrid, Spain

Preliminary results on the electron damage of a CH_3OH (methanol) layer adsorbed on Nb(110) studied by ESD, ELS and AES are presented. After methanol dosing and before the e^- bombardment the ESD spectrum reveals H^+ surface ions. Under e^- bombardment the H^+ surface ion current decreases exponentially with time to a constant value while the e^- beam induced ESD O^+ surface ion current increases linearly with time. Total ESD cross section determined from the H^+ decay is 5.5×10^{-19} cm^2. The ion energy distribution experiments reveal two kinds of ions with most probable kinetic energy of 8 eV and 5 eV. The high energy peak shifts to the lower one as a function of e^- dosing. The initial layer, mainly formed by methoxy species, is completely dissociated as O, H and C as a consequence of the electron bombardment.

1. Introduction

As is well known electronic transitions induced by electron impact of adsorbed complex molecules on metal surfaces yield few ionic species as compared with the rich cracking pattern yielded by the same molecule in the gas phase. The difference is mainly due to the available high number of electrons in the metal. A complex molecule like methanol which is molecularly or dissociatively adsorbed mainly as methoxy species(Ref.[1] and cited literature) should be a good example to study how e^- bombardment interacts with this radical, what are the surface ions with electron stimulated desorption (ESD), and what is the e^- damage. Another interesting aspect to be studied is whether the methanol species would yield OH hydroxyl radicals upon adsorption and/or oxidized states under electron bombardment, as in the case of another molecule containing hydroxyl radicals as H_2O on the same surface [2].

The aim of the work reported here has been to study the electron interaction with methanol adsorbed on the Nb(110) surface by using the techniques of ESD, electron energy loss spectroscopy (EELS) and Auger electron spectroscopy (AES).

2. Experimental

The Nb(110) sample was located in the focus of an hemispherical retarding grid system and in the line of sight of the entrance of a quadrupole mass spectrometer (QMS) [3]. The QMS can detect either volume or surface phase ions by switching on or off the filament of the ion chamber. The grid retarding system is used for experiments of ion energy distribution (IED). The electron

Springer Series in Surface Sciences, Vol. 19
Desorption Induced by Electronic Transitions, DIET IV
Editors: G. Betz · P. Varga

gun outside the grid system sends the electron flux at 45° from the sample normal. The ultimate pressure of the ultrahigh vacuum system was $5x10^{-11}$ mbar with the sputter-ion pump cryopanel on. Adsorption experiments were performed at methanol total pressure of $1.2x10^{-7}$ mbar (N_2 equivalent). The H_2 contamination was 4% of methanol pressure and no other contaminants were detected. The EELS spectra were recorded with a single pass CMA and an electron energy of 250 eV. The electron current density at the sample in the electron damage experiments was 10^{-4} $A.cm^{-2}$ with the same incident electron energy.

3. Results and discussion

The experiments to date have been performed with the surface at 240 K. Figure 1 shows the ESD O^+ and H^+ surface ion currents as a function of the e^- beam dosing after exposing the surface to 250 L of methanol(1 Langmuir= 10^{-6} mbar.s). The surface species yielding H^+ surface ions decrease exponentially with time down to a constant value while species yielding ESD O^+ ion current increase linearly with time. This suggests that readsorption from the residual methanol pressure is important. (Note that after closing the valve the pressure decreased rather slowly.) From the exponential decay of the ESD H^+ surface ion current a total cross section of $5.5x10^{-19}$ cm^2 has been determined. Figure 2 curve (a) shows the IED after 0.03 $C.cm^{-2}$ e^- beam dosing of the initial adsorbed layer. Several interesting features can be observed. Firstly, we note the two peaks in the H^+ surface ions with maxima at 4.8 eV and 8 eV, and secondly, the decrease in intensity of the high energy peak after 0.24 $C.cm^{-2}$ e^- beam dosing, which remains as a shoulder on the high energy side of the low energy peak. Experiments performed at less than 1 L of methanol exposure show that the IED curve is characterized by a peak with the maximum at 8 eV and a shoulder at about 4.8 eV energy [1]. It is also noticeable that the width in energy extends to about 16 eV. Curve (c) shows the IED of the ESD O^+ surface ions. Note that its width is smaller than for the H^+ ions. The coincidence of this distribution with the high energy peak of the H^+ surface ions is also noticeable, but no explanation is available at this time.

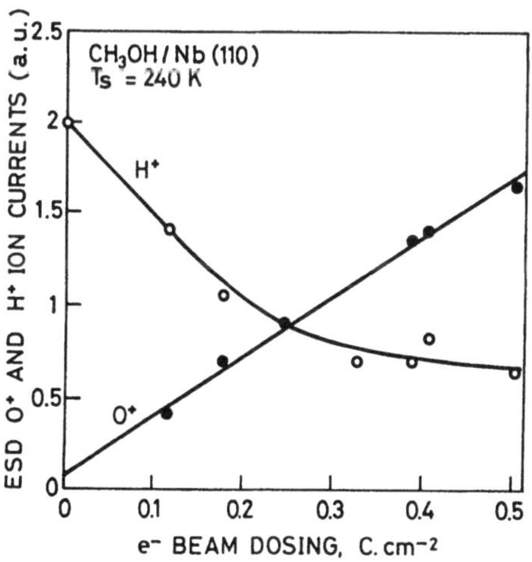

Figure 1. ESD H^+ and O^+ surface ions as a function of the e^- beam dosing after exposing the Nb(110) surface to about 250 L of methanol at 240 K. e^- beam 300 eV, 1.2 μA.

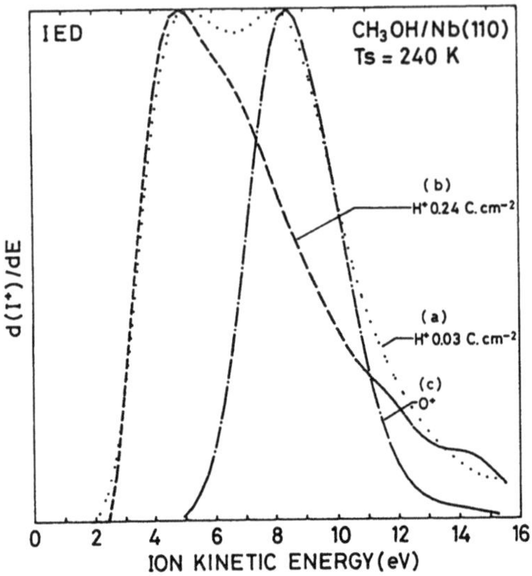

Figure 2. Ion energy distribution (IED) of the ESD H^+ and O^+ surface ions in the conditions specified in Fig.1. (a) IED H^+ surface ion after 0.03 $C.cm^{-2}$ e^- beam dosing. (b) After 0.24 $C.cm^{-2}$ e^- beam dosing. (c) IED O^+ surface ions as in (b).

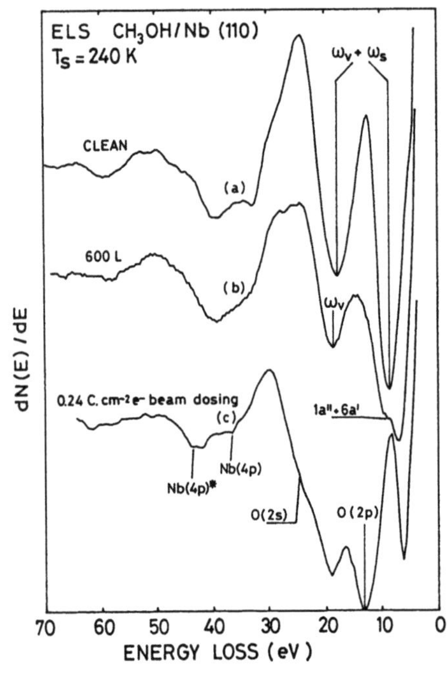

Figure 3. ELS spectra of the Nb(110) surface taken at 250 eV. (a) Clean surface. (b) After exposing to 600 L of methanol at 240 K. (c) After 0.24 $C.cm^{-2}$ e^- beam dosing.

The initial layer of methanol is characterized by the ELS spectrum shown in curve (b) of Fig.3. For comparison the clean surface spectrum, curve (a), is also shown. This spectrum has been discussed in detail in another work on the adsorption of methanol on the same Nb(110) surface [4] where it was concluded that the observed characteristic structure at 9.2 eV was due to the excitation of the 1a"+6a' orbital of the methoxy species. Thus, the initial layer was mainly formed by this methoxy species with the oxygen atom bonded to the Nb surface atom and the hydrogens of the methyl radical out of the surface. In this configuration the ESD H^+ surface ions of the initial layer have their origin in the C-H bond. This is in agreement with the results on the methanol adsorbed on a Ti surface [5].

Thus, from the evidence presented here that e^- beam damage of the initial methoxy adsorbed species produces surface species yielding ESD O^+ and H^+, the questions of what these products are and how they are bonded to the Nb surfaces atoms arise. The oxygen atoms would be directly bonded to the Nb surface atoms, probably in a full valence oxide. This is confirmed by the ELS spectrum shown in curve (c) of Fig.3 after 0.24 $C.cm^{-2}$ e^- beam dosing, where a chemical shift of 3.5 eV of the Nb 4p shallow core level and its resonance, Nb 4p*, towards higher energies can be observed. Even though this value is higher than the reported values for oxidized Nb [6,7], it is clear evidence of a high degree of oxidation. The loss at 13 eV energy can be identified with excitation of an electron from the O 2p orbital to the 3s level. The O 2s to valence band transition at about 23 eV energy overlaps with the structure at 19.5 eV of the volume plasmon. This wide structure is also observed when the surface is carbon contaminated [4]. The C(KLL) AES spectra revealed a structure corresponding with graphite, which demonstrates that the C atoms are not chemically bound to either O or H atoms. The ESD H^+ surface ions induced by e^- bombardment of the initial methoxy species could have their origin in either the O-H radical or in the Nb-H bond. In the first hypothesis ESD O^+ surface ions would be produced as a consequence of the H and H^+ desorption by the electron bombardment of the O-H radical. However, the EELS spectra at different stages of the e^- beam dosing did not reveal the characteristic loss at 10.2 eV energy corresponding to the 3σ molecular orbital of the O-H radical. Thus, the presence of this radical was discarded. At this point it can be concluded that e^- damage produces total dissociation of the initial adsorbed methoxy species as C, O and H. The ESD O^+ and H^+ surface ions should be ejected as a consequence of the strong repulsion of the two-hole (2h) final state of the H-Nb and O-Nb bonds [8,9].

In summary, we have presented preliminary experiments on electron damage of methanol adsorbed on Nb (110) at 240K. The probable methoxy initial layer yields ESD H^+ surface ions which under electron bombardment change to the species yielding O^+ and H^+ surface ions. The electron bombardment induces ESD H^+ surface ions with 8 eV and 5 eV most probable kinetic energy. The yield of high energy ions decreases as a function of electron dosing. The most probable kinetic energy of the O^+ surface ions at 8 eV energy does not change with the electron bombardment. From the ESD and EELS experiments it can be concluded that electron bombardment produces the complete dissociation of the methoxy species as O, C, and H atoms. These O atoms present the structure of a high valence oxidized state.

This research has been performed under the auspices of the Spain-USA Joint Committee for Scientific and Technological Co-Operation, grant CCA 8510/051, and the Comisión Interministerial de Ciencia y Tecnología.

References

1. S.Rey,I.Colera and J.L.de Segovia, 11th Intl. Vac. Congress and 7th Intl.Conf. on Sol.Surfaces, Köln, Sept 1989. To be published in Vacuum.
2. S.Rey, E. Román and J.L. de Segovia, J.Vac.Sci.Technol, **A5**, 562 (1987).
3. J.P. Adrados and J.L. de Segovia, Vacuum, **34**, 737 (1984).
4. I.Colera, S.Rey and J.L.de Segovia, 11th Intl.Vac.Congress and Intl. Conf. on Sol. Surfaces, Köln, Sept. 1989. To be published in Vacuum.
5. R. Stockbauer, E. Bertel and T. E. Madey, J. Chem. Phys., **76**, 5639 (1982)
6. C.N.R. Rao, D.D.Sarma, S.Vasudevan and M.S. Heede, Proc. R. Soc. Lond., A **367** (1979).
7. T.W. Haas, J.T. Grant and G.J. Dooley, in Adsorption-Desorption Phenomena, Ed. F. Rica (Academic Press, N.Y., 1972) p. 369
8. P.J. Feibelman and M.L. Knotek, Phys. Rev., **B18**, 6531 (1978) E. Ramaker, Chem. Phys., **80**, 103 1983)
9. E.M. Wiliams and J.L de Segovia, Vacuum, **39**, 633 (1989)

Role of Core Levels' Ionization in the Electron Induced Dissociation of Silicon Dioxide

L. Calliari, M. Dapor, L. Gonzo, and F. Marchetti

Divisione Scienza dei Materiali, Istituto per la Ricerca Scientifica e Tecnologica (IRST), I-38050 Povo (Trento), Italy

1. Introduction

Desorption of ions and neutrals from surfaces under low energy (a few keV) electron or photon irradiation has long been recognized as due to electronic excitations not only from the outer valence levels involved in the bonding, but also from the inner levels. The latter observation goes back to the work of Knotek and Feibelman [1]. In the original formulation of the model proposed by these authors (KF model), the states leading to desorption are generated via inter-atomic Auger transitions following core ionizations. As such, the model is strictly applicable only to maximal valency ionic compounds, that is when the cation is nominally ionised down to the noble gas configuration. However, it was soon recognized that possible desorptive states can be generated via Auger transitions in covalent systems too [2]: intra-atomic transitions are involved and correlated final state valence holes are required to make Auger Induced Desorption effective in this case [3].

In this work, we present an Auger investigation of electron beam induced compositional changes (damage) at the silicon dioxide surface. This widely studied oxide is known to decompose under electron irradiation [4-8], leading to a depletion of oxygen in the surface region, which in turn becomes enriched in elemental silicon. Such a modification can be due to both desorption of ionic and neutral species and to diffusion phenomena, the latter eventually enhanced by electric fields.

Desorption from stoichiometric oxides, whether they are maximal or submaximal valency compounds, is commonly understood on the basis of the above mentioned Auger model [3]. It makes sense therefore to ask which electronic excitations are the most important primary events in this respect. To this aim, a spectroscopic use of stimulated desorption, that is ion yield measurements as a function of ionizing radiation energy, has been proven to be highly successful. In particular, studies concerning the silicon/oxygen system have shown O^+ electron stimulated desorption from oxidized Si(111) [9] with thresholds near 18 and 31eV, the latter most likely deriving from the O2s excitation. Furthermore, a strong increase in the O^+ ion yield from electron irradiated SiO_2 above the Si2p ionization threshold has been observed [10]. Much less attention has however been devoted to an analogous "spectroscopic use" of the radiation induced surface chemistry, even though some Auger work has been performed in this respect on SiO_2 claiming the importance of the O1s ionization for the oxide decomposition [11,12].

Springer Series in Surface Sciences, Vol. 19
Desorption Induced by Electronic Transitions, DIET IV
Editors: G. Betz · P. Varga

The aim of this investigation is to determine which electronic excitations are responsible for the induced damage. This has been evaluated, after a fixed irradiation dose, as a function of the primary electron energy and has been then compared to the measured energy dependence of electron impact ionization cross sections for the Si2s, Si2p and O1s core levels. The latter have been experimentally determined through an intensity measurement of Auger transitions initiated by such ionizations, i.e. $SiL_1L_{2,3}V$, $SiL_{2,3}VV$ and $OKL_{2,3}L_{2,3}$ respectively. It is shown that the Si L-shell ionization is by far the most important primary event for the electron beam induced SiO_2 decomposition.

2. Experimental

The measurements have been performed in an Ultra High Vacuum (PHI model 4200) system (base pressure 10^{-10}Torr). The analyzer used is a Cylindrical Mirror Analyzer with a coaxial electron gun. The electron beam has been rastered in order to better control the current density at the sample surface. Integral spectra have been acquired in the Voltage to Frequency mode with the analyzer resolution set to 0.6% and have been subsequently computer differentiated.

The samples were thermally grown SiO_2 films (1033Å) on a Si substrate. They have been irradiated with electron beam primary energies ranging from 300eV to 8000eV and up to a dose of ≈7C/cm^2. During each irradiation, the time evolution of the $SiL_{2,3}VV$ Auger transition has been monitored. In order to obtain comparable clean surfaces, a short ion beam etching has been performed before every irradiation. Though this generally leads to oxide reduction [5], it has been checked that no appreciable surface modification was induced by the gentle treatment performed here.

The Auger intensity measurements needed to determine the energy dependencies of electron impact ionization cross sections have been performed in the following way:
i) on a sputter cleaned silicon wafer for the $SiL_{2,3}VV$ and $SiL_1L_{2,3}V$ transitions.
ii) on SiO_2 samples for the $OKL_{2,3}L_{2,3}$ transition. To avoid electron beam induced oxide reduction, the electron current density has been reduced by a factor of ≈100 by rastering the electron beam over a large area; furthermore, the surface status has been periodically checked by looking at the $SiL_{2,3}VV$ transition.

In every case, the incident electron beam current has been calibrated against the energy with a Faraday cup and the Auger intensity measurements have been then divided by the electron beam current.

3. Results and discussion

As mentioned above, a depletion of oxygen, together with the appearance of elemental silicon, is observed in the SiO_2 surface region as a consequence of electron irradiation. By focusing on the $SiL_{2,3}VV$ Auger transition, which allows to distinguish the two components associated with oxidized and elemental silicon, this modification can be easily followed. As a measure of the electron beam induced damage, the ratio $R=h(Si)/h(SiO_2)$ at a fixed irradiation dose has been considered.

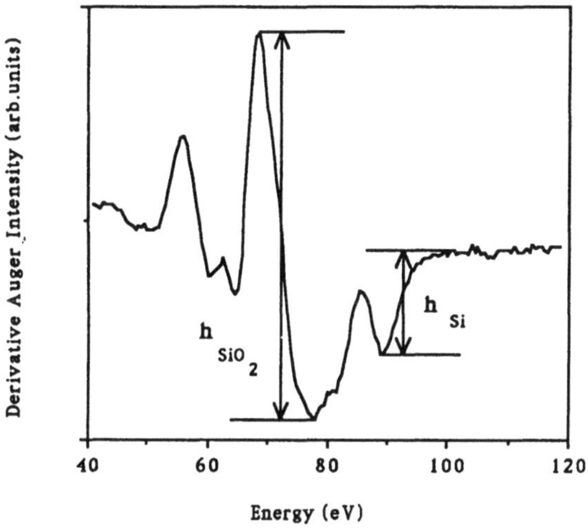

Fig.1: Derivative $SiL_{2,3}VV$ Auger signal in reduced SiO_2.

Here h(Si) and h(SiO_2) are the peak heights, in a derivative $SiL_{2,3}VV$ spectrum, of elemental and oxidized silicon respectively, whose definition is shown in fig.1. By choosing R as the ratio between two Auger signals generated by the same core level ionization, the factors in the Auger intensity, which depend on the primary electron energy, i.e. primary core level ionization cross section and backscattering factor, cancel out.

The ratio R, normalized to 1 at its maximum, is shown at the bottom of fig.2 as a function of the primary beam energy. As one can see, there is a very steep increase on going from 300eV to about 600eV, followed by a slower monotonic decrease up to 8000eV.

In order to understand this energy behaviour and in particular to clarify the relationship between electron beam induced SiO_2 decomposition and core level ionizations, the ionization cross section energy dependence of Si2p, Si2s and O1s has been investigated. This has been done through an intensity measurement of Auger transitions initiated by these primary ionizations, namely the $SiL_{2,3}VV$, $SiL_1L_{2,3}V$ and $OKL_{2,3}L_{2,3}$. These intensities are related to total ionization cross sections, i.e. due to both primary and backscattered electrons. To be compared to the R(E) curve, they are more interesting than calculated electron impact ionization cross sections, which only consider primary electron ionization. The measured curves, each normalized to 1 at its maximum, are also presented in fig.2. They show the typical trend of electron impact ionization cross sections, namely a steep increase above the ionization threshold and a slower decrease after reaching a maximum.

A good correlation between the energy behaviour of the SiO_2 electron induced surface damage and that of both the $SiL_{2,3}VV$ and $SiL_1L_{2,3}V$ transitions is immediately evident, this showing that the SiO_2 decomposition is closely connected to the Si L-shell ionization. More precisely, the SiO_2 decomposition rate results to be proportional to the Si L-shell ionization rate. No distinction can however be done, within our experimental uncertainty, between Si2s and Si2p ionization.

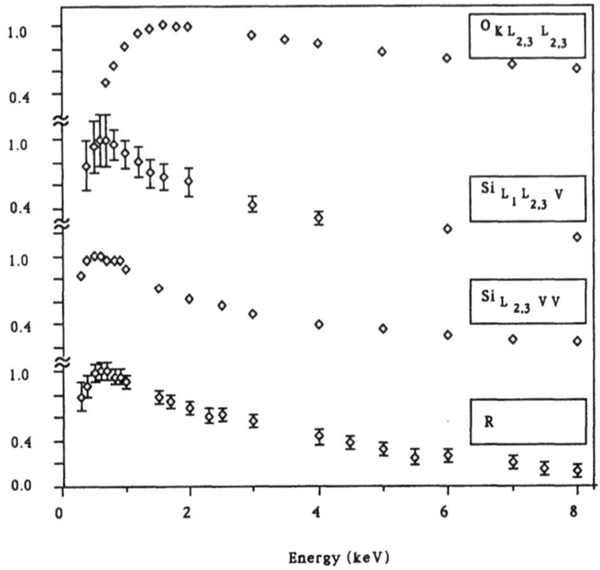

Fig.2: Energy dependence of the ratio R and the intensities of the $SiL_{2,3}VV$, $SiL_1L_{2,3}V$ and $OKL_{2,3}L_{2,3}$ Auger transitions.

On the contrary, the $OKL_{2,3}L_{2,3}$ transition energy dependence is clearly not in agreement with the R(E) curve. In particular, a strong SiO_2 surface damage is observed below the O1s ionization threshold (531eV) and, in addition, no feature is shown in the R(E) curve, which can be related to the maximum of the $OKL_{2,3}L_{2,3}$ transition energy dependence. This means that, once available, the ionization of the O1s core level plays, if any, a negligible role in promoting SiO_2 surface decomposition. As mentioned above, in contradiction with our results, previous Auger work on SiO_2 asserts the predominance of the O1s ionization in this respect [11,12]. We notice however that measuring the electron beam induced damage at low energies (below 1000eV) can be quite critical, due to the difficulty in keeping a constant and sufficiently high electron current density at the sample surface, so that a failure in observing an induced damage below 500eV [11] could be interpreted as due to an unwanted decrease in the electron current density.

Absolute values of electron impact ionization cross sections are worthwhile to be considered at this point beside energy dependencies. By using the Gryzinski theoretical expression [13], the following values have been derived, at the curve maximum, for the three levels considered here:

Si2p: $8.4\ 10^{-18}\ cm^2$ at $Ep \approx 400$ eV
Si2s: $1.3\ 10^{-18}\ cm^2$ at $Ep \approx 550$ eV
O 1s: $1.0\ 10^{-19}\ cm^2$ at $Ep \approx 2050$ eV

This much lower value of the O K-shell ionization cross section with respect to that of the Si L-shell one persists over the whole energy range considered here. Our R(E) curve provides therefore evidence of the fact that Auger final states generated by

O K-shell ionization do not have a much higher oxide decomposition efficiency with respect to those generated by Si L-shell ionization. This would be in fact the only way for contributions associated with O K-shell ionization to be observable in the damage vs. energy curve. A theoretical basis for the interpretation of this result is given by an Auger lineshape investigation concerning SiO_2 [14]. By considering the $SiL_{2,3}VV$ and OKVV Auger transitions from SiO_2, an estimate has been done in fact of what fraction of the final state hole configurations generated by Auger decays should lead to oxide decomposition. By assuming that decomposition arises only from states with two holes localized in a bonding orbital, it has been found that this fraction is approximately 3% of the OKVV Auger signal and 13% of the $SiL_{2,3}VV$ one. Si2p ionization results therefore to be more efficient (by a factor of ≈4) than O1s ionization with respect to damage generation in silicon dioxide. Therefore, both the electron impact ionization cross section and the efficiency for damage generation greatly favour the Si L-shell with respect to the O K-shell, this being in perfect agreement with our finding of the damage vs. energy curve modulated by just the Si L-shell ionization cross section energy behaviour.

4. Conclusions

The energy dependence of the electron beam induced damage at the SiO_2 surface has been compared to that of measured core level ionization cross sections. This allows to single out the Si L-shell ionization as the most important primary event for the oxide decomposition. The result agrees well both with calculated values of electron impact ionization cross sections for the core levels considered here and with estimates of the oxide decomposition efficiency associated with these core level ionizations.

Acknowledgements

The technical assistance of Victor Micheli during the acquisition of the Auger measurements is greatly acknowledged.

References

1. M.L. Knotek and P.J. Feibelman, Phys. Rev. Lett. **40**, 964 (1978)
2. R. Franchy and D. Menzel, Phys. Rev. Lett. **43**, 865 (1979)
3. M. L. Knotek, Rep. Prog. Phys. **47**, 1499 (1984)
4. S. Thomas, J. Appl. Phys. **45**, 161 (1974)
5. J. S. Johannessen, W. E. Spicer and Y. E. Strausser, J. Appl. Phys. **47**, 3028 (1976)
6. B. Carriere and B. Lang, Surf. Sci. **64**, 209 (1977)
7. S. Ichimura and R. Shimizu, J. Appl. Phys. **50**(9), 6020 (1979)

8. B. Lang, Appl. Surf. Sci. **37**, 63 (1989)
9. M.L. Knotek and J.E. Houston, J. Vac. Sci. Technol. **20**(3), 544 (1982)
10. D.E. Ramaker, J. Vac. Sci. Technol. **A1**(2), 1137 (1983)
11. V.V. Nikitin, K.F. Komolova, I.V. Razumovskaya and N.F. Shevtsova, Bull. Acad. Sci. USSR, Phys. Ser. (USA) **48**, 89 (1984)
12. P.Juliet and B. Blanchard, Surf. Int. Analys. **9**, 331 (1986)
13. M. Gryzinski, Phys. Rev. **138**, A336 (1965)
14. D. E. Ramaker, C. T. White and J. S. Murday, Phys. Lett. **89A**, 211 (1982)

DIET in High-Temperature Superconductors

*R.A. Rosenberg and C.-R. Wen**

Synchrotron Radiation Center, University of Wisconsin-Madison,
3731 Schneider Drive, Stoughton, WI53589, USA
*Also affiliated with Department of Physics, University of Wisconsin, Madison
Present address: Department of Electronic Engineering,
National Taiwan University of Marine Science and Technology,
Keelung, Taiwan, Republic of China

Abstract. We present results of photon-stimulated desorption (PSD) measurements from the high-T_c, 123 superconductors, $MBa_2Cu_3O_{7-x}$ where M = Y or Dy. The predominant desorbing species is O_2. Excitation spectra in the energy range 60-140 eV show structure due to excitation of Cu(3p) and Ba(4d) electrons. In the energy range 10-30 eV spectral features may be related to excitation of Ba(5p) and O(2s) electrons with possible contributions from bulk and surface plasmons.

1. Introduction

The properties of high-T_c superconductors such as $YBa_2Cu_3O_{7-x}$ (YBCO) and $DyBa_2Cu_3O_{7-x}$ (DBCO) are very sensitive to the oxygen content [1-3] (an orthorombic to tetragonal phase transition occurs at x = 0.5). Many recent papers have addressed the issue of oxygen loss in the high-T_c superconductors as a result of thermal [4] or radiation-induced processes [5-8]. The general conclusion is that these materials have a high propensity to lose oxygen.

Previous studies have examined the radiation-induced oxygen loss processes in YBCO indirectly, by examining changes in the electronic structure as the result of photon or electron irradiation using photoelectron spectroscopy [5,6] (PES), inverse photoelectron spectroscopy [6,7] (IPES), or electron-energy loss spectroscopy [8] (EELS). These studies found radiation-induced changes in the spectra in periods ranging from 10 min to several hours.

In the first study to examine the radiation-induced desorption product directly we reported on the time dependence of the photodesorbed O_2 upon exposure of a YBCO surface to intense, unmonochromatized synchrotron radiation (SR) [9]. From these measurements a cross section for desorption of weakly-bound O_2 of 3.5±3 Mb was derived. Subsequently we observed photon-stimulated desorption (PSD) of a neutral molecule (O_2) by core-level excitation from YBCO [10] and DBCO [11]. We also reported O_2 PSD data involving low-energy ($\leq$30 eV) excitation and on the yield of the only desorbing ion, H^+ following core-level excitation [11].

In this paper we will present results that emphasize the role of DIET in these materials. In particular we present data on the yield of O_2 following both core- and valence-level excitation. The probable electronic states and mechanisms involved in each process will be discussed.

Springer Series in Surface Sciences, Vol. 19
Desorption Induced by Electronic Transitions, DIET IV
Editors: G. Betz · P. Varga

2. Core-Level Excitation

Results of experiments using monochromatized SR [10] are shown in Fig. 1. This is the first data to show desorption of a neutral molecule from a surface as the result of core-level excitation. Fig. 1(a) shows the yield of O_2 from a scraped YBCO sample as a function of photon energy in the range 60-140 eV. In Fig. 1(b) is the total-electron yield (TEY, yield of electrons of all kinetic energies) spectrum, which is representative of the bulk photoabsorption [12], from the same sample. The features in the TEY spectrum are assigned as: Cu(3p) excitation (threshold = 73.8 eV); Ba(4d) to discrete level excitation (peak = 95 eV); Ba(4d) to continuum excitation (threshold = 99 eV, peaks at 107 and 120 eV) [13].

Two features dominate the O_2 PSD data: the lower energy feature has a threshold of ~80 eV and a peak at 92 eV, while the higher energy feature has a threshold at 102 eV and a sharp peak at 110 eV. We assign the lower-energy feature to excitation of Cu(3p) electrons and the higher-energy structure is a result of Ba(4d) excitation. In Fig. 1(c) is shown the ratio of the O_2 PSD spectrum to the TEY spectrum. The resulting spectrum is proportional to the number of O_2 molecules produced per photon absorbed. It shows that photoabsorption by Cu atoms is much more effective at producing O_2 than absorption by Ba atoms (Cu peak area/Ba peak area > 3:1). This coincides with the fact that in the ab planes of the YBCO unit cell there are 5 oxygen atoms bound directly to Cu atoms and only 2 oxygen atoms bound directly to Ba atoms.

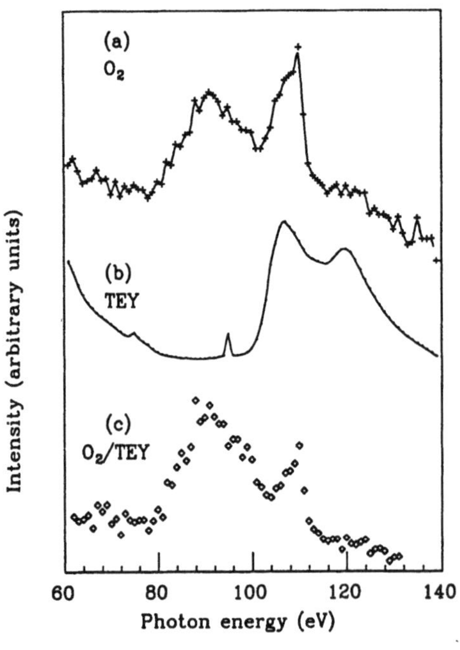

Fig. 1. Comparison of the O_2 PSD spectrum (a) with the TEY (photo-absorption) spectrum (b) of YBCO. In (c) is shown the ratio of the O_2 PSD to the TEY.

The nature of the precursor states involved in the core-level O_2 PSD data of Fig. 1 has been discussed previously [10], so only a brief discussion will be given here. At the Cu(3p) edge, excitation of a Cu^{+2} 3p satellite level results in O_2 PSD. Since this state lies at 7-8 eV higher binding energy than the main Cu^{+2} 3p level [14], the threshold for O_2 PSD is shifted to higher energies relative to the Cu(3p) photoabsorption edge (73.8 eV) as given by the TEY data. The satellite level is thought to arise from a bonding to antibonding, σ to σ^*, transition, which in an ionic picture is effectively a metal-to-ligand charge transfer. Auger decay of the satellite state will result in final states with 2 or 3 holes plus 1 (excited) electron in an antibonding level (2h1e or 3h1e states). Such multi-hole, excited-electron configurations have been found to be very effective pathways for PSD of ions [15], particularly when the holes are in bonding levels and the excited electron is in an antibonding orbital. In the case of YBCO a 3h1e state localized on an O^{-2} ligand or a 2h1e state localized on an O^- ligand could result in desorption of neutral O, provided the state remained localized for a sufficient time for desorption to occur ($\sim 10^{-13}$ sec). These repulsive states could then lead directly to desorption of O_2 or to production of ground or excited-state oxygen atoms. These oxygen atoms would have a high probability for reactive scattering with surface or near-surface oxygen to form O_2, since the binding energy of oxygen in O_2 (5.1 eV) is much higher than the free energy barrier for oxygen desorption (0.9-1.3 eV) [4].

The structure, with a peak at 110 eV in the O_2 PSD spectrum of Fig. 1 is due to excitation of Ba(4d) electrons. The energy position of the peak in the O_2 PSD spectrum is close to the energy of the first strong peak observed in the photoabsorption (TEY) data; however, there is a noticeable lack of structure in the O_2 PSD spectrum at higher energies (120 eV) corresponding to the second strong peak in the TEY data.

As discussed previously [10] we assign the peak at 110 eV in the O_2 PSD spectrum to be a result of a localized Ba^{+2} (4d$\rightarrow$4f) excitation. The energy position of this peak is close to that of the 107 eV peak in the photoabsorption data. There is little or no structure in the O_2 PSD data which corresponds to the 120 eV peak in the photoabsorption data of Fig. 1(b). The reason for this may be that the states reached upon decay of the excitations leading to the 110 eV and 120 eV peaks are different. Some insight into this process may be derived by examining the Ba(5p) CIS data of YBCO [16]. This data gives the relative partial photoionization cross section for production of Ba^+ with a hole in the 5p orbital. The similarity of the O_2 PSD data and the Ba(5p) CIS data suggests that production of a Ba(5p) hole may be an important step in the O_2 desorption process. Data on valence-level O_2 PSD, which will be presented in the following section, in which the Ba(5p) level is directly excited, give further credence to this argument.

3. Valence-Level Excitation

Figure 2 shows the yield of O_2 as a function of photon energy in the range 10-30 eV [11]. The solid curve is the raw data and the dashed curve shows the data

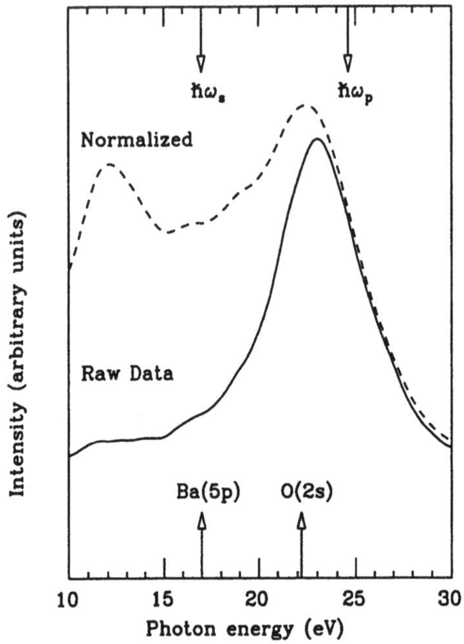

Fig. 2. O_2 PSD spectrum from DBCO in the energy range 10-30 eV presented as raw data (solid line) and normalized to incident photon flux (dashed line). The positions of Ba(5p), O(2s), surface and bulk plasmon excitation energies, as determined by EELS, are marked [8,15].

normalized to photon flux. There are two principal features: a weak threshold at ~15 eV, and a strong structure which peaks at ~23 eV. The peak at 11.5 eV, which is most evident in the normalized data, is a result of excitation of the 23 eV peak by second-order light.

The first important result obtained from the data of Fig. 2 is that the upper limit for the threshold for oxygen desorption is 15 eV. In order to determine if this is the true threshold, the measurements need to be extended to lower photon energies. This is not easily done due to the presence of large amounts of second-order light at lower energies. If 15 eV is the actual threshold then electronic excitations of energies less than 15 eV will not cause oxygen desorption.

In order to explain the structure in Fig. 2, we must try to understand what possible electronic excitations could lead to O_2 desorption. The deep core-level data of Fig.2 can be successfully interpreted in terms of excitation of Cu(3p) and Ba(4d) electrons. Similarly it is possible that excitation of shallow core levels may lead to the structure observed in Fig. 2. The most likely candidates in this energy region are the Ba(5p) and O(2s) levels. Arrows on the bottom of Fig. 2 mark the positions of Ba(5p) and O(2s) to conduction band transitions as revealed by electron-energy loss spectroscopy (EELS) [8,17]. Since the energy positions of these transitions coincide with the structure in

Fig. 2, it is possible that excitation of Ba(5p) and O(2s) electrons may lead to an antibonding state and desorption of O_2. The argument for participation of a Ba(5p) hole in the desorption process is strengthened by the fact that the Ba(5p) constant ionic state (CIS) data [16], is similar to the O_2 PSD yield at the Ba(4d) edge.

There is an alternative explanation for the data of Fig. 2. Recently it has been shown that laser-light excitation of surface plasmons on small metal particles can lead to desorption of atoms [18]. In order to see if such a mechanism could be involved in desorption of O_2 from DBCO, we have marked the energy positions observed for excitation of surface and bulk plasmons in YBCO [17]. The proximity of the surface and bulk plasmon excitation energies to the structure in Fig. 2 gives strength to the argument that plasmon excitation could lead to oxygen desorption. The oxygen observed in our measurements does not necessarily originate from the surface. Atomic or molecular oxygen produced in the near-surface region should have a high probability for escaping the surface. In fact some of the observed O_2 may be formed by reactive scattering of atomic oxygen originating in the bulk. Therefore, excitation of bulk plasmons, as well as surface plasmons, could contribute to O_2 PSD.

4. Conclusions

In this paper we have presented results of PSD measurements on high-T_c superconductors, which emphasize the role of DIET in these materials. We are able to interpret the O_2 PSD results in terms of core-level excitation, with the possible involvement of bulk and surface plasmons in the lower-energy (10-30 eV) data. Besides the obvious technological importance of these materials, the fact that they have such high PSD yields of a neutral molecule make them attractive for DIET experimental studies. However, at this stage of our understanding it is difficult to gain further insight into the desorption process without some theoretical input.

5. Acknowledgements

We would like to thank David Larbalestier and Jeff Seuntjens for supplying the samples used in these studies. The experiments were performed at the Wisconsin Synchrotron Radiation Center, a national facility supported by the National Science Foundation.

References

1. *"Chemistry of High Temperature Superconductors"*, D. L. Nelson, M. S. Whittingham, and T. F. George Eds., ACS Symposium Series **351**, Washington, DC (1988).
2. *"Thin Film Processing and Characterization of High-Temperature Superconductors"*, J. A. Harper, R. J. Colton and L. C. Feldman Eds., AIP Conference Proc. **165** (1988).

3. J. D. Jorgensen, M.A. Beno, D.G. Hinks, L. Soderholm, K.J. Volin, R.L. Hitterman, J.D. Grace, I.K. Schuller, C.U. Segre, K. Zhang, and M.S. Kleefisch, Phys. Rev. B **36**, 3608 (1987).
4. H. Strauven, J.P. Locquet, O.B. Verbeke, and Y. Bruynseraede, Solid State Comm. **65**, 293 (1988).
5. N. G. Stoffel, J.M. Tarascon, Y. Chang, M. Onellion, D. W. Niles, and G. Margaritondo, Phys. Rev. B **36**, 3986 (1987).
6. T. J. Wagener, Y. Gao, H.M. Meyer III, I.M. Vitomirov, C.M. Aldao, D.M. Hill, J.H. Weaver, B. Flandermeyer, and D.W. Capone II, in Ref. 2, p. 368.
7. Y. Gao, T.J. Wagener, J.H. Weaver, A.J. Arko, B. Flandermeyer, and D.W. Capone II, Phys. Rev. B **36**, 3971 (1987).
8. J. Yuan, L. M. Brown, and W. Y. Liang, J. Phys. C **21**, 517 (1988).
9. R. A. Rosenberg and C.-R. Wen, Phys. Rev. B **37**, 5841 (1988).
10. R. A. Rosenberg and C.-R. Wen, Phys. Rev. B **37**, 9852 (1988).
11. R.A. Rosenberg and C.-R. Wen, Phys. Rev. B **39**, 6630 (1989).
12. A. Bianconi, Appl. Surf. Sci. **6**, 392 (1980).
13. M. M. Hecht and I. Lindau, Phys. Rev. Lett. **47**, 821 (1981). M. M. Hecht, Ph.D. Thesis, Stanford Synchrotron Radiation Laboratory Report No. 82/07, 1982 (unpublished), and references therein.
14. K.S. Kim, J. Elec. Spec. Rel. Phen. **3**, 217 (1974).
15. *"Desorption Induced by Electronic Transitions, DIET I"*, N. H. Tolk, M. M. Traum, J. C. Tully, and T. E. Madey, Eds. (Springer, New York, 1983).
16. M. Onellion, Y. Chang, D.W. Niles, R. Joynt, G. Margaritondo, N.G. Stoffel, and J.M. Tarascon, Phys. Rev. B **36**, 819 (1987).
17. Y. Chang, M. Onellion, D.W. Niles, R. Joynt, G. Margaritondo, N.G. Stoffel, and J.M. Tarascon, Solid State Comm. **63**,717 1987).
18. W. Hoheisel, K. Jungmann, M. Vollmer, R. Weidenauer, F. Träger, Phys. Rev. Lett. **60**, 1649 (1988).

A Study of the Initial Adsorption Processes of Oxygen on the Ti(1010) Surface by TOF-Type Electron Stimulated Desorption

K. Ueda and A. Takano

Department of Applied Physics, Osaka University, Suita, Osaka 565, Japan

Adsorption processes of oxygen on a titanium surface were studied time-sequentially by means of TOF-type electron stimulated desorption. We obtained ion energy distribution and relative ion yield data in the initial oxidation stage for both oxygen ions and molecular ions.

1. Introduction

Electron stimulated desorption (ESD) techniques have been widely used in the investigation of adsorption and desorption processes [1,2]. In particular, electron stimulated ion angular distribution (ESDIAD) [3] is a powerful technique for the investigation of adsorbate configurations on the surface. In these studies it is important to identify the desorbed species in order to infer the behaviour of adsorbed atoms or molecules on the surface. There are several methods for mass analysis of desorbed species. We have adopted a time-of-flight (TOF) spectroscopy for ESD study using a pulsed counting technique. It has the advantage of minimizing damage by electron impact.

2. Experiment

Our experimental setup has been described previously [4,5] and here we will describe it only briefly. We have recently incorporated the TOF spectrometer into an UHV chamber equipped with ordinary LEED-AES and RHEED apparatuses.

A schematic drawing of the TOF-ESD system is shown in Fig.1. The electron gun is an off-axial-type LEED gun. A pulsed electron beam of 350eV

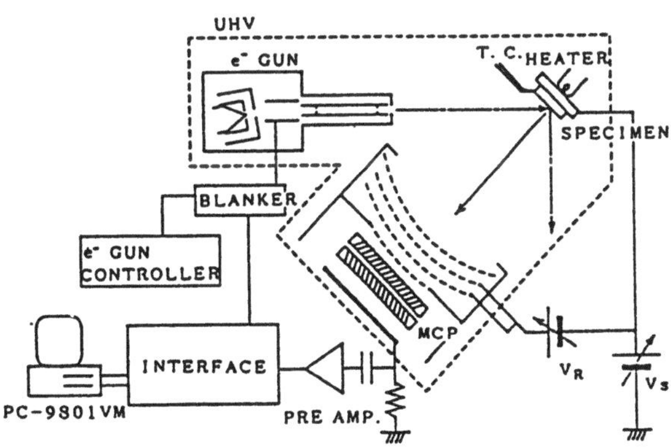

Fig.1
Schematic drawing of the TOF-ESD apparatus. E_p is 350eV, and the pulsed electron width is less than 200ns.

Springer Series in Surface Sciences, Vol. 19
Desorption Induced by Electronic Transitions, DIET IV
Editors: G. Betz · P. Varga

hits the specimen every 2.5ms, and a counting system controlled by a personal computer counts ions between pulses. The sample is mounted on the heater assembly of a Varian manipulator. The pressure of the experimental chamber reaches $1x10^{-10}$ Torr after 8h of bake-out. The sample is cleaned conventionally by: flash heating, Ar ion bombardment and annealing in UHV. The concentrations of impurities of carbon and sulfur on the surface are monitored by Auger analysis. The incident angle of the electron beam is 45deg from the surface normal. The effective diameter of the micro-channel plate (MCP) is 40mm. The flight path of the desorbed ions between the specimen and the detector is 150mm. The orientation of the sample surface was determined by RHEED.

The initial oxidation process of the titanium $(10\bar{1}0)$ surface was measured in the following manner: After cleaning the sample surface, oxygen gas was introduced into the chamber up to a constant pressure which was monitored by a B-A type ionization gauge. During the exposure of oxygen TOF measurements were performed repeatedly. The TOF spectra were measured alternately with and without positive sample bias during oxygen exposure. The spectra measured without a sample bias were used for the conversion to energy distribution curves as described later on. In the TOF spectra with zero sample bias, we obtain a high resolution in the ionic energy, but the S/N ratio is poor in this case. On the other hand with positive sample bias we obtain a high mass resolution and a high S/N ratio.

3.Experimental results

Figure 2(a) shows a series of TOF spectra obtained with +5V sample bias during oxygen exposure of a titanium surface at room temperature. In this figure, exposure/L is presented as the integrated value at the end of measurements in an experimental run, and P denotes the photoelectron peak

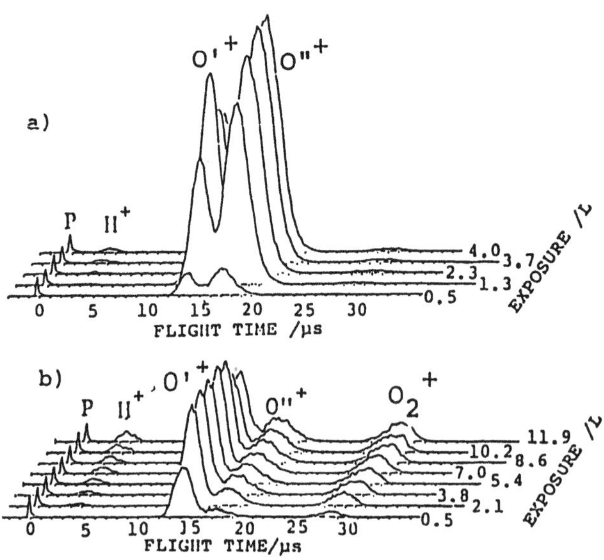

Fig.2 Change of TOF-ESD spectrum during oxygen exposure at room temperature on (a) the fairly clean surface, (due to the alternate measurements with zero sample bias,values of exposure are not linear) and (b) the about 1/3 ML sulfur segregated surface.

produced by soft X-rays generated by beam irradiation. The H^+ peak is due to protons from adsorbed water, O'^+ is the ESD signal of oxygen ions in the high energy state, and O''^+ corresponds to the low energy state. This is the first time that two distinct states for adsorbed oxygen on titanium have been recognized [6]. In the case of a clean surface, the O'^+ species is less stable than the O''^+ species and its signal intensity starts decreasing as soon as oxygen dosing is stopped. In contrast, on a sulfur segregated surface with 1/3 sulfur monolayer, the O'^+ ESD signal intensity increases with oxygen exposure, and simultaneously the signal for molecular oxygen ions (O_2^+) appears (Fig.2(b)). No S^+ signal was observed from the sulfur covered surface without oxygen exposure. We note that the signal ratio of O'^+ to O''^+ for the sulfur segregated surface is larger than the one for the clean titanium surface.

The TOF spectra with zero sample bias can be converted directly to energy distribution curves using the classical equation of motion [7]. In order to avoid complications due to electric fields in the path of desorbed ions, a zero bias TOF spectrum was used for conversion curves showing energy distributions. The converted energy distribution curves are consistent with the ones obtained from a retarding potential method. The converted curves are shown in Fig.3(a) for the clean titanium surface and in Fig.3(b) for the sulfur segregated surface, both for the maximum coverage as shown in Fig.2(a) and (b), respectively. In Fig.3 both of the energy distribution curves were normalized to their maximum intensity. It is remarkable that the O''^+ peak intensity is considerably stronger than the O'^+ peak for the oxygen exposure of the clean surface. On the other hand, the O''^+ peak intensity is almost a quarter of the O'^+ peak height on the sulfur segregated surface.

A similar set of measurements were made at a substrate temperature of 150°C. The converted energy distribution curves are shown in Fig.4, where the spectra are plotted for various exposure times. Similarly to the experiments at room temperature the O''^+ signal increases with increasing

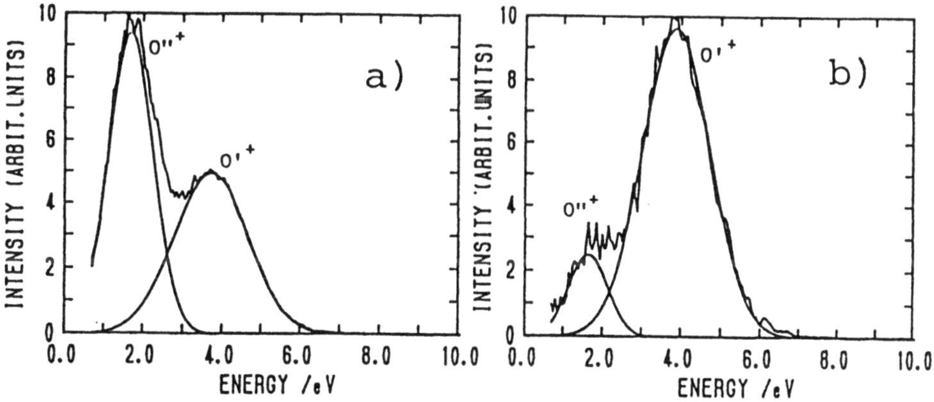

Fig.3 (a) Converted energy distribution curve obtained from $3.0L-O_2$ covered Ti(10$\bar{1}$0) surface. (b) The same for a $3.2L-O_2$ coverage on a sulfur segregated surface.

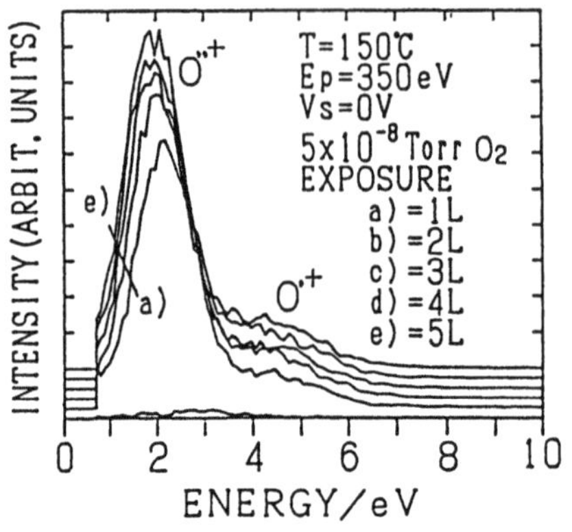

Fig.4 A series of converted energy distribution curves during oxygen exposure at 150°C. Zero levels are shifted.

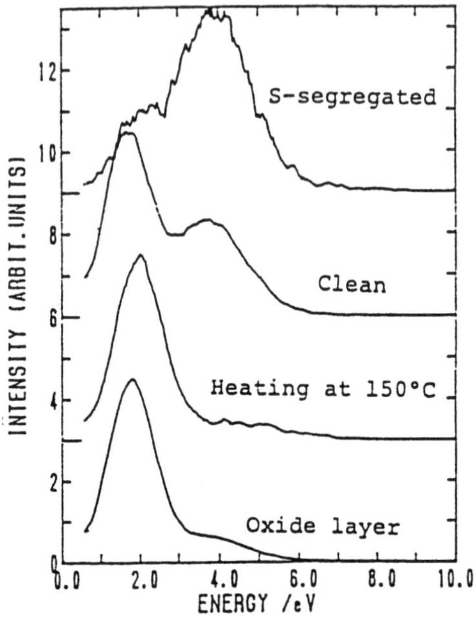

Fig.5 Difference of E.D.curves converted from TOF spectra for different surface preparations before oxygen exposure.

oxygen exposure, but the O'^{+} signal does not. In spite of the large kinetic energy of O'^{+}, for the warmer surface (150°C) oxygen hardly adsorbs in this energy state. As mentioned before this is consistent with a metastable behaviour of O'^{+}.

4.Discussion

As mentioned above, the study of the initial stage of oxygen adsorption on the titanium surface has been investigated by TOF-ESD measurement. As shown in Fig.2(a), the O''^{+} signal increases initially with increasing oxygen exposure prior to the O'^{+} signal. However, on the sulfur segregated surface (Fig.(2b)), the O'^{+} signal intensity grows faster than the O''^{+} signal. From this fact we infer that the adsorption in the O''^{+} state is inhibited by the sulfur layer due to its high electron affinity. In addition to this, Fig.4 indicates that oxygen cannot adsorb into the O'^{+} state, but it adsorbs easily into the O''^{+} state even at 150°C. These facts suggest that in spite of the higher kinetic energy, the O'^{+} signal comes from an unstable state. Oxygen is adsorbed on the sulfur covered titanium surface as O_2 molecules while it is atomically adsorbed on the clean surface. The difference in the kinetic energy between the O'^{+} and O''^{+} species arises from different binding energies for the two cases. The O''^{+}(2eV) is stimulated by the Knotek-Feibelman mechanism [8] because it is almost the same energy as O^+ from the oxidized layer of titanium, while the O'^+ may be ionized directly by the electron beam from molecular oxygen adsorbed in the precursor state on the surface. Generally, desorbing ions stemming from molecular oxygen on the metal surface have a kinetic energy of several electron volts [9].

Molecular oxygen ions were detected from the sulfur segregated surface with a kinetic energy less than 1eV. This is a new discovery. The O_2^+ ESD signal from a sulfur segregated surface can be detected during oxygen exposure only with a positive bias voltage larger than about 5V. When the oxygen flow is stopped, the O_2^+ signal abruptly decreases, furthermore, in the separate experimental runs its flight time depends on the sample bias voltage. Thus, it is clear that the O_2^+ comes from desorption from the sulfur covered surface and not from the ambient gaseous O_2. It has been shown before that oxygen is adsorbed in molecular form on a Pt surface [10].

In conclusion, 1) we have found that two distinct energy peaks appear in the energy distribution of desorbed oxygen species from titanium surfaces during oxygen exposure. 2) The low energy peak of about 2eV (O''^+) is attributed to the chemisorbed oxygen and it is stable after stopping the exposure. The high energy peak of about 4eV (O'^+) is attributed to adsorbed oxygen in molecular form on the titanium surface. 3) On the sulfur segregated surface the O'^+ signal appears prior to the O''^+ due to passivation from the dissociation by a sulfur layer. 4) From the sulfur segregated titanium surface molecular ions of oxygen are observed during oxygen exposure.

Acknowledgments

The authors wish to thank Professor S.Ushioda in Tohoku University, Sendai, for fruitful discussions. This work was partially supported by a Grant-in-Aid for Scientific Research on Priority Area by the Ministry of Education, Science and Culture.

References

[1]D.Lichtman,R.B.McQuistan and T.R.Kirst,Surf.Sci. 5 (1966) 120.

[2]J.J.Czyzewski,T.E.Madey and J.T.Yates,Jr.,Phys.Rev.Lett. 32 (1974) 777.

[3]T.E.Madey and J.T.Yates,Jr., Surf.Sci. 63 (1977) 203.

[4]K.Ueda and A.Takano,Technl. Rept. Osaka Univ. 38 (1988) 217.

[5]K.Ueda and A.Takano, Japan J.Appl.Phys. 27 (1988) L2029.

[6]Oxygen ions from TiO_2 have a single peak in the energy distribution curve, see e.g. R.L.Kurz, Surf.Sci. 177 (1986) 526.

[7]I.G.Newsham,J.V.Hogue and D.R.Sandstrom,J.Vac.Sci.Technl. 9 (1972) 596.

[8]M.L.Knotek and P.J.Feibelman, Phys. Rev. Lett. 40 (1980) 964.

[9]P.A.Redhead,J.P.Hobson and E.V.Kornelsen "The Physical Basis of Ultrahigh Vacuum" , Chapman and Hall Ltd. (London 1968).

[10]J.L.Grand,B.A.Sexton and G.Fisher, Surf.Sci. 95 (1980) 587.

Index of Contributors

MIX
Papier aus verantwortungsvollen Quellen
Paper from responsible sources
FSC® C105338

If you have any concerns about our products,
you can contact us on
ProductSafety@springernature.com

In case Publisher is established outside the EU,
the EU authorized representative is:
Springer Nature Customer Service Center GmbH
Europaplatz 3, 69115 Heidelberg, Germany

Printed by Libri Plureos GmbH
in Hamburg, Germany